Understanding Digital Signal Processing with MATLAB® and Solutions

THE ELECTRICAL ENGINEERING
AND APPLIED SIGNAL PROCESSING SERIES
Edited by Alexander D. Poularikas

Understanding Digital Signal Processing with MATLAB® and Solutions

Alexander D. Poularikas

CRC Press
Taylor & Francis Group
Boca Raton London New York

CRC Press is an imprint of the
Taylor & Francis Group, an **informa** business

MATLAB® is a trademark of The MathWorks, Inc. and is used with permission. The MathWorks does not warrant the accuracy of the text or exercises in this book. This book's use or discussion of MATLAB® software or related products does not constitute endorsement or sponsorship by The MathWorks of a particular pedagogical approach or particular use of the MATLAB® software.

CRC Press
Taylor & Francis Group
6000 Broken Sound Parkway NW, Suite 300
Boca Raton, FL 33487-2742

© 2018 by Taylor & Francis Group, LLC
CRC Press is an imprint of Taylor & Francis Group, an Informa business

No claim to original U.S. Government works

Printed on acid-free paper

International Standard Book Number-13: 978-1-138-08143-7 (Hardback)

Visit the Taylor & Francis Web site at
http://www.taylorandfrancis.com

and the CRC Press Web site at
http://www.crcpress.com

Contents

Abbreviations

ACF	autocorrelation function
ACS	autorrelation sequence
AIC	Akaike information criterion
AR	autoregressive process, $AR(p)$ of order p
ARMA	autoregressive moving average process, $ARMA(p, q)$ of order (p, q)
ARMAX	autoregressive moving average with exogenous source
ARX	autoregressive moving average with exogenous input
BT	Blakman–Tukey
CDF	cumulative distribution function
CRLB	Cramer–Rao lower bound
DFT	discrete Fourier transform
DTFT	discrete-time Fourier transform
ENSS	error normalized step size
FFT	fast Fourier transform of data given by the vector x
FIR	finite impulse response of a discrete system
FPE	final prediction error
FT	Fourier transform
IDFT	inverse discrete Fourier transform
IDTFT	inverse discrete-time Fourier transform
IFFT	inverse fast Fourier transform
IFT	inverse Fourier transform
ID	identically distributed
IID	independent and identically distributed
IIR	infinite impulse response of a discrete system
KVL	Kirchhoff voltage law
LMF	least-mean forth
LMMN	least mean mixed norm
LMS	least mean square
LTI	linear time invariant
LS	least squares
LSE	least-squares error
MA	moving average process, $MA(q)$
MEM	maximum entropy method
MMSE	minimum mean square error
MSE	mean square error
MV	minimum variance
MVUE	minimum variance unbiased estimator
N(m, v)	Normal (Gaussian) distribution with mean m and variance v
NLMS	normalized LMS
PDF	probability density function
PNLMS	power normalized LMS
PSD	power spectral density
QAM	quadradure amplitude modulation
ROC	region of convergence
RV	random variable
RVs	random variables
RVSS	robust variable step-size (algorithm)

RW	random walk
SCLMS	self-correcting LMS
SCWF	self-correcting wiener filter
TDLMS	transform domain LMS
VSLMS	variable step-size LMS
WGN	white Gaussian noise
W–K	Wiener–Khintchine
WN	white noise
WSS	wide-sense stationary process
YW	Yule–Walker equations

Author

Alexander D. Poularikas received his PhD from the University of Arkansas and was a professor at the University of Rhode Island. He became chairman of the Engineering Department at the University of Denver and then served as chairman of the Electrical and Computer Engineering Department at the University of Alabama in Huntsville. He has published, co-authored and edited 14 books. Dr. Poularikas served as editor-in-chief of the *Signal Processing Series* (1993–1997) with ARTECH HOUSE and is now editor-in-chief of the *Electrical Engineering and Applied Signal Processing Series* as well as the *Engineering and Science Primers Series* (1998–) with Taylor & Francis Group. He was a Fulbright scholar, is a life-long senior member of IEEE, and a member of Tau Beta Pi, Sigma Nu and Sigma Pi. In 1990 and 1996, he received the Outstanding Educators Award of the IEEE, Huntsville section.

1 Continuous and Discrete Signals

1.1 CONTINUOUS DETERMINISTIC SIGNALS

PERIODIC SIGNALS

The most fundamental periodic signal is the sinusoidal one. Any periodic signal is defined by

$$f(t \pm nT_p) = f(t) \qquad n = 0, \pm 1, \pm 2, \dots \qquad T_p = \text{period} \tag{1.1}$$

This means that the values of the function at $t, t+T_p, t+2T_p, \dots, t+nT_p, \dots$ are identical.

Two sinusoidal functions with different amplitudes are

$$f(t) = a \sin(\omega t + \theta) \text{ or } f(t) = b \cos(\omega t + \theta), T_p = \frac{2\pi}{\omega}, \quad \omega = 2\pi f, \quad f = \frac{1}{T_p} \tag{1.2}$$

where a and b are the amplitudes, θ the phase, T_p the period in seconds, ω frequency in radians per second (rad/s), and f frequency in cycles per second, Hz.

If we use the Maclaurin series expansion $\exp(x) = 1 + (x/1!) + (x^2/2!) + (x^3/3!) + \cdots$, where (!) indicates **factorial**, e.g., $4! = 1 \times 2 \times 3 \times 4$, substituting $j\omega$ for x in the expansion, we obtain

$$e^{j\omega t} = 1 + j\omega t + \frac{(j\omega t)^2}{2!} + \cdots + \frac{(\omega t)^n}{n!} + \cdots$$

$$= \underbrace{\left[1 - \frac{(j\omega t)^2}{2!} + \frac{(j\omega t)^4}{4!} - \cdots \right]}_{\cos \omega t} + j \underbrace{\left[\omega t - \frac{(j\omega t)^3}{3!} + \frac{(j\omega t)^5}{5!} - \cdots \right]}_{\sin \omega t} = \cos \omega t + j \sin \omega t \tag{1.3}$$

From the above equation we obtain the Euler's equation:

$$\cos \omega t = \frac{e^{j\omega t} + e^{-j\omega t}}{2}, \sin \omega t = \frac{e^{j\omega t} - e^{-j\omega t}}{2j}, \quad \text{Re}\left\{ e^{j\omega t} \right\} = \cos \omega t, \text{Im}\left\{ e^{j\omega t} \right\} = \sin \omega t \tag{1.4}$$

Figure 1.1 shows four different deterministic types of periodic signals. Figure 1.1d includes the following four musical signals: (1) the dotted one (...) is the A note $f(t) = \sin(2\pi 440t)$, (2) the dash-dot line (_ . _ . _) is the B note $f(t) = \sin(2\pi 494t)$, (3) the dash line (----) is the E note $f(t)\sin(2\pi 659t)$, and (4) the continuous line is another periodic signal which is equal to the sum of the three notes.

NON-PERIODIC CONTINUOUS SIGNALS

The following are some of the most fundamental functions which we find in practice and in analysis of systems.

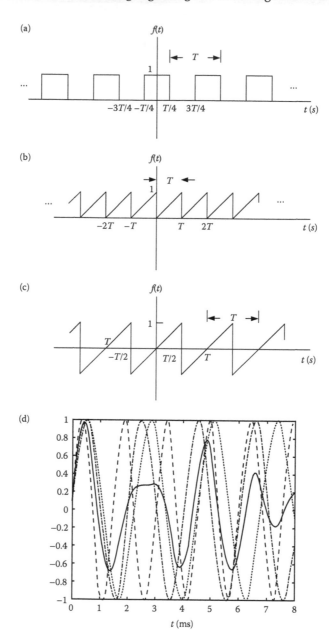

FIGURE 1.1

Unit Step Functions

$$u(t) = \begin{cases} 1 & t > 0 \\ 0 & t < 0 \end{cases} \qquad u(t - t_0) = \begin{cases} 1 & t - t_0 > 0 \\ 0 & t - t_0 < 0 \end{cases} \qquad (1.5)$$

To plot the function we introduce different values of t. For positive t the value is one and for negative values of t the value is zero. We can take the value equal to one for $t=0$. Therefore, the function $-u(t+2)$ is a unit step function which starts from $t=-2$, and its value is negative one for $t > -2$.

Ramp Function

$$r(at) \triangleq ar(t) = \begin{cases} at & t \geq 0 \\ 0 & t < 0 \end{cases} \qquad r(t) = \int_0^t u(x)\,dx = \int_0^t dx = t \quad t \geq 0 \qquad (1.6)$$

Rectangular Function

$$p_a(t) = [u(t+a) - u(t-a)] = \begin{cases} 0 & |t| > a \\ 1 & |t| < a \end{cases}$$

$$p_a(t-t_0) = [u(t-t_0+a) - u(t-t_0-a)] = \begin{cases} 0 & |t-t_0| > a \\ 1 & |t-t_0| < a \end{cases} \qquad (1.7)$$

Triangular Pulse Function

$$\Lambda_a = \begin{cases} 1 - \dfrac{|t|}{a} & |t| < a \\ 0 & |t| > a \end{cases} \qquad (1.8)$$

Signum Function

$$\operatorname{sgn}(t) = \begin{cases} 1 & t > 0 \\ 0 & t = 0 \\ -1 & t < 0 \end{cases} \qquad \operatorname{sgn}(t) = -1 + 2u(t) \qquad (1.9)$$

Sinc Function

$$\sin c_a(t) = \frac{\sin at}{t} \quad -\infty < t < \infty \qquad (1.10)$$

To obtain its value at $t=0$, since we obtain the undefined expression 0/0, we use L'Hôpital's rule, which says: take the derivative of the numerator and denominator and then set $t=0$. For example, for the function $\sin(a(t-2))/(3(t-2))$ we obtain $a\cos(a(t-2))/3|_{t=2} = a/3$. To plot the sinc function we must use the expression, for example, $t=-5:0.01:5$; $\sin(0.5 * t+\text{eps})/(t+\text{eps})$; plot(s).

Gaussian Function

$$f(t) = e^{-at^2} \quad -\infty < t < \infty \quad a > 0 \qquad (1.11)$$

Error Function

$$\operatorname{erf}(t) = \frac{2}{\sqrt{\pi}} \int_0^t e^{-x^2}\,dx = \frac{2}{\sqrt{\pi}} \sum_{n=0}^{\infty} \frac{(-1)^n t^{2n+1}}{n!(2n+1)}$$

$$\text{properties:} \operatorname{erf}(\infty) = 1, \operatorname{erf}(0) = 0, \operatorname{erf}(-t) = -\operatorname{erf}(t) \qquad (1.12)$$

$$\operatorname{erfc}(t) = \text{complementary error function} = 1 - \operatorname{erf}(t) = \frac{2}{\sqrt{\pi}} \int_t^{\infty} e^{-x^2}\,dx$$

Exponential and Double Exponential Functions

$$f(t) = e^{-t}u(t) \quad t \geq 0, \quad f(t) = e^{-|t|} \quad -\infty < t < \infty \tag{1.13}$$

Type of Signals—Even, Odd, Energy, and Power

$f(t) = f(-t)$ is even function; $f(t) = -f(t)$ is odd function

$$\int_{-\infty}^{\infty} f^2(t)\,dt < \infty \text{ is energy function;} \quad 0 \leq \lim_{T \to \infty} \frac{1}{2T} \int_{-T}^{T} f^2(t)\,dt < \infty \text{ is power function} \tag{1.14}$$

The even and odd parts of a general function are given, respectively, by

$$f_e(t) = \frac{f(t) + f(-t)}{2}, \quad f_o(t) = \frac{f(t) - f(-t)}{2}$$

Example 1.1.1

Find the function which describes the graph in Figure 1.2.

Solution: the function is $r(t) - r(t-2) - u(t-2) - u(t-4)$.   ∎

Example 1.1.2

Find the value of the error function for $t = 1.8$ using the summation approximation for ranges from 0 to 5, 0 to 8 and 0 to 20.

Solution: We use the following Book MATLAB® program:

```
>>t=1.80;
>>for 0:5;
>>er(n+1)=(2/sqrt(pi))*(-1)^n*(t)^(2*n+1)/factorial(n)*(2*n+1);
>>        %factorial() is a MATLAB function;
```

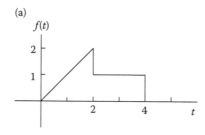

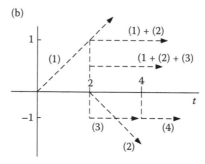

FIGURE 1.2

Similarly, we obtain the remaining two cases. The results were: (1) for $n=0:5$ sum(er)$=0.8117$, (2) for the case $n=0:8$ sum(er)$=0.9980$, (3) for the case $n=0:20$ sum(er)$=0.9891$ which is the same as the exact value to the fourth digit approximation. sum(er) is a MATLAB function which sums the elements of the vector er. ∎

Example 1.1.3

Plot the sinc function $f(t) = \sin(at)/t$ for $a=0.5$, 1, and 2.5.

Solution: The following Book m-file was used: *ex_1_1_3*

```
%m-file:ex_1_1_3;
t=-12:0.05:12;%this creates a vector
    %with values of t from -12 to 12
    %in steps of 0.05;
s1=sin(0.5*(t+eps))./(t+eps);%eps is a small
    %number, and it is used to avoid warning
    %from MATLAB that a division with zero
    %is present, MATLAB would have plotted
    %the signal with one point missing at 0/0;
    %the period in the numerator is an instruction that
    %the division is to be made element by element of
    %the two vectors(numerator and denominator);
s2=sin(t+eps)./(t+eps);
s3=sin(2.5*(t+eps))./(t+eps);
plot(t,s1,'k');hold on;plot(t,s2,'k');hold on;
plot(t,s3,'k');grid on;xlabel('t');
ylabel('sin(at)/t');
```

See Figure 1.3.

Exercise 1.1.1

Show whether the signals (1) $f(t) = \exp(-t)u(t)$ and (2) $f(t) = u(t)$ are power signals or not. ▲

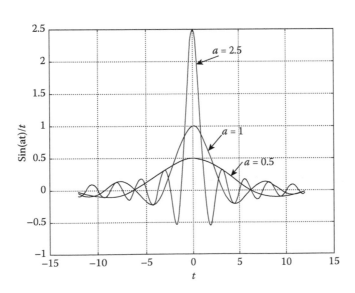

FIGURE 1.3

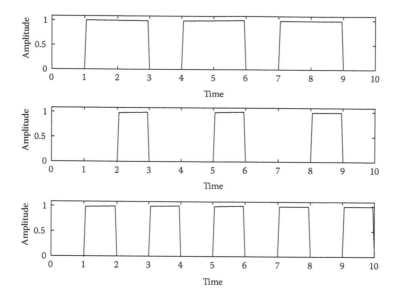

FIGURE 1.4

Exercise 1.1.2

(1) Plot a decaying or increasing sinusoidal signal. (2) Plot a time-varying sinusoidal.　　▲

Example 1.1.4

Plot a periodic square wave.

Solution: The following Book MATLAB program produces Figure 1.4.

BOOK MATLAB PROGRAM

```
T=0:0.05:10; subplot(3,1,1); plot(t,mod(1,3)>1,'k'); xlabel('time');
ylabel('amplitude');axis([0 10 0 1.1]);
subplot(3,1,2); plot(t,mod(t,3)>2,'k'); xlabel('time');
ylabel('amplitude');axis([0 10 0 1.1]);
subplot(3,1,3); plot(t,mod(t,2)>1,'k'); xlabel('time');
ylabel('amplitude');axis([0 10 0 1.1]);
```

Here the modulus (MATLAB function mod) produces a time series of zeros and ones that can vary in width. axis([]) is also a MATLAB function which determines the axis.　　■

1.2　SAMPLING OF CONTINUOUS SIGNALS-DISCRETE SIGNALS

A **discrete-time signal** is a function whose domain is a set of integers, e.g., the values of a stock at each closing day or the height of school children in a school. Therefore, this type of signal is a sequence of numbers denoting by $\{x(n)\}$. It is understood that the discrete-time signal is often formed by **sampling** a continuous-time signal $x(t)$. Under this case, and for equidistance samples, we write

$$x(n) = x(nT) \qquad T = \text{sampling interval} \tag{1.15}$$

Figure 1.5 shows a transformation from a continuous-time to a discrete-time signal. How we decide the value of the time sampling T will be discussed in detail in the next chapter. Table 1.1 gives some analog and discrete functions.

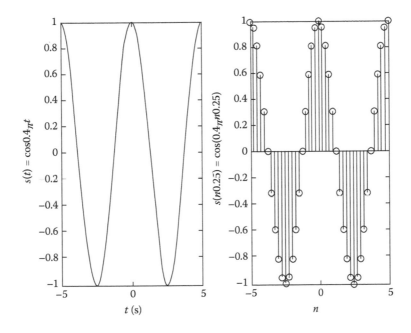

FIGURE 1.5

TABLE 1.1

Some Useful Functions in Analog and Discrete Forms

1. Signum Function

$$sgn(t) = \begin{cases} 1 & t > 0 \\ 0 & t = 0 \\ -1 & t < 1 \end{cases} \quad ; \quad sgn(nT) = \begin{cases} 1 & nT > 0 \\ 0 & nT = 0 \\ -1 & nT < 0 \end{cases}$$

2. Step Function

$$u(t) = \frac{1}{2} + \frac{1}{2}sgn(t) = \begin{cases} 1 & t > 0 \\ 0 & t < 0 \end{cases} \quad ; \quad u(nT) = \begin{cases} 1 & nT > 0 \\ 0 & nT < 0 \end{cases}$$

3. Ramp Function

$$r(t) = \int_{-\infty}^{t} u(x)\,dx = tu(t); \quad r(nT) = nTu(nT)$$

4. Pulse Function

$$p_a(t) = u(t+a) - u(t-a) = \begin{cases} 1 & |t| < a \\ 0 & |t| > a \end{cases} \quad ; \quad p_a(nT) = u(nT + mT) - u(nT - mT)$$

(Continued)

TABLE 1.1 (*Continued*)
Some Useful Functions in Analog and Discrete Forms

5. Triangular Pulse

$$\Lambda_a(t) = \begin{cases} 1 - \dfrac{|t|}{a} & |t| < a \\ 0 & |t| > a \end{cases} \quad ; \quad \Lambda_a(nT) = \begin{cases} 1 - \dfrac{|nT|}{mT} & |nT| < mT \\ 0 & |nT| > mT \end{cases}$$

6. Sinc Function

$$\text{sinc}_a(t) = \frac{\sin at}{t} \quad -\infty < t < \infty; \qquad \text{sinc}_a(nT) = \frac{\sin anT}{nT} \quad -\infty < n < \infty$$

7. Gaussian Function

$$g_a(t) = e^{-at^2} \quad -\infty < t < \infty; \qquad g_a(nT) = e^{-a(nT)^2} \quad -\infty < n < \infty$$

8. Error Function

$$\text{erf}(t) = \frac{2}{\sqrt{\pi}} \int_0^t e^{-x^2}\,dx = \frac{2}{\sqrt{\pi}} \sum_{n=0}^{\infty} \frac{(-1)^n t^{2n+1}}{n!(2n+1)}$$

properties: $\text{erf}(\infty) = 1, \text{erf}(0) = 0, \text{erf}(-t) = -\text{erf}(t)$

$$\text{erfc}(t) = \text{complementary error function} = 1 - \text{erf}(t) = \frac{2}{\sqrt{\pi}} \int_t^{\infty} e^{-x^2}\,dx$$

Exponential and Double Exponential

$$f(t) = e^{-t}u(t) \qquad t \geq 0; \qquad f(t) = e^{-|t|} \qquad -\infty < t < \infty$$
$$f(nT) = e^{-nT}u(nT) \qquad nT \geq 0; \qquad f(nT) = e^{-|nT|} \qquad -\infty < nT < \infty$$

APPROXIMATION OF THE DERIVATIVE AND INTEGRAL

From Figure 1.6 we observe that we can approximate the sampled function $y(nT)$ of the derivative $y(t) = x'(t)$ of the signal $x(t)$ for sufficient small sampling time T as follows:

$$y(t) \triangleq x'(t) \cong \frac{x(t) - x(t-T)}{T} \tag{1.16}$$

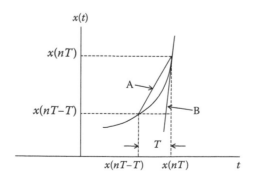

FIGURE 1.6

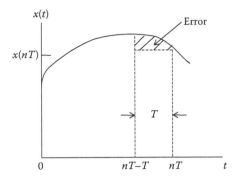

FIGURE 1.7

Therefore, its sampled form becomes

$$y(nT) \triangleq x'(nT) = \frac{x(nT) - x(nT - T)}{T} = \frac{1}{T}\Delta x(nT) \tag{1.17}$$

The approximation of the integral in its discrete form is shown in Figure 1.7. Therefore, we write

$$y(nT) = \int_0^{nT-T} x(t)\,dt + \int_{nT-T}^{nT} x(t)\,dt = y(nT - T) + \int_{nT-T}^{nT} x(t)\,dt$$

$$y(nT) \cong y(nT - T) + Tx(nT) \quad n = 0,1,2,3,\ldots$$

$$y(n) \cong y(n-1) + x(n) \quad n = 0,1,2,3,\ldots \quad T = 1 \tag{1.18}$$

Observe that the last integral was approximated by the rectangle $Tx(nT)$.

Exercise 1.2.1

Obtain the area under the function $f(t) = \exp(-t)$ from 0 to 2 and for $T = 0.5$ and $T = 0.1$ using the help of MATLAB. ▲

IMPULSE (DELTA) FUNCTION

The graphic representation of the **impulse** function, also known as the Dirac's **delta** function, is shown in Figure 1.8. Table 1.2 shows the delta functions properties. The value 1 indicates the area of the function. Its basic definition is

$$\begin{cases} \displaystyle\int_{-\infty}^{\infty} \delta(t)\,dt = 1 \\[2mm] \delta(t) = 0 \quad t \neq 0 \end{cases}$$

$$\int_{-\infty}^{\infty} f(t)\delta(t \pm t_0)\,dt = f(\mp t_0) \tag{1.19}$$

In words, the above formula tells us that the area under the delta function is 1 and its value on all points, besides the $t = 0$, is zero.

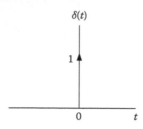

FIGURE 1.8

TABLE 1.2
Basic Delta Function Properties

$$\delta(at) = \frac{1}{|a|}\delta(t)$$

$$\delta\left(\frac{t - t_0}{a}\right) = |a|\,\delta(t - t_0)$$

$$\delta(-t) = \delta(t), \quad \delta(-t + t_0) = \delta(t - t_0) \qquad \delta(t) = \text{even function}$$

$$\int_{-\infty}^{\infty} \delta(t) f(t)\, dt = f(0) \qquad \int_{-\infty}^{\infty} \delta(t - t_0) f(t)\, dt = f(t_0)$$

$$f(t)\delta(t) = f(0)\delta(t) \qquad f(t)\delta(t - t_0) = f(t_0)\delta(t - t_0)$$

$$f(t) * \delta(t) = \text{convolution} = \int_{-\infty}^{\infty} f(t - x)\delta(x)\, dx = f(t), \quad f(t) * \delta(t - t_0) = f(t - t_0)$$

Note: The convolution of a function with a delta reproduces the function at the point where the delta function is located

$$\int_{-\infty}^{\infty} \frac{d\delta(t)}{dt}\, dt = 0 \qquad \frac{d\delta(t)}{dt} = \text{odd function}$$

$$\delta(t) = \frac{du(t)}{dt} \qquad u(t) = \text{unit step function}$$

$$\delta(t) = \lim_{\varepsilon \to 0} \frac{e^{-t^2/\varepsilon}}{\sqrt{\varepsilon\pi}}, \quad \delta(t) = \lim_{\omega \to \infty} \frac{\sin \omega t}{\pi t}, \quad \delta(t) = \lim_{\varepsilon \to 0} \frac{1}{\pi}\frac{\varepsilon}{t^2 + \varepsilon^2} \quad \text{(approximation of the delta function by a limited process of}$$

series of continuous functions)

$$\text{comb}_T(t) = \sum_{n=-\infty}^{\infty} \delta(t - nT) \qquad f(t)\text{comb}_T(t) = \sum_{n=-\infty}^{\infty} f(nT)\delta(t - nT)$$

Exercise 1.2.2

Plot the approximation of the delta function using the representation.

$$\delta(t) = \lim_{\varepsilon \to 0} \frac{e^{-t^2/\varepsilon}}{\sqrt{\varepsilon\pi}}$$

▲

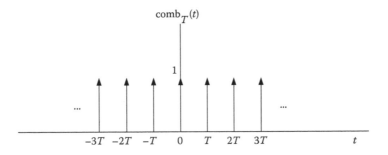

FIGURE 1.9

THE COMB FUNCTION

The **comb** function is represented mathematically as follows:

$$\text{comb}_T(t) = \sum_{n=-\infty}^{\infty} \delta(t - nT) \tag{1.20}$$

The function is represented graphically in Figure 1.9, and it is used for ideal sampling of a continuous signal.

Exercise 1.2.3

Evaluate the integrals:

$$(1) \int_{-2}^{2} (3t^2 + 1)[\delta(t) + 2\delta(t+1)]\,dt \quad (2) \int_{-2}^{2} e^{-2t}\cos 3t[\delta(t) + 2\delta(t-1)]\,dt \quad \blacktriangle$$

1.3   SIGNAL CONDITIONING AND MANIPULATION

MODULATION

To be able to transmit signals, such as speech in space as transmitted via radio signals, we must multiply the signal with a high frequency sinusoidal one, which is equivalent to saying that its wavelength is small, so that the transmitting antenna is manageable as far as its length is concerned. This is connected with the most efficient transmission of signals which is dictated by the antenna height which must be about 1/2 of the wavelength. Therefore, the transmitted signal takes the form

$$y(t) = f(t)\cos\omega_c t \quad f_c = \frac{1}{T_c} = \frac{\omega_c}{2\pi} \equiv \text{carrier frequency in cycles/second (Hz = hertz)} \tag{1.21}$$

Exercise 1.3.1

Find the signal which is received if the signal $s = \cos(100t)$ is modulated by $sm = \cos(2000t)$. $\blacktriangle$

SHIFTING AND FLIPPING

If the function is $f(t)$, then $f(-t)$ is its reflection with respect to the vertical axis. If the function is $f(t)$, then the function $f(t-t_0)$ is the same but shifted to the right by t_0. The function $f(t+t_0)$ is the same but shifted to the left by t_0.

TIME SCALING

The compression or expansion of a signal is known as **time scaling**. Let us assume that the original signal is

$$f(t) = \begin{cases} t+1 & -1 \leq t \leq 0 \\ -\dfrac{1}{2}(t-2) & 0 \leq t \leq 2 \end{cases} \tag{1.22}$$

If, now, we are asked to find the function $f(2.5t)$, we first insert the value $2.5t$ for each t in the above equation, and then write the time range inequalities as shown below:

$$f(2.5t) = \begin{cases} 2.5t+1 & -1 \leq 2.5t \leq 0 \\ -\dfrac{1}{2}(2.5t-2) & 0 \leq 2.5t \leq 2 \end{cases} \quad , \quad f(2.5t) = \begin{cases} 2.5t+1 & -\dfrac{1}{2.5} \leq t \leq 0 \\ -\dfrac{1}{2}(2.5t-2) & 0 \leq t \leq \dfrac{2}{2.5} \end{cases} \tag{1.23}$$

In the above case the function was compressed. If we had asked to find the function $f(0.65t)$, we would have found that this substitution creates an expansion to the original function.

WINDOWING OF SIGNALS

Since in practice, for example, we cannot take the Fourier transform for long signals we, in general, stop the signal between two points. This action is equivalent to multiplying the original signal by a rectangular window of unit height. This approach produces unwanted distortions in the spectra. To reduce these types of artifacts many types of windows have been proposed (Table 1.3).

TABLE 1.3
Windows for Continuous Signal Processing

Windows $w(t)$		First Side-Lobe Level in dB of its Fourier Transform
1. Rectangular window: $w_{rect} = \begin{cases} 1 \\ 0 \end{cases}$	$\begin{aligned} -\dfrac{T}{2} \leq t \leq \dfrac{T}{2} \\ \text{otherwise} \end{aligned}$	-13.3
2. Triangular (Bartlett): $w_{tr}(t) = 1 - \dfrac{\lvert t \rvert}{T}$	$-T \leq t \leq T$	-26.5
3. Hanning: $w_{hn}(t) = 0.5\left[1 + \cos\left(\dfrac{2\pi t}{T}\right)\right]$	$-\dfrac{T}{2} \leq t \leq \dfrac{T}{2}$	-31.5
4. Hamming: $w_{hm}(t) = 0.54 + 0.46\cos\left(\dfrac{2\pi t}{T}\right)$	$-\dfrac{T}{2} \leq t \leq \dfrac{T}{2}$	-42.7
5. Blackman: $w_{bl}(t) = 0.42 + 0.5\cos\left(\dfrac{2\pi t}{T}\right) + 0.08\cos\left(\dfrac{2\pi t}{T}\right)$	$-\dfrac{T}{2} \leq t \leq \dfrac{T}{2}$	-58.1

1.4   CONVOLUTION OF ANALOG AND DISCRETE SIGNALS

ANALOG SIGNALS

Convolution is a mathematical operation, and for any two real-valued functions their convolution is commonly indicated by an asterisk between the functions. Mathematically the convolution is given by

$$g(t) \triangleq f(t) * h(t) = \int_{-\infty}^{\infty} f(\tau)h(t-\tau)d\tau = \int_{-\infty}^{\infty} f(t-\tau)h(\tau)d\tau \qquad (1.24)$$

Note: *Equation 1.24 tells us the following: given two functions in the time domain t we find their convolution g(t) by doing the following steps: (1) re-write one of the functions in the τ domain by just replacing all instances of t with the variable τ; the shape of the function, and its position is identical to that in the t domain; (2) in the second function we substitute t−τ wherever we see t; this produces a function in the τ domain which is flipped (the minus sign in front the τ) and shifted by t (positive values of t shift the function to the right, and negative values shift the function to the left); (3) multiply these two functions, and find another function of τ, since t is a parameter and constant as far as the integration is concerned; (4) next find the area under the product function whose value is equal to the output of the convolution at t (in our case here is g(t)). By introducing the infinite values of t's, from minus infinity to infinity, we obtain the output function g(t).*

From the convolution integral we observe that one of the functions does not change when it is mapped from the t to τ domain. The second function is reversed or folded over (mirrored with respect to the vertical axis) in the τ domain, and it is shifted by an amount t, which is just a parameter in the integrand. Figure 1.10a and b show two functions in the t and τ domains, respectively. We now write

$$g(t) = f(t) * h(t) = \int_{-\infty}^{\infty} e^{-\tau}u(\tau)e^{-0.5(t-\tau)}u(t-\tau)d\tau = \int_{0}^{t} e^{-\tau}e^{-0.5(t-\tau)}\,d\tau$$

$$= e^{-0.5t}\int_{0}^{t} e^{-0.5\tau}\,d\tau = 2(-e^{-t}) = 2e^{-0.5t} - 2e^{-t} \qquad (1.25)$$

Figure 1.10c shows the results of the convolution.

Exercise 1.4.1

Find the convolution of the following two signals: $f(t) = u(t) - u(t-2)$ and $h(t) = e^{-(t-1)}u(t-1)$.   ▲

DISCRETE SIGNALS

The convolution of two continuous functions, as was presented in Equation 1.24, can be written as follows:

$$(1)\ g(t) = \int_{-\infty}^{\infty} f(x)h(t-x)dx = \sum_{m=-\infty}^{\infty}\int_{mT-T}^{mT} f(x)h(t-x)dx \quad \text{or}$$

$$(2)\ g(nT) = T\sum_{m=-\infty}^{\infty} f(mT)h(nT-mT) \quad \text{for} \quad n = 0, \pm 1, \pm 2, \dots \quad m = 0, \pm 1, \pm 2, \dots \qquad (1.26)$$

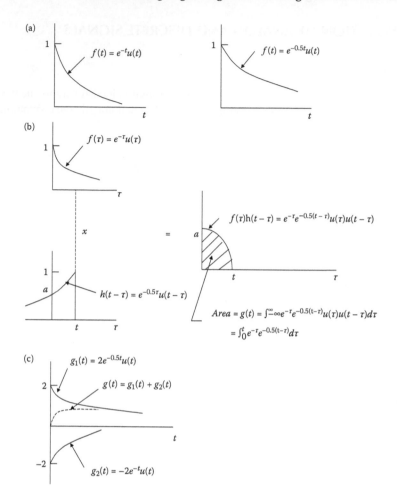

FIGURE 1.10

The result of the discretized convolution approaches the continuous case in the limit $T \to 0$.
   For $T = 1$, the above convolution equation becomes

$$g(n) = \sum_{m=-\infty}^{\infty} f(m)h(n-m) \quad n = 0, \pm 1, \pm 2, \ldots \quad m = 0, \pm 1, \pm 2, \ldots \tag{1.27}$$

If the input function of the system, $f(nT)$, is the delta function

$$f(nT) = \delta(nT) = \begin{cases} 1 & n = 0 \\ 0 & n \neq 0 \end{cases} \qquad \delta(nT - mT) = \begin{cases} 1 & n = m \\ 0 & n \neq m \end{cases} \tag{1.28}$$

with $T = 1$, then Equation 1.27 gives

$$g(n) = \sum_{m=-\infty}^{\infty} \delta(m)h(n-m) = \cdots + \delta(-1)h(n+1) + \delta(0)h(n-0) + \delta(1)h(n-1) + \cdots$$

$$= \cdots + 0h(n+1) + 1h(n) + 0h(n-1) + \cdots = h(n)$$

Example 1.4.1

It is desired to find the convolution of the two functions $f(n)=u(n-5)$ and $h(n)=\exp(-0.2(n+1))$ $u(n+1)$. An equivalent continuous type are the functions $f(t)=u(t-5)$ and $h(t)=\exp(-0.2(t+1))$ $u(t+1)$.

Solution: For the discrete type signals we use Equation 1.27. If we accept the $h(n)$ function unchanged in the m-domain, and the $f(n)$ function shifted and flipped, we produce the following two functions: $h(m)=\exp(-0.2(m+1))u(m+1)$ and $f(n-m)=u(n-m-5)$. For $n<4$, there is no overlapping of the two functions and, hence, the convolution is zero. For $n>4$ there will be an overlap from $m=-1$ to $m=n-5$ and hence

$$g(n)=\sum_{m=-1}^{n-5}e^{-0.2(m+1)} \tag{1.29}$$

For the continuous case and for $t>4$ we obtain (see Figure 1.11)

$$g(t)=\int_{-1}^{t-5}e^{-0.2(x+1)}\,dx=-\frac{e^{-0.2}}{0.2}\left(e^{-0.2(t-5)}-e^{0.2}\right) \tag{1.30}$$

The following Book MATLAB program was used to produce Figure 1.12.

```
>> t=4:0.2:20;
>>g=(-exp(-0.2)/0.2)*(exp(-0.2*(t-5))-exp(0.2));
>>for n=4:20
>>    for m=-1:n-5
>>            gd(n,m+2)=exp(-0.2)*exp(-0.2*m);%we added 2 because
>>            %MATLAB does not accept negative integers or zeros;
>>    end
>>end
>>gdo=sum(gd,2);%sum(v) sums all the elements of a vector v
>>              %sum(A,1) sums the rows of matrix A,sum(A,2)
>>              %sums the columns of the matrix A;
>>plot(t,g,'k');hold on; stem(gdo,'k');
>>xlabel('t,n'); ylabel('g(t), g(n)');
```

We observe that the two results are different. This is due to the fact that we adopted the sampling time $T=1$. If we proceeded to decrease the sampling time, however the result of the discrete case will tend to approach the exact value of the continuous case (Table 1.4). ∎

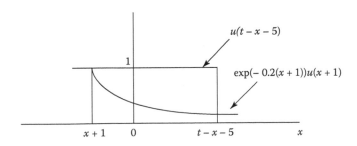

FIGURE 1.11

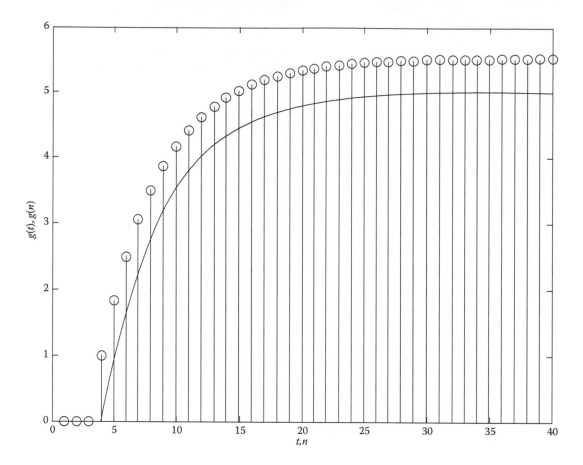

FIGURE 1.12

TABLE 1.4

Basic Convolution Properties

1. Commutative

$$g(t) = \int_{-\infty}^{\infty} f(x)h(t-x)\,dx = \int_{-\infty}^{\infty} f(t-x)h(x)\,dx$$

2. Distributive

$$g(t) = f(t) * [y(t) + z(t)] = f(t) * y(t) + f(t) * z(t)$$

3. Associative

$$[f(t) * y(t)] * z(t) = f(t) * [y(t) * z(t)]$$

4. Fourier transform

$$\mathcal{F}((t) * h(t)) = F(\omega)H(\omega)$$

5. Inverse FT

$$\frac{1}{2\pi} \int_{-\infty}^{\infty} F(\omega)H(\omega)e^{j\omega t}\,d\omega$$

6. Discrete-time

$$x(n) * y(n) = \sum_{m=-\infty}^{\infty} x(n-m)y(m)$$

7. Sampled

$$x(nT) * y(nT) = T \sum_{m=-\infty}^{\infty} x(nT-mT)y(mT)$$

1.5 MATLAB USE FOR VECTORS AND ARRAYS (MATRICES)

An example of numbers, vector, and matrices presentation in MATLAB format is

```
>>x=5;
>>y=[4,2,9];
>>z=[2  3  4; 5  6  7];
```

A **variable** is a single element which may be integer, real number or complex one. A **vector** is an ordered sequence of variables, like the y above. A **matrix** is an ordered sequence of vectors, like z above. The vector elements in MATLAB are separated by comma or by space. Matrix z is a 2 by 3 matrix.

EXAMPLES OF ARRAY OPERATIONS

Sum or difference + or −

$$\begin{bmatrix} 2 & 3 \\ 4 & 5 \end{bmatrix} + \begin{bmatrix} 1 & -4 \\ 4 & 1 \end{bmatrix} = \begin{bmatrix} 3 & -1 \\ 8 & 6 \end{bmatrix} ; \quad \begin{bmatrix} 2 \\ 3 \end{bmatrix} - 2 = \begin{bmatrix} 2 \\ 3 \end{bmatrix} - 2 \begin{bmatrix} 1 \\ 1 \end{bmatrix} = \begin{bmatrix} 2 \\ 3 \end{bmatrix} - \begin{bmatrix} 2 \\ 2 \end{bmatrix} = \begin{bmatrix} 0 \\ 1 \end{bmatrix}$$

Dimensions must be the same unless one variable is a scalar.

*Matrix product**

$$\begin{bmatrix} 1 & 2 & 3 \\ 4 & 5 & 6 \end{bmatrix}_{2\times3} * \begin{bmatrix} 0 & 0 \\ 1 & 1 \\ 2 & -2 \end{bmatrix}_{3\times2} = \begin{bmatrix} 1\times0+2\times1+3\times2 & 1\times0+2\times1+3\times(-2) \\ 4\times0+5\times1+6\times2 & 4\times0+5\times1+6\times(-2) \end{bmatrix}_{2\times2} = \begin{bmatrix} 7 & -4 \\ 17 & -7 \end{bmatrix}$$

Inner dimensions must be the same unless one variable is scalar. The resulting matrix has the dimensions of the outer numbers.

*Array product.**

$$\begin{bmatrix} 1 & 2 \\ -3 & -4 \end{bmatrix} .* \begin{bmatrix} 0 & 2 \\ 1 & -4 \end{bmatrix} = \begin{bmatrix} 0 & 4 \\ -3 & 16 \end{bmatrix} ; \quad 2.* \begin{bmatrix} 4 & 5 \\ 6 & 7 \end{bmatrix} = \begin{bmatrix} 8 & 10 \\ 12 & 14 \end{bmatrix}$$

Dimensions must be the same unless one variable is a scalar.

Transpose'

$$\begin{bmatrix} 1 & 2 & 3 \\ 4 & 5 & 6 \end{bmatrix}' = \begin{bmatrix} 1 & 4 \\ 2 & 5 \\ 3 & 6 \end{bmatrix} ; \quad [\ 1+j6 \quad 3-j2\ ]' = \begin{bmatrix} 1-j6 \\ 3+j2 \end{bmatrix} ; \quad \begin{bmatrix} 1 & 2 & 3 \\ 4 & 5 & 6 \end{bmatrix} .' = \begin{bmatrix} 1 & 4 \\ 2 & 5 \\ 3 & 6 \end{bmatrix} ;$$

$$[\ 1+j6 \quad 3-j2\ ].' = \begin{bmatrix} 1+j6 \\ 3-j2 \end{bmatrix}$$

Exponentiation

$$2.^{\wedge}[\ 1 \quad 2 \quad 3\ ] = [\ 2 \quad 4 \quad 8\ ]; \quad [\ 1 \quad 2 \quad 3\ ].^{\wedge}3 = [\ 1 \quad 8 \quad 27\ ];$$

$$\begin{bmatrix} 0 & 2 & 1 \\ 3 & 4 & 5 \end{bmatrix} .^{\wedge} \begin{bmatrix} 1 & 0 & 2 \\ 2 & 1 & 2 \end{bmatrix} = \begin{bmatrix} 0 & 1 & 1 \\ 9 & 4 & 25 \end{bmatrix}$$

Dimensions must be the same unless one variable is a scalar

Inverse

$$A = \begin{bmatrix} 1 & 2 \\ 3 & 4 \end{bmatrix}, B = \operatorname{inv}(A) = \begin{bmatrix} -2.0000 & 1.0000 \\ 1.5000 & -0.5000 \end{bmatrix}, A*B = \begin{bmatrix} 1.0000 & 0 \\ 0 & 1.0000 \end{bmatrix} = I$$

inv(.) is a MATLAB function that gives the inverse of a matrix. The product of a matrix with its inverse gives the **identity** matrix which has unit values along the main diagonal, and all the other elements are of zero value.

Exercise 1.5.1

Verify the above results using MATLAB.   ▲

HINTS–SUGGESTIONS–SOLUTIONS OF THE EXERCISES

1.1.1

1. $$P = \lim_{T \to \infty} \frac{1}{2T} \int_{-T}^{T} |f(t)|^2 \, dt = \lim_{T \to \infty} \frac{1}{2T} \int_{0}^{T} e^{-2t} \, dt = \lim_{T \to \infty} \frac{1}{4T}\left(1 - e^{-2T}\right) = 0 < \infty \Rightarrow \text{power}$$

2. $$\lim_{T \to \infty} \frac{1}{2T} \int_{0}^{T} dt = \lim_{T \to \infty} \frac{1}{2T}(T - 0) = \frac{1}{2} < \infty \Rightarrow \text{power signal}$$

1.1.2

Use the following Book MATLAB programs:

1. ```
>>a=0.98;n=0:0.2:100;x=a.^n.*sin(0.3*pi*n);plot(x);%decreasing
%sinusoidal;for increasing we can use, for example, a=1.01;
```

2. ```
>>t=0:0.001:5;n=length(t);s=sin(2*pi.*linspace(0,5,n).*t);
>>plot(t,s);%linspace(.) is a MATLAB function;
```

You can also try: $t=0{:}0.001{:}5$; $s = \sin(\exp(t))$; plot(s);

1.2.1

The suggested MATLAB Book program is

```
>>T=0.1; y(1)=0;
>>for n=0:1:2/T; %2/T must be integer;
>>y(n+2)=y(n+1)+T*exp(-T*n);
>>end
```

We can change T, at our will, to obtain the approximation we desire. For $T = 0.5$ the approximate area is 1.1664 and for $T = 0.1$ is 0.9222. The exact value is 0.8647.

1.2.2

Use the following Book MATLAB program:

```
>>t=-6:0.1:6; e=[1,0.2,0.01];
>>for i=1:3
        f(i,:)=exp(-t.^2/e(i))./sqrt(e(i)*pi);
        plot(t,f(i,:));hold on;
end
```

1.2.3

(1) $(3 \times 0^2 + 1) + 2(3(-1)^2 + 1)$,   (2) $e^{-2 \times 0} \cos(3 \times 0) + e^{-2} \cos(3)$

1.3.1

Two signals equal to the sum and difference of the original signals.

1.4.1

In this case we flip the exponential function. Because the exponential function was given at a shifted position at $t = 1$ then its flipped position on the τ-domain will be at -1 point. Hence

Range of t:

$$t < 0 \quad g(t) = 0$$

$$1 < t < 3$$

$$\int_0^{t-1} e^{-(t-\tau-1)} \, d\tau = e^{-(t-1)} \int_0^{t-1} e^{\tau} \, d\tau = e^{-(t-1)}(e^{(t-1)} - 1) = 1 - e^{-(t-1)}$$

$$3 < t < \infty$$

$$\int_0^2 e^{-(t-\tau-1)} \, d\tau = e^{-(t-1)}(e^2 - 1)$$

2 Fourier Analysis of Continuous and Discrete Signals

2.1 INTRODUCTION

Among the most fundamental information about any signal is its cyclic or oscillating activity. Our main interest is to determine the sinusoidal components of the signal, the range of their amplitude and frequency values.

Fourier analysis provides us with different tools, which are appropriate for particular types of signals. If the signal is periodic and deterministic, we use Fourier series analysis. If the signal is finite and deterministic, we use the Fourier transform. For discrete-time signals, the discrete Fourier transform is used, whereas for random signals need special approaches for their frequency content determination which will be explained in a later chapter.

2.2 FOURIER TRANSFORM (FT) OF DETERMINISTIC SIGNALS

The Fourier transform is used to find the frequency content of those signals that, at least, satisfy the Dirichlet conditions, which are: (1) the signal has finite number of its maximums and minimums within a finite interval, (2) the signal has finite number of discontinuities and (3) $\int_{-\infty}^{\infty} |f(t)| \, dt < \infty$, absolute integrative. The Fourier transform pair is $\omega = 2\pi f$, $f = \omega / 2\pi = 1/T_p, T_p =$ period in seconds, ω has units in rad/s, and f is the frequency having units cycles/s = Hz

$$F(\omega) \triangleq \mathscr{F}\{f(t)\} = \int_{-\infty}^{\infty} f(t)e^{-j\omega t} \, dt, \quad f(t) = \frac{1}{2\pi} \int_{-\infty}^{\infty} F(\omega)e^{j\omega t} \, d\omega \tag{2.1}$$

The above set of equations, which constitute a pair, gives the FT and its inverse FT (IFT). The imaginary factor $j = \sqrt{-1}$ is considered as a constant during the integration process.

Example 2.2.1

Find the FT, the amplitude and phase spectra of the functions (1) $f(t) = \exp(-t)u(t)$, (2) $f(t) = \exp(-(t-4))u(t-4)$, (3) $f(t) = \delta(t)$, (4) $\delta(t \pm 3)$.

Solution:

1. $\mathscr{F}\{e^{-t}u(t)\} = \int_{-\infty}^{\infty} e^{-t}u(t)e^{-j\omega t} \, dt = \int_{0}^{\infty} e^{-t}e^{-j\omega t} \, dt = -\frac{1}{1+j\omega}\left[e^{-t}e^{-j\omega t}\right]_{0}^{\infty} = \frac{1}{1+j\omega}$

$= \frac{1}{\sqrt{(1+\omega^2)}e^{j\tan^{-1}(\omega/1)}} = \frac{1}{\sqrt{(1+\omega^2)}}e^{-j\tan^{-1}(\omega)} = |F(\omega)|e^{j\varphi(\omega)}$

2. $\mathcal{F}\{e^{-(t-4)}u(t-4)\} = \int\limits_{4}^{\infty} e^{-(t-4)}e^{-j\omega t}\,dt = e^4\int\limits_{4}^{\infty} e^{-(1+j\omega)t}\,dt = -\dfrac{e^4}{1+j\omega}\Big[e^{-(1+j\omega)t}\Big]_{4}^{\infty}$

$$= \dfrac{1}{\sqrt{1+\omega^2}}e^{-j\tan^{-1}\omega}e^{-j4\omega} = \dfrac{1}{\sqrt{1+\omega^2}}e^{-j(\tan^{-1}\omega+4\omega)}$$

3. $\mathcal{F}\{\delta(t)\} = \int\limits_{-\infty}^{\infty}\delta(t)e^{-j\omega t}\,dt = e^{-j\omega 0} = 1$

4. $\mathcal{F}\{\delta(t\pm 3)\} = \int\limits_{-\infty}^{\infty}\delta(t\pm 3)e^{-j\omega t}\,dt = e^{-j\omega(\mp 3)} = e^{\pm j3\omega}$

$$\left|e^{\pm j3\omega}\right| = 1, \quad \text{angle}(e^{\pm j3\omega}) = \pm 3\omega \qquad\qquad \blacksquare$$

Example 2.2.2

Find the FT of the comb function, $\text{comb}_T(t)$, shown in Figure 2.1a.

Solution: To carry out this problem we must represent the comb function in Fourier series first. Hence, we find

$$\text{comb}_T(t) = \sum_{n=-\infty}^{\infty}\delta(t-nT) \triangleq \sum_{n=-\infty}^{\infty} a_n e^{j\omega_s t} = \dfrac{1}{T}\sum_{n=-\infty}^{\infty} e^{j\omega_s t}$$

$$a_n = \dfrac{1}{T}\int\limits_{-T/2}^{T/2}\delta(t)e^{-jn\omega_s t}\,dt = \dfrac{1}{T}e^{-jn\omega_s 0} = \dfrac{1}{T} \qquad \omega_s = \dfrac{2\pi}{T} \qquad (2.2)$$

The FT is then

$$\mathcal{F}\{\text{comb}_T(t)\} = \dfrac{1}{T}\sum_{n=-\infty}^{\infty}\mathcal{F}\{e^{j\omega_s t}\} = \dfrac{1}{T}\sum_{n=-\infty}^{\infty} 2\pi\delta(\omega-n\omega_s)$$

$$= \dfrac{2\pi}{T}\sum_{n=-\infty}^{\infty}\delta\left(\omega-n\dfrac{2\pi}{T}\right) \triangleq \dfrac{2\pi}{T}\text{COMB}_{\omega_s}(\omega) \qquad \omega_s = \dfrac{2\pi}{T} \qquad (2.3)$$

Figure 2.1b shows the FT of the $\text{comb}_T(t)$ function. $\blacksquare$

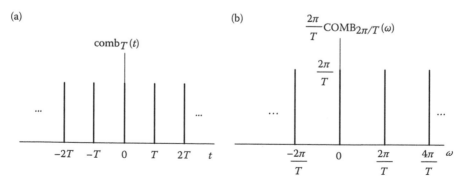

FIGURE 2.1

The FT properties are given in a Table in Appendix 2.1, and some fundamental pairs of FTs are given in a Table in Appendix 2.2

Example 2.2.3

Find the FT of the function sign(t).

Solution: We present the signum function by an approximation, and then take the limiting process. Hence, we write $sign(t) = \lim_{\varepsilon \to 0} \exp(-\varepsilon|t|)\,\text{sgn}(t)$. It is instructive to plot $u(t)$ for different values of ε. Hence, the FT is

$$\mathcal{F}\{\text{sgn}(t)\} = \mathcal{F}\left\{\lim_{\varepsilon \to 0} e^{-\varepsilon|t|}\text{sign}(t)\right\} = \lim_{\varepsilon \to 0} \int_{-\infty}^{\infty} e^{-\varepsilon|t|}\text{sign}(t)e^{-j\omega t}\,dt$$

$$= \lim_{\varepsilon \to 0}\left[\int_{-\infty}^{0} -e^{(\varepsilon - j\omega)t}\,dt + \int_{0}^{\infty} e^{-(\varepsilon + j\omega)t}\,dt\right]$$

$$= \lim_{\varepsilon \to 0}\left(-\frac{1}{\varepsilon - j\omega} + \frac{1}{\varepsilon + j\omega}\right) = \frac{2}{j\omega} = \frac{2}{\omega}e^{-j\pi/2} \triangleq \text{SGN}(\omega) \tag{2.4}$$

∎

Example 2.2.4

Find the FT of the of the unit step function $u(t)$.

Solution: We begin by writing $u(t)$ in its equivalent form:

$$u(t) = \frac{1}{2} + \frac{1}{2}\text{sgn}(t) \tag{2.5}$$

Therefore, the FT is

$$U(\omega) \triangleq \mathcal{F}\{u(t)\} = \frac{2\pi}{2}\delta(\omega) + \frac{2}{2j\omega} = \pi\delta(\omega) + \frac{1}{j\omega} \tag{2.6}$$

In this case, we cannot plot the amplitude and phase spectra but the real and imaginary parts of the spectrum. ∎

Note: *The FT operation on a deterministic continuous signal in the time domain transforms it to a continuous signal in the frequency domain. We must have in mind that only the positive frequencies are physically realizable. Negative frequencies are involved to complete mathematical operations in complex format.*

Exercise 2.2.1

Indicate, by an example, what is the effect of a time-shifted function on its FT spectrum. ▲

Exercise 2.2.2

Finding the FT of the function $f(t) = \exp(-2|t|)$, identify what type of spectrum this function is producing.   ▲

Exercise 2.2.3

Find the FT of the functions (1) $f(t) \odot f(t)$, autocorrelation, (2) $\dfrac{d^2 f(t)}{dt^2}$. The function $f(t)$, and its derivative is zero at minus and plus infinite.   ▲

2.3   SAMPLING OF SIGNALS

The ideal sampling, exact values of the continuous function at equal time distances, is accomplished by multiplying the signal with the $\text{comb}_T(t)$ function. The resulting sampled function is made up of delta functions, equally spaced apart. Figure 2.2a shows the exponential continuous signal, Figure 2.2b shows the $\text{comb}_T(t)$ function, and Figure 2.2c shows the resulting exponential sampled function. The sampled function is given mathematically as follows:

$$f_s(t) = f(t)\text{comb}_T(t) = f(t)\sum_{n=-\infty}^{\infty} \delta(t - nT) = \sum_{n=-\infty}^{\infty} f(nT)\delta(t - nT) \qquad n = 0, \pm 1, \pm 2, \dots \quad (2.7)$$

Since we are always forced to sample continuous functions to be processed by computers, it is natural to ask the question whether or not the spectrum of the original signal is modified by the sampling process. Hence, we write

$$F_s(\omega) = \mathcal{F}\{f(t)\text{comb}_T(t)\} = \frac{1}{2\pi}F(\omega) * \mathcal{F}\{\text{comb}_T(t)\} = \frac{1}{2\pi}F(\omega) * \left[\frac{2\pi}{T}\text{COMB}_{\omega_s}(\omega)\right]$$

$$= \frac{1}{2\pi}F(\omega) * \left[\frac{2\pi}{T}\sum_{n=-\infty}^{\infty} \delta(\omega - n\omega_s)\right]$$

$$= \frac{1}{T}\sum_{n=-\infty}^{\infty}\int_{-\infty}^{\infty} F(x)\delta(\omega - x - n\omega_s)dx$$

$$= \frac{1}{T}\sum_{n=-\infty}^{\infty} F(\omega - n\omega_s) \qquad \omega_s = \frac{2\pi}{T} \tag{2.8}$$

The above equation is based on the following FT property (see Appendix 2.1): the FT of the multiplication of two time functions is equal to the convolution of their FTs. Second, we used Equation 2.3 for the form of the FT of the $\text{comb}_T(t)$ function.

   Note: *Since the spectrum of the sampled function is the sum of shifted spectra of the signal, it is obvious that the resulting spectrum is distorted. If, however, we increase the value of the sampling frequency, or equivalently, we decrease the sampling time T then the distortion becomes less and less. Therefore, if we select a very small sampling time, the distortion becomes negligible and, hence, we will be able to filter the spectrum and recover the signal basically undistorted.*

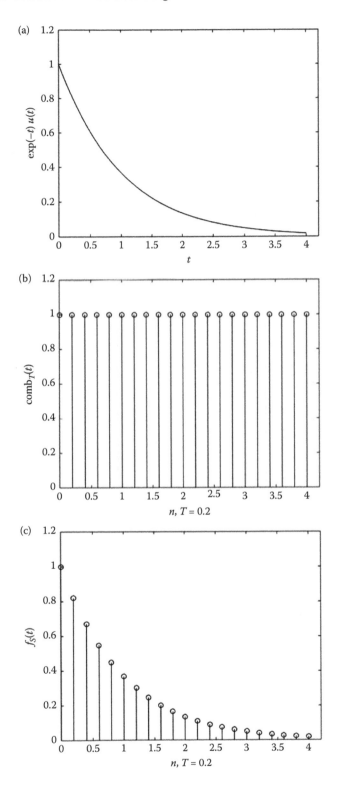

FIGURE 2.2

Sampling Theorem: *For any band-limited signal (signal with finite spectrum), with its highest frequency ω_N (Nyquist frequency), can be recovered completely from its sampled function if the sampling frequency ω_s is at least twice the Nyquist frequency ω_N, or the sampling time must be at least half the Nyquist sampling time $2\pi/\omega_N$. Basically, every detected signal, by any type of transducer, e.g., voltmeter, is band limited since any transducer stops responding at very high frequencies.*

In this text we will consider that the sampling time is small enough so that we neglect any distortion in the recovered signals.

Example 2.3.1

Find the FT of the following sampled continuous functions: (1) $f_s(t) = e^{-|t|}\text{comb}_T(t)$, (2) $f_s(t) = \text{sign}(t)\text{comb}_T(t)$.

Solution:

1. The FT of the function $f(t)$ is

$$\int_{-\infty}^{\infty} e^{-|t|}e^{-j\omega t}\,dt = \int_{-\infty}^{0} e^{t}e^{-j\omega t}\,dt + \int_{0}^{\infty} e^{-t}e^{-j\omega t}\,dt = \frac{1}{1-j\omega} + \frac{1}{1+j\omega} = \frac{2}{1+\omega^2}$$

Therefore, the FT of the sampled function is

$$\mathcal{F}\{e^{-|t|}\text{comb}_T(t)\} = F_s(\omega) = \frac{1}{T}\sum_{n=-\infty}^{\infty} \frac{2}{1+(\omega+n\omega_s)^2} \qquad \omega_s = \frac{2\pi}{T} \tag{2.9}$$

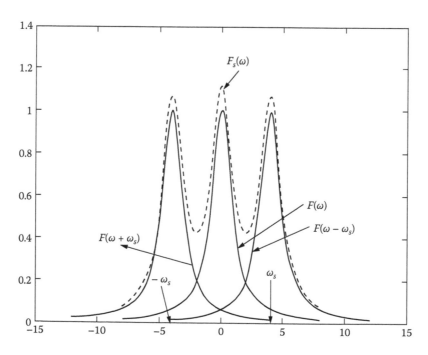

FIGURE 2.3

2. The FT of the sampled function (see Equation 2.4) is

$$\mathscr{F}\{\text{sign}(t)\text{comb}_T(t)\} = \frac{1}{T}\sum_{n=-\infty}^{\infty}\frac{2}{j(\omega-\omega_s)} \qquad (2.10)$$

∎

The FT of the sampled function given in Equation 2.9 is shown in Figure 2.3 for three of the infinite number of spectrums. We observe that the spectrum of the sampled function is periodic and expands from minus infinity to infinity.

From Figure 2.3 we observe that if the spectrum of a signal is a finite, **band limited** signal, we can always select the sampling time T small enough so that the spectrum of the sampled function will be constituted of discrete non overlapping spectra of the unsampled function. This gives us the opportunity to filter the spectrum and recapture the signal with all its details.

Exercise 2.3.1

Use MATLAB to find the inverse FT of the function $\exp(-|t|)$ using the ranges (1) $-2 \leq \omega \leq 2$
(2) $-20 \leq \omega \leq 20$.   ▲

2.4   DISCRETE-TIME FOURIER TRANSFORM (DTFT)

As $T \to 0$, we can approximate the FT integral to obtain an approximate FT known as the DTFT of an ideal sampled signal. Hence, we have

$$F(\Omega) = \mathscr{F}\{f(t)\} = \int_{-\infty}^{\infty} f(t)e^{-j\Omega t}\,dt = \sum_{n=-\infty}^{\infty}\int_{nT}^{nT+T} f(t)e^{-j\Omega t}\,dt$$

$$\cong \sum_{n=-\infty}^{\infty} Tf(nT)e^{-j\Omega nT} \quad T \to 0 \quad \Omega \text{ rad/s} \qquad (2.11)$$

If we set

$$f(n) = Tf(nT), \quad \omega = \Omega T \text{ rad} \qquad (2.12)$$

in the above equation, we obtain the DTFT with $T=1$ to be

$$\mathscr{F}_{DT}\{f(n)\} \triangleq F(\omega) \triangleq \left|F(e^{j\omega})\right|e^{j\varphi(\omega)} \triangleq F(e^{j\omega}) = \sum_{n=-\infty}^{\infty} f(n)e^{-j\omega n} \qquad (2.13)$$

Note: *The discrete frequency ω is equal to ΩT and has the units of radians, where Ω has units of radians per second. If T is unity time, then the discrete frequency and the continuous frequency have the same values but different units. Both frequencies, Ω and ω are continuous independent variables.*

Therefore, the following steps can be taken to approximate the continuous-time Fourier transform:

1. Select the time sampling T such that $F(\Omega) \cong 0$ for all $|\Omega| > \pi/T$.
2. Sample $f(t)$ at times nT to obtain $f(nT)$.
3. Compute the DTFT using the sequence $\{Tf(nT)\}$.
4. The resulting approximation is then $F(\Omega) \cong F(\omega)$ for $-\dfrac{\pi}{T} \leq \Omega \leq \dfrac{\pi}{T}$.

The result $\exp(-j(\omega+(2\pi/T))nT)=\exp(-j\omega nT)\exp(-j2\pi n)=\exp(-j\omega nT)$ indicates that the spectrum of the sampled function with sampling time T is periodic with period $2\pi/T$. Therefore, the **inverse** DTFT is given by

$$f(nT)\triangleq\mathscr{F}_{DT}^{-1}\{F(\omega)\}=\frac{1}{2\pi}\int_{-\pi/T}^{\pi/T}F(\omega)e^{j\omega nT} \tag{2.14}$$

Therefore, Equations 2.13 and 2.14 constitute the DTFT pair (see Appendix 2.3).

Example 2.4.1

Determine the DTFT of the signal $f(t)=\exp(-t)u(t)$, for $T=1$ and $T=0.1$.

Solution:

1. From Equation 2.13, we obtain

$$F(e^{j\omega})\triangleq F(\omega)=\sum_{n=0}^{\infty}e^{-n}e^{-j\omega n}=\sum_{n=0}^{\infty}e^{-(1+j\omega)n}=\sum_{n=0}^{\infty}r^n=\frac{1}{1-e^{-(1+j\omega)}},\quad r=e^{-(1+j\omega)}$$

since the summation is an infinite geometric series (see Appendix 2.1).

2. Taking into consideration the sampling time in Equation 2.13, we find

$$F(e^{j\Omega T})=T\sum_{n=0}^{\infty}f(nT)e^{-j\Omega Tn}=0.1\sum_{n=0}^{\infty}e^{-nT}e^{-j\Omega Tn}=0.1\sum_{n=0}^{\infty}e^{-(0.1+j\Omega 0.1)n}$$

$$=\frac{0.1}{1-e^{-0.1}e^{-j\Omega 0.1}}=\frac{0.1}{1-e^{-0.1}e^{-j\omega}}\qquad \Omega=\frac{\omega}{T}=\frac{\omega}{0.1}$$

The magnitude and phase of the above equation are

$$F(e^{j\Omega 0.1})=F(e^{j\omega})=\frac{0.1}{(1-e^{-0.1}\cos 0.1\Omega)+j(e^{-0.1}\sin 0.1\Omega)}$$

$$=\frac{0.1}{\sqrt{(1-e^{-0.1}\cos\omega)^2+(e^{-0.1}\sin\omega)^2}}e^{-j\tan^{-1}\frac{e^{-0.1}\sin\omega}{1-e^{-0.1}\cos\omega}}$$

If we set $\omega=-\omega$ in the equation above, we observe that the magnitude of the spectrum is **even**, and the phase spectrum is an **odd** function.

Note: *For real functions the amplitude spectrum is even, and the phase spectrum is an odd function. Thus the representation of the frequency spectrum within the range* $0\le\frac{\omega}{T}\le 2\pi$ *or* $-\pi\le\frac{\omega}{T}\le\pi$ *will suffice. The smaller the T the longer the spectrum of the discretized function will approximate the exact value of the continuous function.*

The reader should plot the magnitude of the spectrums given above to observe the differences. Furthermore, if we set $T=0.01$, the spectrum will closely approximate the exact one to

$\pi/0.01 = 100\pi$. For example, its magnitude at zero frequency is $0.01/(1 - \exp(-0.01)) = 1.0050$, which is very close to 1, as it should be. ∎

Note: *For continuous signals, we first discretize the signal every T. Next, we ignore the sampling time T and proceed to find the DTFT with equivalent T = 1. The spectrum is important from $-\pi/2$ to $\pi/2$ or from 0 to π and the unit is in radians per. To translate the spectrum to the equivalent one, corresponding to the continuous signal, we divide π by T. Therefore, we create a rad/s axis. This means that the range from 0 to π/T is the useful spectrum after the discretization of the continuous signal. Therefore, the smaller the T the longer the spectrum of the continuous signal is useful.*

Example 2.4.2

Find the DTFT of the function $g(n) = 0.9^n u(n) \cos 0.1\pi n$.

Solution: We observe that $0.9^n u(n)$ is a modulated function. Modulation is the process of multiplying any signal with a cosine or sine function. This is the most useful process in speech broadcasting in radio and television. Hence,

$$\mathcal{F}\{0.9^n u(n)\cos 0.1\pi n\} = \mathcal{F}\left\{ \frac{0.9^n u(n)e^{j0.1\pi n}}{2} + \frac{0.9^n u(n)e^{-j0.1\pi n}}{2} \right\}$$

$$= \frac{1}{2}\sum_{n=0}^{\infty}(0.9e^{-j(\omega-0.1\pi)})^n + \frac{1}{2}\sum_{n=0}^{\infty}(0.9e^{-j(\omega+0.1\pi)})^n = \frac{1}{2}\frac{1}{1-0.9e^{-j(\omega-0.1\pi)}}$$

$$+ \frac{1}{2}\frac{1}{1-0.9e^{-j(\omega+0.1\pi)}} = \frac{1}{2}G(e^{j(\omega-0.1\pi)}) + \frac{1}{2}G(e^{j(\omega+0.1\pi)})$$

We must always have in mind that the spectrum of discrete signals is infinitely periodic. Therefore, the magnitude of $G(e^{j\omega})$ is a double periodic structure, one shifted to the left by 0.1π, and the other shifted to the right by 0.1π. Since in this case $T = 1$, the spectrum can be plotted from $-\pi/1$ to $\pi/1$. To plot the spectrum the reader can use the following Book m-file.

Book MATLAB m-file: ex2_4_2

```
%ex2_4_2 is an m-file for illustrating Ex 2.4.2
w=-pi:0.01:pi;
f1=abs(0.5*(1./(1-0.9*exp(-j*(w-0.1*pi)))));
f2=abs(0.5*(1./(1-0.9*exp(-j*(w+0.1*pi)))));
      %exp(.) is a MATLAB function;
f=f1+f2;
plot(w,f,'k');
grid on;
```

If we execute the program, by simply writing in the command window ex2_4_2, we observe that the peak is at $0.1 \times 3.14 = 0.314$ rad/unit, as it should be. Before the execution, we must be sure that the MATLAB path includes the director with the Book MATLAB functions of this book, see Appendix 2.1. ∎

Exercise 2.4.1

Verify Parseval's formula using the function

$$F(e^{j\omega}) = \begin{cases} 1 & -\pi < \omega < \pi \\ 0 & \text{otherwise} \end{cases} \qquad \blacktriangle$$

Exercise 2.4.2

Verify the accuracy of the DFT, using $T=1$ and $T=0.2$ and for the time length of the function $f(t) = \exp(-|t|)$ from −5 to 5, with respect to FT. $\qquad \blacktriangle$

2.5 DTFT OF FINITE-TIME SEQUENCES

Practical considerations usually dictate that we deal with truncated series. Therefore, the spectrum of a truncated series requires special attention. When we observe the values of a function at a finite number of data points, $f(0), f(1), f(2), \ldots, f(N-1)$, how do we account for the unobserved time series elements that lie outside the measured interval $0 \le n \le N-1$? We must consider the effect of the missing data outside the given finite range. That is when all the data of the function are given to have an exact DTFT of the discrete band-limited function.

The one-sided truncated DTFT is defined by

$$F_N(e^{j\omega}) = \sum_{n=0}^{N-1} f(n)e^{-j\omega n} \qquad (2.15)$$

We introduce the DTFT of $f(n)$ in to above equation to obtain

$$F_N(e^{j\omega}) = \sum_{n=0}^{N-1} \left[\frac{1}{2\pi} \int_{-\pi}^{\pi} F(e^{j\omega'})e^{j\omega'n}\, d\omega' \right] e^{-j\omega n}$$

$$= \frac{1}{2\pi} \int_{-\pi}^{\pi} F(e^{j\omega'}) \sum_{n=0}^{N-1} e^{j(\omega-\omega')n}\, d\omega' = \frac{1}{2\pi} \int_{-\pi}^{\pi} F(e^{j\omega'})W(e^{j(\omega-\omega')})d\omega'$$

$$= \frac{1}{2\pi} F(e^{j\omega}) * W(e^{j\omega}) \qquad (2.16)$$

$$W(e^{j\omega}) = \sum_{n=0}^{N-1} e^{-j\omega n} = \frac{1-(e^{-j\omega})^N}{1-e^{-j\omega}} = e^{-j\omega(N-1)/2} \frac{\sin(\omega N/2)}{\sin(\omega/2)}$$

where the finite geometric series formula was used (see Appendix 2.1), and the sign $*$ indicates the convolution in the frequency domain. The transform function $W(e^{j\omega})$ is the DTFT of the rectangular window, since it is the transform of the time function $w(n) = u(n) - u(n-N) = p_{N/2}(n-(N/2))$ for even N. We observe that with a finite sequence a convolution appears in the frequency domain. From Equation 2.16 we observe that to find the exact spectrum $F(e^{j\omega})$ we require a Fourier-transformed window $W(e^{j\omega})$ equal to a delta function $\delta(\omega)$ in the interval $-\pi \le \omega \le \pi$. However, the magnitude of $|W(e^{j\omega})| = \sin(\omega N/2)/\sin(\omega/2)$ has the properties of a delta function which approaches it as

$N \to \infty$. Therefore, the longer the time-data sequence we observe, the less distortion will occur in the spectrum of the signal, $F(e^{j\omega})$. This is because the convolution of a function with a delta function produces the function at the point where the delta is located.

We observe that

$$\cos \omega_0 n = \mathscr{F}_{DT}^{-1}\{\pi\delta(\omega - \omega_0) + \pi\delta(\omega + \omega_0)\} \tag{2.17}$$

Exercise 2.5.1

Verify Equation 2.17. ▲

Therefore, the DTFT of the function (see Equation 2.17)

$$f(n) = \cos(\omega_0 n) + \cos[(\omega_0 + \Delta\omega_0)n] \tag{2.18}$$

is

$$F(e^{j\omega}) = \pi[\delta(\omega - \omega_0) + \delta(\omega + \omega_0) + \delta(\omega - \omega_0 - \Delta\omega_0) + \delta(\omega + \omega_0 + \Delta\omega_0)] \tag{2.19}$$

The convolution of Equation 2.19 with Equation 2.16 gives the following expression:

$$F_N(e^{j\omega}) = \frac{1}{2} e^{-j(\omega-\omega_0)(N-1)/2} \frac{\sin\frac{\omega-\omega_0}{2}N}{\sin\frac{\omega-\omega_0}{2}} + \frac{1}{2} e^{-j(\omega+\omega_0)(N-1)/2} \frac{\sin\frac{\omega+\omega_0}{2}N}{\sin\frac{\omega+\omega_0}{2}}$$

$$+ \frac{1}{2} e^{-j(\omega-\omega_0-\Delta\omega_0)(N-1)/2} \frac{\sin\frac{\omega-\omega_0-\Delta\omega_0}{2}N}{\sin\frac{\omega-\omega_0-\Delta\omega_0}{2}}$$

$$+ \frac{1}{2} e^{-j(\omega+\omega_0+\Delta\omega_0)(N-1)/2} \frac{\sin\frac{\omega+\omega_0+\Delta\omega_0}{2}N}{\sin\frac{\omega+\omega_0+\Delta\omega_0}{2}} \tag{2.20}$$

If we set $\omega_0 = 0.2\pi$ and $\Delta\omega_0 = 0.1\pi$ and plot $\left|F_N(e^{j\omega})\right|$ in the range $0 \le \omega \le \pi$ for $N=50$, $N=25$ and $N=10$ we will observe the following: (1) $\Delta\omega_0 = 0.1\pi \gg \frac{2\pi}{N} = \frac{2\pi}{50}$, and we can observe that we can see two peaks. This implies that we can differentiate two sinusoidal signals that are close in frequency, (2) If we use $N=25$, then $\Delta\omega_0 = 0.1\pi > \frac{2\pi}{N} = \frac{2\pi}{25}$, and the peaks barely can be seen, (3) If we use $N=10$, then $\Delta\omega_0 = 0.1\pi < \frac{2\pi}{N} = \frac{2\pi}{10}$, and we will observe only one peak. This implies that we cannot differentiate the two sinusoidal frequencies that are close to each other. Since the limiting value is $2\pi/N$, then as $N \to \infty$, the value $2\pi/N$ approaches zero. This indicates that we can separate two sinusoids with infinitesimal differences in their frequencies if we take enough signal values. The reader should use MATLAB to verify the above.

WINDOWING

It turns out that the Fourier-transform spectrum will depend on the type of window used. It is appropriate that the window functions used are such that the smoothed version of the spectrum resembles the exact spectrum as closely as possible. Typically it is found that for a given value N, the smoothing effect is directly proportional to the width of the main lobe of the window, and the effects of ripple decrease as the relative amplitude of the main lobe and the largest side-lobe diverge. A few important discrete windows are as follows:

1. von Hann or Hanning:

$$w(n) = \frac{1}{2}\left[1 - \cos\left(\frac{2\pi n}{N-1}\right)\right] \qquad 0 \leq n < N \tag{2.21}$$

2. Bartlett:

$$w(n) = \begin{cases} \dfrac{2n}{N-1} & 0 \leq n \leq \dfrac{N-1}{2} \\ 2 - \dfrac{2n}{N-1} & \dfrac{N-1}{2} \leq n < N \end{cases} \tag{2.22}$$

3. Hamming:

$$w(n) = 0.54 - 0.46\cos\left(\frac{2\pi n}{N-1}\right) \qquad 0 \leq n < N \tag{2.23}$$

4. Blackman:

$$w(n) = 0.4 - 0.5\cos\left(\frac{2\pi n}{N-1}\right) + 0.08\cos\left(\frac{4\pi n}{N-1}\right) \qquad 0 \leq n < N \tag{2.24}$$

Example 2.5.1

Find he Fourier transform of the discrete function $f(n)=0.95^n u(n)$ using $N=21$, and compare the spectrums of the rectangular window and that of the Hamming one.

Solution: Figure 2.4a shows the spectrum using the rectangular window, Figure 2.4b shows the spectrum using the Hamming window, and Figure 2.4c shows the exact spectrum. The following Book m-file was used:

Book m-file: ex2_5_1

```
%ex2_5_1 is a book m-file that illustrates Ex. 2.5.1
N=21;n=0:N-1;w=0:0.01:pi;
wh=0.54-0.46*cos(2*pi*n'/(N-1));%Nx1 vector
f1=abs((0.95.^n')'*exp(-j*n'*w));% f1=Nx(length(w))matrix;
f1s=sum(f1,1);%sums the rows of the
    %Nxlength(w) matrix f1;
f2h=abs((0.95.^n'.*wh)'*exp(-j*n'*w));
f2hs=sum(f2h,1);
fwe=abs(1./(1-0.95*exp(-j*w)));%see Ex. 2.4.1a;
subplot(2,3,1);plot(w,f1s,'k');xlabel('\omega');
ylabel('Magnitude');
```

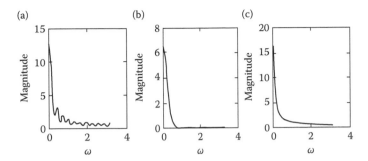

FIGURE 2.4

```
subplot(2,3,2);plot(w,f2hs,'k');xlabel('\omega');
ylabel('Magnitude');
subplot(2,3,3);plot(w,fwe,'k');xlabel('\omega');
ylabel('Magnitude');
    %ex2_5_1 is a book m-file that illustrates Ex. 2.5.1
```

It is recommended to plot the above figures in normalized form. Use e.g plot(w,f1s/abs(max(f1s))). ∎

2.6 THE DISCRETE FOURIER TRANSFORM (DFT)

As we have seen above, if a time function is sampled uniformly in time, its Fourier spectrum is a periodic function. Therefore, corresponding to any sampled function in the frequency domain, a periodic function exists in the time domain. As a result, the sampled signal values can be related in both domains.

As a practical matter, we are only able to manipulate a certain length of a signal. That is, suppose that the data sequence is available only within a finite-time window from $n=0$ to $n=N-1$. The transform is discretized for N values by taking samples at the frequencies $\Omega_b = 2\pi / NT$, where T is the time interval between sample points in the time domain (sampling time). Hence, we define the **discrete Fourier transform** (DFT) of a sequence of N samples $\{f(nT)\}$ by the relation

$$F(k\Omega_b) \triangleq F(e^{jk\Omega_b}) = \mathscr{F}_D\{f(nT)\} = T\sum_{n=0}^{N-1} f(nT)e^{-j\Omega_b Tkn}$$

$$\Omega_b = \frac{2\pi}{NT}; \quad \omega_s = \frac{2\pi}{T} = \text{sampling frequency}; \quad 0 \leq n \leq N-1 \qquad (2.25)$$

where

N = number of samples (even number)
T = sampling time interval
$(N-1)T$ = signal length in time domain
$\Omega_b = \dfrac{2\pi}{NT} = \dfrac{\omega_s}{N} =$ the frequency sampling interval (frequency bin) equal to sampling
frequency divided by the number of the sequence elements N
$e^{-j\Omega_b T} = N$th principal root unity.

The Inverse DFT (IDFT)

The inverse DFT (IDFT) is related to DFT in much the same way that the Fourier transform is related to its inverse transform. IDFT is given by

$$f(nT) \triangleq F_D^{-1}\{F(k\Omega_b)\} = \frac{1}{NT} \sum_{k=0}^{N-1} F(k\Omega_b)e^{j\Omega_b Tnk}$$

$$\Omega_b = \frac{2\pi}{NT} = \frac{\omega_s}{N}; \quad 0 \le k \le N-1 \tag{2.26}$$

It turns out that both sequences are **periodic** with period N. This stems from the fact that the exponential function is periodic as shown below:

$$e^{\pm 2\pi nk/N} = e^{\pm 2\pi(n+N)k/N}, e^{\pm 2\pi nk/N} = e^{\pm 2\pi n(k+N)/N}, \quad k, n = 0, 1, 2, \ldots, N-1 \tag{2.27}$$

This stems from the fact that $\exp(\pm 2\pi nk) = 1$ for any integer n and k.

Exercise 2.6.1

Verify Equation 2.26.   ▲

Exercise 2.6.2

(1) Find the DFT of the frequency varying (chirp) sinusoidal; (2) Find the DFT of a time varying signal (increasing or decreasing). The results are shown in Figure 2.19.   ▲

2.7   PROPERTIES OF DFT

To simplify the proofs, without any loss of generality, we set $T = 1$ and $k2\pi / N = k$.

Linearity

The DFT of the function $f(n) = ax(n) + by(n)$ is

$$F(k) = \mathcal{F}_D\{ax(n) + by(n)\} = a\mathcal{F}_D\{x(n)\} + b\mathcal{F}_D\{y(n)\} = aX(k) + bY(k) \tag{2.28}$$

This property is the direct result of Equation 2.25.

Symmetry

If $f(n)$ and $F(k)$ are a DFT pair, then

$$\mathcal{F}\left\{\frac{1}{N}F(n)\right\} = f(-k) \tag{2.29}$$

Exercise 2.7.1

Verify Equation 2.29.   ▲

TIME SHIFTING

For any real integer m

$$\mathscr{F}_D\{f(n-m)\} = F(k)e^{-j2\pi mk/N} \tag{2.30}$$

The above equation indicates that if the function is shifted in the time domain by m units to the right its spectrum magnitude does not change. However, its phase spectrum is changed by the addition of a linear phase factor equal to $-2\pi m/N$.

Exercise 2.7.2

Verify Equation 2.30.   ▲

FREQUENCY SHIFTING

For any integer m

$$f(n)e^{j2\pi nm} = \mathscr{F}_D^{-1}\{F(k-m)\} \tag{2.31}$$

TIME CONVOLUTION

Based on the discrete-time convolution (see Chapter 1), and due to the fact that both of the signals are periodic with the sample period N,

$$f(n) = f(n+pN) \quad p = 0,\pm1,\pm2,\ldots, \quad h(n) = h(n+pN) \quad p = 0,\pm1,\pm2,\ldots \tag{2.32}$$

the convolution is known as **circular** or **cyclic** one. The DFT of the convolution expression yields

$$Y(k) = \mathscr{F}_D\{f(n)*h(n)\} = F(k)H(k) \tag{2.33}$$

Exercise 2.7.3

Verify Equation 2.33.   ▲

Figure 2.5 shows graphically the convolution of the following two periodic sequences, $f(n) = \{1, 2, 3\}$ and $h = \{1, 1, 0\}$. Figure 2.5a shows the two sequences at 0 shifting, and the result is given by the sum of the products: $1\times1+2\times0+3\times1 = 4$. Next, we shift the inside circle by one step counterclockwise, and the output is given by the sum of the following products: $1\times1+2\times1+3\times0 = 3$ and so on. Note that $y(n)$ is periodic, and, since $F(k)$ and $H(k)$ are periodic, $Y(k)$ is also periodic.

Another approach to the evaluation of the circular convolution is to write the convolution equation in the form given below for the particular case when each sequence has three terms:

(a)

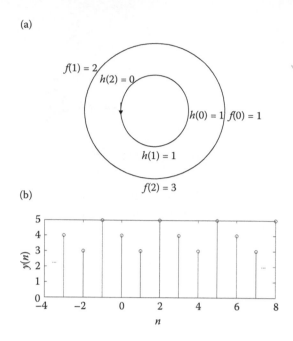

(b)

FIGURE 2.5

$$y(0) = f(0)h(0) + f(1)h(2) + f(2)h(1)$$

$$y(1) = f(0)h(1) + f(1)h(0) + f(2)h(2)$$

$$y(2) = f(0)h(2) + f(1)h(1) + f(2)h(0) \qquad (2.34)$$

This set of equation can be written in matrix from as follows:

$$[y]^T = [\ y(0) \quad y(1) \quad y(2)\ ] = [\ f(0) \quad f(1) \quad f(2)\ ] \begin{bmatrix} h(0) & h(1) & h(2) \\ h(2) & h(0) & h(1) \\ h(1) & h(2) & h(0) \end{bmatrix} \qquad (2.35)$$

If we want to produce a linear convolution, using the matrix format, we must pad the sequences with zeros so that their total length must be equal to $N = F + H - 1$, where F and H are the number of the elements of the corresponding sequences.

Example 2.7.1

Consider the two periodic sequences $f(n) = [1, -1, 4]$ and $h(n) = [0, 1, 3]$. Verify the convolution property for $n = 2$.

Solution: In this case $T = 1$, $\Omega_b = 2\pi/3$, and

$$y(2) = \sum_{m=0}^{N-1} f(m)h(n-m) = \mathscr{F}_D^{-1}\{F(k)H(k)\} = \frac{1}{N} \sum_{k=0}^{N-1} F(k)H(k)e^{j2\pi k2/N}$$

First, we find the summation:

$$\sum_{m=0}^{2} f(m)h(2-m) = f(0)h(2) + f(1)h(1) + f(2)h(0) = 1 \times 3 + (-1) \times 1 + 4 \times 0 = 2$$

Next, we obtain $F(k)$ and $H(k)$:

$$F(0) = \sum_{n=0}^{2} f(n)e^{-j2\pi 0 n/3} = f(0) + f(1) + f(2) = 1 - 1 + 4 = 4$$

$$F(1) = \sum_{n=0}^{2} f(n)e^{-j2\pi 1 n/3} = f(0) + f(1)e^{-j2\pi/3} + f(2)e^{-j2\pi 2/3} = -0.5 + j5 \times 0.866$$

$$F(2) = \sum_{n=0}^{2} f(n)e^{-j2\pi 2 n/3} = f(0) + f(1)e^{-j2\pi 2/3} + f(2)e^{-j2\pi 4/3} = -0.5 - j5 \times 0.866$$

Similarly, we obtain

$$H(0) = 4 \quad H(1) = -2 + j2 \times 0.866 \quad H(2) = -2 - j2 \times 0.866$$

The second summation given above becomes

$$\frac{1}{3}\sum_{k=0}^{2} F(k)H(k)e^{j2\pi 2k/3} = \frac{1}{3}[16 + 11.55e^{j115.7} + 11.55e^{j604.3}] = 2$$

This result shows the validity of the DFT of the convolution.   ■

FREQUENCY CONVOLUTION

The frequency convolution property is given by

$$\mathscr{F}_D\{g(n)\} = \mathscr{F}_D\{f(n)h(n)\} = \frac{1}{N}\sum_{m=0}^{N-1} F(m)H(k-m) \tag{2.36}$$

where $g(n) = f(n)h(n)$.

PARSEVAL'S THEOREM

$$\sum_{n=0}^{N-1} f^2(n) = \frac{1}{N}\sum_{k=0}^{N-1} |F(k)|^2 \tag{2.37}$$

2.8   EFFECT OF SAMPLING TIME *T*

Let us deduce the DFT of the continuous function

$$f(t) = e^{-t} \quad 0 \le t \le 1 \tag{2.38}$$

for the following three cases:

1. $T=0.5, NT=8$
2. $T=0.5, NT=16$
3. $T=0.1, NT=8$

Observe that we change the values of two constants. The first is the time sampling T which defines the frequency resolution in the frequency domain. This means that two frequencies close together can be identified in the frequency domain when T is small enough. The other constant identifies the frequency bins.

Case a: In this case we find $N=8/T=8/0.5=16$ and $\Omega_b = 2\pi/(NT) = 2\pi/8$. The fold over frequency for this case is at $N/2=8$, or at $8\times\Omega_b = 8\times[2\pi/(0.5\times16)] = 2\pi$ rad/s. The Fourier amplitude spectrums for the continuous and discrete case are shown in Figure 2.6a. The exact spectrum is found to be equal to: $F(\omega) = (1-\exp(-(1+j\omega)))/(1+j\omega)$. Although the spectrum of the continuous signal is continuous and infinite in length, we only present the values which correspond to the spectrum of the discrete case. We observe that the errors exist between the discrete and continuous spectrums up to the fold over frequency, and these errors become larger for the rest of the range. The Book MATLAB program given below was used to find the exact and DFT magnitude values of the spectrum and were plotted in Figure 2.6a. Note that the number of function values used in the calculations were three, $\{f(0), f(0.5), f(1)\}$. However, since $N=16$ the MATLAB function **fft**(.) added 13 zeros (fft means Fast Fourier Transform), thus creating a vector with 16 values. The m-file below produces Figure 2.6a. Similarly, we graphed Figure 2.6b and c.

Book MATLAB m-file: eff_of_sampl_time_a

```
%Book MATLAB m-file:eff_of_sampl_time_a
w1=0:pi/4:4*pi-(pi/4);
fw1=abs((1-exp(-(1+j*w1)))./(1+j*w1));
t1=0:0.5:1;
```

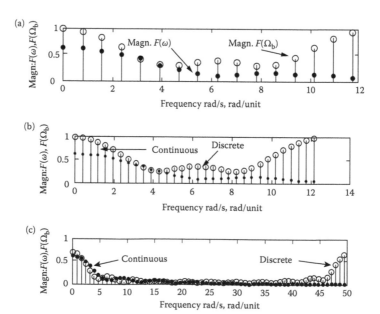

FIGURE 2.6

```
ft1=exp(-t1);
dft1=0.5*abs(fft(ft1,16));%fft(.) is the
    %fast Fourier transform MATLAB function;
subplot(3,1,1);stem(w1,dft1,'k');%'k' instructs
    %MATLAB to produce black graphs;
hold on;plot(w1,fw1,'.k','markersize',15);
    %produces black dots of magnitude 15;
xlabel('a) Frequency rad/s, rad/unit')
ylabel('Magn:F(\omega),F(\Omega_b)');
```

Calling from the Book m-files: eff_of_sampl_time_b and eff_of_sampl_time_c will produce Figure 2.6b and c corresponding to Cases (b) and (c) discussed below.

Case b: In this case we do not change the sampling time, $T = 0.5$, but we double the number $NT = 16$. From these data we obtain $N = 32$. Hence, the folding over frequency is $16 \times 2\pi/16 = 2\pi$, which is identical to that found in Case (a) above. Therefore, the discrete spectrum is periodic having period 2π. We observe in Figure 2.6b that although we doubled NT, doubling the number of points in the frequency domain for the same range of frequencies, the accuracy of the discrete spectrum has not improved, and, in the same time, the largest frequency which approximates the exact spectrum is the same, $\pi/0.5 = 2\pi$.

Case c: In this case $NT = 8$, but the sampling time was decreased to 0.1 value. This decrease of sampling time resulted in the increase of function sampling points from 3 to 11. From Figure 2.6c we observe that the accuracy has increased, and the approximation to the maximum frequency has also increased to $\pi/0.1 = 10\pi$. The frequency bins are equal to $2\pi/8 = \pi/4$, and the fold over frequency is equal to $(\pi/4)40 = 10\pi$. The total number of points were $N = 8/0.1 = 80$.

Note: *Based on the above results we conclude: (1) Keeping the sampling time T constant and increasing the number of values of the sampled function by padding it with zeros increases the number of frequency bins and, thus, makes the spectrum better defined but its **accuracy does not increase. (2) Decreasing the sampling time T results in better accuracy of the spectrum.** It further extends the range in simulating the exact spectrum using the DFT approach, and this range is π/T.*

It is suggested to the reader, using the above Book MATLAB program, to find their spectrum for different other functions.

2.9   EFFECT OF TRUNCATION

Because the DFT uses a finite number of samples, we must be concerned about the effect that the truncation has on the Fourier spectrum, even if the original function extends to infinity. Specifically, if the signal $f(t)$ extends beyond the total sampling period NT, the resulting frequency spectrum is an approximation to the exact one. If, for example, we take the DFT of a truncated sinusoidal signal, we find that the Fourier transform consists of additional lines that are the result of the truncation process. Therefore, if N is small and the sampling covers neither a large number nor an integral number of cycles of the signal, a large error in spectral representation can occur. This phenomenon is known as **leakage** and is the direct result of truncation. Since the truncated portion of the signal is equal to $f(t)p_a(t)$, the distortion is due to the presence of the rectangular window. This becomes obvious when we apply the Fourier transform property:

$$\mathscr{F}_D\{f(nT)p_a(nT)\} = F(k\Omega_b) * P_a(k\Omega_b).$$

Exercise 2.9.1

Show the leakage phenomenon by example.   ▲

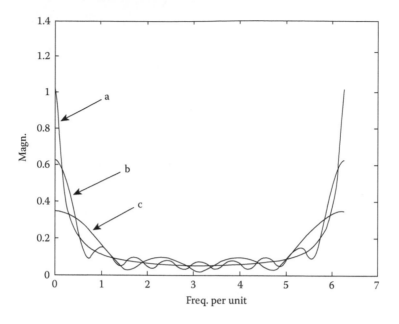

FIGURE 2.7

This expression indicates that the exact spectrum is modified due to the convolution operation. It turns out that as the width of the pulse becomes very large, its spectrum resembles a delta function, and, hence, the output of the convolution is close to the exact spectrum. Figure 2.7 shows the effect of truncation. For this case we used the exponential function $\exp(-t)u(t)$ and applied the fft to obtain the spectrum. Furthermore, we plotted in a continuous curve for a better inspection. Curve (c) is the result of using only the function from 0 to 0.3 s. This means that we multiplied the function by a rectangular window of the same length, Curve (b) is the result of using only the function from 0 to 0.8 s and, finally, for Curve (a) we used the function from 0 to 5. We observe that as the rectangular window becomes larger, the spectrum approaches the exact one. This is due to the fact that a rectangular window of infinite length has a delta function spectrum, and, therefore, the convolution in the frequency domain gives back exactly the spectrum of the windowed function.

WINDOWING

The reader must run a few examples using different types of windows which are given in Section 2.5. At the same time the reader should also find the spectrums of the windows and get familiar with level of their side-lobes and, in particular, the level of the first side-lobe. The plots should be done in a log-linear scale. Furthermore, the width of the main lobe plays an important role during the convolution process in the frequency domain. The thinner the main lobe is the better approximation we obtain of the desired spectrum. Both the main lobe and the level of the first side-lobe play an important role in the attempt to produce the smallest distortion of the spectrum of the truncated signal.

2.10 RESOLUTION

One interesting question is about the ability to resolve two frequencies which are close together. The main lobe of the window plays an important effect in the resolution of frequencies. Figure 2.8 shows the effect of the window that affects the resolution. The function we used was $f(n) = \cos(0.2\pi n) + \cos(0.4\pi n) + \cos(0.45\pi n)$. The resolution is understood by comparing the values $\Delta\omega$ with the unit bin in the frequency domain, which is equal to $2\pi/NT$. In this case we used

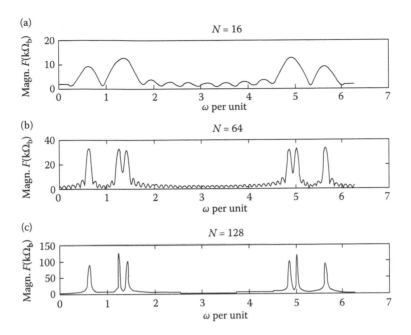

FIGURE 2.8

$T=1$. In Figure 2.8a we find that $(0.45-0.4)\pi \ll 2\pi/16$, and it is obvious that the two closed spaced frequencies cannot be resolved. In Figure 2.8b we find that 0.05π is about equal to $2\pi/32$, and it is apparent that a separation of the two frequencies is starting to be resolved. However, in the third case, where $0.05\pi \gg 2\pi/128$, the resolution is perfect. The Book MATLAB m-file below plots Figure 2.8. We plotted the curves in continuous format for a better observation of the resolution phenomenon.

Book m-File: non_random_resolu

```
%resolution m-file: non_rand_resolu.m
n1=0:15;
f1=cos(0.2*pi*n1)+cos(0.4*pi*n1)+cos(0.45*pi*n1);
n2=0:63;
f2=cos(0.2*pi*n2)+cos(0.4*pi*n2)+cos(0.45*pi*n2);
n3=0:255;
f3=cos(0.2*pi*n3)+cos(0.4*pi*n3)+cos(0.45*pi*n3);
ftf1=fft(f1,256);ftf2=fft(f2,256);ftf3=fft(f3,256);
w=0:2*pi/256:2*pi-(2*pi/256);
subplot(3,1,1);plot(w,abs(ftf1),'k');xlabel('\omega per unit');
ylabel('Magn. F(k\Omega_b)');title('N=16, a)');
subplot(3,1,2);plot(w,abs(ftf2),'k');xlabel('\omega per unit');
ylabel('Magn. F(k\Omega_b)');title('N=64, b)');
subplot(3,1,3);plot(w,abs(ftf3),'k');xlabel('\omega per unit');
ylabel('Magn. F(k\Omega_b)');title('N=128, c)');
```

2.11   DISCRETE SYSTEMS

When we deal with continuous systems, our interest is in the relationship between an input signal $v(t)$ and output signal $g(t)$. Hence, we were interested in finding the appropriate operator $\odot$ that, when operating on the input, would yield the relation between the input and output. Similarly, a **discrete**, or **digital system** establishes a relationship between two discrete signals: an input $v(n)$ and output $g(n)$. The values of n are integers. However, we will also deal with discrete signals whose values will not be apart by a unit time but by a fraction of time, e.g., nT, where T is equal to a value in the range, $0 < T < 1$. An actual discrete system is a combination of many different electronic components, such as amplifiers, shift registers, and gates. However, it is represented here in simple block diagram form.

In discrete-time systems, a **discrete source** produces discrete-time signals that are pulses of assumed zero width and finite height. A discrete signal sequence may arise by pulse sampling a continuous-time excitation function, usually at uniform time intervals. It might represent a sequence of narrow pulses generated in a pulse-generating source. As we mentioned above, the generation of discrete signals having zero width, and finite height is not possible using any physical system. Therefore, there is no physical source that is able to produce discrete signals. Another element of a discrete system is the **scalar multiplier**, a component that produces pulses at the output that are proportional to input pulses. There is also a **delay element** that produces an output identical with that at its input, but delayed by a predetermined number of time units or a multiple-fraction of a time unit. The symbol z^{-1} will be used to denote a delay of a unit time or a fraction of unit time (T): two delay units will be denoted by z^{-2} and for k delay units by z^{-k}.

A group of such connected discrete elements and **adders** (summing points or summers) with appropriate **pickoff points** (observe that the function appears identical in shape and magnitude in all the directions) constitute a digital system. Discrete sources are its inputs, and the resulting signals in its various parts are the outputs, or responses, of the system. The analysis of the behavior of such systems will parallel, to some extent, the systems analysis procedures in continuous-time systems. The main connected elements of discrete systems are shown in Figure 2.9.

Example 2.11.1

Find the output of the system shown in Figure 2.10.

Solution: Clearly, the output is the sum of the input signal plus the input signal shifted by two units of time. The result and graphic representation is shown in Figure 2.10. ∎

A system consisting of discrete-time elements with a discrete-time input signal sequence is described by a **difference equation**. This description is to be contrasted with the description of linear, lumped, time-invariant systems that are described by ordinary differential equations. Depending on the number and character of the elements in the discrete system, the system is described by difference equation of different order. The following examples will help us understand how difference equations are developed. The structure layout of the discrete elements will produce a specific type of difference equations, which will be identified with specific names. Furthermore, the iterative method of solution will be adopted at present, and other methods will be considered in later chapters.

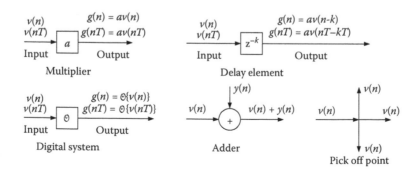

FIGURE 2.9

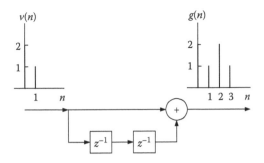

FIGURE 2.10

Example 2.11.2

Consider the system shown in Figure 2.11a. Find the difference equation that describes the system, and deduce $g(n)$ if the input is $v(n)=u(n)$, the unit step function. The system is relaxed at $n=0$, which implies that $g(-1)=0$.

Solution: From the figure we find

$$g(n) = v(n) - 2g(n-1) \quad \text{or} \quad g(n) + 2g(n-1) = v(n)$$

The above equation can be put in the general form:

$$y(n) + a_1(n-1) = b_0 x(n) \tag{2.39}$$

Since the difference between the independent variables of the output of the system is $(n-(n-1))=1$, the difference equation is of the **first order**. Furthermore, since the output is given in its unshifted and shifted format but the input is only given by its zero shift position, the system is of the **Infinite Impulse Response (IIR)** system. Observe that this type of discrete systems produce a feedback-type block diagram form.

Note: *The infinite impulse response discrete system (IIR) consists of the output plus additional delayed outputs multiplied by constants. This type of system has only one input and not delayed ones. The block diagram structure is a feedback type.*

To find a numerical solution, we proceed by successively introducing values of n into the difference equation, starting with $n=0$ since the input function starts at time $n=0$. We proceed iteratively thereafter. Hence,

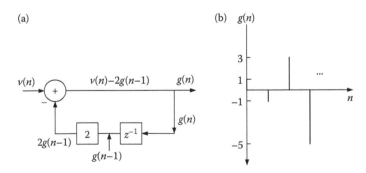

FIGURE 2.11

$$g(0) = v(0) - 2g(-1) = 1 - 2 \times 0 = 1$$

$$g(1) = v(1) - 2g(0) = 1 - 2 \times 1 = -1$$

$$g(2) = v(2) - 2g(1) = 1 - 2 \times (-1) = 3$$

$$\vdots$$

The output has been plotted in Figure 2.11b. The following Book m-function can solve any first-order IIR discrete system. ∎

Book m-Function for an IIR System: urdsp_firstorder_iir(a1,b0,N,x)

```
function[y]=urdsp_firstorder_iir(a1,b0,N,x)
y(1)=0;% if the initial condition is different than 0
       %the new value must be introduced here,x=input;
for n=0:N% N=the number of the output;
    y(n+2)=-a1*y(n+1)+b0*x(n+1);
end;
    % since matlab starts from 1, everything must
    %be shifted by two; the answer will also
    %contain the value of the initial condition
    %at n=1; to have the exact plot we must set:
    %m=0:N+1;
m=0:N+1;
stem(m,y,'k');
```

Example 2.11.3

Find the difference equation that describes the system shown in Figure 2.12a. Determine the output $g(n)$ if its input is $v(n)=u(n)$. The system is relaxed at $n=0$, which implies that $g(-1)=0$.

Solution: From the figure we deduce that

$$g(n) = v(n) + 2v(n-1) - 4v(n-2)$$

The above equation may be put in the general form:

$$y(n) = b_0 x(n) + b_1 x(n-1) + b_2 x(n-2) \tag{2.40}$$

Note that the output is expressed only in terms of the input function. Such a system is called a **nonrecursive, transversal** or **finite duration impulse response (FIR)**. To find the solution, we solve the equation recursively. Hence,

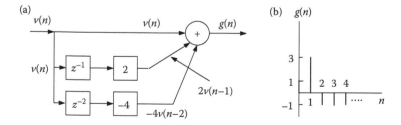

FIGURE 2.12

$$g(0) = v(0) + 2v(-1) - 4v(-2) = 1 + 2 \times 0 - 4 \times 0 = 1$$
$$g(1) = v(1) + 2v(0) - 4v(-1) = 1 + 2 \times 1 - 4 \times 0 = 3$$
$$g(2) = v(2) + 2v(1) - 4v(0) = 1 + 2 \times 1 - 4 \times 1 = -1$$
$$\vdots$$

The Book m-File Function for Four-Term FIR System: function[y]=urdsp_fir4term(x,b)

```
function[y]=urdsp_fir4term(x,b)
N=length(x)-4;
for n=0:N
    y(n+1)=[x(n+4) x(n+3) x(n+2) x(n+1)]*b';
end;
    %the input vector x must have its first three
    %elements zero for a relaxed system;
k=0:N;
stem(k,y,'k');
```

Note: *The finite impulse respond system has only one output and additional delayed inputs multiplied by constants. The output of the system is the sum of both the input and the delayed inputs. The block diagram form is of the forward format.*

Example 2.11.4

Find the difference equation that describes the system shown in Figure 2.13a. Determine the output $g(n)$ for an input $v(n)=u(n)$. The system is relaxed at $n=0$, which implies that $g(-1)=g(-2)=0$.

Solution: From the figure we see that

$$g(n) = v(n-1) - 2v(n-2) - 2g(n-2) \tag{2.41}$$

The general second-order mixed (IIR and FIR) discrete system is given by

$$y(n) + a_1 y(n-1) + a_2 y(n-2) = b_0 x(n) + b_1 x(n-1) + b_2 x(n-2) \tag{2.42}$$

Because the difference between n and $n-2$ belonging to the dependent variable $g(n)$ and $g(n-2)$ is equal to 2, the difference equation is of the **second order**. We note that the output is the sum of the inputs and the delayed outputs with their proportionality factors. The solution of Equation 2.41 is

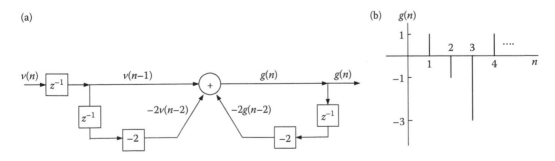

FIGURE 2.13

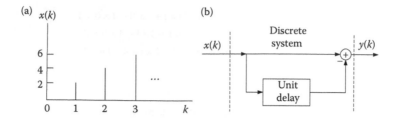

FIGURE 2.14

$$g(0) = v(-1) - 2v(-2) - 2g(-2) = 0 - 2 \times 0 - 2 \times 0 = 0$$

$$g(1) = v(0) - 2v(-1) - 2g(-1) = 1 - 2 \times 0 - 2 \times 0 = 1$$

$$g(2) = v(1) - 2v(0) - 2g(0) = 1 - 2 \times 1 - 2 \times 0 = -1$$

$$g(3) = v(2) - 2v(1) - 2g(1) = 1 - 2 \times 1 - 2 \times 1 = -3$$

$$\vdots$$

The output is plotted in Figure 2.13b. ■

Exercise 2.11.1

Find the output for the system shown in Figure 2.14 if the input is the ramp function. ▲

2.12 DIGITAL SIMULATION OF ANALOG SYSTEMS

If we wish to carry out continuous-time systems analysis using digital computers, a procedure which will be using very often in practice, it is required that we determine a digital system whose output is equivalent to that of the continuous time or analog system. We refer to the digital equivalent as a **digital simulator** and requires, of course, that the digital output closely approximates the output of the corresponding analog system. The objective of digital simulation is to determine a simulator of an analog system and to determine the class of signals that can be processed.

Because the analog systems are described by ordinary differential equations, the digital simulators will correspond to a recursive equation, different equations, that can be represented by delay elements and scalar multipliers. To accomplish these objectives, we must study how to create the digital simulators from their analog counterparts.

Example 2.12.1

To understand the process, let us study the analog system shown in Figure 2.15a. It is desired to find the output when the input is a unit step function $v(t) = u(t)$, and the system is initially relaxed at $t = 0$ or equivalently the system has zero initial conditions.

Solution: The differential equation that describes the system is

$$\frac{di(t)}{dt} + 0.5i(t) = u(t), \quad v_0(t) = 0.5i(t)$$

After multiplying the above equation by 0.5, we obtain

$$\frac{dv_0(t)}{dt} + 0.5v_0(t) = 0.5u(t) \tag{2.43}$$

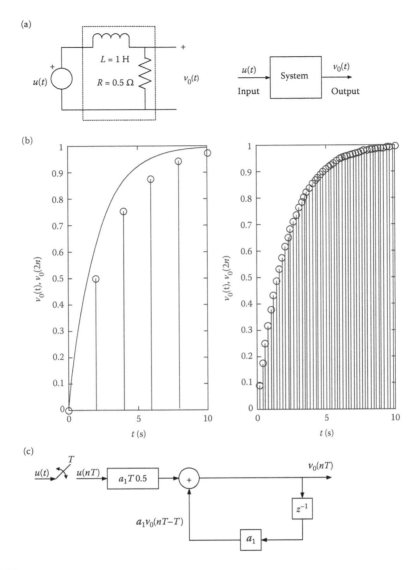

FIGURE 2.15

The solution to the homogeneous equation is

$$\frac{dv_0(t)}{dt} + 0.5v_0(t) = 0$$

is $v_{0h}(t) = Ae^{-0.5t}$. The reader can verify it by introducing the solution to the above homogeneous equation. Since the input is a constant, we set the relation $v_{0p}(t) = B$ for this particular solution. Introducing this solution in Equation 2.43 we obtain

$$0 + 0.5v_{0p}(t) = 0.5 \Rightarrow v_{0p}(t) = 1$$

Therefore, the total solution is $v(t) = v_{0h}(t) + v_{0p}(t) = A\exp(-0.5t) + 1$. Setting $t = 0$ and zero initial condition, $v(0) = 0$, we find that $A = -1$. Hence the solution is

$$v(t) = (1 - e^{-0.5t})u(t) \tag{2.44}$$

Next, we discretize Equation 2.43 using the approximation of the derivative given in Chapter 1. Hence, we obtain

$$\frac{v_0(nT) - v_0(nT - T)}{T} + 0.5v_0(nT) = 0.5u(nT)$$

or

$$v_0(nT) - a_1 v_0(nT) = a_1 T 0.5 u(nT) \quad a_1 = \frac{1}{1 + 0.5T} \tag{2.45}$$

To have zero value at $n=0$, we introduce $n=0$ in Equation 2.45 to obtain the elation:

$$v_0(0T) - a_1 v_0(-T) = 0 - a_1 v_0(-T) = a_1 T 0.5 \Rightarrow v_0(-T) = -T 0.5 \tag{2.46}$$

Next, we solve Equation 2.45 by iteration, taking into consideration the obtained initial condition, to obtain the general form of solution, which is

$$n = 0 \quad v_o(0T) = a_1 v_o(-T) + a_1 T 0.5 = -a_1 T 0.5 + a_1 T 0.5 = 0$$

$$n = 1 \quad v_o(1T) = a_1 v_o(0T) + a_1 T 0.5 = 0 + a_1 T = a_1 T 0.5$$

$$n = 2 \quad v_o(2T) = a_1 v_o(1T) + a_1 T 0.5 = a_1^2 T 0.5 + a_1 T 0.5$$

$$n = 3 \quad v_o(3T) = a_1 v_o(2T) + a_1 T 0.5 = a_1^3 T 0.5 + a_1^2 T 0.5 + a_1 T 0.5$$

$$n = 4 \quad v_o(4T) = a_1 v_o(3T) + a_1 T 0.5 = a_1^4 T 0.5 + a_1^3 T 0.5 + a_1^2 T 0.5 + a_1 T 0.5$$

$$= a_1 T 0.5 (1 + a_1 + a_1^2 + a_1^3)$$

$$\vdots$$

Observe that if $n = N$ then the last parenthesis will have the form $(1 + a_1 + a_1^2 + a_1^3 + \cdots + a_1^{N-1})$. This expression is a finite-term geometric series, and, thus, the output voltage is given for any n by

$$v_0(nT) = a_1 T 0.5 \frac{1 - a_1^n}{1 - a_1} \quad a_1 = \frac{1}{1 + 0.5T} \tag{2.47}$$

The following Book m-function produces Figure 2.15b.

Book m-Function: function[v,vc]=urdsp_ex2_12_1(T,t1)

```
function[v,vc]=urdsp_ex2_12_1(T,t1)
    %the name of this file is: Ex2_12_1;
    %this is an m-function
    %that produces outputs for the
    %Ex.2.12.1;the reader
    %can change the value of T and get
    %different desired results;
    %T=sampling time;
    %t1=time on the time axis we
    %would like to plot
    %the functions;t1/T
    %must be an integer;
N=t1/T;
a1=1/(1+0.5*T);
n=0:N;
v=a1*T*0.5*(1-a1.^n)/(1-a1);
stem(n*T,v,'k');
t=0:0.2:10;
```

```
vc=1-exp(-0.5*t);
hold on;
plot(t,vc,'k');
```

Figure 2.15b gives plots of the analog system output and the output from its digital simulator. We observe that the smaller the sampling time is introduced the better approximation we obtain. Figure 2.15c shows the **block** diagram representation of the simulated continuous system for any sampling time T. ∎

Example 2.12.2

Find the output of the analog and equivalent discrete form of the system shown in Figure 2.16a. The input to the system is exponential function $v_i(t)=\exp(-t)u(t)$.

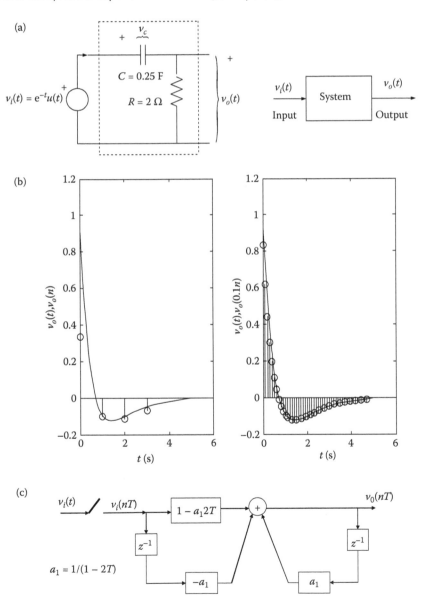

FIGURE 2.16

Solution: The intergrodifferential equation describing the system is

$$4\int i(t)\,dt + 2i(t) = v_i(t) \tag{2.48}$$

We make use of the fact that the current, $i(t)$, through the capacitor is $C\,dv_c/dt$. This current is the same through the circuit. Substituting this expression in the above equation, and taking into consideration the properties of the integration of a derivative, we obtain the equation

$$v_c(t) + 0.5\frac{dv_c}{dt} = v_i(t) \tag{2.49}$$

For an exponential decaying input function and relaxed initial conditions, the solution is easily found to be

$$v_c(t) = 2\left(e^{-t} - e^{-2t}\right) \quad t > 0 \tag{2.50}$$

Further, the output voltage is

$$v_o(t) = 2e^{-2t} - e^{-t} \quad t > 0 \tag{2.51}$$

A sketch of the function is given in Figure 2.16b by the solid line. For comparison, a digital simulation with sampling times of $T=1$ and $T=0.1$ is also shown. As we have seen in the previous example the smaller the sampling time the better approximation we produce.

  Next, we approximate (Equation 2.49) in its discrete form (see Chapter 1). The resulting equation is

$$v_c(nT) - a_1 v_c(nT - T) = a_1 2T v_i(nT) \quad a_1 = 1/(1 + 2T) \tag{2.52}$$

Also, by inspection of Figure 2.16a, we write Kirchhoff's voltage law (KVL):

$$v_i(nT) - v_c(nT) - v_o(nT) = 0 \tag{2.53}$$

Combining these two equations by eliminating $v_c(nT)$ and $v_c(nT - T)$, the latter by appropriately changing the time variable nT to $nT - T$ in Equation 2.53. The result is found to be

$$v_o(nT) = a_1 v_o(nT) + (1 - a2T)v_i(nT) - a_1 v_i(nT - T) \tag{2.54}$$

For the particular case of $T = 0.5\,$s, the solution is calculated as follows:

$$n = 0 \quad v_o(0) = 0.5v_o(-0.5) + (1 - 0.5)v_i(0) - 0.5v_i(-0.5)$$

$$= 0.5 \times 0 + 0.5 \times 1 - 0.5 \times 0 = 0.5$$

$$n = 1 \quad v_o(0.5) = 0.5v_o(0) + (1 - 0.5)v_i(0.5) - 0.5v_i(0)$$

$$= 0.25 + 0.5 \times 0.6065 - 0.5 \times 1 = 0.0533$$

$$n = 2 \quad v_o(1) = 0.5v_o(0.5) + (1 - 0.5)v_i(1) - 0.5v_i(0.5)$$

$$= 0.5 \times 0.0533 + 0.5 \times 0.3679 - 0.5 \times 0.6065 = -0.927$$

$$\vdots$$

Figure 2.16b has been plotted using the Book m-Function given below for two different sampling times.

Book m-File for the example 2.12.2: [vo]=urdsp_ex2_12_2(tl,T)

```
function[vo]=urdsp_ex2_12_2(tl,T)
N=tl/T; %tl/N must be integer;
     %tl=is the number of time points;

a1=1/(1+2*T);
v(1)=0;
n=2:N;
vi1=exp(-(n-2)*T);
vi=[0   vi1];
for m=2:N
     v(m)=a1*v(m-1)+(1-a1*2*T)*vi(m)...
          -a1*vi(m-1);
end;
vo=v(1,2:N);
t=0:0.2:tl;
vco=2*exp(-2*t)-exp(-t);
m=0:N-2;
stem(m*T,vo,'k');
hold on;
plot(t,vco,'k');
```

Figure 2.16c gives the block diagram of the discrete systems which represents the simulated continuous system. ∎

Exercise 2.12.1

Find the solution directly from Equation 2.48. ▲

Exercise 2.12.2

Simulate the system shown in Figure 2.17, and find its discrete-time solution. Assume $T=1$, $v(k)=u(k)$ and zero initial equation, $y(-1)=0$. ▲

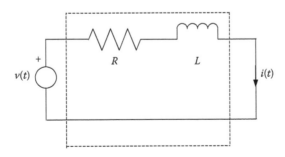

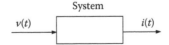

FIGURE 2.17

2.12.1   SECOND-ORDER DIFFERENTIAL EQUATIONS

Because more complicated linear systems can be described by higher-order linear differential equations, it is desired to develop the equivalent digital representations of different-order derivatives. This development yields

$$\frac{dy(t)}{dt} \cong \frac{y(nT) - y(nT - T)}{T} \qquad (2.55)$$

$$\frac{d^2(y(t))}{dt^2} = \frac{d}{dt}\frac{dy(t)}{dt} \cong \frac{y(nT) - 2y(nT - T) + y(nT - 2T)}{T^2} \qquad (2.56)$$

The values of initial conditions are determined from those values specified for the continuous system. For example, given the values of $y(t)$ and $dy(t)/dt$ at $t = 0$, the required values of $y(-T)$ can be obtained approximately from the relationship

$$\left.\frac{dy(t)}{dt}\right|_{t=0} = \frac{dy(0)}{dt} \cong \frac{y(0T) - y(0T - T)}{T}$$

or

$$y(-T) = y(0) - T\frac{dy(0)}{dt} \qquad (2.57)$$

Following such a procedure, the value of $y(-nT)$ can be obtained from $d^n y(0)/dt^n$ by proceeding from the lower- to higher-order derivatives using values already found for $y(0T), y(-T), \ldots, y(-nT + T)$.

Example 2.12.1.1

Find the current in the circuit shown in Figure 2.18a for both continuous and discrete cases. The following data are given: $L=1$, $R=5$, $C=0.25$, $v_i(t)=2\exp(-2t)$, $q(0)=0$, $dq(0)/dt=0$.

Solution: The application of the Kirchhoff's voltage law for the circuit gives the relation:

$$Ri(t) + L\frac{di(t)}{dt} + \frac{1}{C}\int i(t)\,dt = v(t) \qquad (2.58)$$

Since the charge in the capacitor is related to the current by $i(t)=dq(t)/dt$, the above equation becomes

$$\frac{d^2 q(t)}{dt^2} + 5\frac{dq(t)}{dt} + 4q(t) = 2e^{-2t} \quad t > 0 \qquad (2.59)$$

The homogeneous, the particular and the total solutions, which is equal to the sum of the particular and homogeneous equations, to the above equation are

$$q_h(t) = Ae^{-t} + Be^{-4t}, \quad q_p(t) = -e^{-2t}, \quad q(t) = \frac{2}{3}e^{-t} + \frac{1}{3}e^{-4t} - e^{-2t} \qquad (2.60)$$

The current is found by taking the derivative of the charge $q(t)$.
    The digital simulation of Equation 2.59 is deducted by employing Equations 2.55 and 2.56 in this expression. We obtain

$$\frac{q(nT) - 2q(nT - T) + q(nT - 2T)}{T^2} + 5\frac{q(nT) - q(nT - T)}{T} + 4q(nT) = 2e^{-2nT} \quad n = 0,1,2,\ldots \quad (2.61)$$

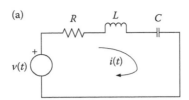

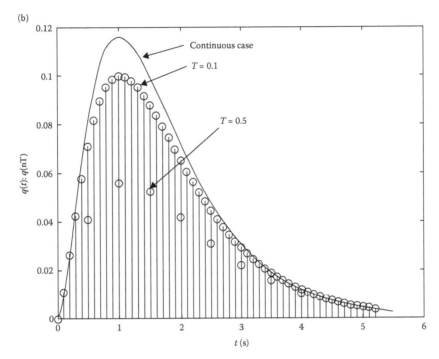

FIGURE 2.18

After rearrangement, the above equation becomes

$$q(nT) = a(2+5T)q(nT-T) - aq(nT-2T) + a2T^2 e^{-2nT}$$

$$a = \frac{1}{1+5T+4T^2}, \quad n = 0,1,2,3,\ldots \tag{2.62}$$

Using Equation 2.57, we find that $q(-T)=0$. Next, introducing this value and initial $q(0T)=0$ in Equation 2.62, we obtain $q(-2T)=2T^2$. Therefore, introducing the initial conditions just found we proceed to simulate the discrete case using Equation 2.62.

Book m-Function: function[q]=urdsp_ex2_12_1_1(t1,T)

```
function[q1]=urdsp_ex2_12_1_1(t1,T)
N=t1/T;       %t1/N must be integer;
a=1/(1+5*T+4*T^2);
q(2)=0;
q(1)=2*T*T;
n=3:N;
for m=0:N
    q(m+3)=a*(2+5*T)*q(m+2)-a*q(m+1)...
        +a*2*T^2*exp(-2*m*T);
end;
```

```
q1=q(1,3:N);
t=0:0.1:tl;
qc=(2/3)*exp(-t)+(1/3)*exp(-4*t)-exp(-2*t);
m=0:N-3;
stem(m*T,q1,'k');
hold on;
plot(t,qc,'k');
```

Figure 2.18b shows both the continuous and discrete cases for sampling times $T = 0.5$ and $T = 0.1$. We observe that the smaller the sampling time the closer the approximation to the exact continuous case. ∎

Exercise 2.12.1.1

Find the current for the continuous and the discrete case for the Example 2.12.1.1.　▲

HINTS–SUGGESTIONS–SOLUTIONS OF THE EXERCISES

2.2.1

$$\int_2^\infty e^{-(t-2)}e^{-j\omega t}\,dt = e^2\int_2^\infty e^{-(1+j\omega)t}\,dt = e^2(-)\frac{1}{1+j\omega}\left(-e^{-(1+j\omega)2}\right) = \frac{1}{\sqrt{1+\omega^2}}e^{-j(\tan^{-1}\omega+2\omega)}$$

This shows that shifting does not change the amplitude spectrum but only changes the phase spectrum.

2.2.2

$$\int_{-\infty}^\infty e^{-2|t|}e^{-j\omega t}\,dt = \int_{-\infty}^0 e^{2t}e^{-j\omega t}\,dt + \int_0^\infty e^{-2t}e^{-j\omega t}\,dt, t = -y \Rightarrow \int_0^\infty e^{(-2+j\omega)y}\,dy + \int_0^\infty e^{-(2+j\omega)t}\,dt$$

$$= \frac{1}{-2+j\omega}e^{(-2+j\omega)y}\Big|_0^\infty + \frac{1}{-(2+j\omega)}e^{-(2+j\omega)t}\Big|_0^\infty = -\frac{1}{-2+j\omega} + \frac{1}{2+j\omega} = \frac{4}{4+\omega^2}$$

The spectrum is real and symmetric.

2.2.3

$$(1)\int_{-\infty}^\infty\int_{-\infty}^\infty f(x)f^*(x-t)e^{-j\omega t}\,dt\,dx, x-t = y, t = -\infty \Rightarrow y = \infty, t = \infty$$

$$\Rightarrow \int_{-\infty}^\infty\int_{-\infty}^\infty f(x)f^*(y)e^{-j\omega(x-y)}\,dy\,dx = \int_{-\infty}^\infty f(x)e^{-j\omega x}\,dx\int_\infty^{-\infty}(-)f^*(y)e^{j\omega y}\,dy$$

$$= F(\omega)\int_\infty^{-\infty}(-)f^*(y)e^{j\omega y}\,dy = F(\omega)\int_{-\infty}^\infty f^*(y)e^{j\omega y}\,dy = F(\omega)\left(\int_{-\infty}^\infty f(y)e^{-j\omega y}\,dy\right)^*$$

$$= F(\omega)F^*(\omega) = |F(\omega)|^2 = \text{The answer is the same for real functions}$$

$$(2)\,(j\omega)^2 F(\omega)$$

2.3.1

$$\frac{1}{2\pi}\int_{-\infty}^{\infty}F(\omega)e^{-j\omega t}\,d\omega = \frac{1}{2\pi}\int_{-\infty}^{\infty}\frac{2}{1+\omega^2}e^{-j\omega t}\,d\omega = \frac{1}{2\pi}\int_{-\infty}^{\infty}\frac{2}{1+\omega^2}\cos\omega t\,d\omega$$

$$-j\frac{1}{2\pi}\int_{-\infty}^{\infty}\frac{2}{1+\omega^2}\sin\omega t\,d\omega = \frac{1}{2\pi}\int_{-\infty}^{\infty}\frac{2}{1+\omega^2}\cos\omega t\,d\omega$$

The integral with the sine function is zero since the $F(\omega)$ is an even function, and the sine is odd, and, therefore, the product is odd, and the integral is zero. The integration is done using the following Book MATLAB program with varying length of integration. As the length increases substantially we recapture the time function.

We use the following Book MATLAB m-file:

```
%exerc2_3_1
w=-2:0.01:2;
for t=-5000:5000
    f(5001+t)=sum(((2./(1+w.^2)).*cos(w*t*0.01)*0.01)/(2*pi));
end; %observe the places of the dots; f(a), a must be an %integer;
w=-20:0.01:20;
for t=-5000:5000
    f1(5001+t)=sum(((2./(1+w.^2)).*cos(w*t*0.01)*0.01)/(2*pi));
end
subplot(2,1,1);plot(f(1,4000:6000),'k');
subplot(2,1,2);plot(f1(1,4000:6000),'k');
```

2.3.2

(1) we observe one line as it should repeating every sampling frequency; (2) the same happens at sampling frequency at 19 since it is equal to Nyquist one; (3) at 9.5 we observe that the first peak is at $\omega_0 - \omega_s$, the second at ω_0, the third at $\omega_0 + \omega_s$ and the forth is at $2\omega_0$, etc.

2.4.1

$$f(n) = \frac{1}{2\pi}\int_{-\pi}^{\pi}e^{j\omega n}\,d\omega = \frac{\sin n\pi}{n\pi}, \sum_{n=0}^{\infty}\left(\frac{\sin n\pi}{n\pi}\right)^2 = \frac{1}{\pi^2}(\pi^2+0+0+\cdots)=1, \frac{1}{2\pi}\int_{-\pi}^{\pi}d\omega = 1$$

2.4.2

In MATLAB we write: >>n=−5:5; F=fft(exp(−abs(n)),512);% sampled every second

>>n1 =−5:0.2:5; F1=fft(exp(−abs(n1)),512);%sampled every 0.2 s;
>>w=0:0.1:4; Fw=2./(1+w.^2);%this is the FT of the time function from 0 to 4 rad/s;

Next we find the areas under the curves to compare results. For the exact case we have selected the distance in rad/s equal to 4. Therefore, we must select the same lengths for the DTFTs.

For the first case the frequency bin is $2*\mathrm{pi}/512$ ($2\pi/NT$). For this case, the corresponding distance (number of bins) can be found by the ratio $\dfrac{\mathrm{pi}}{4} = \dfrac{256}{x}$ or $x = \dfrac{256*4}{\mathrm{pi}} \cong 326$.

Hence the area is given by $(2*pi/512)*sum(abs(F(1,1:326))) = 3.5694$. Similarly, for the second case we find that the area is: $(2*pi/512/0.2)*sum(abs(F1(1,1:65))) = 2.7342$. For the exact case, we have: $sum(0.1*Fw) = 2.7575$. This shows that the smaller we set the sampling time the closer we arrive to the exact value.

2.5.1

$$\frac{1}{2\pi}\int_{-\pi}^{\pi} \pi\delta(\omega-\omega_0)e^{j\omega n}\,d\omega + \frac{1}{2\pi}\int_{-\pi}^{\pi} \pi\delta(\omega+\omega_0)e^{j\omega n}\,d\omega = \frac{1}{2}\left(e^{j\omega_0 n}+e^{-j\omega_0 n}\right)=\cos\omega_0 n$$

where we used Euler's identity and the linearity of the integration.

2.6.1

$$\frac{1}{NT}\sum_{k=0}^{N-1}F(k\Omega_b)e^{j\Omega_b Tnk} = \frac{1}{NT_s}\sum_{k=0}^{N-1}\left[T\sum_{m=0}^{N-1}f(mT)e^{-j\Omega_b Tmk}\right]e^{j\Omega_b Tnk}$$

$$= \frac{1}{N}\sum_{m=0}^{N-1}f(mT)\sum_{k=0}^{N-1}e^{-j\Omega_b T(m-n)k};\text{ but }\sum_{k=0}^{N-1}e^{-j\Omega_b T(m-n)k}=\begin{cases} N & m=n \\ 0 & m\neq n \end{cases}$$

The last identity is the result of using the finite sum of the geometric series.

2.6.2

The following Book MATLAB functions can be used (Figure 2.19):

1.

```
>>s=sin(0.5*pi*[0:0.001:5].*[0:0.001:5]);
>>sf=abs(fft(s));plot(sf(1,1:100));
```

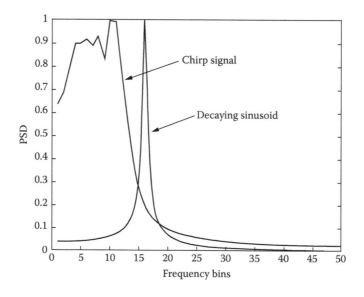

FIGURE 2.19

2.
```
>>x=0.98.^[0:0.1:100].*sin(0.3*pi*[0:0.1:100]);
>>xf=abs(fft(x)); subplot(2,1,1);plot(x);subplot(2,1,2);plot
(xf(1,1:100));
```

2.7.1

Re-write the inverse DFT in the form

$$f(-n) = \frac{1}{N}\sum_{k=0}^{N-1} F(k)e^{j\pi(-n)k/N}, \text{ interchange } n \text{ and } k \Rightarrow f(-k) = \frac{1}{N}\sum_{n=0}^{N-1} F(n)e^{-j\pi kn/N}$$

2.7.2

We write the IDFT in the form

$$f(r) = \frac{1}{N}\sum_{k=0}^{N-1} F(k)e^{j2\pi rk/N}; f(n-m) = \frac{1}{N}\sum_{k=0}^{N-1} F(k)e^{j2\pi k(n-m)/N}$$

$$= \left[\frac{1}{N}\sum_{k=0}^{N-1} F(k)e^{-j2\pi km}\right]e^{j2\pi nk/N} = F_D^{-1}\left\{F(k)e^{-j2\pi mk/N}\right\}$$

2.7.3

$$y(n) = \sum_{m=0}^{N-1} f(m)h(n-m) = \sum_{m=0}^{N-1}\frac{1}{N}\sum_{k=0}^{N-1} F(k)e^{j2\pi mk/N} \times \frac{1}{N}\sum_{r=0}^{N-1} H(r)e^{j2\pi r(n-m)/N}$$

$$= \frac{1}{N}\sum_{k=0}^{N-1}\sum_{r=0}^{N-1} F(k)H(r)e^{j2\pi rn/N}\left[\frac{1}{N}\sum_{m=0}^{N-1} e^{j2\pi km/N}e^{-j2\pi mr/N}\right]$$

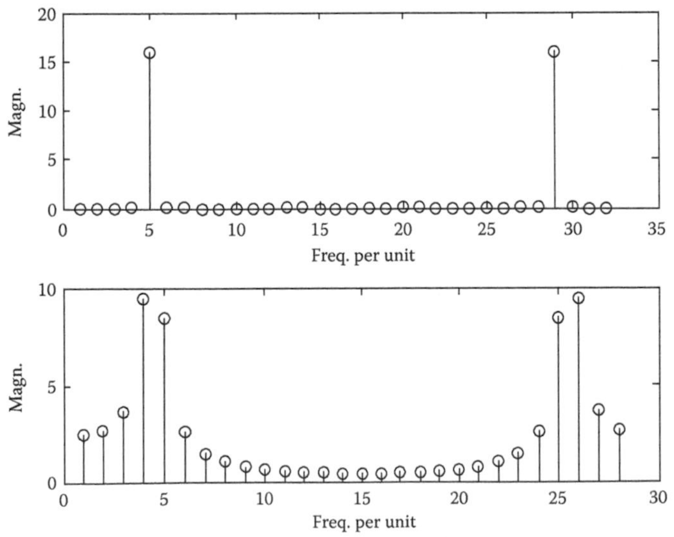

FIGURE 2.20

The expression in the bracket is equal to 1 if $k=r$ and 0 for $k \neq r$. To show this, use the finite geometric series formula. Hence, for $r=k$ in the second sum, we obtain the desired results:

$$y(n) = \sum_{m=0}^{N-1} f(m)h(n-m) = \frac{1}{N}\sum_{k=0}^{N-1} F(k)H(k)e^{j2\pi kn/N} = F_D^{-1}\{F(k)H(k)\}$$

2.9.1

If we use the continuous function $f(t) = \sin(2\pi f t)$ and we set $f = 1/8$ and $T = 1$, we use the following Book MATLAB program to produce Figure 2.20. From the figure we observe that if we find the DFT for a complete cycle(s) we find the spectrum to be a delta function as it should be. However, when we use a nonintegral cycle(s) from the signal, we observe that the spectrum broadens, and we call this phenomenon the leakage.

BOOK MATLAB PROGRAM

```
>>n=0:31;% creates complete cycles;
>>x=sin(2*pi*n/8);
>>xf=fft(x);
>>m=0:27; % creates noncomplete cycles;
>>xm=sin(2*pi*m/8);
>>xmf=fft(xm);
>>subplot(2,1,1); stem(abs(xf),'k');
>>xlabel('Freq. per unit'); ylabel('Magn.');
>>subplot(2,1,2);stem(abs(xmf),'k');
>>xlabel('Freq. per unit'); ylabel('Magn.')
```

2.11.1

The system is a differentiator.

2.12.1

We proceed as follows:

$$4\int_{0}^{nT-T} i(t)\,dt + 4\int_{nT-T}^{nT} i(t)\,dt + 2i(t) = v_i(t) \text{ or } 4q(nT-T) + 4Ti(nT) + 2i(nT) = v_i(nT)$$

But $2i(.)$ is equal to the voltage output $v_o(.)$, and the accumulated charge on the capacitor divided by the capacitance is equal to the voltage across the capacitor, $v_c(t)$. Hence, the above equation becomes

$$4 \times 0.25\frac{q(nT-T)}{0.25} + (1+2T)[2i(nT)] = v_i(nT) \text{ or } v_c(nT-T) + (1+2T)v_o(nT) = v_i(nT)$$

or $v_c(nT-T) + (1+2T)v_o(nT) = v_i(nT)$

But from Kirchhoff's voltage law we have $v_c(nT) = v_i(nT) - v_o(nT)$. Therefore, shifting the above equation we obtain $v_c(nT-T) = v_i(nT-T) - v_o(nT-T)$. The combination of the last two equations we obtain the solution.

2.12.3

Using Kirchhoff's voltage law, we obtain

$$\frac{di(t)}{dt} + \frac{R}{L}i(t) = \frac{1}{L}v(t) \tag{1}$$

The difference equation approximation of the above equation is

$$\frac{i(kT) - i(kT - T)}{T} = \frac{1}{L}v(kT) - \frac{R}{L}i(kT) \tag{2}$$

Rearrange this expression to the form

$$i(kT) = \frac{\frac{T}{L}}{1 + \frac{R}{L}T}v(kT) + \frac{1}{1 + \frac{R}{L}T}i(kT - T)$$

Set $T = 1$ for simplicity, and this equation becomes

$$i(k) = \frac{\frac{1}{L}}{1 + \frac{R}{L}}v(k) + \frac{1}{1 + \frac{R}{L}}i(k - 1) \tag{3}$$

This equation can be set in the form of the first-order difference equation:

$$y(k) = \beta_0 x(k) + \alpha_1 y(k-1), \quad \beta_0 = \frac{(1/L)}{1 + (R/L)} \quad \alpha_1 = \frac{1}{1 + (R/L)} \tag{4}$$

We wish to study the response of this first-order system to a unit step sequence:

$$v(k) = u(k) = \begin{cases} 1 & k = 0,1,2,\ldots \\ 0 & \text{negative} \end{cases}$$

Further, the system is assumed to be initially relaxed so that $y(-1) = 0$. Now we proceed systematically as follows:

$k = 0 \qquad y(0) = \beta_0 x(0) + \alpha_1 y(-1) = \beta_0 + 0$

$k = 1 \qquad y(1) = \beta_0 x(1) + \alpha_1 y(0) = \beta_0 \cdot 1 + \alpha_1 \beta_0 = (1 + \alpha_1)\beta_0$

$k = 2 \qquad y(2) = \beta_0 x(2) + \alpha_1 y(1)$

$\qquad\qquad\qquad = \beta_0 \cdot 1 + \alpha_1(1 + \alpha_1)\beta_0 = (1 + \alpha_1 + \alpha_1^2)\beta_0$

$k = 3 \qquad y(3) = \beta_0 x(3) + \alpha_1 y(2)$

$\qquad\qquad\qquad = \beta_0 \cdot 1 + \alpha_1(1 + \alpha_1 + \alpha_1^2)\beta_0 = (1 + \alpha_1 + \alpha_1^2 + \alpha_1^3)\beta_0$

$\qquad \vdots \qquad\qquad \vdots$

By induction, we write

$$y(k) = (1 + \alpha_1 + \alpha_1^2 + \alpha_1^3 + \cdots + \alpha_1^k)\beta_0$$

By the finite series, we can write

$$1 + \alpha_1 + \alpha_1^2 + \alpha_1^3 + \cdots + \alpha_1^k = \frac{1 + \alpha_1^{k+1}}{1 - \alpha_1} \quad \text{for} \quad \alpha_1 \neq 1$$

so that finally

$$y(k) = \frac{1 - \alpha_1^{k+1}}{\alpha_1} \beta_0 \tag{5}$$

For values of $\alpha_1 > 1$, the factor $(1 - \alpha_1^{k+1})$ becomes arbitrarily large as k increases; this indicates a condition of **instability**. For $|\alpha_1| < 1, (1 - \alpha_1^{k+1})$ approaches 1 as k increases, and the unit step response approaches the value

$$y(k) = \frac{\beta_0}{1 - \alpha_1} \quad k \text{ large} \tag{6}$$

Note that for our particular circuit specified by Equation 4, we obtain

$$y(k) \triangleq i(k) = \frac{1}{R} \quad k \text{ large}$$

which is the steady state current for this physical example.

2.12.1.1

Use the Book m-Function q1=urdsp_ex2_12_1_1(tl,T), and then use first the following MATLAB program:

```
>>for m=1:length(q1)
>>       i(m)=(q1(m+1)-q1(m))/T;
>>end
```

Then plot i. The continuous case current is found from the equation:

$$i(t) = \frac{dq}{dt} = -\frac{2}{3} e^{-t} - \frac{4}{3} e^{-4t} + 2e^{-2t}.$$

APPENDIX 2.1
Fourier Transform Properties

Linearity

$$af(t) + bh(t) \overset{\mathscr{F}}{\leftrightarrow} aF(\omega) + bH(\omega)$$

Time shifting

$$f(t \pm t_0) \overset{\mathscr{F}}{\leftrightarrow} e^{\pm j\omega t_0} F(\omega)$$

Symmetry

$$\begin{cases} F(t) & \overset{\mathscr{F}}{\leftrightarrow} 2\pi f(-\omega) \\ 1 & \overset{\mathscr{F}}{\leftrightarrow} 2\pi\delta(-\omega) = 2\pi\delta(\omega) \end{cases}$$

Time scaling

$$f(at) \overset{\mathscr{F}}{\leftrightarrow} \frac{1}{|a|} F\left(\frac{\omega}{a}\right)$$

Time reversal

$$f(-t) \overset{\mathscr{F}}{\leftrightarrow} F(-\omega)$$

Frequency shifting

$$e^{\pm j\omega_0 t} f(t) \overset{\mathscr{F}}{\leftrightarrow} F(\omega \mp \omega_0)$$

Modulation

$$\begin{cases} f(t)\cos\omega_0 t \overset{\mathscr{F}}{\leftrightarrow} \frac{1}{2}[F(\omega + \omega_0) + F(\omega - \omega_0)] \\ f(t)\sin\omega_0 t \overset{\mathscr{F}}{\leftrightarrow} \frac{1}{2j}[F(\omega - \omega_0) - F(\omega + \omega_0)] \end{cases}$$

Time differentiation

$$\frac{d^n f(t)}{dt^n} \overset{\mathscr{F}}{\leftrightarrow} (j\omega)^n F(\omega)$$

Frequency differentiation

$$\begin{cases} (-jt)f(t) \overset{\mathscr{F}}{\leftrightarrow} \frac{dF(\omega)}{d\omega} \\ (-jt)^n f(t) \overset{\mathscr{F}}{\leftrightarrow} \frac{d^n F(\omega)}{d\omega^n} \end{cases}$$

Time convolution

$$f(t) * h(t) = \int_{-\infty}^{\infty} f(x)h(t-x)\,dx \overset{\mathscr{F}}{\leftrightarrow} F(\omega)H(\omega)$$

Frequency convolution

$$f(t)h(t) \overset{\mathscr{F}}{\leftrightarrow} \frac{1}{2\pi} F(\omega) * H(\omega) = \frac{1}{2\pi}\int_{-\infty}^{\infty} F(x)H(\omega - x)\,dx$$

Autocorrelation

$$f(t) \odot f(t) = \int_{-\infty}^{\infty} f(x)f^*(x-t)\,dx \overset{\mathscr{F}}{\leftrightarrow} F(\omega)F^*(\omega)$$

$$= |F(\omega)|^2$$

Central ordinate

$$f(0) = \frac{1}{2\pi}\int_{-\infty}^{\infty} F(\omega)\,d\omega \quad F(0) = \int_{-\infty}^{\infty} f(t)\,dt$$

Parseval's theorem

$$E = \int_{-\infty}^{\infty} |f(t)|^2\,dt \quad E = \frac{1}{2\pi}\int_{-\infty}^{\infty} |F(\omega)|^2\,d\omega$$

The letter $\mathscr{F}$ stands for FT.

APPENDIX 2.2
Fourier Transform Pairs

$$f(t) = \frac{1}{2\pi} \int_{-\infty}^{\infty} F(\omega)e^{j\omega t} d\omega \qquad\qquad F(\omega) = \int_{-\infty}^{\infty} f(t)e^{-j\omega t} dt$$

1. $$f(t) = \begin{cases} 1 & |t| \le a \\ 0 & \text{otherwise} \end{cases} \qquad\qquad F(\omega) = 2\frac{\sin a\omega}{\omega}$$

2. $$f(t) = Ae^{-at}u(t) \qquad\qquad F(\omega) = \frac{A}{a+j\omega}$$

3. $$f(t) = Ae^{-a|t|} \qquad\qquad F(\omega) = \frac{2aA}{a^2+\omega^2}$$

4. $$f(t) = \begin{cases} A\left(1-\frac{|t|}{a}\right) & |t| \le a \\ 0 & \text{otherwise} \end{cases} \qquad\qquad F(\omega) = Aa\left[\frac{\sin(a\omega/2)}{(a\omega/2)}\right]^2$$

5. $$f(t) = Ae^{-a^2 t^2} \qquad\qquad F(\omega) = A\frac{\sqrt{\pi}}{a}e^{-(\omega/2a)^2}$$

6. $$f(t) = \begin{cases} A\cos\omega_0 t e^{-at} & t \ge 0 \\ 0 & t < 0 \end{cases} \qquad\qquad F(\omega) = A\frac{a+j\omega}{(a+j\omega)^2+\omega_0^2}$$

7. $$f(t) = \begin{cases} A\sin\omega_0 t e^{-at} & t \ge 0 \\ 0 & t < 0 \end{cases} \qquad\qquad F(\omega) = \frac{A\omega_0}{(a+j\omega)^2+\omega_0^2}$$

8. $$f(t) = \begin{cases} A\cos\omega_0 t & |t| \le a \\ 0 & \text{otherwise} \end{cases} \qquad\qquad \begin{aligned} F(\omega) &= A\frac{\sin a(\omega-\omega_0)}{\omega-\omega_0} \\ &+ A\frac{\sin a(\omega+\omega_0)}{\omega+\omega_0} \end{aligned}$$

9. $$f(t) = A\delta(t) \qquad\qquad F(\omega) = A$$

10. $$f(t) = \begin{cases} A & t > 0 \\ 0 & \text{otherwise} \end{cases} \qquad\qquad F(\omega) = A\left[\pi\delta(\omega) - j\frac{1}{\omega}\right]$$

11. $$f(t) \begin{cases} A & t > 0 \\ 0 & t = 0 \\ -A & t < 0 \end{cases} \qquad\qquad F(\omega) = -j2A\frac{1}{\omega}$$

12. $$f(t) = A \qquad\qquad F(\omega) = 2\pi A\delta(\omega)$$

(Continued)

APPENDIX 2.2 (*Continued*)
Fourier Transform Pairs

13. $\quad f(t) = A\cos\omega_0 t \qquad\qquad F(\omega) = \pi A[\delta(\omega - \omega_0) + \delta(\omega + \omega_0)]$

14. $\quad f(t) = A\displaystyle\sum_{n=-\infty}^{\infty}\delta(t - nT) \qquad\qquad F(\omega) = \dfrac{2\pi A}{T}\displaystyle\sum_{n=-\infty}^{\infty}\delta\left(\omega - n\dfrac{2\pi}{T}\right)$

15. $\quad f(t) = \dfrac{\sin at}{\pi t} \qquad\qquad F(\omega) = p_a(\omega)$

16. $\quad f(t) = \dfrac{2\sin^2(at/2)}{\pi t^2} \qquad\qquad F(\omega) = \begin{cases} 1 - \dfrac{|\omega|}{a} & |\omega| \le a \\ 0 & \text{otherwise} \end{cases}$

APPENDIX 2.3
DTFT Properties

Linearity	$af(n) + bh(n) \overset{\mathcal{F}_{DT}}{\leftrightarrow} aF\left(e^{j\omega}\right) + bH\left(e^{j\omega}\right)$
Time shifting	$f(n - m) \overset{\mathcal{F}_{DT}}{\leftrightarrow} e^{-j\omega m}F\left(e^{j\omega}\right)$
Time reversal	$f(-n) \overset{\mathcal{F}_{DT}}{\leftrightarrow} F\left(e^{-j\omega}\right)$
Convolution	$f(n) * h(n) \overset{\mathcal{F}_{DT}}{\leftrightarrow} F\left(e^{j\omega}\right)H\left(e^{j\omega}\right)$
Frequency shifting	$e^{j\omega_0 n}f(n) \overset{\mathcal{F}_{DT}}{\leftrightarrow} F\left(e^{j(\omega - \omega_0)}\right)$
Modulation	$f(n)\cos\omega_0 n \overset{\mathcal{F}_{DT}}{\leftrightarrow} \dfrac{1}{2}F\left(e^{j(\omega + \omega_0)}\right) + \dfrac{1}{2}F\left(e^{j(\omega - \omega_0)}\right)$
Correlation	$f(n) \odot h(n) \overset{\mathcal{F}_{DT}}{\leftrightarrow} F\left(e^{j\omega}\right)H\left(e^{-j\omega}\right)$
Parseval's formula	$\displaystyle\sum_{n=-\infty}^{\infty}|f(n)|^2 = \dfrac{1}{2\pi}\int_{-\pi}^{\pi}|F(e^{j\omega})|^2 \, d\omega$

APPENDIX 2.4
DFT Properties

Linearity

$$af(n) + bh(n) \overset{\mathscr{F}_{DF}}{\longleftrightarrow} aF(k) + bH(k)$$

Symmetry

$$\frac{1}{N} F(n) \overset{\mathscr{F}_{DF}}{\longleftrightarrow} f(-k)$$

Time shifting

$$f(n-i) \overset{\mathscr{F}_{DF}}{\longleftrightarrow} F(k)e^{-j2\pi ki/N} = F(k)W^{ki}$$

Frequency shifting

$$f(n)e^{jni} \overset{\mathscr{F}_{DF}}{\longleftrightarrow} F(k-i)$$

Time convolution

$$y(n) \triangleq f(n) * h(n) \overset{\mathscr{F}_{DF}}{\longleftrightarrow} F(k)H(k)$$

Frequency convolution

$$f(n)h(n) \overset{\mathscr{F}_{DF}}{\longleftrightarrow} \frac{1}{N} \sum_{x=0}^{N-1} F(x)H(n-x)$$

Parseval's theorem

$$\sum_{n=0}^{N-1} |f(n)|^2 = \frac{1}{N} \sum_{k=0}^{N-1} |F(k)|^2$$

Time reversal

$$f(-n) \overset{\mathscr{F}_{DF}}{\longleftrightarrow} F(-k)$$

Delta function

$$\delta(n) \overset{\mathscr{F}_{DF}}{\longleftrightarrow} 1$$

Central ordinate

$$f(0) = \frac{1}{N} \sum_{k=0}^{N-1} F(k); \quad F(0) = \sum_{n=0}^{N-1} f(n)$$

3 The z-Transform, Difference Equations, and Discrete Systems

3.1 THE z-TRANSFORM

To understand the essential features of the z-transform, consider a **one-sided sequence** of numbers $\{y(n)\}$ taken at uniform time intervals. This sequence might be the values of a continuous function that has been sampled at uniform time intervals; it could, of course, be a number sequence, e.g., the values of the amount that are present in a bank account at the beginning of each month that includes the interest. This number sequence is written as

$$\{y(n)\} = [y(0), y(1), y(2); \ldots, y(n), \ldots] \tag{3.1}$$

We now create the series,

$$Y(z) = \frac{y(0)}{z^0} + \frac{y(1)}{z^1} + \frac{y(2)}{z^2} + \cdots = y(0) + y(1)z^{-1} + y(2)z^{-2} + \cdots \tag{3.2}$$

In this expression, z denotes the general complex variable and $Y(z)$ denotes the z-transform of the sequence $\{y(n)\}$. In this general form, the **one-sided** z-transform of a sequence $\{y(n)\}$ is written as follows:

$$Y(z) \triangleq \mathfrak{Z}\{y(n)\} = \sum_{n=0}^{\infty} y(n)z^{-n} \tag{3.3}$$

This expression can be taken as the definition of the one-sided z-transform.

Since the exponent of the z's is equal to the distance from the sequence element $y(0)$, we identify the negative exponents as the amount of delay. This interpretation is not explicit in the mathematical form of Equation 3.3, but it is implied when the shifting properties of functions are considered. This same concept will occur when we apply the z-transform to the solution of difference equations. Initially, however, we study the mathematics of the z-transform.

When the sequence of numbers is obtained by sampling a function $y(t)$ every T seconds—for example, by using an analog-to-digital (A/D) converter, the numbers represent sample values $y(nT)$ for $n = 0, 1, 2,\ldots$.This suggests that there is a relationship between the Laplace transform of a continuous function and the z-transform of a sequence of samples of the function at the time constants $\ldots - nT, -(n-1)T, \ldots - T, 0, T, 2T, \ldots, nT, (n+1)T, \ldots$. To show that there is such a relationship, let $y(t)$ represent a function that is sampled at time constants T, seconds apart. The sampled function is (see Equation 2.7)

$$y_s(t) = y(t)\text{comb}_T(t) = \sum_{n=-\infty}^{\infty} y(nT)\delta(t - nT) \tag{3.4}$$

The Laplace transform of this equation is

$$Y_s(s) \triangleq \mathcal{L}\{y_s(t)\} = \mathcal{L}\left\{\sum_{n=-\infty}^{\infty} y(nT)\delta(t - nT)\right\} = \sum_{n=-\infty}^{\infty} y(nT)\mathcal{L}\{\delta(t - nT)\}$$

$$= \sum_{n=-\infty}^{\infty} y(nT)e^{-nTs} \tag{3.5}$$

Note that the Laplace operator operates on time t and s is a complex variable. If we make the substitution $z = e^{sT}$, then

$$Y_s(s)\Big|_{z=e^{sT}} = \sum_{n=-\infty}^{\infty} y(nT)z^{-n} = Y(z) \tag{3.6}$$

Hence, $Y(z)$ is the z-transform of the sequence of samples of $y(t)$, namely, $y(nT)$, with $n = 0, 1, 2,\dots$. From this discussion, we observe that the z-transform may be viewed as the Laplace transform of the sampled time function $y_s(t)$, with an appropriate change of variables. This interpretation is in addition to that given in Equation 3.3, which, as already noted, specifies that $Y(z)$ is the z-transform of the number sequence $\{y(nT)\}$ for $n = \dots-2, -1, 0, 1, 2,\dots$. The above equation is the two-sided z-transform.

Example 3.1.1

Deduce the z-transform of the function $f(t) = A\exp(-at)$ $t \geq 0$ sampled every T seconds.

Solution: The sampled values of the function are

$$\{f(nT)\} = A, Ae^{-aT}, Ae^{-a2T}, Ae^{-a3T},\dots$$

The z-transform of this sequence is written as

$$F(z) = A\left[1 + \frac{e^{-aT}}{z} + \left(\frac{e^{-aT}}{z}\right)^2 + \left(\frac{e^{-aT}}{z}\right)^3 + \cdots\right]$$

The series can be written in closed form, recalling that the infinite geometric series is given by $[1/(1-x)] = 1 + x^1 + x^2 + x^3 + \cdots$ $|x| < 1$. Thus, we have

$$\mathcal{Z}\{e^{-aT}u(nT)\} \triangleq F(z) = \frac{A}{1 - \dfrac{e^{-aT}}{z}} = \frac{A}{1 - e^{-aT}z^{-1}} = \frac{Az}{z - e^{-aT}} \tag{3.7}$$

Keeping in mind that z is a complex variable, the convergence is satisfied if

$$|e^{-aT}z^{-1}| = e^{-aT}|z^{-1}| < 1 \Rightarrow |z| > e^{-aT}$$

Exercise 3.1.1

Find the z-transform of the given functions when sampled every T seconds:
(1) $f(t) = u(t)$, (2) $f(t) = tu(t)$, (3) $f(t) = \exp(-bt)u(t)...b > 0$, (4) $f(t) = \sin\omega t u(t)$. ▲

3.2 PROPERTIES OF THE z-TRANSFORM

The most important properties of the z-transform, for one-sided sequences, are given in Table 3.1.

TABLE 3.1
Summary of z-Transform Properties

1. *Linearity*

$$\Im\{ax(n) + by(n)\} = a\Im\{x(n)\} + b\Im\{y(n)\}$$

$$\Im\{ax(nT) + by(nT)\} = a\Im\{x(nT)\} + b\Im\{y(nT)\}$$

2. *Right-shifting property*

a. $\Im\{y(n-m)\} = z^{-m}\sum_{n=0}^{\infty} y(nT)z^{-n} = z^{-m}Y(z)$ zero initial conditions

     $\Im\{y(nT - mT)\} = z^{-m}\sum_{n=0}^{\infty} y(nT)z^{-n} = z^{-m}Y(z)$ zero initial conditions

b. $\Im\{y(n-m)\} = z^{-m}Y(z) + \sum_{i=0}^{m-1} y(i-m)z^{-i}$ initial conditions present

     $\Im\{y(n-m)\} = z^{-m}Y(z) + \sum_{i=0}^{m-1} y(i-m)z^{-i}$ initial conditions present

3. *Left-shifting property*

$$\Im\{y(n+m)\} = z^{m}Y(z) - \sum_{n=0}^{m-1} y(n)z^{m-n}$$

$$\Im\{y(nT + mT)\} = z^{m}Y(z) - z^{m}\sum_{n=0}^{m-1} y(nT)z^{m-n}$$

4. *Time scaling*

$$\Im\{a^{n}y(n)\} = Y(a^{-1}z) = \sum_{n=0}^{\infty} (a^{-1}z)^{-n}$$

$$\Im\{a^{nT}y(nT)\} = Y(a^{-T}z) = \sum_{n=0}^{\infty} (a^{-T}z)^{-n}$$

(Continued)

TABLE 3.1 (*Continued*)
Summary of z-Transform Properties

5. *Periodic sequence*

$$\mathcal{Z}\{y(n)\} = \frac{z^N}{z^N - 1}\mathcal{Z}\{y_{(1)}(n)\}, \qquad N = \text{number of time units in a period}$$

$$y_{(1)}(n) = \text{first period}$$

$$\mathcal{Z}\{y(nT)\} = \frac{z^N}{z^N - 1}\mathcal{Z}\{y_{(1)}(nT)\}, \quad N = \text{number of time units in a period}$$

$$y_{(1)}(nT) = \text{first period}$$

6. *Multiplication by n*

$$\mathcal{Z}\{ny(n)\} = -z\frac{dY(z)}{dz}$$

$$\mathcal{Z}\{nTy(n)\} = -Tz\frac{dY(z)}{dz}$$

7. *Initial value*

$$y(n_0) = z^{n_0}\,Y(z)\big|_{z\to\infty} \qquad y(n) = 0 \text{ for } n < n_0$$

$$y(n_0T) = z^{n_0}\,Y(z)\big|_{z\to\infty} \qquad y(nT) = 0 \text{ for } n < n_0$$

8. *Final value*

$$\lim_{n\to\infty} y(n) = \lim_{n\to 1}(1 - z^{-1})Y(z) \qquad \text{provided } y(\infty) \text{ exists}$$

$$\lim_{n\to\infty} y(nT) = \lim_{n\to 1}(1 - z^{-1})Y(z) \qquad \text{provided } y(\infty) \text{ exists}$$

9. *Convolution*

$$\mathcal{Z}\{y(n)\} = \mathcal{Z}\{h(n) * x(n)\} = \mathcal{Z}\left\{\sum_{m=0}^{\infty} h(n-m)x(m)\right\} = H(z)X(z)$$

$$\mathcal{Z}\{y(nT)\} = \mathcal{Z}\{h(nT) * x(nT)\} = \mathcal{Z}\left\{\sum_{m=0}^{\infty} h(nT - mT)x(mT)\right\} = H(z)X(z)$$

10. *Bilateral convolution*

$$\mathcal{Z}\{y(n)\} = \mathcal{Z}\{h(n) * x(n)\} = \mathcal{Z}\left\{\sum_{m=-\infty}^{\infty} h(n-m)x(m)\right\} = H(z)X(z)$$

$$\mathcal{Z}\{y(nT)\} = \mathcal{Z}\{h(nT) * x(nT)\} = \mathcal{Z}\left\{\sum_{m=-\infty}^{\infty} h(nT - mT)x(mT)\right\} = H(z)X(z)$$

Exercise 3.2.1

Prove the following z-transform properties: (1) left-shift property, (2) time scaling, (3) multiplication by n, and (4) convolution. ▲

Example 3.2.1

Find the output of the relaxed system shown in Figure 3.1a if the input is that shown in Figure 3.1b. Express the system in its discrete form.

Solution: A direct application of the KVL yields the equation

$$\frac{di(t)}{dt} + i(t) = v(t), \qquad \frac{dv_o(t)}{dt} + v_o(t) = v(t)$$

The second equation above follows from the first since $v_o(t) = Ri(t)$ with $R = 1\ \Omega$. If we assume the sampling time $T = 1$, then from Equation 1.16 we find that $dv_o(t)/dt \cong v_o(n) - v_o(n-1)$, and thus, the above equation takes the following discretized form:

$$v_o(n) = \frac{1}{2}v(n) + \frac{1}{2}v_o(n-1)$$

Note: *The impulse response of a system is found by introducing a delta function for its input. The z-transform of the output due to a delta function input to the system is equal to the system function (or transfer function) H(z). Its inverse z-transform is its impulse response h(n).*

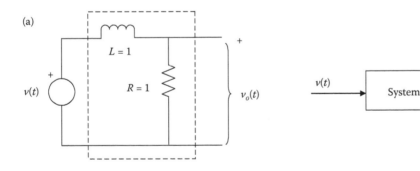

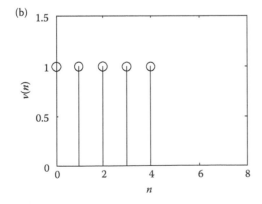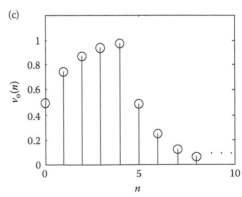

FIGURE 3.1

Therefore, $H(z)$ is given by

$$\mathcal{Z}\{v_o(n)\} = \frac{1}{2}\mathcal{Z}\{\delta(n)\} + \frac{1}{2}\mathcal{Z}\{v_o(n-1)\} \text{ or } H(z) = \frac{V_o(z)}{V(z)} = \frac{1}{2}\frac{1}{1-\frac{1}{2}z^{-1}}$$

$$= \frac{1}{2}\left[1 + \left(\frac{1}{2}\right)z^{-1} + \left(\frac{1}{2}\right)^2 z^{-2} + \cdots\right], \quad V(z) = \mathcal{Z}\{v(n)\},$$

$$\mathcal{Z}\{\delta(n)\} = \sum_{n=0}^{\infty}\delta(n)z^{-n} = \delta(0)z^{-0} + \delta(1)z^{-1} + \cdots = 1 + 0z^{-1} + 0 \cdots = 1$$

From the definition of the z-transform (Equation 3.2), we obtain the inverse z-transform of $H(z)$, its impulse response, which is

$$\mathcal{Z}^{-1}\{H(z)\} = h(n) = \frac{1}{2}\left(\frac{1}{2}\right)^n \quad n \geq 0$$

Note: *The output of a system is equal to the convolution of its input and its impulse response.*

The convolution in this case is

$$v_o(n) = \sum_{m=0}^{n} h(n-m)v(m)$$

Therefore, the output at successive time steps for a unit step function is

$$v_o(0) = h(0)v(0) = \frac{1}{2} \times 1 = \frac{1}{2}$$

$$v_o(1) = h(1)v(0) + h(0)v(1) = \frac{1}{4} \times 1 + \frac{1}{2} \times 1 = \frac{3}{4}$$

$$v_o(2) = h(2)v(0) + h(1)v(1) + h(0)v(2) = \frac{1}{8} \times 1 + \frac{1}{4} \times 1 + \frac{1}{2} \times 1 = \frac{7}{8}$$

$$\vdots \qquad\qquad \vdots$$

The output is shown in Figure 3.1c. The reader should use inputs of different lengths to obtain different outputs.

Book m-Function: [vo]=urdsp3_2_1(v)

```
function[vo]=urdspex3_2_1(v)
n=0:length(v)-1;
h=0.5*((0.5).^n);
vo=conv(v,h);% conv() is a MATLAB function;
m=0:length(vo)-1;
stem(m,vo,'k');
```

Example 3.2.2

Find the z-transform of the *RL* series circuit, with a voltage input and the current output. The initial condition is $i(0) = 2$ and the input voltage is $v(t) = \exp(-t)u(t)$. Discretize the analog system with values $R = 2$ and $L = 1$.

Solution: The Kirchhoff voltage law (KVL) is

$$L\frac{di(t)}{dt} + Ri(t) = v(t) \quad \text{or} \quad \frac{di(t)}{dt} + \frac{R}{L}i(t) = \frac{1}{L}v(t)$$

The homogeneous solution is $i_h(t) = A\exp(-(R/L)t)$ which verifies the homogeneous equation $\frac{di(t)}{dt} + \frac{R}{L}i(t) = 0$. For an exponential function input voltage $v(t) = \exp(-t)u(t)$, the particular solution is $i_p(t) = B\exp(-t)$. Introducing this assumed solution, $i_p(t) = B\exp(-t)u(t)$, in the nonhomogeneous equation, we obtain $B = (1/L)(1/[(R/L) - 1])$. The solution is

$$i(t) = i_h(t) + i_p(t) = Ae^{-\frac{R}{L}t} + \frac{1}{L}\frac{1}{\frac{R}{L}-1}e^{-t}$$

Introducing the initial condition $i(0) = 2$ in the above equation, we obtain $A = 2 - (1/L)[1/((R/L) - 1)]$ and, therefore, the solution is

$$i(t) = \left(2 - \frac{1}{L}\frac{1}{\frac{R}{L}-1}\right)e^{-\frac{R}{L}t} + \frac{1}{L}\frac{1}{\frac{R}{L}-1}e^{-t}$$

The discretized form of the above equation is given by

$$L\frac{i(nT) - i(nT-T)}{T} + Ri(nT) = v(nT)$$

or

$$i(nT) - \frac{1}{1+T\frac{R}{L}}i(nT-T) = \frac{T}{L}\frac{1}{1+T\frac{R}{L}}v(nT)$$

Taking into consideration the linearity and time shifting properties (see Table 3.1), we find

$$I(z) - a\left[z^{-1}I(z) + i(0-0)z^{-0}\right] = \frac{T}{L}aV(z) \quad a = \frac{1}{1+T\frac{R}{L}}$$

Finally, we obtain the algebraic relation

$$I(z) = \frac{2a}{1-az^{-1}} + \frac{T}{L}\frac{a}{1-az^{-1}}V(z) = 2a\frac{z}{z-a} + \frac{aT}{L}\frac{z}{z-a}V(z)$$

Since $v(nT) = \exp(-nT)u(nT)$, then using the z-transform definition, we obtain

$$V(z) = \sum_{n=0}^{\infty} e^{-nT}z^{-n} = 1 + e^{-T}z^{-1} + e^{-2T}z^{-2} + e^{-3T}z^{-3} + \cdots = \frac{1}{1-e^{-T}z^{-1}}$$

Therefore, the current in z-transform format is

$$I(z) = 2a\frac{z}{z-a} + \frac{aT}{L}\frac{z}{z-a}\frac{z}{z-e^{-T}}$$

We use the following identity from the last term:

$$\frac{z}{z-a}\frac{z}{z-e^{-T}} = \frac{Az}{z-a} + \frac{Bz}{z-e^{-T}}$$

If we first multiply both sides by $z - a$ and set in both sides $z = a$, and, second, we multiply both sides of the equation by $z - \exp(-T)$ and set $z = \exp(-T)$ in both sides, we obtain A and B, which are

$$A = \frac{a}{a-e^{-T}}, \quad B = \frac{e^{-T}}{e^{-T}-a}$$

Therefore, the current in its final form is

$$I(z) = 2a\frac{z}{z-a} + \frac{aT}{L}\frac{a}{a-e^{-T}}\frac{z}{z-a} + \frac{aT}{L}\frac{e^{-T}}{e^{-T}-a}\frac{z}{z-e^{-T}}$$

$$= \left(2a+\frac{a^2T}{L}\frac{1}{a-e^{-T}}\right)\frac{z}{z-a} + \frac{aT}{L}\frac{e^{-T}}{e^{-T}-a}\frac{z}{z-e^{-T}}$$

Based on the z-transform definition, the inverse of the $I(z)$ is given by

$$i(nT) = \left(2a+\frac{Ta^2}{a-e^{-T}}\right)a^n + \frac{Tae^{-T}}{e^{-T}-a}e^{-nT}$$

$$i(nT) = \left(2a+\frac{a^2T}{L}\frac{1}{a-e^{-T}}\right)a^n + \frac{aT}{L}\frac{e^{-T}}{e^{-T}-a}e^{-nT}$$

Figure 3.2 shows the results for the continuous case and the discrete case for two different sampling times $T = 0.5$ and $T = 0.1$.

The following Book MATLAB program was used to produce Figure 3.2.

BOOK MATLAB PROGRAM

```
>>t=0:0.01:3; i=exp(-2*t)+exp(-t);
>>T=0.5; n=0:3/0.5;
>>i5=((2*a+a^2*T*1/(a-exp(-T))))*a.^n…
+(a*T*exp(-T)*exp(-n*T))/(exp(-T)-a);
>>T=0.1; m=0:3/0.1;
>>i1=((2*a+a^2*T*1/(a-exp(-T))))*a.^m…
+(a*T*exp(-T)*exp(-m*T))/(exp(-T)-a);
>>plot(t,i,'k'); hold on; stem(n*0.5, i5, 'k');
>>hold on; stem(m*0.1,i1,'xk');
>>xlabel('t (s), n'); ylabel('i(t), i(nT)');            ■
```

Exercise 3.2.2

Find the current of an RL relaxed circuit in series if its input voltage is a unit step function with the following values: $R = 2$ and $L = 0.5$.   ▲

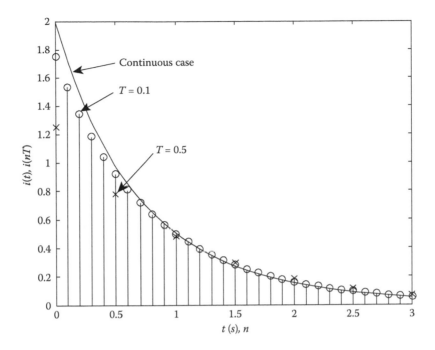

FIGURE 3.2

3.3   INVERSE z-TRANSFORM

As already discussed in our studies, we assume that an $F(z)$ corresponds to a sequence $\{f(n)\}$ that is bounded as $n \to \infty$. To find the inverse z-transform, we cast the transform function into a form that is amenable to simple tabular lookup using Table 3.2. The approach parallels that followed in performing similar operations using the Laplace transform. The functions that will concern us are rational functions of z which are the ratio of two polynomials. Ordinarily, these are **proper fractions** since the degree of the numerator polynomial is less than the degree of the denominator polynomial. If the functions are not proper functions, we perform long division till the degree of the numerator is less than that of the denominator. This result is a power series and a proper fraction.

Example 3.3.1

Determine the inverse z-transform of the functions: (1) $F(z) = \dfrac{1}{1 - 0.1z^{-1}}$, (2) $F(z) = \dfrac{z^2}{(z - 0.2)(z + 0.2)}$, (3) $F(z) = \dfrac{z^3}{z - 0.2}$.

Solution:

1. We can also write $F(z) = \dfrac{z}{z - 0.1}$ and with the help of Table 3.2, we obtain $f(n) = (0.1)^n \quad n \geq 0$.

2. We use the splitting of the function as a sum of functions as follows:

$$F(z) = \frac{z^2}{(z - 0.2)(z + 0.2)} = \frac{Az}{z - 0.2} + \frac{Bz}{z + 0.2}$$

TABLE 3.2
Common z-Transform Pairs

Entry $f(n)$, $f(nT)$ for Number $n \geq 0$	$F(z) = \sum\limits_{n=0}^{\infty} f(n)z^{-n}$	Radius of Convergence $\|z\| > R$
1. $\delta(n)$	1	0
2. $\delta(n-m)$	z^{-m}	0
3. 1	$\dfrac{z}{z-1}$	1
4. n	$\dfrac{z}{(z-1)^2}$	1
5. n^2	$\dfrac{z(z+1)}{(z-1)^3}$	1
6. n^3	$\dfrac{z(z^2+4z+1)}{(z-1)^4}$	1
7. a^n	$\dfrac{z}{z-a}$	$\|a\|$
8. na^n	$\dfrac{az}{(z-a)^2}$	$\|a\|$
9. $n^2 a^n$	$\dfrac{az(z+a)}{(z-a)^3}$	$\|a\|$
10. $\dfrac{a^n}{n!}$	$e^{a/z}$	0
11. $(n+1)a^n$	$\dfrac{z^2}{(z-a)^2}$	$\|a\|$
12. $\dfrac{(n+1)(n+2)a^n}{2!}$	$\dfrac{z^3}{(z-a)^3}$	$\|a\|$
13. $\dfrac{(n+1)(n+2)\cdots(n+m)a^n}{m!}$	$\dfrac{z^{m+1}}{(z-a)^{m+1}}$	$\|a\|$
14. $\sin n\omega T$	$\dfrac{z\sin\omega T}{z^2 - 2z\cos\omega T + 1}$	1
15. $\cos n\omega T$	$\dfrac{z(z-\cos\omega T)}{z^2 - 2z\cos\omega T + 1}$	1
16. $a^n \sin n\omega T$	$\dfrac{az\sin\omega T}{z^2 - 2az\cos\omega T + a^2}$	$\|a\|^{-1}$
17. $a^{nT} \sin n\omega T$	$\dfrac{a^T z\sin\omega T}{z^2 - 2a^T z\cos\omega T + a^{2T}}$	$\|a\|^{-T}$
18. $a^n \cos n\omega T$	$\dfrac{z(z-a\cos\omega T)}{z^2 - 2az\cos\omega T + a^2}$	$\|a\|^{-1}$
19. $e^{-anT} \sin n\omega T$	$\dfrac{ze^{-aT}\sin\omega T}{z^2 - 2e^{-aT}z\cos\omega T + e^{-2aT}}$	$\|z\| > \|e^{-aT}\|$
20. $e^{-anT} \cos n\omega T$	$\dfrac{z(z-e^{-aT}\cos\omega T)}{z^2 - 2e^{-aT}z\cos\omega T + e^{-2aT}}$	$\|z\| > \|e^{-aT}\|$

(Continued)

TABLE 3.2 (*Continued*)
Common z-Transform Pairs

Entry $f(n)$, $f(nT)$ for Number $n \geq 0$	$F(z) = \sum_{n=0}^{\infty} f(n)z^{-n}$	Radius of Convergence $\|z\| > R$
21. $\dfrac{n(n-1)}{2!}$	$\dfrac{z}{(z-1)^3}$	1
22. $\dfrac{n(n-1)(n-2)}{3!}$	$\dfrac{z}{(z-1)^4}$	1
23. $\dfrac{n(n-1)(n-2)\cdots(n-m+1)}{m!}a^{n-m}$	$\dfrac{z}{(z-a)^{m+1}}$	1
24. e^{-anT}	$\dfrac{z}{z-e^{-aT}}$	$\left\| e^{-aT} \right\|$
25. ne^{-anT}	$\dfrac{ze^{-aT}}{(z-e^{-aT})^2}$	$\left\| e^{-aT} \right\|$

Multiplying by $z - 0.2$ and divide by z both sides of the equation and setting $z = 0.2$, we obtain $A = 0.5$. Next, we multiply both sides by $(z + 0.2)$ and set $z = -0.2$ and find $B = 0.5$.

$$F(z) = 0.5\frac{z}{z-0.2} + 0.5\frac{z}{z+0.2} \implies \mathcal{Z}^{-1}\{F(z)\} = 0.5\mathcal{Z}^{-1}\left\{\frac{z}{z-0.2}\right\} + 0.5\mathcal{Z}^{-1}\left\{\frac{z}{z+0.2}\right\}$$

or

$$f(n) = 0.5\left[0.2^n + (-0.2)^n\right]$$

3. The function can be written in the form

$$F(z) = \frac{z^3}{z-0.2} = z^2 + 0.2z + 0.2^2\frac{z}{z-0.2}$$

Using Table 3.2, the inverse z-transform is

$$f(n) = \delta(n+2) + 0.2\delta(n+1) + (0.2)^2(0.2)^n$$

where the last term is applicable for $n \geq 0$. ∎

Example 3.3.2

Find the inverse z-transform of the function

$$F(z) = \frac{z^2 - 3z + 8}{(z-2)(z+2)(z+3)} = \frac{z^2 - 3z + 8}{z^3 + 3z^2 - 4z - 12}$$

Solution: If we execute the division, we obtain $z^{-1}(1 - 6z^{-1} + 30z^{-2} - 102z^{-3} + \cdots)$. The z^{-1} indicates a shifted sequence to the right and MATLAB ignores the shifting. To proceed, we multiply both sides by z^{-1} to find

$$F(z)z^{-1} = \frac{z^2 - 3z + 8}{z(z-2)(z+2)(z+3)} = \frac{A}{z} + \frac{B}{z-2} + \frac{C}{z+2} + \frac{D}{z+3}$$

Following the same procedure as earlier, we obtain: $A = -(2/3)$, $B = (3/20)$, $C = (9/4)$, and $D = -(26/15)$. Therefore, the inverse z-transform is

$$f(n) = \frac{2}{3}\delta(n) + \frac{3}{20}2^n + \frac{9}{4}(-2)^n - \frac{26}{15}(-3)^n$$

We can use MATLAB to find the inverse z-transform. We can use the following program:

Book MATLAB Program

```
>>num=[0   1   -3   8]; den=[conv(conv([1   -2],[1   2]),[1   3])];
>>      %we can also write den=[1   3   -4   -12] instead of the
>>      %double convolution;
>>[r,p,k]=residue(num,den);%residue() is a MATLAB function;
r=[5.2000    -4.5000   0.3000];
p=[-3.0000   -2.0000   2.0000];
k=[];
```

Based on the MATLAB results, we write

$$F(z) = 5.2\frac{z}{z+3} - 4.5\frac{z}{z+2} + 0.3\frac{z}{z-2}$$

Note: *Observe the corresponding numbers between r and p vectors.*

The inverse z-transform of the above equation is

$$f(n) = 5.2(-3)^n - 4.5(-2)^n + 0.3(2)^n$$

Although the two results look different, the sequence $\{f(n)\}$ is the same in the two cases. We must remember that any shifting of the sequence is ignored by MATLAB.   ∎

Example 3.3.3

Find the inverse z-transform of the function

$$F(z) = \frac{z^2 - 9}{(z-1)(z-2)^2}$$

Observe that the function has both single- and second-order poles. In addition, since the denominator is lager by one unit it means that the signal is shifted to the right by one unit time, which MATLAB ignores.

Solution: The function $F(z)/z$ is expanded in partial fraction form as follows:

$$\frac{F(z)}{z} = \frac{z^2 - 9}{z(z-1)(z-2)^2} = \frac{A}{z} + \frac{B}{z-1} + \frac{C}{z-2} + \frac{D}{(z-2)^2} \tag{3.8}$$

Following the same procedure as in the above examples we obtain $A = (9/4)$, $B = -8$, $D = -(5/2)$. Next, we set $z = 3$ in both sides of the equation, or any other value besides any one of the poles, and we obtain $C = (23/4)$. Introducing these constants in Equation 3.8, multiplying both sides by z and then taking the inverse z-transform, we find

$$f(n) = \frac{9}{4}\delta(n) - 8u(n) + \frac{23}{4}2^n - \frac{5}{2}n2^{n-1}$$

For $n = 0, 1, 2, 3$, and 4, the value of the sequence is $\{f(n) = \{\; 0 \quad 1 \quad 5 \quad 8 \quad 4 \;\}\}$.

Book MATLAB Program

```
>>num=[0   1   0   -9];
>>den=[1   -5   8   -4];
>>f=dimpulse(num,den,5);%the number 5 indicates the number
>>              %of the values desired for the f(n);
>>              %dimpulse() is a MATLAB function;
```

The values found using the Book MATLAB program are the same values as found above. ∎

Note: *We note that to find the inverse z-transform, we must*

1. *initially ignore any factor of the form z^n, where n is an integer*
2. *expand the remaining part into partial fraction form*
3. *use z-transform tables or properties to obtain z-transform of each term in the expansion*
4. *combine the results and perform the necessary shifting required by z^n omitted in the first step*

3.4 TRANSFER FUNCTION

It is desired to obtain the transfer function for the system shown in Figure 3.3a. The reader must note that, in finding the transfer function, the initial conditions are ignored. In other words, we assume that the system is in a relaxing condition. To proceed, we consider initially the portion of the system (subsystem) shown between $x(n)$ and $y_1(n)$ described by the difference equation

$$y_1(n) = bx(n) + a_1 y_1(n-1)$$

The z-transform of the expression is

$$Y_1(z) = bX(z) + a_1 z^{-1} Y_1(z) \quad \text{or} \quad Y_1(z) = \frac{b}{1 - a_1 z^{-1}} X(z) = H_1(z)X(z)$$

The portion of the system between $y_1(n)$ and $y(n)$ (a cascaded subsystem within the first subsystem) is described by a similar expression whose z-transform is

$$Y(z) = \frac{b}{1 - a_1 z^{-1}} Y_1(z) = H_2(z)Y_1(z)$$

(a)

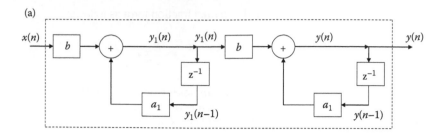

(b)

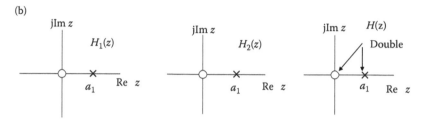

FIGURE 3.3

Substituting the known expression for $Y_1(z)$ into this final expression, we obtain

$$Y(z) = \frac{b}{1 - a_1 z^{-1}} \frac{b}{1 - a_1 z^{-1}} X(z) = H_1(z)H_1(z)X(z) = H(z)X(z)$$

where

$$H(z) = H_1(z)H_1(z) = \left(\frac{b}{1 - a_1 z^{-1}}\right)^2$$

The pole-zero configurations for each subsystem and for the combined system are shown in Figure 3.3b.

Exercise 3.4.1

Find the transfer function for the system shown in Figure 3.4 which is a combination of an FIR (**finite impulse response**) and an IIR (**infinite impulse response**) systems.   ▲

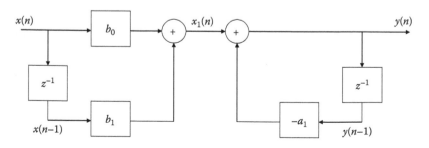

FIGURE 3.4

HIGHER-ORDER TRANSFER FUNCTIONS

The general case of a system is described by the following difference equation:

$$y(n)+a_1 y(n-1)+a_2 y(n-2)+\cdots+a_p y(n-p)=b_0 x(n)+b_1 x(n-1)$$

$$+b_2 x(n-2)+\cdots+b_q x(n-q) \qquad (3.9)$$

Taking the z-transform of both sides of the above equation and solving for the ratio $Y(z)/X(z)$ (output/input), we obtain the transfer function

$$H(z) \triangleq \frac{Y(z)}{X(z)} = \frac{\text{output}}{\text{input}} = \frac{b_0 + b_1 z^{-1} + \cdots + b_q z^{-q}}{1 + a_1 z^{-1} + \cdots + a_p z^{-p}} = \frac{\displaystyle\sum_{n=0}^{q} b_n z^{-n}}{1 + \displaystyle\sum_{n=1}^{p} a_n z^{-n}} \qquad (3.10)$$

This equation indicates that if we know the transfer function $H(z)$, then the output $Y(z)$ to any input $X(z)$ (or equivalently, $x(n)$) can be determined.

If we set $a_1, a_2, \ldots, a_p$ equal to zero, Equation 3.9 becomes

$$y(n) = b_0 x(n)+b_1 x(n-1)+b_2 x(n-2)+\cdots+b_q x(n-q) \qquad (3.11)$$

This expression defines a **qth-order IIR filter** (system).

For the case $b_0 = 1$ and the rest of b_i's are zero, the difference Equation 3.9 becomes

$$y(n)+a_1 y(n-1)+a_2 y(n-2)+\cdots+a_p y(n-p)=x(n) \qquad (3.12)$$

This equation defines a **pth-order IIR filter**. Finally, if none of the constants is zero in Equation 3.9, the block diagram representation of the combined FIR and IIR system is shown in Figure 3.5. The block diagrams for FIR and IIR filters can be obtained from Figure 3.5 by introducing the appropriate constants equaling zero.

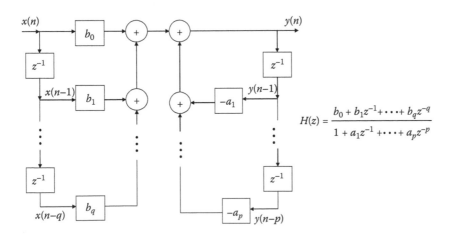

FIGURE 3.5

Exercise 3.4.2

From Figure 3.5, set the appropriate constants equal to zero and produce the block diagram of an IIR filter. ▲

3.5   FREQUENCY RESPONSE OF DISCRETE SYSTEMS

To obtain the frequency response of a system, we introduce into its transfer function the identity $z = e^{-j\omega}$. Hence, for a first-order system we have

$$H(e^{j\omega}) = \frac{b_0 + b_1 e^{-j\omega}}{1 + a_1 e^{-j\omega}} \tag{3.13}$$

which is the **frequency response function**.

If we set $\omega = \omega + 2\pi$ in $H(.)$, we find that $H(e^{j(\omega+2\pi)}) = H(e^{j\omega}e^{j2\pi}) = H(e^{j\omega})$ which indicates that the frequency response function is periodic with period 2π. If, on the other hand, we had introduced $z = e^{j\omega T}$ (T = sampling time), then the frequency response function would be of the form

$$H(e^{j\omega T}) = \frac{b_0 + b_1 e^{-j\omega T}}{1 + a_1 e^{-j\omega T}} \tag{3.14}$$

If we set

$$\omega = \omega + \frac{2\pi}{T}$$

in $H(.)$, we obtain

$$H\left(e^{j(\omega+\frac{2\pi}{T})T}\right) = H\left(e^{j\omega T}e^{j2\pi}\right) = H\left(e^{j\omega T}\right)$$

which indicates that the frequency response function is periodic with period $2\pi/T$.

Note: *Discrete systems with unit sampling time (T = 1) have periodic frequency response functions with period 2π, and those with time sampling equal to T have periodic frequency response functions with period $2\pi/T$.*

Example 3.5.1

Find the frequency characteristics of the system described by the difference equation

$$y(n) = 4x(n) + 4x(n-1) + x(n-2)$$

Solution: The system function is

$$H(z) \triangleq \frac{Y(z)}{X(z)} = 4 + 4e^{-j\omega T} + e^{-j2\omega T}$$

The frequency responses are obtained from the relationships

$$H(z) = Ae^{j\varphi} = 4 + 4e^{-j\omega T} + e^{-j2\omega T}$$

$$A = \left[H(z)H*(z)_{z=e^{j\omega T}} \right]^{1/2} = \left[H(z)H(z^{-1})\Big|_{z=e^{j\omega T}} \right]^{1/2} = (33 + +40\cos\omega T + 8\cos 2\omega T)^{1/2}$$

$$\varphi = \tan^{-1}\left(\frac{-4\sin\omega T - \sin 2\omega T}{4 + 4\cos\omega T + \cos 2\omega T} \right)$$

The frequency characteristics of the system for $T = 0.5$ and ω from 0 to $2\pi/0.5 = 4\pi$ are obtained using the Book MATLAB program:

```
>>w=0:0.1:4*pi;
>>A=abs(4+4*exp(-j*w*0.5)+exp(-2*w*0.5);
>>ef=angle(4+4*exp(-j*w*0.5)+exp(-2*w*0.5);
>>subplot(2,1,1);plot(w,A);subplot(2,1,2);plot(w,ef);
```                    ■

Exercise 3.5.1

From the results above for the amplitude A and the phase ef, state a general conclusion about the spectra.   ▲

Frequency Response of Higher-Order Digital Systems

$$H(e^{j\omega T}) = \frac{b_0 + b_1 e^{-j\omega T} + b_2 e^{-j2\omega T} + \cdots + b_q e^{-jq\omega T}}{a_0 + a_1 e^{-j\omega T} + a_2 e^{-j2\omega T} + \cdots + a_p e^{-jp\omega T}} = H(z)\Big|_{z=e^{j\omega T}} \qquad (3.15)$$

Example 3.5.2

Find the frequency response of the digital system shown in Figure 3.6.

Solution: From the figure we obtain the two difference equations

$$x_1(n-1) + 2x_1(n) = y(n)$$

$$-[y(n-1) + 2y(n)] + x(n) = x_1(n)$$

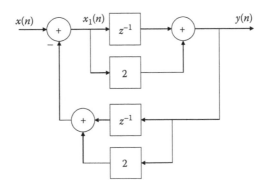

FIGURE 3.6

Substituting $x_1(n)$ from the second to the first equation, we obtain

$$5y(n) + 4y(n-1) + y(n-2) = 2x(n) + x(n-1)$$

By taking the z-transform of the above equation and solving for the ratio of the output to input, we obtain the transfer function

$$H(z)\Big|_{z=e^{j\omega}} \triangleq \frac{Y(z)}{X(z)}\Big|_{z=e^{j\omega}} = \frac{2+z^{-1}}{5+4z^{-1}+z^{-2}} = \frac{2+e^{-j\omega}}{5+4e^{-j\omega}+e^{-j2\omega}}$$

$$= \frac{2e^{j2\omega}+e^{j\omega}}{5e^{j2\omega}+4e^{j\omega}+1}$$

Another way to find the frequency response of a transfer function is to use the following Book MATLAB program. For this particular example it is

```
>>w=0:2*pi/256:2*pi-(2*pi/256);
>>b=[2   1   0];
>>a=[5   4   1];
>>H=fft(b,256)./fft(a,256);
>>subplot(2,1,1);stem(w,abs(H));
>>subplot(2,1,2);stem(w,angle(H));
```
∎

3.6 z-TRANSFORM SOLUTION OF DIFFERENCE EQUATIONS

The following examples will explain how to use the z-transform method for the solution of difference equations.

Example 3.6.1

Solve the discrete-time problem defined by the equation

$$y(n) + 2y(n-1) = 3.5u(n) \tag{3.16}$$

with $y(-1) = 0$ and, $u(n)$ is the discrete unit step function.

Solution: Taking the z-transform of the above equation (see Tables 3.1 and 3.2), we obtain

$$Y(z) + 2z^{-1}Y(z) = 3.5\frac{z}{z-1}$$

Therefore, we proceed as follows:

$$Y(z) = 3.5\frac{z}{z-1}\frac{z}{z+2} = \frac{A}{z-1} + \frac{B}{z+2}$$

Multiplying both sides by $(z-1)$ and setting $z=1$, we obtain $A = 6/7$. Similarly, we obtain $B = 7/3$. Therefore, the inverse z-transform is

$$y(n) = \frac{7}{6}u(n) + \frac{7}{3}(-2)^n \quad n = 0,1,2,\dots$$
∎

Exercise 3.6.1

Obtain the results in Example 3.6.1 using MATLAB.   ▲

Exercise 3.6.2

Repeat Exercise 3.6.1, but now with the initial condition $y(-1) = 4$.   ▲

Example 3.6.2

Determine the output of the discrete approximation of the system shown in Figure 3.7a for a sampling time *T*. The output for $T = 0.2$ and for $T = 1$ are to be plotted, and the results are compared with the output of the continuous system. The input is a unit step current source $i(t) = u(t)$, and an initial condition $v_o(0) = 2V$.

Solution: The differential equation describing the system is

$$\frac{dv_o(t)}{dt} + \frac{v_o(t)}{0.5} = i(t)$$

The analogous discrete form of this equation is

$$\frac{v_o(nT) - v_o(nT - T)}{T} + 2v_o(nT) = i(nT)$$

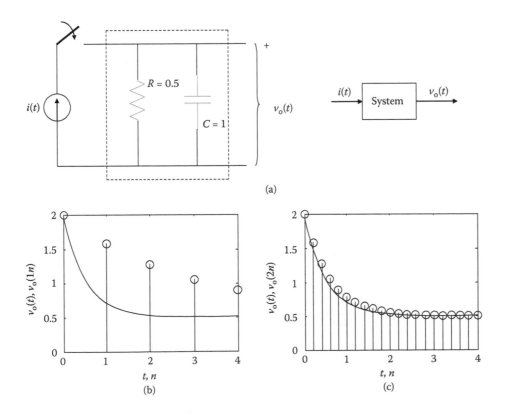

(a)

(b)

(c)

FIGURE 3.7

From this,

$$v_o(nT) = \frac{T}{1+2T}i(nT) + \frac{1}{1+2T}v_o(nT-T) = b_0 i(nT) + a_1 v_o(nT-T)$$

The z-transform of this equation gives

$$V_o(z) = \underbrace{\frac{b_0}{1-a_1 z^{-1}}I(z)}_{\substack{\text{zero-state}\\\text{response}}} + \underbrace{\frac{a_1}{1-a_1 z^{-1}}v_o(-T)}_{\substack{\text{zero-input}\\\text{response}}}$$

Since in the continuous case $v_o(0) = 2$, we must refer back to the difference equation and set the appropriate values to find the value of $v_o(-T)$. Hence, we find

$$v_o(0) = b_0 i(0) + a_1 v_o(T) \quad n=0, \text{ or } v_o(-T) = (2-b_0)/a_1$$

Thus, we obtain

$$\text{zero-input response} = \mathcal{Z}^{-1}\left\{ b_0 \frac{z^2}{(z-1)(z-a_1)} \right\} = \mathcal{Z}^{-1}\left\{ \frac{b_0}{1-a_1}\frac{z}{z-1} + \frac{b_0 a_1}{a_1-1}\frac{z}{z-a_1} \right\}$$

$$= \frac{b_0}{1-a_1}u(n) + \frac{b_0 a_1}{a_1-1}a_1^n u(n)$$

$$\text{zero-state response} = \mathcal{Z}^{-1}\left\{ \frac{(2-b_0)z}{(z-a_1)} \right\} = (2-b_0)a_1^n u(n)$$

The solution is the sum of the **zero-input** and **zero-state** solutions. The solution of the continuous case is easily found to be equal to

$$v_o(t) = 0.5 + 1.5e^{-2t} \quad t \geq 0$$

The result for the continuous and discrete case $T = 1$ and $T = 0.2$ are shown in Figure 3.7b. ∎

HINTS–SUGGESTIONS–SOLUTIONS OF THE EXERCISES

3.1.1

1. $\mathcal{Z}\{f(nT)\} = \mathcal{Z}\{u(nT)\} = \sum_{n=0}^{\infty} u(nT)z^{-n} = \left(1 + z^{-1} + z^{-2} + \cdots\right)$

$$= \frac{1}{1-z^{-1}} = \frac{z}{z-1}$$

To have convergence of the series, we must have $\left|z^{-1}\right| < 1$. This leads us to conclude that the region of convergence (ROC) in the complex plane is the region beyond the radius $z > 1$.

2. $\mathfrak{Z}\{f(nT)\} = \mathfrak{Z}\{nTu(nT)\} = \sum_{n=0}^{\infty} nTu(nT)z^{-n} = Tz^{-1} + 2Tz^{-2} + 3Tz^{-3} + \cdots$

$$= -Tz\frac{d}{dz}\left(z^{-1} + z^{-2} + z^{-3} + \cdots\right) = -Tz\frac{d}{dz}\left[z^{-1}\left(z^{-1} + z^{-2} + z^{-3} + \cdots\right)\right]$$

$$= -Tz\frac{d}{dz}\left[z^{-1}\frac{z}{z-1}\right] = \frac{Tz}{(z-1)^2}, \quad \text{ROC}|z| > 1$$

3. $\mathfrak{Z}\{f(nT)\} = \mathfrak{Z}\{e^{-bnT}u(nT)\} = \mathfrak{Z}\{u(nT)c^{-n}\}; \, c = e^{bT}$

$$= \sum_{n=0}^{\infty} u(nT)c^{-n}z^{-n} = 1 + c^{-1}z^{-1} + c^{-2}z^{-2} \mp \cdots = \frac{1}{1 - c^{-1}z^{-1}}$$

$$= \frac{cz}{cz-1} = \frac{ze^{bT}}{ze^{bT}-1} = \frac{z}{z - e^{-bT}} \quad \text{ROC}|c^{-1}z^{-1}| < 1 \text{ or } |z| > e^{-bT}$$

4. $\mathfrak{Z}\{f(nT)\} = \mathfrak{Z}\{u(nT)\sin \omega nT\} = \mathfrak{Z}\left\{u(nT)\dfrac{e^{j\omega nT} - e^{-j\omega nT}}{2j}\right\}$

$$= \sum_{n=0}^{\infty} \frac{u(nT)}{2j}c_1^{-n}z^{-n} - \sum_{n=0}^{\infty} \frac{u(nT)}{2j}c_2^{-n}z^{-n}; \, c_1 = e^{-j\omega nT}, c_2 = e^{j\omega nT}$$

$$= \frac{1}{2j}\left[\frac{c_1 z}{c_1 z - 1} - \frac{c_2 z}{c_2 z - 1}\right] = \frac{z\sin \omega T}{z^2 - 2z\cos \omega T + 1} \qquad |z| > 1$$

3.2.1

1. From the left-shift property, we have $\mathfrak{Z}\{y(n+1)\} = \sum_{n=0}^{\infty} y(n+1)z^{-n}$. Next, setting $n + 1 = m$ we obtain

$$\mathfrak{Z}\{y(m)\} = \sum_{m=1}^{\infty} y(m)z^{-m+1} = z\sum_{m=1}^{\infty} y(m)z^{-m} = z\sum_{m=0}^{\infty} y(m)z^{-m} - zy(0) = zY(z) - zy(0)$$

By a similar procedure, we can show that

$$\mathfrak{Z}\{y(n+m)\} = z^m Y(z) - z^m y(0) - z^{m-1}y(1) - \cdots - zy(m-1) = z^m Y(z) - \sum_{n=0}^{m-1} y(n)z^{m-n}$$

2. The time scaling property is proved using the z-transform definition as follows:

$$\mathfrak{Z}\{a^n y(n)\} = \sum_{n=0}^{\infty} a^n y(n)z^{-n} = \sum_{n=0}^{\infty} y(n)(a^{-1}z)^{-n} = Y(a^{-1}z)$$

The above result indicates that in the z-transform of $y(n)$, wherever we see z we substitute it with $a^{-1}z$.

3. Multiplication by n property is proved from the basic definition as follows:

$$\Im\{ny(n)\} = \sum_{n=0}^{\infty} y(n)nz^{-n} = z\sum_{n=0}^{\infty} y(n)(nz^{-n-1}) = z\sum_{n=0}^{\infty} y(n)\left[-\frac{d}{dz}z^{-n}\right]$$

$$= -z\frac{d}{dz}\sum_{n=0}^{\infty} y(n)z^{-n} = -z\frac{dY(z)}{dz}$$

4. The convolution property is shown as follows:

$$\Im\{y(n)\} \triangleq Y(z) = \sum_{n=0}^{\infty} z^{-n}\sum_{m=0}^{n} h(n-m)x(m) = \sum_{n=0}^{\infty} z^{-n}\sum_{m=0}^{\infty} h(n-m)x(m)$$

We substitute $m = \infty$ in the second summation since h (negative argument) is zero. Next, write $n - m = q$ and invert the order of summation

$$Y(z) = \sum_{m=0}^{\infty} x(m)\sum_{n=0}^{n} h(n-m)z^{-n} = \sum_{m=0}^{\infty} x(m)z^{-m}\sum_{q=-m}^{\infty} h(q)z^{-q}$$

$$= \sum_{m=0}^{\infty} x(m)z^{-m}\sum_{q=0}^{\infty} h(q)z^{-q} = Z(z)H(z)$$

since $h(q) = 0$ for $q < 0$.

3.2.2

$$\frac{di(t)}{dt} + 4i(t) = 2v(t), \quad \frac{i(nT) - i(nT - T)}{T} + 4i(nT) = 2v(nT), \quad i(nT) - ai(nT - T) = 2Tav(nT)$$

$$a = 1/(1 + 4T), \Rightarrow I(z) - az^{-1}I(z) = 2aTV(z) \Rightarrow I(z) = 2aT\frac{1}{1 - az^{-1}}V(z),$$

$$V(z) = 1 + z^{-1} + z^{-2} + \cdots = \frac{1}{1 - z^{-1}} = \frac{z}{z - 1}, \Rightarrow \Im^{-1}\{I(z)\} = \Im^{-1}\left\{2aT\frac{z}{z - a}\frac{z}{z - 1}\right\},$$

$$\frac{z}{z - a}\frac{z}{z - 1} = \frac{Az}{z - a} + \frac{Bz}{z - 1} \text{ multiply both sides by}(z - a) \text{ and set } z = a \text{ and next multiply both}$$

sides by$(z - 1)$ and set $z = 1$ we obtain $A = \dfrac{a}{a - 1}, B = \dfrac{1}{1 - a} \Rightarrow \Im^{-1}\{I(z)\} = \Im^{-1}\left\{2aT\dfrac{z}{z - a}\dfrac{z}{z - 1}\right\}$

$$= 2aT\Im^{-1}\left\{\frac{a}{a - 1}\frac{z}{z - a} + \frac{1}{1 - a}\frac{z}{z - 1}\right\} \Rightarrow i(nT) = \frac{2a^2T}{a - 1}a^n + \frac{2aT}{1 - a}u(nT), \text{ continuous case:}$$

$$i_h(t) = Ae^{-4t}, i_p(t) = B \Rightarrow 0 + 4B = 2 \Rightarrow B = 0.5, i(t) = i_h(t) + i_p(t) = Ae^{-4t} + 0.5, \Rightarrow$$

$$i(0) = A + 0.5 \Rightarrow A = -0.5 \Rightarrow i(t) = -0.5e^{-4t} + 0.5$$

3.3.1

The difference equation describing the total system is found by the following two equations, which are obtained by inspection of Figure 3.4:

$$x_1(n) - a_1 y(n-1) = y(n) \qquad x_1(n) = b_0 x(n) + b_1 x(n-1)$$

Hence,

$$y(n) + a_1 y(n-1) = b_0 x(n) + b_1 x(n-1)$$

The a's and b's can take either positive or negative values. Taking the z-transform of both sides of the above equation, we obtain the transfer function

$$H(z) \triangleq \frac{Y(z)}{X(z)} = \frac{b_0 + b_1 z^{-1}}{1 + a_1 z^{-1}}$$

3.3.2

See Figure 3.8.

3.5.1

The amplitude functions are even functions of the frequency, and the phase functions are odd functions of the frequency.

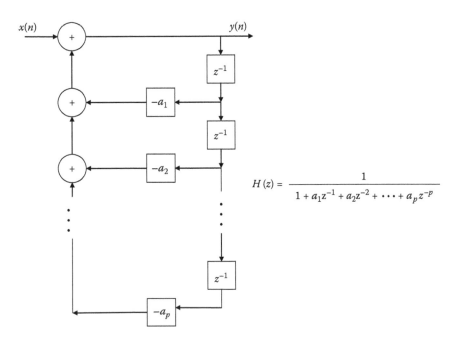

$$H(z) = \frac{1}{1 + a_1 z^{-1} + a_2 z^{-2} + \cdots + a_p z^{-p}}$$

FIGURE 3.8

3.6.1

$\dfrac{Y(z)}{3.5z} = \dfrac{z}{z^2 + z - 2}$, and, hence, we use the following Book MATLAB program:

>>[r, p, k] = residue ([0   1   0],[1   1   −2]);

$r = 0.6667, 0.3333;\ p = -2, 1;\ k = 0$

Therefore,

$$Y(z) = 3.5 \times 0.6667 \dfrac{z}{z+2} + 3.5 \times 0.3333 \dfrac{z}{z-1} = \dfrac{7}{3}\dfrac{z}{z+2} + \dfrac{7}{6}\dfrac{z}{z-1} \text{ or } y(n) = \dfrac{7}{3}(-2)^n + \dfrac{7}{6}u(n)$$

3.6.2

$$Y(z) + 2z^{-1}Y(z) + 2y(-1) = 3.5U(z) \text{ or } Y(z) = \underbrace{\dfrac{3.5}{1+2z^{-1}}U(z)}_{\substack{\text{zero-state}\\\text{response}}} - \underbrace{\dfrac{2y(-1)}{1+2z^{-1}}}_{\substack{\text{zero-input}\\\text{response}}}$$

The inverse z-transform is

$$y(n) = \underbrace{4(-2)^{n+1}}_{\substack{\text{zero-input}\\\text{response}}} + \underbrace{\dfrac{7}{6}u(n) + \dfrac{7}{3}(-2)^2}_{\substack{\text{zero-state}\\\text{response}}}$$

4 Finite Impulse Response (FIR) Digital Filter Design

4.1 INTRODUCTION

In this chapter, we discuss the simplest digital finite impulse response (FIR) and infinite impulse response design techniques. Digital filter design often stems from analog filters of the low-pass or high-pass class that meet design specifications by the use of transformations that yield the equivalent z-plane for a given analog description in the s-plane or in the time domain. In essence, in this procedure, we establish a roughly equivalent sampled $H(z)$ for a given analog function $H(s)$. A number of different transformations exist. We will study the impulse invariant and bilinear transformations in some detail as well the Fourier series, discrete Fourier transform (DFT), and window methods.

4.2 FINITE IMPULSE RESPONSE (FIR) FILTERS

FIR filters (nonrecursive) are filters whose present output is determined from the present and past inputs, but is independent of the previous outputs. Because no feedback is present, FIR filters are stable. Furthermore, such filters are associated with zero phase or linear phase characteristics, and so no phase distortion occurs in the output. In such filters the output is a time-delayed version of the input signal.

Linear phase FIR digital filters have many advantages, such as guaranteed stability, freedom from phase distortion, and low coefficient sensitivity. Such filters are used where frequency dispersion in the pass band is harmful. Several methods have been proposed in the literature for reducing the complexity of sharp filters.

The design of FIR filters is approached from two points of view: (1) by employing the DFT series and (2) by employing the DFT. In both cases, $H(z)$ is used to obtain the appropriate $h(k)$ that is used in the design process. Because of the truncation of $H(z)$, ripples appear that are closely eliminated using windows.

In this chapter, the capital omega, Ω (rad/s), represents the frequency of continuous functions in their FT, and the lower case omega, ω (rad), represents the frequency of discrete signals.

DISCRETE FOURIER-SERIES METHOD

Low-pass filter. In this procedure, the assumed analog filter response function $H(j\Omega)$ is considered to be periodic function to allow the function to be expressed by a Fourier series, but with the restriction that the series representation is constrained to the range of the original function. Then, we approximate the continuous-time function by a sampled function that adequately represents the original function in sampled form. This procedure involves replacing the infinite integral by a finite summation, the number of terms in the expansion being limited to a value N that is sufficiently large to limit the aliasing errors in the respective coefficients.

For the desired periodic digital filter,

$$H\left(e^{j\Omega T}\right) \triangleq H\left(e^{j\omega}\right) = \sum_{k=-\infty}^{\infty} h(kT)e^{-j\Omega Tk} \qquad \Omega = \text{rad/s}, \quad \omega = \Omega T = \text{rad} \tag{4.1}$$

where the time function $h(t)$ is taken at $t=kT$, so that

$$h(kT) = \frac{1}{\Omega_s} \int_{-\Omega_s/2}^{\Omega_s/2} H\left(e^{j\Omega T}\right) e^{jk\Omega T} d\Omega \qquad \Omega_s = \frac{2\pi}{T} \qquad \omega = \Omega T \tag{4.2}$$

If we set $z = e^{j\Omega T}$ in Equation 4.1, we obtain

$$H(z) = \sum_{k=-\infty}^{\infty} h(kT) z^{-k} \tag{4.3}$$

A typical and ideal low-pass filter description is illustrated in Figure 4.1. The description requires that

$$h(kT) = 0 \quad \text{for } k > \frac{N-1}{2} \text{ and } k < -\frac{N-1}{2} \tag{4.4}$$

and Equation 4.3 becomes

$$H'(z) = h(0) + \sum_{k=1}^{(N-1)/2} \left[h(-kT)z^k + h(kT)z^{-z} \right] \tag{4.5}$$

This sequence is noncausal because of the term z^k. To obtain a causal filter, we multiply $H'(z)$ by $z^{-(N-1)/2}$ to obtain the expression

$$H(z) = z^{-(N-1)/2} H'(z) \tag{4.6}$$

The factor $z^{-(N-1)/2}$ introduces a delay, or phase, factor proportional to frequency Ω. Consequently, a filter with an initially zero phase for all frequencies is converted into one that has a linear phase with frequency. Therefore, it introduces no distortion in the output. Observe also that if $h(kT)$ is real, then $H'(z)$ is real.

For the case when $h(kT) = h(-kT)$, an even function (Equation 4.5) becomes

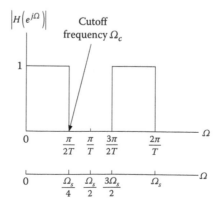

FIGURE 4.1

$$H'(z) = h(0) + \sum_{k=1}^{(N-1)/2} h(kT)\left(z^k + z^{-k}\right)$$ (4.7)

For $z = e^{j\Omega T}$, this equation becomes

$$H'\left(e^{j\Omega T}\right) = h(0) + 2\sum_{k=1}^{(N-1)/2} h(kT)\cos\left(\frac{2\pi k\Omega}{\Omega_s}\right) d\Omega$$ (4.8)

Under the conditions of Equation 4.2, Equation 4.8 becomes

$$h(kT) = \frac{1}{\Omega_s}\int_{-\Omega_s/2}^{\Omega_s/2} H'\left(e^{j\Omega T}\right)\cos\left(\frac{2\pi k\Omega}{\Omega_s}\right) d\Omega$$ (4.9)

The following steps are required in the design of an FIR filter using Fourier-series approach:

1. From the given amplitude–frequency characteristic of the filter, we obtain $\{h(kT)\}$ using Equation 4.2.
2. Truncate the sequence $\{h(kT)\}$ by choosing $-(N-1)/2 < k < (N-1)/2$.
3. Use Equation 4.5 to find $H'(z)$, and then use Equation 4.6 to determine the desired filter $H(z)$.
4. Plot $\left|H\left(e^{j\omega}\right)\right|$ versus ω.

Let the input to the filter be z^k. Then, using the convolution property of the system response, the output is given by

$$y(k) = z^k * h(k) = \sum_{n=0}^{\infty} h(n)z^{k-n} = z^k \sum_{n=0}^{\infty} h(n)z^{-n} = z^k H(z) \text{ or } y(k) = e^{j\omega k}H\left(e^{j\omega k}\right)$$ (4.10)

by setting $z = e^{j\omega}$.

Consider a causal nonrecursive system (FIR) defined by the difference equation

$$y(k) = \sum_{n=0}^{N-1} b_n x(k-n)$$ (4.11)

If the input is given by $x(k) = e^{jk\omega}$, the output (as shown in Equation 4.10) is given by

$$y(k) = e^{jk\omega} H\left(e^{j\omega}\right)$$ (4.12)

where $e^{j\omega}$ is the **eigenfunction** and $H\left(e^{j\omega}\right)$ is the associated **eigenvalue**. Substitute Equation 4.12 in the left of Equation 4.11 and set $x(k) = e^{jk\omega}$ in the right of this equation, and multiply by $e^{-jk\omega}$. The result is

$$H\left(e^{j\omega}\right) = b_0 + b_1 e^{-j\omega} + b_2 e^{-j2\omega} + \cdots + b_{N-1} e^{-j(N-1)\omega} = \sum_{n=0}^{N-1} b_n e^{-jn\omega} \tag{4.13}$$

For a causal filter, $h(k)=0$ for $k < 0$, and a finite number of delays, Equation 4.11 becomes

$$H\left(e^{j\omega}\right) = \sum_{n=0}^{N-1} h(n) e^{-jn\omega} \tag{4.14}$$

From the last two equations, we obtain the important relation

$$h(n) = b_n \tag{4.15}$$

Example 4.2.1

Determine the FIR filter that approximates the ideal filter shown in Figure 4.2a using the Fourier-series method. Sketch the resulting $\left|H\left(e^{j\omega}\right)\right|$. Choose $N=5$ and $N=15$.

Solution: By Equation 4.2, we have

$$h(kT) = \frac{T}{2\pi} \int_{-\pi/4T}^{\pi/4T} e^{jk\Omega T} d\Omega = \frac{1}{\pi k} \sin\left(\frac{k\pi}{4}\right)$$

The sampling frequency is $\Omega_s = \pi / T$. The sampled values $\{h(k)\}$ are as follows:

$k =$	0	1	2	3	4	5	6	7	8
$\{h(k)\}$		$\frac{1}{\pi}0.707$	$\frac{1}{2\pi}1$	$\frac{1}{3\pi}0.707$	0	$-\frac{1}{5\pi}0.707$	$-\frac{1}{6\pi}0.707$	$-\frac{1}{7\pi}0.707$	0
	0.25	0.225	0.159	0.075	0	−0.045	−0.053	−0.032	0

Observe that the sequence is symmetric about the midpoint; hence, the transfer function for $N=5$ can be written per Equation 4.5,

$$H'(z) = h(0) + \sum_{k=1}^{(N-1)/2} \left[h(-k)z^k + h(k)z^{-k} \right]$$

$$= 0.250 + 0.225z + 0.225z^{-1} + 0.159z^2 + 0.159z^{-2}$$

Thus, we obtain

$$H(z) = z^{-2} H'(z) = 0.159 + 0.225z^{-1} + 0.25z^{-2} + 0.225z^{-3} + 0.195z^{-4}$$

Setting $z = e^{j\omega}$ in the above equation and using the Book MATLAB function, we obtain Figure 4.2b.

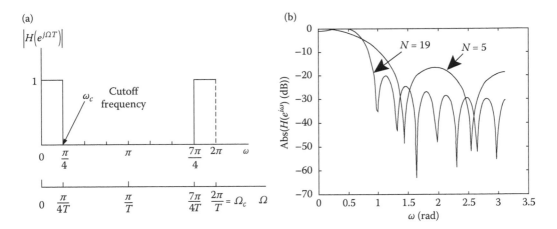

FIGURE 4.2

Book MATLAB Function: [hf2,hf]=urdsp_ex4_2_1(N,h,T,h0)

```
function[hf2,hf]=urdsp_ex4_2_1(N,h,T,h0)
    %h=impulse response of the system;
    %N=odd number of elements of h; h0= is the
    %value of h at n=0 and must be found manually
    %since in case 0/0 MATLAB will always give 1;
for m=0:127
    for k=1:(N-1)/2
        hf(m+1,k)=h(k)*(exp(j*k*m*pi*T/128)+...
            exp(-j*k*m*pi*T/128));
    end
    hf2=(h0+sum(hf,2))*exp(-j*k*m*pi*T*(N-1)/2);
end
    %wh0 is the h function at k=0 multilied by w(0);
    %sum(hf,2) a MATLAB function
    %which sums the columns of hf;
w=0:pi/128:pi-(pi/128);
plot(w,20*log10(abs(hf2)/max(abs(hf2))),'k');
```

We have plotted the filter response in dB and the frequency in radians. We have selected $T=1$ and, therefore, the equivalent in rad/s is obtained by dividing π by the sampling time T which was selected different than 1. ∎

Observe that the truncation of the series representation of the ideal magnitude function causes the appearance of ripples in the amplitude characteristics; the smaller the value of N, the wider is the spread of the ripples. If we view the series termination of the magnitude function as a multiplication of the original sequence by a rectangular window, the oscillations in the frequency domain arise because of leakage. The leakage can be reduced by multiplying the sequence by a more appropriate window or weighing sequence.

Therefore, if we denote the window function as $w(kT)$, the modified truncated sequence of the Fourier coefficients is specified by

$$h(kT) = \begin{cases} h(kT)w(kT) & 0 \leq k \leq N-1 \\ 0 & \text{otherwise} \end{cases}$$

$$w(kT) = \begin{cases} w(kT) & 0 \leq k \leq N-1 \\ 0 & \text{otherwise} \end{cases}$$

(4.16)

COMMONLY USED WINDOWS

Three commonly used windows functions with $T = 1$ are as follows:

1. Hamming window:

$$w(k) = 0.54 + 0.46\cos\left(\frac{k\pi}{K}\right) \qquad |k| \leq K = \frac{N-1}{2} \tag{4.17}$$

2. Blackman window:

$$w(k) = 0.42 + 0.5\cos\left(\frac{k\pi}{K}\right) + 0.08\cos\left(\frac{k\pi}{K}\right) \qquad |k| \leq K = \frac{N-1}{2} \tag{4.18}$$

3. Hann (or Hanning) window:

$$w(k) = 0.5 + 0.5\cos\left(\frac{k\pi}{K}\right) \qquad |k| \leq K = \frac{N-1}{2} \tag{4.19}$$

The following example illustrates the use of window functions.

Example 4.2.2

Consider again the ideal filter characteristics shown in Figure 4.2a. Examine the effect of a Hamming window on the FIR approximation. Choose $N = 19$ and plot $|H(e^{j\omega})|$ versus ω.

Solution: We modify the Book m-function urdsp_ex4_2_1 given earlier, and the results are plotted in Figure 4.3. The Book m-function is

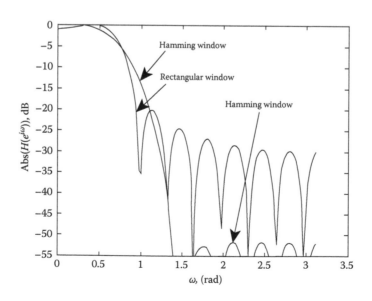

FIGURE 4.3

Book m-function: [hf2,hf]=urdsp_ex4_2_2(N,wh,T,wh0)

```
function[hf2,hf]=urdsp_ex4_2_2(N,wh,T,wh0)
    %h=impulse response of the system;
    %N=odd number of elements of h; wh0=is the
    %value of h at n=0 and must be found manually
    %since in case 0/0 MATLAB will always give 1,
    %multiplied by the w(0) of the window; wh is h
    %multiplied by the window;
for w=0:127
    for k=1:(N-1)/2
        hf(w+1,k)=wh(k)*(exp(j*k*w*pi*T/128)+...
            exp(-j*k*w*pi*T/128));
    end
    hf2=(wh0+sum(hf,2))*exp(-j*k*w*pi*T*(N-1)/2);
end
    %sum(hf,2) a MATLAB function
    %which sums the columns of hf;
w=0:pi/128:pi-(pi/128);
plot(w,20*log10(abs(hf2)/max(abs(hf2))),'k');
```

The Book MATLAB program to obtain Figure 4.3 is:

```
>>m=1:19;
>>h=sin(m*pi/4)./(pi*m);
>>w=0.54+0.46*cos(m*pi/9);
>>h0=0.25; wh0=0.25*1;
>>wh=h.*w;
>>[hf2,hf]=urdsp_ex4_2_1(19,h,1,0.25);
>>hold on
>>[hf2,hf]=urdsp_ex4_2_2(19,wh,0.25);
>>xlabel('\omega  dB');ylabel('Abs(H(e^{j\omega})), dB);
```

Exercise 4.2.1

Compare all four windows.   ▲

A more complete window set with their characteristics and performance is given in Appendix 4.1.

DISCRETE FOURIER TRANSFORM METHOD

To illustrate the use of the DFT method, refer again to Example 4.2.2. For the sampled $h(k)$ and the window function $w(k)$, we write, as before,

$$h_w(k) = h(k)w(k)$$

The DFT of the above function is

$$H_w\left(e^{j\omega}\right) = \sum_{k=-\infty}^{\infty} h(k)w(k)e^{-j\omega k}$$

$$= \sum_{k=-\infty}^{\infty}\left[\frac{1}{2\pi}\int_{-\pi}^{\pi} H\left(e^{jx}\right)e^{jxk}\,dx\right]w(k)e^{-j\omega k}$$

We, next, transpose the summation and integration operations; this is correct for finite and regular functions:

$$H_w\left(e^{j\omega}\right) = \frac{1}{2\pi} \int_{-\pi}^{\pi} H\left(e^{jx}\right) \left[\sum_{k=-\infty}^{\infty} w(k)e^{-j(\omega-x)k}\right] dx$$

$$= \frac{1}{2\pi} \int_{-\pi}^{\pi} H\left(e^{jx}\right) W\left(e^{-j(\omega-x)}\right) dx$$

This function (complex) $H_w\left(e^{j\omega}\right)$ is the result of the convolution of $H\left(e^{j\omega}\right)$ with $W\left(e^{j\omega}\right)$, the transform of the window sequence $\{w(k)\}$. Here, $H\left(e^{j\omega}\right)$ and $W\left(e^{j\omega}\right)$ are periodic, this is a circular convolution. The result is the frequency convolution,

$$H_w\left(e^{j\omega}\right) = H\left(e^{j\omega}\right) * W\left(e^{-j\omega}\right)$$

A common feature of the methods discussed earlier is that the resulting filters generally require a large number of terms in the expansion and, hence, a large number of multiplications in the design procedure.

HIGH-PASS FILTER

We observe from Figure 4.4 that if we set $\omega_c = \omega_a$ and shift the low-pass filter frequency response by π, we obtain the high-pass filter characteristics. We first start with a low-pass frequency response designed for $\omega_c = \omega_a$ or $H_{lp}\left(e^{j\omega}\right)\Big|_{\omega_c \triangleq \omega_a}$. Then, we write

$$H_{hp}\left(e^{j\omega}\right) = H_{lp}\left(e^{j(\omega-\pi)}\right) = \sum_{n=-\infty}^{\infty} h_{lp}(n)e^{-jn(\omega-\pi)}$$

$$= \sum_{n=-\infty}^{\infty} h_{lp}(n)e^{jn\pi}e^{-jn\omega}$$

(4.20)

This indicates that ($e^{j\pi} = \cos\pi - j\sin\pi = -1$)

$$h_{hp}(n) = h_{lp}(n)e^{jn\pi} = (-1)^n h_{lp}(n)$$

(4.21)

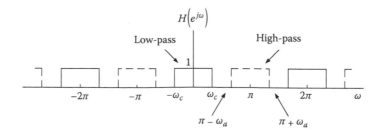

FIGURE 4.4

Example 4.2.3

Design a high-pass filter that passes frequencies greater than 20 kHz. The sampling frequency is 100 kHz. Use Fourier design approach and assume that the pass-band gain is unity.

Solution: The cutoff digital frequency corresponding to the analog 20 kHz is

$$\omega_c = \Omega T = 2\pi \times 2 \times 10^4 \times \frac{1}{f_s} = 2\pi \times 2 \times 10^4 \times \frac{1}{10^5} = 0.4\pi \text{ rad}, \quad f_s = \frac{1}{T}$$

From Figure 4.4, we observe that the cutoff digital frequency for the high-pass filter is $\pi - 0.4\pi = 0.6\pi$. The inverse low-pass filter is given by

$$h_{lp}(n) = \frac{1}{2\pi} \int_{-\omega_c}^{\omega_c} H\left(e^{j\omega}\right) e^{jn\omega} \, d\omega = \frac{1}{2\pi} \int_{-0.6\pi}^{0.6\pi} e^{jn\omega} \, d\omega = \frac{\sin(0.6\pi n)}{\pi n} \qquad n = 0, \pm 1, \pm 2, \ldots \qquad (4.22)$$

This result in conjunction with Equation 4.21 provides the details of the high-pass filter $h_{hp}(n) = (-1)^n h_{lp}(n)$. ■

Exercise 4.2.2

Plot the time and frequency domain of the high- and low-pass filter discussed in Example 4.2.3. Be sure that you substitute the middle point of the impulse response with its exact value, by taking the derivative of the numerator and denominator and setting $n=0$ (assume n as a continuous variable). ▲

Table 4.1 contains the pertinent information for the necessary frequency transformations from the low-pass filter to filters of different frequency response characteristics.

Example 4.2.4

Design a band-pass filter for the frequency band from 20 to 30 kHz with a sampling frequency of 100 kHz (observe that the sampling frequency is quite a bit larger than the range of the filter frequency).

Solution: The digital frequency pass band extends from

$$\omega_l = \frac{2\pi \times 2 \times 10^4}{10^5} = 0.4\pi \quad \omega_u = \frac{2\pi \times 3 \times 10^4}{10^5} = 0.6\pi$$

and

$$\omega_c = \frac{(\omega_u - \omega_l)}{2} = 0.1\pi$$

For a cutoff frequency ω_c for the low-pass filter, we obtain its impulse response

$$h_{lp}(n) = \frac{\sin 0.1\pi n}{\pi n}$$

TABLE 4.1

Frequency Transformations

Low-pass: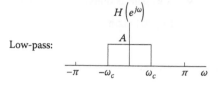

$$h_{lp}(n) = \frac{A}{\pi n}\sin(n\omega_c) \qquad n = 0, \pm 1, \pm 2, \ldots$$

High-pass: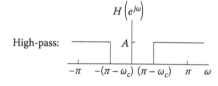

$$h_{hp}(n) = (-1)^n h_{lp}(n) \qquad n = 0, \pm 1, \pm 2, \ldots$$

Band-pass: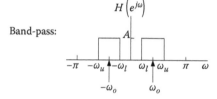

$$h_{bp}(n) = [2\cos(n\omega_o)]h_{lp}(n) \qquad n = 0, \pm 1, \pm 2, \ldots$$

$$\omega_u - \omega_l = 2\omega_c, \qquad \omega_o = \frac{\omega_u + \omega_l}{2}$$

Band-stop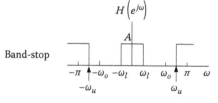

$$h_{bs}(0) = A - h_{bp}(0)$$

$$h_{bs}(n) = -h_{bp}(n) \qquad n = \pm 1, \pm 2, \ldots$$

$$\omega_u - \omega_l = 2\omega_c$$

$$\omega_o = \frac{\omega_u + \omega_l}{2}$$

The centered frequency is

$$\omega_o = \frac{0.6\pi + 0.4\pi}{2} = 0.5\pi$$

Thus, the impulse response of the band-pass filter is

$$h_{bp}(n) = [2\cos(0.5\pi n)]h_{lp}(n) \qquad n = 0, \pm 1, \pm 2, \ldots$$

Figure 4.5 shows the result of the above band-pass filter with $n = -10$ to 10. Remember to set the 0/0 value of the impulse response with its exact value. ■

Exercise 4.2.3

Plot the spectra of Hamming and Blackman windows. ▲

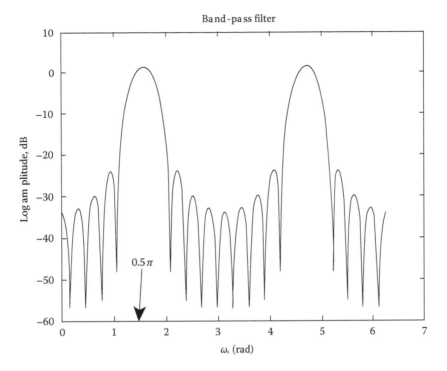

Band-pass filter

FIGURE 4.5

Example 4.2.5

Let us assume that we have the signal $f(t) = 10\sin(2\pi 10t) + 10\sin(2\pi 25t)$, which contains two frequencies, one at 10 Hz and the other at 25 Hz. It is desired to retain the 10 Hz signal using a low-pass filter. We transform all in discrete form with a sampling frequency of 100 Hz.

Solution: The sampling time is $T = 1/100$ and, therefore, the digital frequencies of the signal are $2\pi 10/100 = 0.2\pi$ and $2\pi 25/100 = 0.5\pi$. The sampling frequency is $\Omega_s = \pi/T$. We next take as the cutoff frequency of the low-pass filter to be $\omega_c = 0.5\pi - 0.2\pi = 0.3\pi$. Using Equation 4.2 and Example 4.2.1, we obtain

$$h(kT) = \frac{1}{\Omega_s} \int_{-\Omega_s/2}^{\Omega_s/2} H\left(e^{j\Omega T}\right) e^{jk\Omega T}\, d\Omega = \frac{2\pi}{T} \int_{-0.3\pi/T}^{0.3\pi/T} e^{jk\Omega T}\, d\Omega = \frac{\sin(0.3\pi k)}{k\pi} \tag{4.23}$$

The first row of Figure 4.6 shows the signal and its spectrum. We can identify the sine frequencies at 0.6283 and 1.5708 rad. The second row shows impulse response of the low-pass filter and its frequency spectrum. The first figure of the last row shows the convolution of the impulse response and the signal. The second figure shows the spectrum at the output of the low-pass filter. Based on the results, it was possible to eliminate the second frequency of 25 Hz. ∎

Exercise 4.2.4

Eliminate the low frequency of 10 Hz for the signal given in Example 4.2.5.   ▲

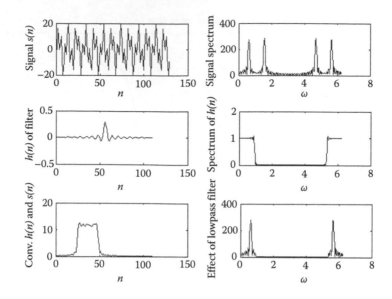

FIGURE 4.6

HINTS–SUGGESTIONS–SOLUTIONS OF THE EXERCISES

4.2.2

We use the following Book MATLAB program:

```
>>n=-10:10;
>>hp=(-1).^n.*(sin(0.6*pi*n)./(n*pi));
>>hp(11)=0.6;
>>hpf=fft(hp,256);
>>stem(hp,'k');
>>w=0:2*pi/256:2*pi-(2*pi/256);
>>plot(w,abs(hpf),'k');
>>title('High-Pass Filter in Frequency Doain');
```

Figure 4.7a presents the high-pass filter in time domain and Figure 4.7b presents the frequency spectrum of the high-pass filter. If we plot the low-pass filter we will observe that the cutoff frequency is at 0.6π.

4.2.3

See Figure 4.8.

4.2.4

From Example 4.2.3 and observing the signal spectrum e set as low-pass frequency, $\omega_c = 2 = 0.64\pi$. Therefore, the low-pass filter impulse response is $h_{lp}(kT) = \sin(0.64\pi k)/(k\pi)$. From Table 4.1, we obtain the high-pass filter impulse response to be $h_{hp}(n) = (-1)^n h_{lp}(n)$ $n = 0,\pm1,\pm2,\ldots$. The following Book MATLAB program produces the desired results and are shown in Figure 4.9.

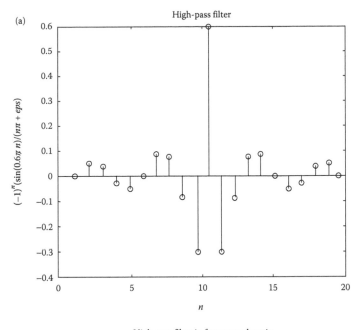

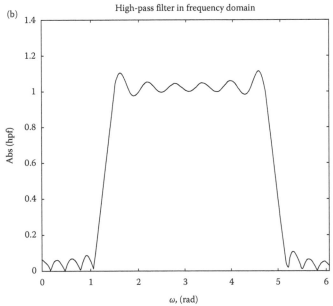

FIGURE 4.7

```
>>ωc=0.64*pi; %low-pass frequency;
>>m=-64:64;
>>hl=(sin(0.64*pi*m)./(m*pi);
>>hl(65)=0.64;
>>n=0:128;
>>hh=(-1).^n.*hl;
>>hhf=fft(hh,256);
```

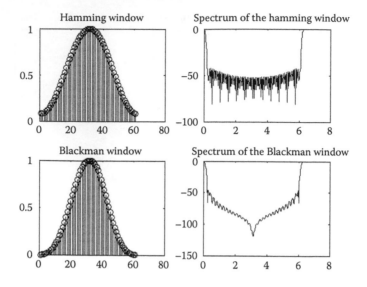

FIGURE 4.8

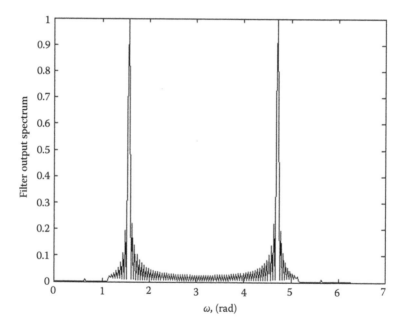

FIGURE 4.9

```
>>q=0:125;
>>f=10*sin(2*pi*10*q/100)+10*sin(2*pi*25*q/100);
>>ff=fft(f,256);
>>hfout=ff.*hhf;
>>w=0:2*pi/256:2*pi-(2*pi/256);
>>plot(w,abs(hfout)/max(abs(hfout)),'k');
>>xlabel('\omega rad')
>>ylabel('Filter output spectrum');
```

Appendix 4.1: Window Characteristics and Performance

Type Function	Transform	Main Lobe Width	Side-Lobe Level (dB)
Rectangular			
$w_R(n) = \begin{cases} 1 & 0 \le n \le N-1 \\ 0 & \text{elsewhere} \end{cases}$,	$W_R(\omega) = e^{-j\omega(N-1)/2}\dfrac{\sin(\omega N/2)}{\sin(\omega/2)}$,	$\dfrac{2\pi}{N}$	-13
Triangular			
$w_T(n) = \begin{cases} \dfrac{2n}{N-1} & 0 \le n \le \dfrac{N-1}{2} \\ 2 - \dfrac{2n}{N-1} & \dfrac{N-1}{2} \le n \le N-1 \end{cases}$,	$W_T(\omega) = \dfrac{1}{N} e^{-j\omega(N-1)/2}\left(\dfrac{\sin(\omega N/2)}{\sin(\omega/2)}\right)^2$,	$\dfrac{4\pi}{N}$	-27
Hanning			
$w_{HAN}(n) = \dfrac{1}{2}\left[1 - \cos\left(\dfrac{2\pi n}{N-1}\right)\right]$, $0 \le n \le N-1$	$W_{HAN}(\omega) = \dfrac{1}{2}W_R(\omega) + \dfrac{1}{4}W_R\left(\omega + \dfrac{2\pi}{N}\right)$ $+ \dfrac{1}{4}W_R\left(\omega - \dfrac{2\pi}{N}\right)$	$\dfrac{4\pi}{N}$	-31
Hamming			
$w_{HAM}(n) = 0.54 - 0.46\cos\left(\dfrac{2\pi n}{N-1}\right)$, $0 \le n \le N-1$	$W_{HAM}(\omega) = 0.54 W_R(\omega) + 0.23 W_R\left(\omega + \dfrac{2\pi}{N}\right)$ $+ 0.23 W_R(\omega - \dfrac{2\pi}{N})$	$\dfrac{4\pi}{N}$	-41
Blackman			
$w_B(n) = 0.42 - 0.5\cos\left(\dfrac{2\pi n}{N-1}\right)$ $+ 0.08\cos\left(\dfrac{4\pi n}{N-1}\right)$, $0 \le n \le N-1$	$W_B(\omega) = 0.42 W_R(\omega)$ $+ 0.25\left[W_R\left(\omega + \dfrac{2\pi}{N}\right) + W_R\left(\omega - \dfrac{2\pi}{N}\right)\right]$, $+ 0.04\left[W_R\left(\omega + \dfrac{2\pi}{N}\right) + W_R\left(\omega - \dfrac{2\pi}{N}\right)\right]$	$\dfrac{6\pi}{N}$	-58

5 Random Variables, Sequences, and Probability Functions

5.1 RANDOM SIGNALS AND DISTRIBUTIONS

Random signal can be described by precise mathematical analysis, the tools of which are contained in the theory of statistical analysis.

A discrete random signal $\{X(n)\}$ (at this point, the capital letter indicates a random variable (RV)) is a sequence of indexed RVs assuming the random values:

$$\{x(0) \quad x(1) \quad x(2) \quad \cdots\} \tag{5.1}$$

This type of random sequence can be thought of as the result of sampling a continuous random signal. The development of random sequences from sampling random continuous signals is natural because every continuous random signal must be processed by digital computers and, as a consequence, random sequences are developed. In our book, we assume that the RV at any time n will take any real value. This type of sequence is also known as **discrete time series**.

A particular RV that is continuous at time n, $X(n)$, is characterized by its **probability density function** (PDF) $f(x(n))$, which is given by

$$f(x(n)) = \frac{\partial F(x(n))}{\partial x(n)} \tag{5.2}$$

and its **cumulative density function** (CDF) $F(x(n))$:

$$F(x(n)) = \mathrm{pr}\{X(n) \le x(n)\} = \int_{-\infty}^{x(n)} f(y(n))\,dy(n) \tag{5.3}$$

The expression $p\{X(n) \le x(n)\}$ is interpreted as the probability that the RV $X(n)$ will take values less than or equal to the value $x(n)$ at time n. As the value of the RV at time n approaches infinity, the CDF $F(x(n))$ approaches unity.

Figure 5.1a presents a time series indicating, for example, some observations as recorded at 20 instants of time. Figure 5.1b shows the histogram, that is, the approximation to the PDF of the RV, which in this case is the observation of some characteristic quantity: the width of a wire as it is produced. For example, the number of magnitude falling between 0 and 0.5 is 4, and between 1 and 1.5 is 6. Finally, Figure 5.1c shows the CDF, where each step is proportional to the number of widths found within the range of width 0.5 divided by the total number of samples, in this case 20.

The empirical CDF of a RV, given the values $x_1(n), x_2(n), \ldots, x_N(n)$, is given by

$$\hat{F}(x \mid x_1, \ldots, x_N) = \frac{\text{Number of samples values } x_1, \ldots, x_N \text{ not greater than } x}{N} \tag{5.4}$$

This function is a histogram of staircase type and its value at x is the percentage of the points $x_1, \ldots, x_N$ that are not larger than x. MATLAB Statistics Toolbox has the function cdfplot(x), which plots the staircase CDF given the vector x of the data.

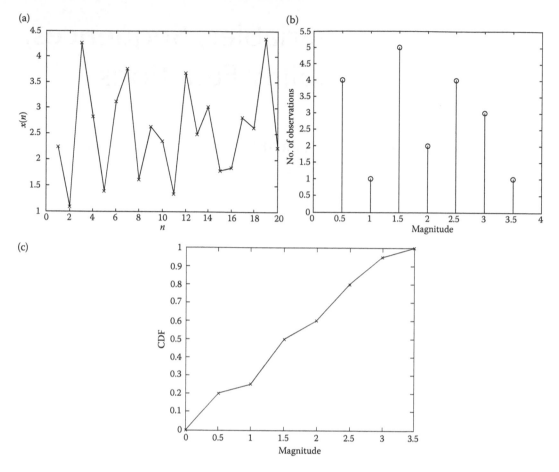

FIGURE 5.1

Similarly, the empirical PDF can be found based on the relation:

$$\hat{f}(x) \cong \frac{p\{x < X \leq x + \Delta x\}}{\Delta x} \tag{5.5}$$

for small Δx. Thus, we write

$$\hat{f}(x \mid x_1, \ldots, x_N) = \frac{\text{Number of the samples } x_1, \ldots, x_N \text{ in } [x, x + \Delta x]}{N \Delta x} \tag{5.6}$$

Similarly, the **multivariate** distributions are given by

$$F(x(n_1), \ldots, x(n_k)) = p\{X(n_1) \leq x(n_1), \ldots, X(n_k) \leq x(n_k)\}$$

$$f(x(n_1), \ldots, x(n_k)) = \frac{\partial^k F(x(n_1), \ldots, x(n_k))}{\partial x(n_1) \cdots \partial x(n_k)} \tag{5.7}$$

Note that here we have used a capital letter to indicate RVs. In general, we would not keep this notation because it will be obvious from the context.

If, for example, we want to check the accuracy of reading a dial by a person, we will have two readings: one due to the person and another due to the instruments. A simultaneous plot of these two readings, each one associated with a different orthogonal axis, will produce a scattering diagram

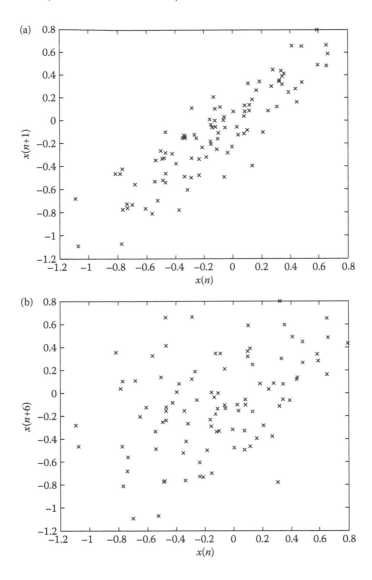

FIGURE 5.2

but with a linear dependence. The closer the points fall on a straight line, the more reliable the person's readings are. This example presents a case of a **bivariate** distribution. Figure 5.2a shows a scatter plot of $x(n)$s versus $x(n-1)$s of a time series $\{x(n)\}$. Figure 5.2b shows similar plot, but for $x(n)$ s versus $x(n-6)$s. It is apparent that the $x(n)$s and $x(n-1)$s are more correlated (tending to be around a straight line) than the second case.

To obtain a formal definition of a discrete-time stochastic process, we consider an experiment with a finite or infinite number of unpredictable outcomes, vectors with random elements, from a sample space, $S(\xi_1, \xi_2, \ldots)$, each one occurring with a probability $p(\xi_i)$. We recall that the **sample space**, S, is a collections of samples corresponding to the set of all possible outcomes, ξ, which are **elements** of the space. Certain subsets of the space S, or collections of outcomes, are called **events**. When we attempt to specify how likely it is that a particular event will occur during an experiment, we define the notion of **probability**. A **probability function**, $p(.)$, is a function defined on a class of events that assigns a number of the likelihood of a particular event occurring during an experiment. We constrain the events to a set of events, B, on which the probability function is defined and satisfy certain properties. We, therefore define an experiment by the triple $\{S, B, P\}$. The following example will elucidate the preceding notions.

Example 5.1.1

Consider the experiment $\{S, B, P\}$ of tossing a fair coin. Therefore, we observe

$$\text{Sample space} \qquad S = \{H, T\}$$

$$\text{Events} \qquad B = \{0, H, T, S\}$$

$$\text{Probability(assumed)} \qquad p(H) = p = 0.4(0 \leq p \leq 1)$$

$$p(T) = 1 - p = 0.6$$

From the preceding, we have
Random variable

$$\text{Random varibles} \qquad \begin{aligned} x\left(\xi_1 = H\right) &= x_1 = 0.4 \\ x\left(\xi_2 = T\right) &= x_2 = 0.6 \end{aligned}$$

$$\text{Distribution} \quad = \begin{cases} 0.4 & 0 \leq x_i \leq 0.4 \\ 1 - 0.4 & 0.4 \leq x_i \leq 1 \\ 0 & x_i < 0 \end{cases}$$

The PDF and CDF are shown in Figure 5.3. We observe that that the sum of the mass function value (probabilities) must be one, and the maximum value of the cumulative distribution is one. ∎

Example 5.1.2

The following Book MATLAB program created Figure 5.4a and b.

BOOK MATLAB M-FILE:EX5_1_2

```
%ex5_1_2
x=2+0.5*randn(1,120);% randn(1,N) is a MATLAB function
    %creating a vector with N rvs normally (Gaussian)
    %distributed, in this case the mean value is 2;
subplot(2,1,1);
hist(x,20);% hist(x,n)is a MATLAB function creating a
    %histogram from the rvs of x, in this case a normal
    %pdf,with n bars, the bars have color blue;
colormap([1 1 1]);% colormap([ 1 1 1]) is a MATLAB function
    %that changes the color to white, if we want to change to
    %black we write colormap([0 0 0]), the colormap([0 1 1])
    %will give light green etc, for any other color go to
    %help colormap;
```

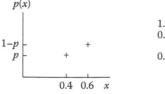 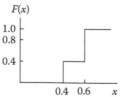

FIGURE 5.3

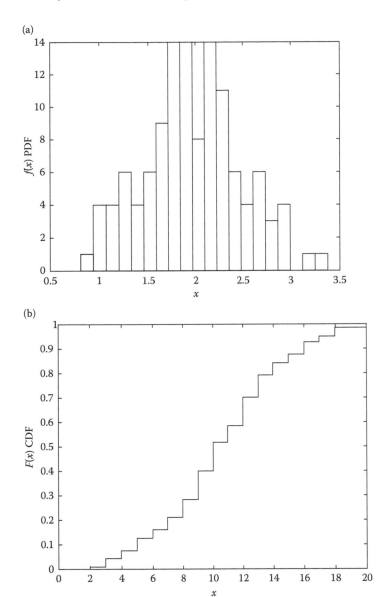

FIGURE 5.4

```
xlabel('x'); ylabel('f(x) pdf');
subplot(2,1,2);
[N,X]=hist(x,20);%the values of x are split in 20 bins, N is a
    %vector with each element assigning a number equal to how
    %many times the value of the rvs fall inside each bin;
y(1)=0;
for m=1:19%observe 19=20-1;
y(m+1)=sum(y(m)+N(m));%sum() is a MATLAB function that sums
    %the elements inside the parenthesis and sum(x) will
    %add all the elements of the vector x;
end;
stairs(y/120,'k');%stairs() is a MATLAB function that creates
    %a staircase, note that vector y is devited by %120=length(x);
xlabel('x');ylabel('F(x) cdf');
```

In case we want to plot the CDF without creating bins of values, as we did previously, we use the following Book MATLAB program for CDF:

```
>>N=120;
>>x=randn(1,N);
>>sx=sort(x);
>>plot(sx,(1:N)./N,'k')
```
■

Exercise 5.1.1

Suppose we toss a fair coin at each time instant, that is,

$$rv\, x(n,\xi_i), \quad n \in \mathbf{I}\,(\text{integers}) = \{0,1,2,\ldots,N-1\}, \quad \xi \in S = \{H,T\}$$

$$x(n,\xi_i) = \begin{cases} K & \xi_1 = H \\ -K & \xi_2 = T \end{cases}$$

We define the corresponding probability mass function as

$$p(x(n,\xi_i) = \pm K) = \frac{1}{2}$$

Define the time function [time series, Random Walk (RW)]

$$y(n,\xi_i) = \sum_{n=0}^{N-1} x(n,\xi_i) \quad i = 1, 2$$

Let the two realizations (set $N=5$) are $\xi_1 = \{HHTHTT\}$ and $\xi_2 = \{THHTTT\}$. Plot the two corresponding time functions. ▲

Exercise 5.1.2

Plot the CDF of the central (normal or Gaussian) PDF: $f(x) = \dfrac{\exp(-0.5x^2)}{\sqrt{2\pi}}$, $\sigma = 1$. Plot it from $x = -3$ to 3. ▲

STOCHASTIC PROCESSES

To obtain a formal definition of a discrete-time stochastic process, we consider an experiment with a finite or infinite number of unpredictable outcomes from a space, $S(\xi_1,\xi_2,\ldots)$, each one occurring with a probability $p(\xi_i)$. Next, by some rule, we assign a deterministic sequence, $x(n,\xi_i)$, $-\infty < n < \infty$, to each element ξ_i of the sample space. The sample space, the probabilities of each outcome and the sequences constitute a **discrete-time stochastic process** or **random sequence**. From this definition, we obtain the following four interpretations:

1. For any n and ξ varying, $x(n,\xi)$ is a random time function
2. For fixed n and ξ varying, $x(n,\xi)$ is a random variable
3. For fixed ξ and varying n, $x(n,\xi)$ is a stochastic process (one sample)
4. For fixed ξ and fixed n, $x(n,\xi)$ is a constant (number)

Therefore, a single discrete random sample signal can be precisely characterized by

$$x_i(n) \triangleq x(n,\xi_i) \tag{5.8}$$

This is a **realization** of a stochastic process and in fact **each** distinct value of time can be interpreted as a RV. Thus, we can consider a stochastic process as simply a sequence of ordered (in time) RVs.

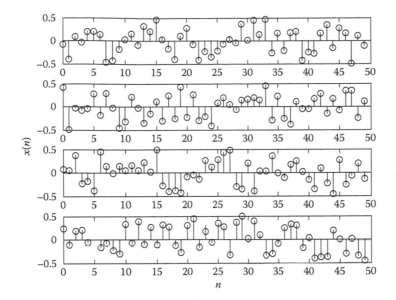

FIGURE 5.5

A collection of realizations of a stochastic process is called an **ensemble**. It is tacitly assumed that every realization is produced under the same exact physical and detection conditions. Figure 5.5 shows four realizations (events) using the following Book MATLAB m-file: realizations.

Book MATLAB m-File: Realizations

```
%Bok MATLAB m-file: realizations
for n=1:4
    x(n,:)=rand(1,50)-0.5;% x=4x50 matrix
        %each row having zero mean and
        %uniform distribution;
end;
m=0:49;
for i=1:4
    subplot(4,1,i);stem(m,x(i,:),'k');
        %plots the four rows of the matrix x;
end;
xlabel('n');ylabel('x(n)');
```

Stationary and Ergodic Processes

Seldom in practice are we able to create an ensemble of a random process with numerous realizations so that we will be able to find some of its statistical characteristics, for example, mean value, variance, and so on. To find these statistical quantities, we need the PDF of the process, which, most of the time, is not possible to produce or guess. Therefore, we will restrict our studies to processes that are easy to study and easy to handle mathematically using one realization.

The process, which produces an ensemble of realizations and whose statistical characteristics do not change, is called **stationary**. For example, the PDF of the RVs $x(n)$ and $x(n+k)$ of the process $\{x(n)\}$ are the same independently of the values of n and k.

Since we will be unable to produce ensemble averages in practice, we are left with only one realization of the stochastic process. To overcome this difficulty, we assume that the process is **ergodic**. This characterization permits us to find the desired statistical characteristics of the process from only one realization at hand. We refer to those statistical values as **sample mean**, **sample variance**, and so on. This assumes that under some conditions, ergodicity is applicable to those statistical characteristics as well.

5.2 AVERAGES

MEAN VALUE

The mean value or expectation value m_x at time n of a continuous RV $x(n)$ at time n having PDF $f(x(n))$ is given by

$$m_x(n) = E\{x(n)\} = \int_{-\infty}^{\infty} x(n) f(x(n)) \, dx(n) \tag{5.9}$$

where $E\{\}$ stands for **expectation operator** and indicates in the same time also that $x(n)$ is an RV as well as when it is written in the form $x(n,\xi)$, where ξ indicates a point from the sample space S. In the situation, we have obtained an ensemble of realizations and we can obtain the mean value using the **frequency interpretation** formula:

$$m_x(n) = \lim_{N \to \infty} \left\{ \frac{1}{N} \sum_{i=1}^{N} x_i(n) \right\} \tag{5.10}$$

where N is the number of realizations and $x_i(n)$ is the ith outcome at sample index n (or time n) of the ith realization. The above formula indicates that the sum is all over the ensemble, different ξ's, at a particular time n. Depending on the type of the RV, the mean value may or may not vary with time.

For an ergodic process, we obtain the sample mean (**estimator of the mean**) using the **time-average** formula (see Exercise 5.2.1):

$$\hat{m}_x = \frac{1}{N} \sum_{n=0}^{N-1} x(n) \tag{5.11}$$

Exercise 5.2.1

Using the continuous-case formulation, find the sample mean value formula 5.10.   ▲

Example 5.2.1

The following results were obtained using MATLAB. The results indicate that, as N goes to infinity, the sample mean value approaches the expectation (population) one:

We first produced a 10,000 uniformly distributed RVs with zero mean value using the following command:

```
x=rand(1,10000)-0.5;
```

Next, we obtained the following mean values:

```
m1=sum(x(1,1:10000))/10000=0.0005731,  m2=sum(x(1,1:1000))/1000=-0.0038,
m3=sum(x(1,1:100))/100=0.0183, and finally m4=sum(x(1,1:10))/10=0.2170.
```  ■

Exercise 5.2.2

Show that the sample mean is equal to the population mean.   ▲

Exercise 5.2.3

Use MATLAB to produce an ensemble of 1,000 realizations of white noise with zero mean. Next, (1) add all the rows and find the mean value for the first 10 elements, (2) add all the rows and find the mean value of the first 100 elements, and (3) add all the rows and find the mean value of all 1,000 elements. Observe and state your conclusions.   ▲

CORRELATION

The **cross-correlation** between two random sequences is defined by

$$r_{xy}(m,n) = E\{x(m), y(n)\} = \iint x(m)y(n)f(x(m),y(n))\,dx(m)\,dy(n) \qquad (5.12)$$

where the integrals are from $-\infty$ to ∞. If $x(n)=y(n)$, the correlation is known as the **autocorrelation**. Having an ensemble of realizations, the frequency interpretation of the autocorrelation function is found using the formula:

$$r_{xx}(m,n) = \lim_{N\to\infty}\left\{\frac{1}{N}\sum_{i=1}^{N} x_i(m)x_i(n)\right\} \qquad (5.13)$$

The Book m-function that produces the mean and autocorrelation using the frequency interpretation formulas is as follows:

BOOK M-FUNCTION: [MX,RX]=MEAN _ AUTOC _ ENSEMBLE(M,N)

```
%Book MATLAB m-function: [mx,rx]=mean_autoc_ensemble
function[mx,rx]=mean_autoc_ensemble(M,N);
    %N=number of time instances; easily modified for
    %other pdf;M=number of realizations; to store
    %the function we write mean_autoc_ensemble;
    %on the command window we write: [a,b]=mean_
    %autoc_ensemble(M,N);
x=randn(M,N);%x=MxN matrix;sum(x,1)=MATLAB function
    %that sums all the rows; sum(x,2)=MATLAB function
    %that adds all the columns;
mx=sum(x,1)/M;
for i=1:N
    rx(i)=sum(x(:,1).*x(:,i))/M;
end;
```

Using the preceding function, we obtained Figure 5.6. The figure indicates clearly that the larger realizations we add, the better results we obtain. The reader should use this Book MATLAB function to obtain similar results with the MATLAB function randn(M,N), which gives random samples normally distributed.

SAMPLE AUTOCORRELATION FUNCTION

The sample autocorrelation function is found from only one realization at hand. The formula is

$$\hat{r}_{xx}(m) = \frac{1}{N-|m|}\sum_{n=0}^{N-|m|-1} x(n)x(n+|m|) \quad m = 0,1,2,\ldots,N-1 \qquad (5.14)$$

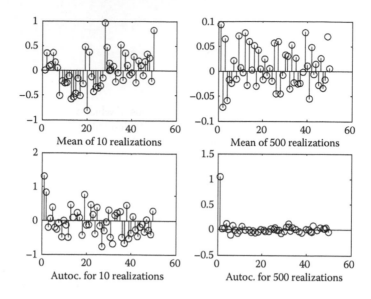

FIGURE 5.6

The absolute value of m ensures the symmetry of the sample autocorrelation sequence at $n=0$. The preceding formula gives an **unbiased** autocorrelation sequence that sometimes produces matrices that do not have inverses. Therefore, it is customary in practice to use anyone of the **biased** autocorrelation formulas:

$$\hat{r}_{xx}(m) = \frac{1}{N} \sum_{n=0}^{N-|m|-1} x(n)x(n+|m|) \quad 0 \le m \le N-1$$

or

$$\hat{r}_{xx}(m) = \frac{1}{N} \sum_{n=0}^{N-1} x(n)x(n-m) \quad\quad 0 \le m \le N-1$$

(5.15)

Book m-Function for Biased Autocorrelation Function: [r]=sample_biased_autoc(x,lg)

```
%Book MATLAB m-function: [r]=sample_biased_autoc
function[r]=sample_biased_autoc(x,lg)
    %the function finds the biased autocorrelation
    %with lag from 0 to lg,it is recommended that lg
    %is 20-30% of N;N=total number of elements of
    %the observed vector x (data);N=length(x);lg
    %stands for lag number;
N=length(x);%length(x)=MATLAB function returning
            %the number of elements of the vector x
for m=1:lg
    for n=1:N+1-m
        x1(n)=x(n-1+m)*x(n);
    end;
    r(m)=sum(x1)/N;
end;
```

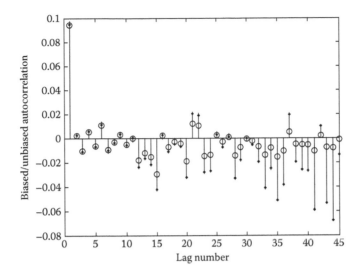

FIGURE 5.7

BOOK M-FUNCTION FOR UNBIASED AUTOCORRELATION FUNCTION

```
%Book MATLAB m-function: [r]=sample_unbiased_autoc(x,lg);
function[r]=sample_unbiased_autoc(x,lg)
N=length(x);
for m=1:lg
    for n=1:N+1-m
        x1(n)=x(n-1+m)*x(n);
    end;
    r(m)=sum(x1)/(N-m);
end;
```

See also the preceding unbiased type for detail explanations.

Figure 5.7 shows the plots of a biased autocorrelation function (shown by circles) and an unbiased autocorrelation function (shown by dots). Note also that for the first 20% lags of the total number of data, the two autocorrelations are identical.

We can also use MATLAB functions to find the biased and unbiased autocorrelation of a signal. The function is

```
>>r=xcorr(x,y,'biased');%for biased case; xcorr() is a MATLAB function;
>>r=xcorr(x,y,'unbiased');%for unbiased case;
>>    %x,y are N length vectors; r is a 2N-1 symmetric
>>    %cross-correlation vector; for unequal vectors we pad
>>    %with zeros the shorter vector;
```

If we do not use any of the words, biased or unbiased, the correlation will be unbiased and the results are not divided by N. The reader will learn more about the xcorr function by entering xcorr in the MATLAB command window help.

COVARIANCE

The covariance (or cross-covariance) of a random ensemble sequences at times m and n is defined by

$$c_x(m,n) = E\{(x(m) - m_m)(x(n) - m_n)\} = E\{x(m)x(n)\} - m_m m_n$$

$$= r_x(m,n) - m_m m_n$$

(5.16)

To find the **variance** (auto-covariance) of the ensemble at one time, we set $m=n$ in Equation 5.16. Hence,

$$c_{xx}(n,n) = \sigma_n^2 = E\left\{(x(n)-m_n)^2\right\} = E\left\{x^2(n)\right\} - m_n^2 \qquad (5.17)$$

If the mean value is 0, the variance and the correlation function are identical.

$$c_{xx}(n,n) = \sigma_n^2 = E\left\{x^2(n)\right\} = r_{xx}(n,n) \qquad (5.18)$$

The estimator, for the biased covariance and variance, due to one sample from the ensemble is given by

$$\hat{\sigma}^2 = \frac{1}{N}\sum_{n=1}^{N}\left(x(n)-\hat{m}\right)^2; \quad \hat{\sigma}^2 = \frac{1}{N}\sum_{n=1}^{N}x(n)^2 \quad \hat{m}=0 \qquad (5.19)$$

The **standard deviation** is defined as the square root of the variance: σ or $\hat{\sigma}$. MATLAB has two functions that give these two quantities for any vector x: σ^2=var(x)=variance, σ=std(x)=standard deviation.

Exercise 5.2.4

Compare the Book autocorrelation function with the MATLAB xcorr() function.   ▲

Exercise 5.2.5

Create an ensemble of 1000 random vectors, each one with 500 elements using uniform distributed (white noise) RVs with zero mean value. Do the following: (1) add 10 rows and find their histogram, (2) add 100 rows and plot the histogram, (3) add 1,000 rows and plot the histogram. Observe and characterize the results.   ▲

Exercise 5.2.6

Create an ensemble of 1,000 random vectors, each one with 1000 elements using uniform distributed (white noise) RVs with zero mean value. Next, (1) add all the 1000 rows and find the standard deviation, std(), (2) for the first 10, 100, and 1000 elements. Plot in log–log scale the STD versus the number used.   ▲

INDEPENDENT AND UNCORRELATED RVS

If the joint PDF of two RVs can be separated into two PDFs, $f_{x,y}(m,n) = f_x(m)f_y(n)$, the RVs are **statistically independent**. Hence,

$$E\{x(m)x(n)\} = E\{x(m)\}E\{x(n)\} = m_x(m)m_x(n) \qquad (5.20)$$

The preceding equation is necessary and sufficient condition for the two RVs $x(m)$, $x(n)$ to be uncorrelated. Note that independent RVs are always uncorrelated. However, the converse is not necessarily true. If the mean value of any two uncorrelated RVs is 0, the RVs are called **orthogonal**. In general, two RVs are called orthogonal if their correlation is 0.

5.3 STATIONARY PROCESSES

It is a **wide-sense** (or **weakly**) **stationary** (WSS) process; the CDF satisfies the following first and second moments:

$$F(x(m)) = F(x(m+k))$$

$$F(x(m),x(n)) = F(x(m+k)x(n+k)) \qquad (5.21)$$

TABLE 5.1

Properties of WSS Processes

1. $m_n = m_{n+k} = m = $ constant
2. The autocorrelation $r_{xx}(m,n)$ depends on the difference $m-n$
3. The variance is less than unity
4. The autocorrelation is symmetric; $r_{xx}(n) = r_{xx}(-n)$
5. For lag $l=0$, $r_{xx}(0) = E\left\{|x(n)|^2\right\} \geq 0$
6. $r_{xx}(0) \geq |r_{xx}(l)|$

for all m, n, and k. The preceding relationships also apply for all the statistical characteristics such as mean value, variance, correlation, and so on. If the preceding relationship is true for any number of RVs of the time series, the process is known as **strictly stationary** process.

The basic **properties** of the wide-sense real stationary process (see Exercise 5.3.1) are listed in Table 5.1.

Exercise 5.3.1

Verify the properties of the WSS processes. ▲

AUTOCORRELATION MATRIX

If $x = [x(0)\,x(1)\cdots x(p)]^T$ is a vector representing a finite random sequence, the autocorrelation matrix is given by ($r_{xx}(m,n) \triangleq r_x(m,n)$)

$$
\mathbf{R}_x = E\{\mathbf{x}\mathbf{x}^T\} = \begin{bmatrix}
E\{x(0)x(0)\} & E\{x(0)x(1)\} & \cdots & E\{x(0)x(p)\} \\
E(x(1)x(0)) & E\{x(1)x(1)\} & \cdots & E\{x(1)x(p)\} \\
\vdots & \vdots & \vdots & \vdots \\
E\{x(p)x(0)\} & E\{x(p)x(1)\} & \cdots & E\{x(p)x(p)\}
\end{bmatrix}
$$

$$
= \begin{bmatrix}
r_x(0) & r_x(-1) & \cdots & r_x(-p) \\
r_x(1) & r_x(0) & \cdots & r_x(-p+1) \\
\vdots & \vdots & \vdots & \vdots \\
r_x(p) & r_x(p-1) & \cdots & r_x(0)
\end{bmatrix}
$$

(5.22)

However, in practical applications, we will have a single sample at hand and this will permit us to find only the estimate correlation matrix $\widehat{\mathbf{R}}_x$. Since, in general and in this text, we will have single realizations to work with, we will not explicitly indicate the estimate with the over bar. The reader will be able to decide if the derived quantity is an estimate or not.

Example 5.3.1

Find the estimate-biased autocorrelation function with lag time 25 for a sequence of 50 terms, which is the single realization of RVs having normal (Gaussian) distribution with zero mean value; create a 4 × 4 autocorrelation matrix.

Solution: At the command window, we write the program:

```
>>x=randn(1,50); r=sample_biased_autoc(x,25);
>>R=toeplitz(r(1,1:4))
R=
```

1.3339	−0.1530	−0.0676	−0.0035
−0.1530	1.3339	−0.1530	−0.0676
−0.0676	−0.1530	1.3339	−0.1530
−0.0035	−0.0676	−0.1530	1.3339

■

Note: *MATLAB gives the following results from a vector or a matrix:*

```
>>y=x(1,m:n);%y is a row vector made up from the mth
>>      %to nth element of the row vector x;
>>z=x(m:n,1);%z is a column vector made up from the mth to
>>      %the nth elements of the column vector x;
>>R=toeplitz(x(1,2:6));% R is a 5x5 square matrix
>>      %made from the 2nd to the 6th element of
>>      %the row vector r;
>>A=R(2:5,4:6);%A is a 5x3 matrix which is produced from
>>      %the row elements 2nd to 5th and from the column elements
>>      %4th and 6th of the matrix R;
```

Example 5.3.2

Let $\{v(n)\}$ be a zero mean, uncorrelated Gaussian random sequence with variance $\sigma_v^2(n) = \sigma^2 = $ constant. (1) Characterize the random sequence $\{v(n)\}$ and (2) determine the mean and the autocorrelation of the sequence $\{x(n)\}$ if $x(n) = v(n) + av(n-1)$, in the range $-\infty < n < \infty$, a is constant.

Solution: (1) The variance of $\{v(n)\}$ is constant and, hence, is independent of the time, n. Since $\{v(n)\}$ is an uncorrelated sequence, it is also independent due to the fact that the sequence is Gaussian distributed. Therefore, we obtain $c_v(l,n) = r_v(l,n) - m_l m_n = r_v(l,n)$ or $\sigma^2 = r_{vv}(n,n)$ is constant. Hence, $r_v(l,n) = \sigma^2 \delta(l-n)$, which implies that $\{v(n)\}$ is a WSS process. (2) $E\{x(n) = 0\}$ since $E\{v(n)\} = E\{v(n-1)\} = 0$. Hence,

$$r_x(l,n) = E\{[v(l)+av(l-1)][v(n)+av(n-1)]\}$$

$$= E\{v(l)v(n)\} + aE\{v(l-1)v(n)\} + aE\{v(l)v(n-1)\} + a^2 E\{v(l-1)v(n-1)\}$$

$$= r_v(l,n) + ar_v(l-1,n) + ar_v(l,n-1) + a^2 r_v(l-1,n-1)$$

$$= \sigma^2 \delta(l-n) + a\sigma^2 \delta(l-n+1) + a^2 \sigma^2 \delta(l-n)$$

$$= (1+a^2)\sigma^2 \delta(m) + a\sigma^2 \delta(m+1), \quad m = l-n$$

Since the mean of $\{x(n)\}$ is 0, a constant, and its autocorrelation is a function of the lag factor $m = l-n$, it is a WSS process.   ■

Exercise 5.3.2

Determine if the following process is WSS: $x(n) = A\cos n + B\sin n$. A and B are uncorrelated, zero mean, RVs with variance σ^2.   ▲

PURELY RANDOM PROCESS (WN)

A discrete process is a **purely random process** which implies that RVs $\{x(n)\}$ are a sequence of mutually **independent and identically distributed** (IID) variables. Since the mean and conv$\{x(m),$ $x(m-k)\}$ do not depend on time, the process is WSS. This process is also known as WS and is given by

$$c_{xx}(k) = \begin{cases} E\{(x(m)-m_x)(x(m-k)-m_x)\} = \sigma_x^2\delta(k)-m_x^2 & k=0 \\ 0 & k=\pm1,\pm2,\ldots \end{cases}$$

$$\delta(k) = \begin{cases} 1 & k=0 \\ 0 & k\neq0 \end{cases} \tag{5.23}$$

$$r_{xx}(k) = c_{xx}(k) \quad m_x = 0$$

Since the process is white, which implies that for the value of k different than zero, $E\{(x(i)-m_x)(x(i-k)-m_x)\} = [E\{x(i)\}-m_x][E\{x(i-k)\}-m_x] = 0$.

RANDOM WALK (RW)

Let $\{x(n)\}$ be a purely random process (IID RVs) with mean m_x and variance σ_x^2. A process $\{y(n)\}$ is a RW if

$$y(n) = y(n-1)+v(n) \quad y(0)=0 \tag{5.24}$$

Therefore, the process takes the form

$$y(n) = \sum_{i=1}^{n} v(i) \tag{5.25}$$

which is found from Equation 5.24 by recursive substitution.

The mean is found to be $E\{y(n)\}=nm_x$ and the $\text{cov}\{y(n),y(n)\} = n\left(E\{x^2\}-m_x^2\right)=n\sigma_x^2$. It is interesting to note that the difference $v(n)=y(n)-y(n-1)$ is purely random and, hence, stationary.

5.4 PROBABILITY DENSITY FUNCTIONS

Some of the important properties and definitions of the PDFs are given in Table 5.2.

UNIFORM DISTRIBUTION

A WSS process of a discrete random sequence of IID elements satisfies the relation

$$f(x(0),x(1),\ldots) = f(x(0))f(x(1))f(x(2))\cdots \tag{5.26}$$

A RV $x(n)$ at time n has a uniform distribution if its PDF is of the form

$$f(x) = \begin{cases} \dfrac{1}{b-a} & a\leq x\leq b \\ 0 & \text{otherwise} \end{cases} \tag{5.27}$$

The MATLAB function rand(1,1000), for example, will provide a random sequence of a 1,000 elements whose sample mean value is 0.5 and its PDF is uniform from 0 to 1 with amplitude 1. To produce the PDF, we can use the MATLAB function hist(x,20). This means that we want to produce the PDF dividing the range of values, in this case from 0 to 1, of the random vector x in 20 equal steps. They are two more useful functions of MATLAB that give the mean value, mean(x), and the variance, var(x). We can also find the standard deviation using the MATLAB function std(x), which is equal to the square root of the variance.

TABLE 5.2
Properties and Definitions

1. The PDF is nonnegative
$$f(x(n)) \geq 0 \quad \text{for all } x$$

2. Integrating the PDF yields CDF
$$F(x(n)) = \int_{-\infty}^{x(n)} f(y(n)) \, dy(n)$$

$$= \text{Area under PDF over the interval } (-\infty, x(n))$$

3. Normalization property
$$\int_{-\infty}^{\infty} f(x(n)) \, dx(n) = F(\infty) = 1$$

4. The area under the PDF $f(x(n))$ over the interval $(x_1(n), x_2(n))$ is equal to
$$p\{x_1(n) < x(n, \xi) \leq x_2(n)\} = \int_{x_1(n)}^{x_2(n)} f(x(n)) \, dx(n)$$

5. The conditional probability of the RV $x_1(n, \xi_1)$, given the value of the second RV, $x_2(n, \xi_2) = x_2(n)$ is given by
$$p\{x_1(n) \mid x_2(n)\} \triangleq p\{x(n, \xi_1) \mid x(n, \xi_2) = x_2(n)\}$$

6. In a simplified way, the Bayes' rule is given by
$$p\{x_1(n) \mid x_2(n)\} = p\{x_2(n) \mid x_1(n)\} \frac{p\{x_1(n)\}}{p\{x_2(n)\}}$$

This means that the probability of $x_1(n)$ given $x_2(n)$ is equal to the probability of $x_2(n)$ given $x_1(n)$ times the ratio of the probabilities.

7. The simplified form, and at time n, of the Markov process is defined as follows:
$$p(x_1, x_2, \ldots, x_N) = p(x_1 \mid x_2) p(x_2 \mid x_3) \cdots p(x_{N-1} \mid x_N) p(x_N)$$

Note that the mean value is given by

$$\int_{-\infty}^{\infty} f(x) \, dx = \int_{a}^{b} \frac{1}{b-a} \, dx = 1$$

and the CDF is

$$F(x) = \int_{-\infty}^{x} f(y) \, dy = \begin{cases} 0 & x < 0 \\ \int_{a}^{x} \frac{1}{b-a} \, dy = \frac{x-a}{b-a} & a \leq x \leq b \\ 1 & x > b \end{cases} \tag{5.28}$$

Other Statistical Characteristics of the Uniform Distribution:

Range: $a \leq x \leq b$

Mean: $\dfrac{(a+b)}{2}$

Variance: $\dfrac{(b-a)^2}{12}$

To find the variance, we use the definition and proceed as follows:

$$\sigma_x^2 = E\{(x - m_x)^2\} = \int_{-\infty}^{\infty} (x - m_x)^2 f(x) \, dx = \frac{1}{b-a} \int_{-\infty}^{\infty} (x - m_x)^2 \, dx = \frac{1}{b-a} \frac{1}{3} (x - m_x)^3 \Big|_{a}^{b}$$

GAUSSIAN (NORMAL) DISTRIBUTION

The PDF of a Gaussian RV $x(n)$ at time n with mean value m_x and variance σ_x^2 is

$$f(x) = \frac{1}{\sqrt{2\pi\sigma_x^2}} e^{-\frac{(x-m_x)^2}{2\sigma_x^2}} = N\left(m_x, \sigma_x^2\right) \tag{5.29}$$

Example 5.4.1

Find the joint PDF of a white Gaussian noise with two elements each one having zero mean value and the same variance. White means that the RVs are independent and identically distributed.

Solution: The joint PDF is

$$f(x, y) = f(x)f(y)$$

$$= \frac{1}{(2\pi)^{1/2}\sigma_x} e^{-[x^2]/2\sigma_x} \cdot \frac{1}{(2\pi)^{1/2}\sigma_y} e^{-[y^2]/2\sigma_y}$$

$$= \frac{1}{(2\pi)^{1/2}\sigma_x^2} \frac{1}{(2\pi)^{1/2}\sigma_y} \exp\left[-\frac{1}{2\sigma_x^2} x^2 \frac{1}{2\sigma_y^2} y^2\right] \tag{5.30}$$

∎

Some properties of the random Gaussian process $\{x(n)\}$ are listed in Table 5.3.

Example 5.4.2

Create a vector of Gaussian distributed 600 elements using MATLAB function randn(). Then, find the approximate mean and variance from the data. Based on these two approximate results, plot the analytical PDF superimposed on the histogram.

Solution: The following Book m-file and Book program produces Figure 5.8.

BOOK M-FILE: [PDFX,BINCENTERS]=HISTOGRAMX(X) (SEE EXERCISE 5.4.1)

```
%Book m-file: [pdfx,bincenters]=histogramx(x)
function[pdfx,bincenters]=histogramx(x)
M=length(x);
bincenters=[min(x)-1:0.2:max(x)+1];
bins=length(bincenters);
h=zeros(1,bins);
```

TABLE 5.3
Properties of a Gaussian Random Process

1. A random process $\{x(n)\}$ is said to be Gaussian if every finite collection of samples $x(n)$ are jointly Gaussian.
2. A Gaussian random process is completely defined by its mean vector and covariance matrix.
3. Any **linear** operation on the Gaussian process $\{x(n)\}$ produces another Gaussian process that only the mean and variance changes.
4. All the higher moments of a Gaussian random process can be expressed by the first and second moments of the distribution.
5. The Gaussian model is motivated by the central limit theorem, which states that the sum of a number of RVs, each one with different PDF, is Gaussian.
6. WN is necessarily generated by IID samples (independence implies uncorrelated RVs and vice versa).

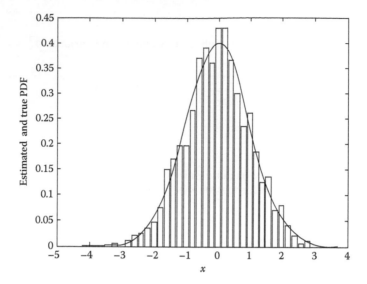

FIGURE 5.8

```
for i=1:length(x)
    for k=1:bins
        if x(i)>bincenters(k)-(0.2/2)...
            & x(i)<bincenters(k)+(0.2/2);
            h(1,k)=h(1,k)+1;
        end
    end
end
pdfx=h/(M*0.2);
```

BOOK PROGRAM

```
m=mean(x); va=var(x);
f=(1/sqrt(2*pi*va))*exp((-(bincenters-m).^2)/2*va);
bar(bincenters,pdfx); colormap([1 1 1]);hold on; plot(bincenters,2*f,'k');   ■
```

Exercise 5.4.1

Based on the PDF definition and the definition of the probability of an event $p(x_0 - \Delta x / 2 < x(\xi) < x_0 + \Delta x / 2)$, verify the previous Book m-function.   ▲

Example 5.4.3

If the PDF $f(x)$ is $N(2.5,2)$, find $p\{x \le 6.5\}$.

   Solution:

$$p\{x \le 6.5\} = \int_{-\infty}^{6.5} \frac{1}{\sqrt{2\pi}} \frac{1}{2} e^{-(x-2.5)^2/8} dx \qquad (1)$$

Set in (1) $y = (x - 2.5)/2$ to obtain: for $x = -\infty \Rightarrow y = -\infty$, for $x = 6.5 \Rightarrow y = 2$ and $dx = 2dy$.

Hence (1) becomes: $p\{x \le 6.5\} = p\{y \le 2\} = \frac{1}{\sqrt{2\pi}} \int_{-\infty}^{0} e^{-y^2/2} dy + \frac{1}{\sqrt{2\pi}} \int_{0}^{2} e^{-y^2/2} dy = 0.5 + \text{errf}(2).$

The error function is tabulated, but MATLAB can give very accurate values by introducing sampling values less than 0.0001. For this sampling value, MATLAB gives the results: $\text{erf}(2) = (1/\text{sqrt}(2*\text{pi}))*0.0001*\text{sum}(-y.^2/2) = 0.47727$, where the vector y is given by $y = 0:0.0001:2$. Tables give $\text{errf}(2) = 0.47724$. Therefore, $p\{x \le 6.5\} = 0.97727$.   ∎

Algorithm Producing Zero Mean and Variance One Normally Distributed

1. Generate two independent RVs, u_1 and u_2, from a uniform distribution $(0,1)$
2. $x_1 = (-2\ln(u_1))^{1/2}\cos(2\pi u_2)$ (or $x_2 = (-2\ln(u_1))^{1/2}\sin(2\pi u_2)$)
3. Keep x_1 and x_2

Book m-Function to Produce N(m,s) RVs

```
function[x]=normallydistr(m,s,N)
    %N=number of elements in x; m=mean
    %value; s=standard deviation;
for i=1:N
    r1=rand;
    r2=rand;
    z(i)=sqrt(-2*log(r1))*cos(2*pi*r2);
end
x=s*z+m;
```

Book m-Function to Produce N(m,s)

```
function[x]=normallydistr1(m,s,N)
z=sqrt(-2*log(rand(1,N))).*cos(2*pi*rand(1,N));
x=s*z+m;
```

Book m-Function to Produce Normal PDF by Addition

```
function[y]=normallydistr_by_summation(m,s,N,M)
    %m=mean value; s=standard deviation;
    %Number of rvs e.g. 2000-5000; M=summable
    %normal variables N(0,1)e.g. 10-20-100;
for n=1:N
    x(n)=sum(randn(1,M))/sqrt(M);%x is normal
    %distributed with std of 1 and mean value 0;
end;
y=m+s*x;
```

To produce the PDF of y, we use the following Book MATLAB program

```
>>[pdf, bincenters]=histogram(y);
>>plot(bincenters,pdf)
```

The summation approach was based on the **central limit theorem**, which states:

For N independent RVs $x(\xi_1), x(\xi_2), \ldots, x(\xi_N)$ ith mean m_i and variance σ_i^2, respectively,

$$y = \frac{\sum_{i=1}^{N}(x_i - m_i)}{\sqrt{\sum_{i=1}^{N}\sigma_i^2}}, \quad \lim N \to \infty \Rightarrow y \equiv N(0,1) \tag{5.31}$$

The central limit theorem has the following interpretation:

> The properly normalized sum of many uniformly small and negligible independent RVs tends to be a standard normal (Gaussian) RV. If a random phenomenon is the cumulative effect of many uniformly small sources of uncertainty, it can be reasonably modeled as normal RV.

Exercise 5.4.2

Show that if the RV is $N(m_x, \sigma_x^2)$ and $\sigma_x^2 > 0$ then the RV $y = (x - m_x)/\sigma_x$ is $N(0,1)$ ▲

Statistical characteristics of the normal distribution

Range: $-\infty < x < \infty$
Mean: m_x
Variance: σ_x^2

TRANSFORMATION TO N(M,SD) FROM N(0,1) OF THE RV x

```
>>z=sd*x+m;%the variance of z is sd² and its mean value is m;
>>              %the variance of x=1 and its mean value is 0;
```

The following m-file creates a histogram for a normal RV (see Figure 5.9). With small modifications, it can be used for any distribution.

BOOK M-FILE: NORMAL_HIST

```
%normal histogram m-File: normal_hist
n=1000;
x=1.5*randn(1,n)+6*ones(1,n);
subplot(1,2,1);plot(x(1,1:200),'k');
xlabel('n');ylabel('x(n)');grid on;
[m,z]=hist(x,10); %calculates counts in each bin,m,
              %and bin coordinates,z, for 10 bins;
w=max(x)/length(z);   %calculates bin width;
pb=m/(n*w);      %probability in each bin;
v=linspace(min(x),max(x));%generates 100 values over
              %range of rv x;
y=(1/(2*sqrt(2*pi)))*exp(-((v-6*ones(size(v))).^2)/4.5);
              %normal pdf;
subplot(1,2,2);
colormap([1 1 1]);%creates white bars, for other colors
              % see >>help colormap;
bar(z,pb); %plots histogram;
hold on;plot(v,y,'k')%superimpose plot of normal pdf;
xlabel('RV value');ylabel('Probability density');
```

EXPONENTIAL DISTRIBUTION

The PDF of an exponential distribution is

$$f(x) = \begin{cases} \dfrac{1}{b} e^{-\frac{x}{b}} & 0 \le x < \infty, b > 0 \\ 0 & \text{elsewhere} \end{cases} \tag{5.32}$$

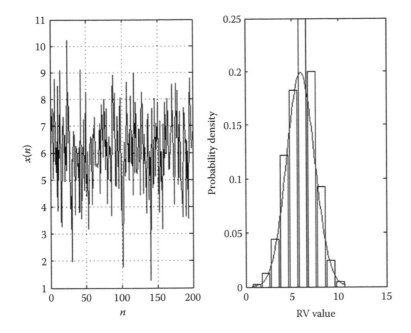

FIGURE 5.9

ALGORITHM TO PRODUCE THE DISTRIBUTION

1. Generate u from a uniform distribution (0,1)
2. $x = -b \ln(u)$
3. Keep x

BOOK M-FUNCTION:[X,M,SD]=EXPONENTIALLYDISTR(B,N)

```
function[x,m,sd]=exponentialpdf(b,N)
for i=1:N
    x(i)=-b*log(rand);
end;
m=mean(x);sd=std(x);
    %value of b=0.2, 0.1 etc, the smaller
    %b the pdf spreads less along the values
    %of x;
```

BOOK M-FUNCTION: [X,M,SD]=EXPONENTIALLYDISTR1(B,N)

```
function[x,m,sd]=exponentialpdf1(b,N)
x=-b*log([rand(1,N)]);
m=mean(x);sd=std(x);
```

Other Statistical Characteristics of the Exponential Distribution

Range: $0 \le x < \infty$
Mean: b
Variance: b^2

LOGNORMAL DISTRIBUTION

Let the RV x be $N(m_x, \sigma^2)$. Then, the RV $y = \exp(x)$ has the lognormal distribution with the following PDF:

$$f(y) = \begin{cases} \dfrac{1}{\sqrt{2\pi}\sigma y} e^{-\frac{\ln y - m_x}{2\sigma^2}} & 0 \le y < \infty \\ 0 & \text{otherwise} \end{cases} \qquad (5.33)$$

The values of σ_x and m_x must take small values to produce a lognormal-type distribution.

ALGORITHM TO PRODUCE THE DISTRIBUTION

1. Generate z from $N(0,1)$
2. $x = m_x + \sigma_x z$ $\left(x \text{ is } N(m_x, \sigma_x^2)\right)$
3. $y = \exp(x)$
4. Keep y

BOOK M-FUNCTION: LOGNORMALLYDISTR(M,SIG,N)

```
function[y]=lognormallydistr(m,sig,N)
    %[y]=lognormallydistr(m,sig,N);
    %m=mean value; sig=standard deviation;
    %N=number of samples;
for i=1:N
    r1=rand;
    r2=rand;
    z(i)=sqrt(-2*log(r1))*cos(2*pi*r2);
end;
x=m+sig*z;
y=exp(x);
```

BOOK M-FUNCTION: LOGNORMALPDF1

```
function[y]=lognormalpdf1(m,sd,N)
n=sqrt(-2*log(rand(1,N))).*cos(2*pi*rand(1,N));
x=m+sd*n;
y=exp(x);
```

Other Statistical Characteristics of the Lognormal Distribution (s_x = standard deviation)

Range: $0 \le x < \infty$
Mean: $e^{m_x} e^{s_x^2/2}$
Variance: $\left(e^{\sigma_x^2} - 1\right) e^{2m_x + \sigma_x^2}$

The reader should create a normal RV with mean 0.7 and standard deviation of 0.2. Next, plot the PDF of the lognormal distribution.

CHI-SQUARE DISTRIBUTION

If $x_1, x_2, \ldots, x_N$ is a random sample of size r from a distribution that is $N(0,1)$, then

$$y = \sum_{n=1}^{r} x_n^2 \qquad (5.34)$$

has a **chi-square** distribution with *r* **degrees of freedom**.

The PDF of the chi-square distribution is

$$f(x) = \frac{1}{\Gamma\left(\dfrac{r}{2}\right) 2^{r/2}} x^{(r/2)-1} e^{-x/2} \quad x \ge 0 \tag{5.35}$$

where *r* is the number of degrees of freedom and $\Gamma()$ is the **gamma** function. MATLAB uses the function gamma( ) for evaluating the gamma function.

BOOK M-FUNCTION FOR CHI-SQUARE DISTRIBUTION: [Y]=CHISQUAREDISTR(N,DF)

```
function[y]=chisquaredistr(n,df)
    %n=number of the ch-square distributed
    %rv y; df=degrees of freedom that must be
    % an EVEN number;
    for m=1:n
        for i=1:df/2
            u(i)=rand;
        end;
        y(m)=-2*log(prod(u));
    %prod(x) gives the product of the elements
    %of the vector x;
    end;
```

To plot the PDF, we just write in the command window >>hist(y,20). We can also use a Monte Carlo approach to find RVs that are chi-square distributed. The following function does this:

BOOK M-FUNCTION: CHISQUAREDISTR1(R,N)

```
function[y]=chisquaredistr1(r,N)
    %N=number of rvs y;
    %r=degrees of freedom, EVEN number;
for n=1:N
    y(n)=sum(randn(1,r).^2);
end;
```

To plot the PDF, we write >>hist(y,20)% 20 bars.

Other Statistical Characteristics of the Chi-square Distribution

Range: $0 \le x < \infty$
Mean: *r*
Variance: $2r$

STUDENT'S DISTRIBUTION

The PDF of the Student's *t* distribution is

$$f(x) = \frac{\Gamma\left(\dfrac{r+1}{2}\right)}{\sqrt{\pi r}\,\Gamma\left(\dfrac{r}{2}\right)\left(1+\dfrac{x^2}{r}\right)^{(r+1)/2}} \frac{1}{} \quad -\infty < x < \infty \tag{5.36}$$

where $r=N-1$ is the **number of degrees of freedom** and *N* is the number of terms in the sequence $\{x(n)\}$.

If z has a standard normal distribution, $N(0,1)$, and y has a chi-square distribution with r degrees of freedom, then

$$x = \frac{z}{\sqrt{y/r}} \tag{5.37}$$

has Student's distribution with r degrees of freedom. To generate x, we first generate z, as described previously for the Gaussian distribution; then, we generate y as described previously for the chi-square distribution and apply Equation 5.37.

BOOK M-FUNCTION: STDISTRIBUTED(N,R)

```
function[t]=stdistributed(N,r)
    %r=degrees of freedom;
    %N=number of iid variables;
z=randn(1,N);
y=chisquarepdf(N,r);
t=z./sqrt(y/r);% t is student distributed;
        % if you write in the command window
        % hist(y,100) will plot the pdf in 100 bins;
```

OTHER STATISTICAL PROPERTIES OF THE *T* DISTRIBUTION

Mean: 0
Variance: $r/(r-2)$, $r > 2$

F DISTRIBUTION

If y_1 is a chi-square RV with r_1 degrees of freedom and y_2 is chi-square RV with r_2 degrees of freedom and both RVs are independent, the RV

$$x = \frac{\dfrac{y_1}{r_1}}{\dfrac{y_2}{r_2}} \qquad 0 \le x < \infty \tag{5.38}$$

is distributed as an F distribution. To create an F distribution, we first generate two chi-square variates and apply Equation 5.38.

The PDF of the F distribution is

$$f(x) = \frac{\Gamma\!\left(\dfrac{r_1+r_2}{2}\right)}{\Gamma\!\left(\dfrac{r_1}{2}\right)\Gamma\!\left(\dfrac{r_2}{2}\right)}\left(\dfrac{r_1}{r_2}\right)^{n/2} \frac{x^{(n/2)-1}}{\left(1+\dfrac{r_1}{r_2}x\right)^{\frac{n+r_2}{2}}} \qquad 0 \le x < \infty \tag{5.39}$$

where $\Gamma()$ is the gamma function.

RAYLEIGH PROBABILITY DENSITY FUNCTION

A radio wave field, which arrives at a receiving point (antenna) after been scattered from a number of scattering points (trees, buildings, and so on), is given by the relation:

$$S = Re^{j\theta} = \sum_{i=1}^{n} A_k e^{j\Phi_k} \tag{5.40}$$

where

Φ_k s is an uniformly distributed phase

A_k s are identically distributed

Resolving S into its real and imaginary components, we have

$$X = \text{Re}\{S\} = R\cos\theta = \sum_{k=1}^{n} A_k \cos\Phi_k = \sum_{k=1}^{n} X_k \tag{5.41}$$

$$Y = \text{Im}\{S\} = R\sin\theta = \sum_{k=1}^{n} A_k \sin\Phi_k = \sum_{k=1}^{n} Y_k \tag{5.42}$$

In general, n is large and, therefore, the central limit theorem dictates that both X and Y are normally distributed. If, in addition, we assume that A_ks and Φ_ks are independent, the ensemble means of X and Y are

$$E\{X\} = \sum_{k=1}^{n} E\{A_k\}E\{\cos\Phi_k\} = \sum_{k=1}^{n} E\{A_k\}\frac{1}{2\pi}\int_{c}^{c+2\pi} \cos\phi_k \, d\phi_k = 0 \tag{5.43}$$

and, similarly,

$$E\{Y\} = \sum_{k=1}^{n} E\{A_k\}E\{\sin\Phi_k\} = \sum_{k=1}^{n} E\{A_k\}\frac{1}{2\pi}\int_{c}^{c+2\pi} \sin\phi_k \, d\phi_k = 0 \tag{5.44}$$

Since $E\{X\}=0$ and $E\{Y\}=0$, the variances are

$$E\{X^2\} = \sum_{k=1}^{n} E\{A_k^2\}E\{\cos^2\Phi_k\} = \frac{1}{2}nE\{A_k^2\} \tag{5.45}$$

$$E\{Y^2\} = \sum_{k=1}^{n} E\{A_k^2\}E\{\sin^2\Phi_k\} = \frac{1}{2}nE\{A_k^2\} \tag{5.46}$$

Setting

$$\frac{1}{2}nE\{A_k^2\} = \sigma^2$$

we obtain the PDFs of X and Y as

$$f(x) = \frac{1}{\sqrt{2\pi\sigma^2}} e^{-x^2/2\sigma^2} \tag{5.47}$$

$$f(y) = \frac{1}{\sqrt{2\pi\sigma^2}} e^{-y^2/2\sigma^2} \tag{5.48}$$

To find the PDF $f(x, y)$, we must find if X and Y are correlated. Therefore, we investigate the ensemble average

$$E\{XY\} = \sum_{k}^{n}\sum_{l}^{n} E\{A_k A_l\}E\{\cos\Phi_k \sin\Phi_l\} = 0 \tag{5.49}$$

since $E\{\cos \Phi_k \sin \Phi_l\} = 0$ for k and l. Therefore, X and Y are uncorrelated and since they are normal, they are also independent. Hence, the combined PDF is

$$f(x,y) = f(x)f(y) = \frac{1}{\sqrt{2\pi\sigma^2}} e^{-x^2/2\sigma^2} \frac{1}{\sqrt{2\pi\sigma^2}} e^{-y^2/2\sigma^2}$$

$$= \frac{1}{2\pi\sigma^2} e^{-(x^2+y^2)/2\sigma^2}$$

(5.50)

Exercise 5.4.3

If the RV x is $N\left(m_x, \sigma^2\right)$ with variance greater than zero, the RV $y = (x - m_x)/\sigma$ is $N(0,1)$.   ▲

Example 5.4.4

Investigate, using MATLAB, the number of $N(1, 0.1)$ RVs that lie within the limits $-2\sigma < x < 2\sigma$.

Solution: The following Book MATLAB program was used:

```
>>x=randn(1,2000);
>>sdx=sqrt(var(x));%sdx=standard deviation, var( ) is a MATLAB
>>    %function given the variance of x;
>>m=0;
>>for n=1:2000
>>    if  x(n)>2*sdx | x(n)<-2*sdx
>>    m=m+1
>>    end
>>end
```

The random vector is shown in Figure 5.10 with two lines at two times the standard deviations: $\pm 2 \times 1.0106 = \pm 2.0212$. From the example, we obtained that 98 times the value of the vector exceeded the ± 2.0212 values. Hence, we obtain a confidence of $(2000 - 98)100/2000 = 95.1\%$ that the value of x will fall within the $\pm 2\sigma$. If we are taken the distance to be 2.5 times the standard deviation, we obtain a 98.4% confidence that the values of x will be between -2.5×1.0106 and 2.5×1.0106.   ■

5.5   TRANSFORMATIONS OF PDFS

Consider the functions:

$$U = U(X,Y) \quad V = V(X,Y)$$

(5.51)

and their inverse functions:

$$X = X(U,V) \quad Y = Y(U,V)$$

(5.52)

If the RV X is near the value x and the RV Y is near the value y, U and V must be near $u = u(x,y)$ and $v = v(x,y)$ given by Equation 5.51. Hence,

$$p\{x < X \le x + dx, y < Y \le y + dy\} = f_{XY}(x,y)dxdy$$

$$= p\{u(x,y) < U \le u(x,y) + du(x,y),$$

(5.53)

$$v(x,y) < V \le v(x,y) + dv(x,y)\} = f_{UV}(u,v)dudv$$

Substituting the inverse functions from Equation 5.52 into preceding equation and dividing by du and dv both sides, we obtain the required PDF to be equal to

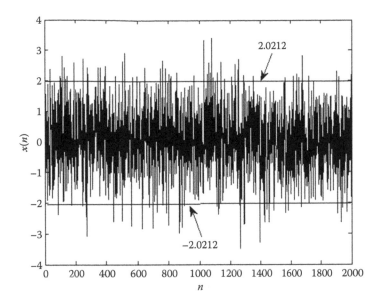

FIGURE 5.10

$$f_{UV}(u,v) = f_{XY}[x(u,v), y(u,v)]\left|\frac{dxdy}{dudv}\right| \qquad (5.54)$$

where the absolute value is necessary to take into consideration both the nondecreasing and non-increasing PDFs. The quantity inside the absolute value symbol is known as the **Jacobian** of the transformation.

The equivalent one-dimensional transformation is given by

$$f_Y(y) = f_X\left[f^{-1}(y)\right]\left|\frac{dx}{dy}\right| \quad y = f(x) \quad x = f^{-1}(y) = \text{inverse function} \qquad (5.55)$$

To transform Equation 5.50 to polar coordinates, we use the following equation (corresponds to Equation 5.51):

$$R(X,Y) = \sqrt{X^2 + Y^2} \quad \Theta(X,Y) = \tan^{-1}(Y/X) \qquad (5.56)$$

with inverse function

$$X(R,\Theta) = R\cos\Theta \quad Y(R,\Theta) - R\sin\Theta \qquad (5.57)$$

The Jacobian of this transformation is

$$\frac{\partial(x,y)}{\partial(r,\theta)} = \begin{vmatrix} \dfrac{\partial x}{\partial r} & \dfrac{\partial x}{\partial \theta} \\ \dfrac{\partial y}{\partial r} & \dfrac{\partial y}{\partial \theta} \end{vmatrix} = \begin{vmatrix} \cos\theta & -r\sin\theta \\ \sin\theta & r\cos\theta \end{vmatrix} = r \qquad (5.58)$$

Substituting Equations 5.57 and 5.58 into Equation 5.54, we obtain

$$f(x,y) = \frac{r}{2\pi\sigma^2}e^{-r^2/2\sigma^2} \quad 0 \le \theta \le 2\pi \quad 0 \le r < \infty \qquad (5.59)$$

If we set $\alpha = 2\sigma^2$ in the previous equation, we obtain the general form of the Rayleigh PDF, which is given by

$$f(r,\theta) = \frac{r}{\pi\alpha} e^{-r^2/\alpha} \tag{5.60}$$

The distribution of θ is, as expected, uniform:

$$f(\theta) = \int_{r=0}^{\infty} f(r,\theta)\, dr = \frac{1}{\pi\alpha} \int_{r=0}^{\infty} r e^{-r^2/\alpha}\, dr = \frac{1}{\pi}\left[-\frac{1}{2} \int_{r=0}^{\infty} \frac{d}{dr} e^{-r^2/\alpha}\, dr \right] \tag{5.61}$$

$$= -\frac{1}{2\pi}\left[e^{-r^2/\alpha} \Big|_{r=0}^{\infty} \right] = -\frac{1}{2\pi}[0-1] = \frac{1}{2\pi} \quad 0 \le \theta \le 2\pi$$

The distribution for the RV R is

$$f(r) = \int_0^{2\pi} \frac{r}{\pi\alpha} e^{-r^2/\alpha}\, d\theta = \frac{2r}{\alpha} e^{-r^2/\alpha} \quad r \ge 0 \tag{5.62}$$

that is, the Rayleigh distribution.

HINTS, SUGGESTIONS, AND SOLUTIONS FOR THE EXERCISES

5.1.1

The results are shown in Figure 5.11.

5.1.2

Use the following Book MATLAB program

```
>>for x=1:120
>>     y=1:120;
>>     F(x)=0.05*sum(exp(-0.5*(y*0.05-3).^2))/...
>>     sqrt(2*pi);%sqrt() is a MATLAB function
>>                %taking the square root of the quantity
>>                %in the parenthesis;
>>end;
>>m=-3:0.05:3-0.05;
>>f=exp(-m.^2/2)/sqrt(2*pi);
>>plot(m,F,m,f);%both curves have blue color, plot(m,F,'g',m,f,'r')
>>                %produces
>>                %one curve green and the other red;
```

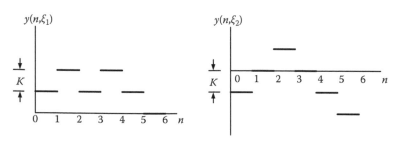

FIGURE 5.11

5.2.1

In this development, we will drop the time n and, hence, it is assumed that we deal with an RV x at time n. Let's divide the real line x into intervals of length Δx with boundary points x_i. We further assume that we want to find the mean value of the function $g(x)$. Hence, the mean of the function $g(x)$ we obtain

$$E\{g(x)\} = \int_{-\infty}^{\infty} g(x)f(x)\,dx \cong \sum_{i=-\infty}^{\infty} g(x_i)f(x_i)\Delta x.$$

To make the transformation from the integral to summation, we made the following important assumption: All the sample values of x have the same PDF, which means that they are identically distributed. Next, we see that $f(x_i)\Delta x \cong p(x_i \le x < x_i + \Delta x)$, where x_i is a value of the RV x at time n (we have suppressed the time here). Using the relative frequency definition of the probability, we obtain

$$E\{g(x)\} \cong \sum_{i=-\infty}^{\infty} g(x_i)p(x_i \le x < x_i + \Delta x) = \sum_{i=-\infty}^{\infty} g(x_i)\frac{N_i}{N}$$

where N_i is the number of measurements in the interval: $x_i \le x < x_i + \Delta x$. Because $g(x_i)N_i$ approximates the sum of values of $g(x)$ for points within the interval i, then $E\{g(x)\} \cong \frac{1}{N}\sum_{i=1}^{N} g(x_i)$. If $g(x)=x$, we obtain $E\{x\} \cong \hat{m}_x = \frac{1}{N}\sum_{i=1}^{N} x_i$.

5.2.2

$$E\{\hat{m}_x\} = E\left\{\frac{1}{N}\sum_{n=0}^{N-1} x(n)\right\} = \frac{1}{N}\sum_{n=0}^{N-1} E\{x(n)\} = \frac{1}{N}\sum_{n=0}^{N-1} m = \frac{Nm}{N} = m$$

5.2.3

From the m-file realizations, we write: for n=1:1000;x(n,:)=rand(1,600)-0.5;end;. Hence, we created a matrix 1000x600. The desired results for this particular realization were (1) m1=sum(x(:,10))/1000=-9.1836 10^{-5},m2=sum(x(:,100))/1000=0.0051,m3=sum(x(:,500))/1000=2.7703 10^{-4}; (2) m1=sum(x(5:24,10))/20=0.0118, m2=sum(x(5:24,100))/20=0.0618, m3=sum(x(5:24,500))/20=0.0611.

5.2.4

Use the following MATLAB program:

```
>>x=randn(1,50);
>>rb=sample_biased_autoc(x);
>>rc=xcorr(x,'biased');
>>subplot(2,1,1);stem(rb(1,1:8));
>>subplot(2,1,2);stem(rc(1,50:57),'g');
```

We have taken from both results about the 20% of the total for comparison, as it should be.

5.2.5

The following Book m-file and Book MATLAB program was used. From Figure 5.12, we conclude that as the added number of ensemble sequences increases, the value of the mean becomes more accurate.

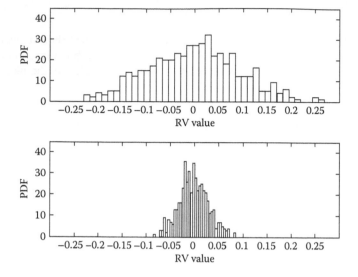

FIGURE 5.12

```
%Bok MATLAB m-function: [A]=ensemble_sequences(M,N)
function[A]=ensemble_sequences(M,N)
for m=1:M%M=number of rows;N=Number of columns;
    for n=1:N
        A(m,n)=rand-0.5;%A=MxN matrix;
    end;
end;
```

Book program

```
>>a=ensemble_sequences(1000,500);
>>a10=a(1:10,:);%a10 is a 10x500 matrix;
>>sa10=sum(a10,1);%sa10 is a vector of 500 elements;
>>a100=a(1:100,:);%a100 is a 100x500 matrix;
>>sa100=sum(a100,1);%sa100 is a vector of 500 elements;
>>subplot(2,1,1);hist(sa10/10,40);colormap([1 1  1]);
>>axis([-0.3  0.3  0  45]);
>>xlabel('rv value');ylabel('PDF');
>>subplot(2,1,2);hist(sa100/100,40);colormap([1 1  1]);
>>axis([-0.3  0.3  0  45]);
>>xlabel('rv value');ylabel('PDF');
```

5.2.6

The plot on a log–log format, dividing the standard deviation by the square root of the number of elements used, $\sqrt{N}$, shows a linear relationship. This indicates that the variation decreases ith the number of RV that are included.

5.3.1

1. $E\{x(n+k)\} = E\{x(m+k)\}$ implies that the mean value must be constant independent of time for the equality to hold or all n, m, and k.

2. $r_x(m,n) = \iint x(m)x(n)f(x(m+k)x(n+k))dx(m)dx(n) = r_x(m+k,n+k) = r_x(m-n,0) = r_x(m-n)$

3. $r_x(k) = E\{x(n+k)x(n)\} = E\{x(n)x(n+k)\} = r_x(n-n-k) = r_x(-k);$

4. $E\{[x(n+k)-x(n)]^2\} = r_x(0) - 2r_x(k) + r_x(0) \geq 0$ (because of the square) or $r_x(0) \geq r_x(k)$

5.3.2

$$m_x(n) = E\{A\}\cos n + E\{B\}\sin n = 0; r_{xx}(n,m) = E\{x(n)x(n+m)\} - m_x^2 = E\{x(n)x(n+m)\}$$

$$= E\{(A\cos n + B\sin n)(A\cos(n+m) + B\sin(n+m))\} = E\{A^2\}\cos n\cos(n+m)$$

$$+E\{B^2\}\sin n\sin(n+m) + E\{AB\}\sin n\cos(n+m) + E\{AB\}\cos n\sin(n+m)$$

$$= \sigma^2\cos(n-n+m) = \sigma^2\cos(m) \Rightarrow WSS, E\{AB\} = E\{A\}E\{B\} = 0$$

5.4.1

The reader will find the desired result by using the following Book MATLAB function:

```
function[f]=exerc5_4_1(x,dx,step,sigma)
    %x=random vector; dx=length of integration section
    %range of pdf; step=the integration step length for each
    %section; sigma=standard deviation;
    r=1;
        for m=round(min(x)):dx:round(max(x));
        y=[m:step:m+dx];
        f(r)=(1/sigma*sqrt(2*pi))*(sum(exp(-y.^2)/(2*sigma^2)))*step;
        r=r+1;
        end

stem(f,'k');
```

5.4.2

$$p(x_0 - \Delta x/2 \leq x(\xi) \leq x_0 + \Delta x/2) = \int_{x_0-\Delta x/2}^{x_0+\Delta x/2} f(x)\,dx \cong f(x_0)\Delta x = p(x_0)\Delta x \text{ or}$$

$$p(x_0) \cong \frac{p(x_0 - \Delta x/2 \leq x(\xi) \leq x_0 + \Delta x/2)}{\Delta x}, p(x_0 - \Delta x/2 \leq x(\xi) \leq x_0 + \Delta x/2)$$

$$\cong \frac{\text{Number of outcomes in } [x_0 - \Delta x/2, x_0 + \Delta x/2]}{M}, M = \text{total outcomes}, \Rightarrow \text{PDF estimator is}$$

$$p(x_0) = \frac{\text{Number of outcomes in } [x_0 - \Delta x/2, x_0 + \Delta x/2]}{M\Delta x}$$

5.4.3

$$F(y) = \text{cdf} = p\left\{\frac{x-m_x}{\sigma} \leq y\right\} = p\{x \leq \sigma y + m_x\} = \int_{-\infty}^{\sigma y+m_x} \frac{1}{\sigma\sqrt{2\pi}}\exp\left[-\frac{x-m_x}{2\sigma^2}\right]dx.$$ If we change the

variable of integration by setting $w = (x-m_x)/\sigma$, then $F(y) = \int_{-\infty}^{y}\left(1/\sqrt{2\pi}\right)e^{-w^2/2}\,dy$. But $f(y)$ is equal to the integrand which is the desired solution.

5.4.4

$$F(y) = \text{cdf} = p\left\{\frac{x - m_x}{\sigma} \leq y\right\} = p\{x \leq y\sigma + m_x = \int\limits_{-\infty}^{y\sigma + m_x} \frac{1}{\sigma\sqrt{2\pi}} \exp\left[-\frac{x - m_x}{2\sigma^2}\right] dx.$$ If we change the

variable of integration by setting $z = (x - m_x)/\sigma$, then $F(y) = \text{cdf} = \int\limits_{-\infty}^{y} \frac{1}{\sigma\sqrt{2\pi}} \exp\left[-z^2 / 2\right] dz.$ But

$f(y) = dF(y)/dy$ and, hence, $f(y)$ is equal to the integrand which is the desired solution.

6 Linear Systems with Random Inputs, Filtering, and Power Spectral Density

6.1 SPECTRAL REPRESENTATION

We first begin by defining the power spectrum of discrete random signals. The discrete-time Fourier transform (DTFT) pair is given by

$$x(n) = \mathscr{F}_{DT}^{-1}\left[X\left(e^{j\omega}\right)\right] = \frac{1}{2\pi}\int_{2\pi} X\left(e^{j\omega}\right)e^{j\omega n} \quad T = 1, \quad \omega = \Omega T\,(\text{rad}),\ \Omega(\text{rad/s})$$

(6.1)

$$X\left(e^{j\omega}\right) = \mathscr{F}_{DT}\{x(n)\} = \sum_{n=-\infty}^{\infty} x(n)e^{-j\omega n} \quad T = 1, \quad \omega = \Omega T\,(\text{rad}),\ \Omega(\text{rad/s})$$

The convergence of $X\left(e^{j\omega}\right)$ is based whether the time function obeys the following

$$\sum_{n=-\infty}^{\infty}|x(n)| < \infty \quad \text{or} \sum_{n=-\infty}^{\infty}|x(n)|^2 < \infty$$

which, in this exist, we will accept as true.

If $x(n)$ is random, $X\left(e^{j\omega}\right)$ is also random because any signal is a realization. Hence,

$$x(n,\xi_i) \overset{\mathscr{F}}{\Leftrightarrow} X\left(e^{j\omega},\xi_i\right) \quad \text{for all } i$$

Therefore, both are simply realizations of the stochastic process over the ensemble generated by i.

A finite-duration realization produces always a convergent transform and is defined as follows:

$$x_N(n) = \begin{cases} x_N(n,\xi_i) & -N \leq n \leq N < \infty \\ 0 & n < -N \text{ and } n > N \end{cases}$$

(6.2)

Exercise 6.1.1

Plot the discrete Fourier transform (DFT) of two sinusoids with added white noise for one realization. Create 200 realizations and average these realizations. Then, find the DFT and plot the spectrum. ▲

The average power of finite sequence $\{x(n)\}$ over the integral $(-N,N)$ is

$$\text{Average power} = \frac{1}{2N+1}\sum_{n=-N}^{N} x^2(n) = \frac{1}{2N+1}\sum_{n=-\infty}^{\infty} x_N^2(n)$$

(6.3)

By the Parseval's theorem for discrete signals, we have

$$\sum_{n=-\infty}^{\infty} x_N^2(n) = \frac{1}{2\pi}\int_{2\pi}\left|X_N\left(e^{j\omega}\right)\right|^2 d\omega = \frac{1}{2\pi}\int_{2\pi} X_N\left(e^{j\omega}\right)X_N^*\left(e^{j\omega}\right)d\omega \tag{6.4}$$

where * means the complex conjugate of the expression. Substituting Equation 6.4 into Equation 6.3, we obtain

$$\text{Average power} = \int_{2\pi}\left(\frac{\left|X_N\left(e^{j\omega}\right)\right|^2}{2N+1}\right)\frac{d\omega}{2\pi} \tag{6.5}$$

The quantity in parentethesis of the preceding equation represents the average power per unit bandwidth and it is called the **power spectral density** (PSD) of $\{x_N(n)\}$. Since the sequence $\{x_N(n)\}$ is a realization of a stochastic process, we must average over the ensemble of realizations. Hence,

$$S_{x_N x_N}\left(e^{j\omega}\right) = E\left\{\frac{\left|X_N\left(e^{j\omega}\right)\right|^2}{2N+1}\right\} \tag{6.6}$$

and since $\{x_N(n)\} \to x$ as $N \to \infty$, we have

$$S_{xx}\left(e^{j\omega}\right) = \lim_{N\to\infty} E\left\{\frac{\left|X_N\left(e^{j\omega}\right)\right|^2}{2N+1}\right\} \tag{6.7}$$

Example 6.1.1

Obtain the PSD using $x(n) = \sin(0.1\times 2\pi n)+0.5\sin(0.3\times 2\pi n)+2\text{randn}$ as a realization, with $N = 30,200$ realizations and using Equation 6.6. The results are shown in Figure 6.1.

Solution: The following Book m-File was used: **ex6_1_1**

```
%m-file:ex6_1_1
n=1:30;
for m=1:200
  x(m,:)=sin(0.2*pi*n)+0.5*sin(0.6*pi*n)+2*randn(1,30);
  X(m,:)=fft(x(m,:),256)/30;
  Xab(m,:)=X(m,:).*conj(X(m,:));%conj( ) is a MATLAB
        %function that gives the conjugate of a complex
        %number;
end
Xt=sum(Xab,1)/200;
w=0:2*pi/256:2*pi-(2*pi/256);
plot(w,Xt,'k')
```

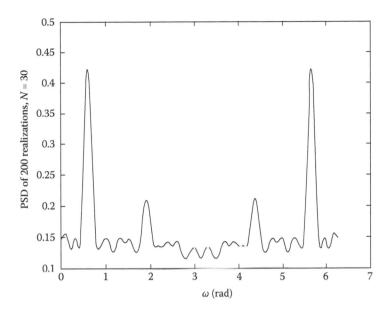

FIGURE 6.1

THE WIENER–KHINTCHINE (W–K) RELATIONS

Having the autocorrelation of a wide-sense stationary (WSS) process $\{x(n)\}$, the PSD pair is given by

$$S_{xx}\left(e^{j\omega}\right) = \sum_{k=-\infty}^{\infty} r_{xx}(k)e^{-j\omega k} \quad \omega = 2\pi, T = 1 \tag{6.8}$$

$$r_{xx}(k) = \frac{1}{2\pi}\int_{-\pi}^{\pi} S_{xx}\left(e^{j\omega}\right)e^{j\omega k}d\omega \tag{6.9}$$

The PSD function is periodic 2π since $\exp[-jk(\omega + 2\pi)] = \exp(-jk\omega)\exp(-jk2\pi) = \exp(-jk\omega)$. Since, for real process, the autocorrelation function is symmetric, $r_{xx}(k) = r_{xx}(-k)$, the PSD is an **even** function. In addition, for a WSS is also non-negative. Hence, in mathematical form, we have

$$S_{xx}\left(e^{j\omega}\right) = S_{xx}\left(e^{-j\omega}\right) = S_{xx}^{*}\left(e^{j\omega}\right)$$

$$S_{xx}\left(e^{j\omega}\right) \geq 0 \tag{6.10}$$

Example 6.1.2

For the following autocorrelation functions, find the PSDs: (1) $r_{xx}(k) = A\delta(k)$, (2) $r_{xx}(k) = A$ for all k and (3) $r_{xx}(k) = Aa^{|k|} \, a < 1$.

Solution: Applying Equation 6.8, we obtain
  1.

$$S_{xx}(e^{j\omega}) = \sum_{k=-\infty}^{\infty} A\delta(k)e^{-j\omega k} = \cdots + A\delta(-1)e^{j\omega} + A\delta(0)e^{-j\omega 0} + A\delta(1)e^{-j\omega 1}\cdots$$

$$= \cdots + 0 + A + 0 + \cdots = A$$

where $\delta(k)$, for k different than zero is zero and for $k = 0$ is one.

2.

$$S_{xx}\left(e^{j\omega}\right) = A\sum_{k=-\infty}^{\infty} e^{-j\omega k} = A\left(\sum_{k=1}^{\infty}\left(e^{j\omega} + e^{-j\omega}\right) + 1\right) = A\left(\sum_{k=1}^{\infty} 2\cos\omega k + 1\right) = A\delta(\omega)$$

The reader can verify the equivalence of the summation to a delta function by using the following Book m-File: **ex6_1_2**

```
%ex6_1_2
N=40;
for k=1:N
     for w=0:1:100
       s(k,w+1)=2*cos(w*0.05*k);%s is a N by 60 matrix;
     end
end
sp=sum(s,1)+1;plot(sp);
```

If we change N to 80, for example, we observe that the pulse at the origin will double in height and the width of the impulse will become narrower, as it should. At the limit, as $k \to \infty$, we will construct the delta function.

3.

$$S_{xx}\left(e^{j\omega}\right) = A\sum_{k=-\infty}^{-1} a^{-k} e^{-j\omega k} + A\sum_{k=1}^{\infty} a^k e^{-j\omega k} + A$$

$$= A\left(\sum_{k=0}^{\infty} a^k e^{j\omega k} - 1 + \sum_{k=0}^{\infty} a^k e^{-j\omega k} - 1\right) + A$$

$$= A\left(\frac{1}{1-ae^{j\omega}} + \frac{1}{1-ae^{-j\omega}} - 1\right) = A\frac{1-a^2}{1+a^2 - 2a\cos\omega}$$

We used the infinite geometric series formula $(1 + x^2 + x^3 + \cdots) = 1/1 - x$ for the two summations. ∎

Example 6.1.3

Find the PSD of the sequence $x(n) = \sin(0.1*2*pi*n) + 1.5*randn(1,32)$ with $n = [0, 1, 2,\ldots, 31]$.

Solution: The following Book m-File produces Figure 6.2. From the figure, it is obvious that a better approximation is accomplished with the correlation method in identifying the signal, in this case, a sine function.

```
%Book MATLAB m-file: ex6_1_3
n=0:31;s=sin(0.18*pi*n);v=randn(1,32);
        %v=white Gaussian noise;
x=s+v;
r=xcorr(x,'biased');%r=biased autocorrelation
                    %function
                    %is divided by N=length(x);
fs=fft(s,32);fr=fft(r,32);fx=fft(x,32);
w=0:2*pi/32:2*pi-(2*pi/32);
subplot(3,2,1);stem(n,s,'k');xlabel('n');
        %ylabel('s(n)');
subplot(3,2,2);stem(w,abs(fs),'k');
        %xlabel('\omega rad');
```

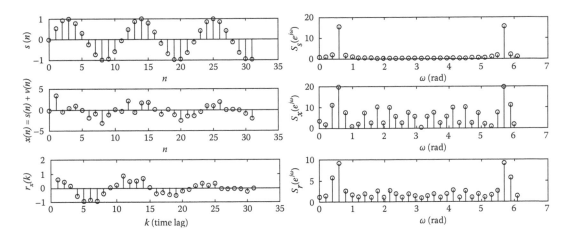

FIGURE 6.2

```
ylabel('S_s(e^{j\omega})');
subplot(3,2,3);stem(n,x,'k');xlabel('n');
        %ylabel('x(n)=s(n)+v(n)');
subplot(3,2,4);stem(w,abs(fx),'k');
        %xlabel('\omega rad');
ylabel('S_x(e^{j\omega})');
subplot(3,2,5);stem(n,r(1,32:63),'k');
        %xlabel('k, time lag');
ylabel('r_x(k)');
subplot(3,2,6);stem(w,abs(fr),'k');
        %xlabel('\omega rad');
ylabel('S_x(e^{j\omega})');                                    ■
```

If we set $z = e^{j\omega}$ in Equation 6.8, we obtain the z-transform of the correlation function instead of the DTFT. Hence,

$$S_{xx}(z) = \sum_{k=-\infty}^{\infty} r_{xx}(k)z^{-k} \tag{6.11}$$

The total power of a zero-mean WSS random process is proportional to the area under the power density curve and is given by

$$r_{xx}(0) = \sigma_x^2 = E\{x^2(n)\} = \frac{1}{2\pi} \int_{-\pi}^{\pi} S_{xx}\left(e^{j\omega}\right)d\omega \tag{6.12}$$

Exercise 6.1.2

Suppose we have a discrete signal given by

$$x(n) = A\sin(\omega_0 n + \varphi)$$

where φ is uniformly random with $\varphi \sim N(0, 2\pi)$. Using the W–K theorem, find the PSD. ▲

6.2   LINEAR SYSTEMS WITH RANDOM INPUTS

Linear time-invariant (LTI) filters (systems) are used in many signal processing applications. Since the input signals of these filters are usually random processes, we need to determine how the statistics of these signals are modified as a result of filtering.

Let $x(n)$, $y(n)$, and $h(n)$ be the filter input, filter output, and the filter impulse response, respectively. It can be shown (see Exercise 6.2.1) that if $x(n)$ is WSS process, the filter output autocorrelation $r_{yy}(k)$ is related to the filter input autocorrelation $r_{xx}(k)$ as follows:

$$r_{yy}(k) = \sum_{l=-\infty}^{\infty} \sum_{m=-\infty}^{\infty} h(l)r_{xx}(m-l+k)h(m)$$

$$= r_{xx}(k) * h(k) * h(-k) = r_{xx}(k) * r_{hh}(k)$$

$$h(k) * h(-k) = \text{convolution between } h(k) \text{ and its reflected form } h(-k)$$

$$= \text{autocorrelation of } h(k) \tag{6.13}$$

Exercise 6.2.1

Find the output autocorrelation function of a system if the input is a WSS process. The input, output, and the impulse response of the system are, respectively, $x(n)$, $y(n)$, and $h(n)$.   ▲

We can also proceed as shown. For linear systems, we have

$$y(n) = h(n) * x(n) = \sum_{k=0}^{\infty} h(k)x(n-k) \tag{6.14}$$

or, taking the DTFT, we have

$$Y\left(e^{j\omega}\right) = H\left(e^{j\omega}\right)X\left(e^{j\omega}\right) \tag{6.15}$$

Since $\{x(n)\}$ is a discrete random signal, we must resort to spectral representation of random processes. Exciting a causal linear system with zero-mean random signal, we obtain the output correlation at lag k as

$$r_{yy}(k) = E\{y(n+k)y(n)\} = E\left\{\left(\sum_{i=0}^{\infty} h(i)x(n+k-i)y(n)\right)\right\}$$

$$= \sum_{i=0}^{\infty} h(i)E\{x(n+k-i)y(n)\} = \sum_{i=0}^{\infty} h(i)r_{xy}(k-i) = h(k) * r_{xy}(k) \tag{6.16}$$

We, next, calculate the output PSD based on the preceding result. Hence,

$$S_{yy}(z) = \sum_{k=-\infty}^{\infty} r_{yy}(k)z^{-k} = \sum_{k=-\infty}^{\infty}\left(\sum_{i=0}^{\infty} h(i)r_{xy}(k-i)\right)z^{-k} \tag{6.17}$$

Multiplying by $z^i z^{-i}$ and interchanging the order of summation, we obtain

$$S_{yy}(z) = \left(\sum_{k=0}^{\infty} h(i)z^{-i} \right) \left(\sum_{i=-\infty}^{\infty} r_{xy}(k-i)z^{-(k-i)} \right) \qquad (6.18)$$

If we let $m = k - i$, we have

$$S_{yy}(z) = \left(\sum_{k=0}^{\infty} h(i)z^{-i} \right) \left(\sum_{m=-\infty}^{\infty} r_{xy}(m)z^{-m} \right) = H(z)S_{xy}(z) \qquad (6.19)$$

Table 6.1 gives the relations between correlations and PSDs.

From the z-transform table (see Chapter 3), we know that the z-transform of the convolution of two functions is equal to the product of their z-transform. Remembering the definition of the z-transform, we obtain the relationship (the order of summation does not change the results)

TABLE 6.1
Summary of Correlation and Spectral Densities

Definitions

Mean value	$m_x = E\{x(n)\}$
Autocorrelation	$r_{xx}(m) = E\{x(n)x^*(n-m)\}$
Auto-covariance	$c_{xx}(m) = E\{[x(n)-m_x][x(n-m)-m_x]^*\}$
Cross-correlation	$r_{xy}(m) = E\{x(n)y^*(n-m)\}$
Cross-covariance	$c_{xy}(m) = E\{[x(n)-m_x][y(n)-m_y]\}$
PSD	$S_{xx}(z) = \displaystyle\sum_{m=-\infty}^{\infty} r_{xx}(m)z^{-m}$
Cross PSD	$S_{xy}(z) = \displaystyle\sum_{m=-\infty}^{\infty} r_{xy}(m)z^{-m}$

Interrelations

$$c_{xx}(m) = r_{xx}(m) - m_x m_x^* = r_{xx}(m) - |m_x|^2$$

$$c_{xy}(m) = r_{xy}(m) - m_x m_x^*$$

Properties

Autocorrelation	**Auto-PSD**
$r_{xx}(m) =$ non-negative definite	$S_{xx}(z) \geq 0$ and real
$r_{xx}(m) = r_{xx}^*(-m)$	$S_{xx}(z) = S_{xx}\left(z^{-1}\right)$ [real $x(n)$]
$|r_{xx}(m)| \leq r_{xx}(0)$	$S_{xx}(z) = S_{xx}^*(1/z^*)$
Cross-Correlation	**Cross-PSD**
$r_{xy}(m) = r_{xy}^*(-m)$	
$|r_{xy}(m)| \leq [r_{xx}(0)r_{yy}(0)]^{1/2} \leq \dfrac{1}{2}[r_{xx}(0)+r_{yy}(0)]$	$S_{xy}(z) = S_{yx}^*(1/z^*)$

Note: $z = e^{j\omega}$.
* is neglected for real sequences.

$$\mathfrak{Z}\{h(-k)\} = \sum_{k=-\infty}^{\infty} h(-k)z^{-k} = \sum_{m=\infty}^{-\infty} h(m)\left(z^{-1}\right)^{-m} = H\left(z^{-1}\right) \tag{6.20}$$

because the summation is the same regardless of the direction in which we sum the series. Therefore, the z-transform of Equation 6.13 becomes

$$R_{yy}(z) = \mathfrak{Z}\{r_{xx}(k) * h(k)\}\mathfrak{Z}\{h(-k)\} = R_{xx}(z)H(z)H\left(z^{-1}\right) \tag{6.21}$$

If we set $z = e^{j\omega}$ in the definition of the z-transform of a function, we obtain the spectrum of the function. Therefore, the W–K theorem, Equation 6.8, becomes

$$S_{yy}\left(e^{j\omega}\right) = S_{xx}\left(e^{j\omega}\right)\left|H\left(e^{j\omega}\right)\right|^2 \tag{6.22}$$

Note: *The preceding equation indicates that the PSD of the output random sequence is equal to the PSD of the input random sequence modified by the square of the absolute value of the spectrum of the filter (system) transfer function.*

Example 6.2.1

One of the first-order **finite impulse response** (FIR) filter (system) is defined in the time domain by the difference equation $y(n) = x(n) + 0.8x(n - 1)$. If the input signal is $N(0,1)$, find the PSD of the filter output.

Solution: Taking into consideration the z-transform properties of linearity and time shifting, the z-transform of the difference equation is $Y(z) = \left(1+0.8z^{-1}\right)X(z)$. Therefore, the transfer function is given by $H(z) = $ output / input $= Y(z)/X(z) = 1+0.8z^{-1}$. The absolute value square of the spectrum of the transfer function is

$$\left|H\left(e^{j\omega}\right)\right|^2 = \left(1+0.8e^{j\omega}\right)\left(1+0.8e^{-j\omega}\right) = 1.64 + 0.8\left(e^{j\omega} + e^{-j\omega}\right)$$

$$= 1.64 + 0.8(\cos\omega + j\sin\omega + \cos\omega - j\sin\omega) = 1.64 + 1.6\cos\omega$$

Therefore, the PSD of the output is given by

$$S_{yy}\left(e^{j\omega}\right) = S_{xx}\left(e^{j\omega}\right)\left|H\left(e^{j\omega}\right)\right|^2 = S_{xx}\left(e^{j\omega}\right)(1.64 + 1.6\cos\omega)$$

Figure 6.3 shows the requested results. Remember that the spectrum is valid in the range $0 \le \omega \le \pi$.

We remind the reader that if the sequence is the sampled form of a continuous function with sampling time T, the spectrum range would have been $0 \le \omega \le \pi/T$. The following Book m-file produces Figure 6.3.

BOOK M-FILE: EX6_2_1

```
%Book m-file: ex6_2_1
x=randn(1,128);%x is N(0,1);
rx=xcorr(x,'biased');
sx=fft(rx(1,128:255),128);
w=0:2*pi/128:2*pi-(2*pi/128);
sy=abs(sx).^2.*(1.64+1.6*cos(w));
```

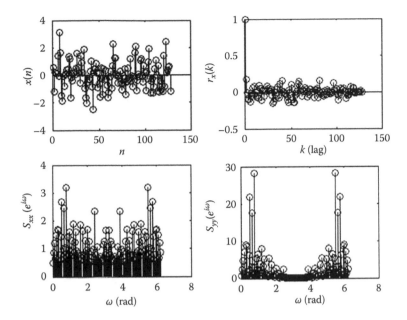

FIGURE 6.3

```
subplot(2,2,1);stem(x,'k');
xlabel('n');ylabel('x(n)')
subplot(2,2,2);stem(rx(1,128:255),'k');
xlabel('k, lag'); ylabel('r_x(k)');
subplot(2,2,3);stem(w,abs(sx),'k');
xlabel('\omega rad');
ylabel('S_{xx}(e^{j\omega})');
subplot(2,2,4);stem(w,abs(sy),'k');
xlabel('\omega rad');
ylabel('S_{yy}(e^{j\omega})');
```

The reader should observe that the filter is a low-pass one because it attenuates the high frequencies (frequencies close to π). ■

Exercise 6.2.2

Let $\{y(n)\}$ be a random process that is generated by filtering white noise $\{v(n)\}$ with a first-order linear shift-invariant filter having a system function:

$$H(z) = \frac{1}{1 - 0.25z^{-1}}$$

The variance of the input noise is equal to one, $\sigma_v^2 = 1$. It is desired to find the correlation of the output, $r_{yy}(k)$. ▲

Exercise 6.2.3

Find and plot the spectra of white with zero-mean value random process. Do this for $N = 64, 256, 512, 1024$. From the plots, state your conclusion. ▲

Example 6.2.2

Generate *M* successive realizations of a white random noise sequence. Plot the spectrum of one realization, the average of 20 realizations. Next, for each of these realizations, plot their histograms.

Solution: The following Book MATLAB program produces Figure 6.4.

BOOK M-FILE: EX6_2_2

```
%Book m-file:ex6_2_2
for m=1:25
    for n=1:256
        x(n,m)=2+randn;
    end;
end;
fx=fft(x,256);
psx=abs(fx)/sqrt(256);
psx256_20=psx(:,1:20);
psxa20=sum(psx256_20,2)/20;
subplot(2,2,1);plot(10*log10(psx(5:256,1)),'.k');
xlabel('Frequency bins');ylabel('Amplitude, dB');
subplot(2,2,2);plot(10*log10(psxa20(5:256,1)),'.k');
axis([0 300 -10 5]);
xlabel('Frequency bins');ylabel('Amplitude, dB')
subplot(2,2,3);hist(psx(5:256,1),20);
colormap([1 1 1]);
xlabel('Amplitude (linear)');
subplot(2,2,4);hist(psxa20(5:256,1),20);
xlabel('Amplitude (linear)');
colormap([1 1 1]);
axis([0 3 0 40]);
```

We observe that averaging produces less random variation of the spectrum. However, the mean of one realization and the average one is close to each other. In this example, mean(psx(:,1)) = 0.9895 and the mean(psxa20) = 1.0111, there is only a difference of 0.0216. ∎

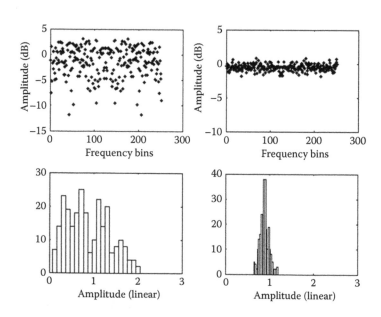

FIGURE 6.4

Exercise 6.2.4

(Noise filtering) Use a Butterworth filter to produce a filtered output. Compare the input unfiltered white nose spectrum with that filtered by a fourth-order Butterworth filter.   ▲

Example 6.2.3

(Noise filtering and windowing) Repeat Exercise 6.2.4 but with the difference of pre-multiplying the data with the Hamming window.

Solution: Figure 6.5 shows the results created by using the following Book m-file.

BOOK M-FILE: EX6_2_3

```
%ex6_2_3
[b,a]=butter(4,0.1*pi);%b and a are two vectors that
    %identify the coefficients of a rational
    %transfer function, b are the numerator coefficient
    %and a the denominator coefficient;
r=randn(1024,50);%x=1024 by 50 matrix of normal rvs;
k=-512:511;
hw=(0.5+0.5*cos(k*pi/512))';a column vector of the
    %Hann window;
c1=hw(:,ones(1,1024));%creates a 1024 by 1024 matrix
    %with identical columns each one the Hann window;
c=c1(:,1:50);%creates a 1024 by 50 matrix with each
    %column the Hann window;
x=r.*c;%multiplies the matrices element by element
    %which means it windows every column of matrix r;
yx=filter(b,a,x);%yx=1024 by 50 matrix that identifies
    %the output of the butterworth filter, MATLAB
    %operates on the columns;
```

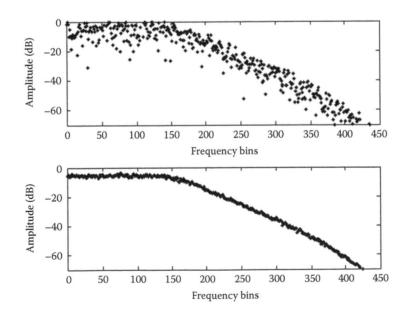

FIGURE 6.5

```
fyx=abs(fft(yx,1024));%fyx=1024 by 50 FFT results, the
    %fft function operates on the columns;
subplot(2,1,1);
plot(20*log10((fyx(:,1))/sqrt(1024)),'.k');
axis([0 450 -70 0]);
xlabel('Frequency bins');ylabel('Amplitude, dB');
subplot(2,1,2);
plot(20*log10((sum(fyx,2)/50)/sqrt(1024)),'.k');
axis([0 450 -70 0]);
xlabel('Frequency bins');ylabel('Amplitude, dB');
```

It is apparent that the windowing extends the attenuation by more than 30 dB.   ∎

Example 6.2.4

Say that a random signal $\{x(n)\}$ passes through some medium and is received m time later. If A is the attenuation of the medium, the received signal is

$$y(n) = Ax(n-m) + v(n)$$

where $\{v(n)\}$ is white noise. The cross-correlation of the zero-mean transmitted and received signals is given by

$$r_{xy}(k) = E\{x(n)y(n+k)\} = AE\{x(n)x(n-m+k)\} + E\{x(n)v(n+k)\}$$

$$r_{xy}(k) = Ar_{xx}(k-m)$$

since $\{v(n)\}$ is white and uncorrelated with $\{x(n)\}$. From the maximum property of the auto-correlation function, we see that the cross-correlation function will achieve a maximum value when $k = m$; that is

$$r_{xy}(m) = Ar_{xx}(0) \quad \text{for } k = m$$

If we take the DTFT of the received signal, we have

$$Y(e^{j\omega}) = AY(e^{j\omega})e^{-j\omega m} + V(e^{j\omega})$$

The corresponding cross-spectrum is given by

$$S_{xy}(e^{j\omega}) = E\{X(e^{j\omega})R*(e^{j\omega})\} = AE\{X(e^{j\omega})X*(e^{j\omega})\}e^{j\omega m} + E\{X(e^{j\omega})V*(e^{j\omega})\}$$

$$S_{xy}(e^{j\omega}) = AS_{xx}(e^{j\omega})e^{j\omega m}$$

Thus, the time delay results in a peak in the cross-correlation function and is characterized by a linear phase component, which can be estimated from the slope of the cross-central phase function.

The following Book m-File produces the cross-correlation of the received signal unshifted and shifted.

BOOK M-FILE: EX6_2_4

```
%ex6_2_4
n=0:99;
x=(0.95).^n.*sin(n*2*pi*0.1);
y=x+0.1*randn(1,100);
```

```
ryy=xcorr(y);
y1=[zeros(1,30) y];
ry1y1=xcorr(y1);
subplot(2,1,1);plot(ryy);
subplot(2,1,2);plot(ry1y1);
```

■

6.3 AUTOREGRESSIVE MOVING AVERAGE PROCESSES (ARMA)

The autoregressive moving average process is defined in the time domain by the difference equation

$$y(n)+a(1)y(n-1)+\cdots+a(p)y(n-p)=b(0)+b(0)x(n-1)+\cdots+b(0)x(n-q) \tag{6.23}$$

Taking the z-transform (see Chapter 3) of both sides of the preceding equation and assuming that the initial conditions are all zero, we obtain the transfer function to be equal to

$$H(z)=\frac{\text{output}}{\text{input}}=\frac{B(z)}{A(z)}=\frac{\displaystyle\sum_{k=0}^{q}b(k)z^{-k}}{1+\displaystyle\sum_{k=1}^{p}A(k)z^{-k}} \tag{6.24}$$

Assuming that the filter (system) is stable, the output process $\{y(n)\}$ will be WSS process, and with input power spectrum, $S_{vv}(z)=\sigma_v^2$, the power output spectrum will be (see Equation 6.22)

$$S_{yy}(z)=\sigma_v^2\frac{B(z)B(1/z)}{A(z)A(1/z)} \tag{6.25}$$

or in terms of ω,

$$S_{yy}\left(e^{j\omega}\right)=\sigma_v^2\frac{B\left(e^{j\omega}\right)B\left(1/e^{j\omega}\right)}{A\left(e^{j\omega}\right)A\left(1/e^{j\omega}\right)}=\sigma_v^2\frac{B\left(e^{j\omega}\right)B\left(e^{-j\omega}\right)}{A\left(e^{j\omega}\right)A\left(e^{-j\omega}\right)}=\sigma_v^2\frac{B\left(e^{j\omega}\right)B\left(e^{j\omega}\right)^*}{A\left(e^{j\omega}\right)A\left(e^{j\omega}\right)^*}$$

$$=\sigma_v^2\frac{\left|B\left(e^{j\omega}\right)\right|^2}{\left|A\left(e^{j\omega}\right)\right|^2} \tag{6.26}$$

Example 6.3.1

Let assume that an ARMA system has two zeros at $z=0.95e^{\pm j\pi/2}$ and two poles at $z=0.9e^{\pm j2\pi/5}$. Obtain the PSD of this system if the input has unit variance.

Solution: To obtain the poles and zeros, we first substitute the exponentials using Euler's expansion $e^{\pm j\theta}=\cos\theta\pm j\sin\theta$. Next, we apply Equation 6.26 to obtain

$$H(z)=\frac{\left(z-0.95e^{j\pi/2}\right)\left(z-0.95e^{j\pi/2}\right)}{\left(z-0.9e^{j2\pi/5}\right)\left(z-0.9e^{j2\pi/5}\right)}=\frac{z^2+0.95^2}{z^2-0.5562z+0.81}=\frac{e^{j2\omega}+0.9025}{e^{j2\omega}-0.5562e^{j\omega}+0.81}$$

For the PSD of the output, here $H(z)$, we use the Book MATLAB program

```
>>w=0:2*pi/256:2*pi-(2*pi/256);
>>psd=abs((exp(j*2*w)+0.9025)./(exp(j*2*w)…
>>-0.5562*exp(j*w)+0.81));
>>plot(w,psd);
```

Next, we plot it. From the plot, we observe a peak close to $\omega = 0.4\pi$ due to the poles in the denominator of $H(z)$ and a null in the spectrum close to $\omega = 0.5\pi$ due to the zeros of $H(z)$. ∎

Exercise 6.3.1

Compare the PSD found in Exercise 6.3.1 and for a transfer function with the same zeros and the poles at $z = 0.98e^{\pm j2\pi/5}$. ▲

Since the output $y(x)$ and the input $v(x)$ are related by a linear constant coefficients difference equation

$$y(n) + \sum_{l=1}^{p} a(l)y(n-l) = \sum_{l=0}^{q} b(l)v(n-l) \tag{6.27}$$

Next, multiply both sides of the preceding equation by $y(n-k)$ and then take the ensemble average. Then, we have

$$r_{yy}(k) + \sum_{l=1}^{p} a(l)r_{yy}(k-l) = \sum_{l=0}^{q} b(l)E\{v(n-l)y(n-k)\} = \sum_{l=0}^{q} b(l)r_{vy}(k-l) \tag{6.28}$$

This is true because $\{v(n)\}$ is a WSS process and since $\{y(n)\}$ is linearly related with $\{v(n)\}$; hence, the output is also WSS process. We also know that the relationship between the input output and system function is

$$y(n) = h(n) * v(n) = \sum_{m=-\infty}^{\infty} v(m)h(n-m) \tag{6.29}$$

We proceed to obtain the cross-correlation of the input and output. Therefore, we find

$$r_{vy}(k-l) = E\{v(n-l)y(n-k)\} = E\left\{ \sum_{m=-\infty}^{\infty} v(n-l)v(m)h(n-k-m) \right\}$$

$$= \sum_{m=-\infty}^{\infty} E\{v(n-l)v(m)\}h(n-k-m) = \sum_{m=-\infty}^{\infty} \sigma_v^2 \delta(n-l-m)h(n-k-m)$$

$$= \sigma_v^2 h(l-k) \tag{6.30}$$

Due to the fact that we have accepted a white noise as input, there exists the following relation: $E\{v(n-l)v(m)\} = \sigma_v^2 \delta(n-l-m)$. Substituting Equation 6.30 in Equation 6.28, we obtain

$$r_{yy}(k) + \sum_{l=1}^{p} a(l)r_{yy}(k-l) = \begin{cases} \sigma_v^2 \sum_{l=k}^{q} b(l)h(l-k) & 0 \le k \le q \\ 0 & k > q \end{cases} \tag{6.31}$$

The reason that the summation on the right side of the equation starts at $l = k$ is due to the fact that for physical systems are causal and, therefore, their impulse response (system function) is zero: $h(-n) = 0$, $n = 0, 1, 2,...$.

The set of the equations depicted by Equation 6.31 are known as the **Yule–Walker equations**. For $k = 0, 1, 2,..., p + q$ in Equation 6.31, we find its matrix form

$$
\begin{bmatrix}
r_{yy}(0) & r_{yy}(-1) & \cdots & r_{yy}(-p) \\
r_{yy}(1) & r_{yy}(0) & \cdots & r_{yy}(-p+1) \\
\vdots & \vdots & & \vdots \\
r_{yy}(q) & r_{yy}(q+p-1) & & r_{yy}(q-p) \\
\hdashline
r_{yy}(q+1) & r_{yy}(q) & \cdots & r_{yy}(q-p+1) \\
\vdots & \vdots & & \vdots \\
r_{yy}(q+p) & r_{yy}(q+p-1) & & r_{yy}(q)
\end{bmatrix}
\begin{bmatrix}
1 \\ a(1) \\ a(2) \\ \vdots \\ a(p)
\end{bmatrix}
= \sigma_v^2
\begin{bmatrix}
c(0) \\ c(1) \\ \vdots \\ c(q) \\ \hdashline 0 \\ \vdots \\ 0
\end{bmatrix}
\tag{6.32}
$$

$$
c(k) = \sum_{l=k}^{q} b(l)h(k-l) = \sum_{l=0}^{q-k} b(l+k)h(l)
$$

One use of the Yule–Walker equations is the extrapolation of the autocorrelation from a finite set of values $r_{yy}(k)$. For example, if $p \geq q$ and if $r_{yy}(0),...,r_{yy}(p-1)$ are known, then $k \geq p$ may be computed recursively using the difference equation

$$
r_{yy}(k) = -\sum_{l=1}^{p} a(l)r_{yy}(k-l)
\tag{6.33}
$$

Furthermore, we observe that the Yule–Walker equations provide also a relationship between the filter (system) coefficients and the autocorrelation of the output sequence. Therefore, Equation 6.31 may be used to estimate the filter coefficients as and bs from the autocorrelation sequence $r_{yy}(k)$. However, due to the product $h(l-k)b(l)$ in Equation 6.32, the Yule–Walker equations are nonlinear in the filter coefficients and solving them for the filter coefficients is, in general, difficult. The same is true for moving average (MA) processes, $p = 0$.

6.4   AUTOREGRESSIVE (AR) PROCESS

A special form of the ARMA (p, q) process results hen $q = 0$ in Equation 6.24 which results into relation $B(z) = b(0)$. In this case, the output $\{y(n)\}$ is generated by filtering white noise with all-pole filter of the form

$$
H(z) = \frac{b(0)}{1 + \displaystyle\sum_{k=1}^{p} a(k)z^{-k}} = \frac{b(0)}{A(z)}
\tag{6.34}
$$

An ARMA $(p, 0)$ process is called an autoregressive process of order p and is referred to as AR (p) process. From the relation

$$S_{yy}(e^{j\omega}) = S_{vv}(e^{j\omega})\left|H\left(e^{j\omega}\right)\right|^2 = S_{vv}(e^{j\omega})H\left(e^{j\omega}\right)H\left(e^{j\omega}\right)$$

$$= S_{vv}(e^{j\omega})H\left(e^{j\omega}\right)H\left(\frac{1}{e^{j\omega}}\right) \quad (6.35)$$

which was developed previously, we obtain

$$S_{yy}(z) = S_{vv}(z)H(z)H\left(\frac{1}{z}\right) \quad (6.36)$$

Since the PSD of the input, white noise, is $S_{vv}(z) = \sigma_v^2$ (the z-transform of the delta function is equal to one), the PSD of the output is

$$S_{yy}(z) = \sigma_v^2 \frac{b(0)^2}{A(z)A\left(\dfrac{1}{z}\right)} \quad (6.37)$$

or, in terms of the frequency variable, $z = \exp(j\omega)$, we find

$$S_{yy}\left(e^{j\omega}\right) = \sigma_v^2 \frac{b(0)^2}{A\left(e^{j\omega}\right)A\left(\dfrac{1}{e^{j\omega}}\right)} = \sigma_v^2 \frac{b(0)^2}{A\left(e^{j\omega}\right)A\left(e^{-j\omega}\right)} = \sigma_v^2 \frac{b(0)^2}{\left|A\left(e^{j\omega}\right)\right|^2} \quad (6.38)$$

Therefore, to find the PSD of the output, we must use the Yule–Walker equation to obtain the *as* and $b(0)$. From Equation 6.32 and $q = 0$, we obtain

$$\begin{bmatrix} r_{yy}(0) & r_{yy}(-1) & \cdots & r_{yy}(-p) \\ r_{yy}(1) & r_{yy}(0) & \cdots & r_{yy}(-p+1) \\ \vdots & \vdots & & \vdots \\ r_{yy}(p) & r_{yy}(p-1) & \cdots & r_{yy}(0) \end{bmatrix}\begin{bmatrix} 1 \\ a(1) \\ \vdots \\ a(p) \end{bmatrix} = \sigma_v^2 b(0)^2 \begin{bmatrix} 1 \\ 0 \\ \vdots \\ 0 \end{bmatrix} \quad (6.39)$$

From Equation 6.32, with $k = 0$ and $l = 0$, we obtain $c(0) = b(0)h(0)$. From Equation 6.34 (see Exercise 6.4.1), we obtain $h(0) = b(0)$ and, hence, $c(0) = b(0)^2$ and it was taken outside the matrix in Equation 6.39.

Exercise 6.4.1

Prove that $h(0) = b(0)$ using Equation 6.34.   ▲

Example 6.4.1

Obtain the PSD of the output of a second-order autoregressive process if the input to the filter is a white noise with zero-mean value and unit variance. The filter has the following denominator:

$$A(z) = 1 + 0.4z^{-1} - 0.2z^{-2}$$

Solution: It is know that the autocorrelation function is symmetric, $r_{yy}(k) = r_{yy}(-k)$. First step is to find the coefficients from the relation that was produced from the preceding equation. Hence, we have

$$
\begin{bmatrix}
r_{yy}(0) & r_{yy}(-1) & \cdots & r_{yy}(-p) \\
r_{yy}(1) & r_{yy}(0) & \cdots & r_{yy}(-p+1) \\
\vdots & \vdots & & \vdots \\
r_{yy}(p) & r_{yy}(p-1) & \cdots & r_{yy}(0)
\end{bmatrix}
\begin{bmatrix}
1 \\
a(1) \\
a(2) \\
\vdots \\
a(p)
\end{bmatrix}
\begin{bmatrix}
b(0)^2 \\
0 \\
0 \\
\vdots \\
0
\end{bmatrix}
\tag{6.40}
$$

The preceding equation can be split into two matrices as follows:

$$
r_{yy}(0) + r_{yy}(-1)a(1) + \cdots + r_{yy}(-p)a(p) = b(0)^2
\tag{6.41a}
$$

$$
\begin{bmatrix}
r_{yy}(-1) & \cdots & r_{yy}(-p) \\
r_{yy}(0) & \cdots & r_{yy}(-p+1) \\
\vdots & & \vdots \\
r_{yy}(p-1) & \cdots & r_{yy}(0)
\end{bmatrix}
\begin{bmatrix}
a(1) \\
a(2) \\
\vdots \\
a(p)
\end{bmatrix}
= -
\begin{bmatrix}
r_{yy}(0) \\
r_{yy}(1) \\
\vdots \\
r_{yy}(p)
\end{bmatrix}
\tag{6.41b}
$$

From Equation 6.41b, we obtain the a's coefficients, and from Equation 6.41a, we obtain the $b(0)^2$. Then we apply Equation 6.38 to obtain the PSD. The following Book MATLAB m-file produces Figure 6.6.

The estimated a's were found to be 0.4232, −0.1561, 0.0074, 0.0313, 0.0127, and 0.0536. This indicates that the approximation is reasonably good and the last four approaching zero as it should be.

BOOK M-FILE: EX6_4_1

```
%Book m-file:ex6_4_1
y(1)=0;y(2)=0;
for n=3:256;
    y(n)=-0.4*y(n-1)+0.2*y(n-2)+randn;
end
```

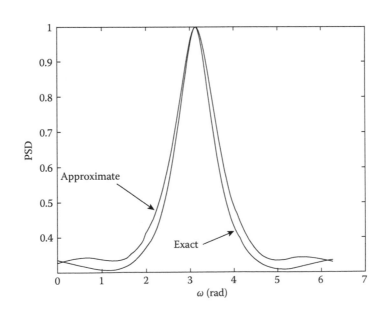

FIGURE 6.6

```
ry=xcorr(y);
Ry=toeplitz(ry(1,256:262));%toeplitz(r) is
    %a MATLAB function creating correlation
    %matrix;
Rya=Ry(2:7,2:7);
a=inv(Rya)*(-Ry(2:7,1));
bsq=sum(Ry(1,:).*[1; a]');
w=0:2*pi/256:2*pi-(2*pi/256);
psd=abs(bsq./(1+a(1)*exp(-j*w)+...
    a(2)*exp(-j*2*w)+a(3)*exp(-j*3*w)+...
    a(4)*exp(-j*4*w)+a(5)*exp(-j*5*w)));
w1=0:2*pi/256:2*pi-(2*pi/256);
plot(w1,psd/max(psd),'k');
```

∎

Exercise 6.4.2

Find the AR coefficients $a(1)$ to $a(5)$ if the autocorrelation of the observed signal $\{y(n)\}$ is given. Let $\{x(n)\}$ be a white noise with mean zero and variance one, which is the input to the filter: $y(n)-0.8y(n-1)+0.4y(n-2)=x(n)$ (Table 6.2). ▲

*6.5   PARAMETRIC REPRESENTATIONS OF STOCHASTIC PROCESSES: ARMA AND ARMAX MODELS

A general model that includes the ARMA model is the **autoregressive moving average with exogenous inputs** (ARMAX), which is represented by the equations

$$y(n)+a_1y(n-1)+\cdots+a_py(n-p)=b_0x(n)+\cdots+b_qx(n-q)$$

$$+v(n\cdots)+c_1v(n-1)+\cdots+c_gv(n-g)$$ (6.42)

In compact form, we write this equation in the form

TABLE 6.2
Linear Systems and Random Signals

Correlation	Spectrum
$r_{yy}(k)=h(k)*h(k)*r_{xx}(k)$	$S_{yy}(z)=H(z)H(z^{-1})S_{xx}(z)=\|H(z)\|^2 S_{xx}(z),\ z=e^{j\omega}$
$r_{yy}(k)=h(k)*r_{xy}(k)$	$S_{yy}(z)=H(z)S_{xy}(z)\quad z=e^{j\omega}$
$r_{yx}(k)=h(k)*r_{xx}(k)$	$S_{yx}(z)=H(z)S_{xx}(z)\quad z=e^{j\omega}$
where	
$r_{yy}(k)=\mathfrak{Z}^{-1}\{S_{yy}(z)\}$	$S_{yy}(z)=\displaystyle\sum_{k=-\infty}^{\infty}r_{yy}(k)z^{-1}$
$r_{xy}(k)=\mathfrak{Z}^{-1}\{S_{xy}(z)\}$	$S_{xy}(z)=\displaystyle\sum_{k=-\infty}^{\infty}r_{xy}(k)z^{-1}$

---

* Means section may be skipped.

$$A(z)y(n) = B(z)x(n) + C(z)v(n)$$

$$A(z) = 1 + a_1 z^{-1} + \cdots + a_p z^{-p}$$

$$B(z) = b_0 + b_1 z^{-1} + \cdots + b_q z^{-q}$$

$$C(z) = 1 + c_1 z^{-1} + \cdots + c_g z^{-g}$$

(6.43)

The abbreviated forms of several models that are developed from the ARMAX(p, q, g) model are as follows:

1. The infinite impulse response of a discrete system (IIR) model: $C(.)$ or ARMAX (p, q, 0)

$$A(z)y(n) = B(z)x(n)$$

2. FIR model: $A(.) = 1$, $C(.) = 0$, or ARMAX(1, q, 0)

$$y(n) = B(z)x(n)$$

3. Autoregressive (AR) model: $B(.) = 0$, $C(.) = 1$, or ARMAX(p, 0, 1)

$$A(z)y(n) = v(n)$$

4. Moving-average (MA) model: $A(.) = 1$, $B(.) = 0$, or ARMAX(1, 0, g)

$$y(n) = C(z)v(n)$$

5. Autoregressive-moving average (ARMA) model: $B(.) = 0$ or ARMAX(p, 0, g)

$$A(z)y(n) = C(z)v(n)$$

6. Autoregressive model with exogenous input (ARX): $C(.) = 1$ or ARMAX(p, q, 1)

$$A(z)y(n) = B(z)x(n) + v(n)$$

The different systems that can be developed from the ARMAX system are shown in Figure 6.7.

Example 6.5.1

Let us use the following model ARMAX(1, 0, 0) with $a = 0.4$, and $\{v(n)\}$ be a $N(0, 1)$. Therefore, the system is given by

$$H(z) = \frac{Y(z)}{V(z)} = \frac{1}{1 + 0.4z^{-1}}$$

Find the PSD of the output.

Solution: After cross multiplying, we obtain

$$Y(z)(1 + 0.4z^{-1}) = V(z) \text{ or } Y(z) + 0.5Y(z)z^{-1} = V(z)$$

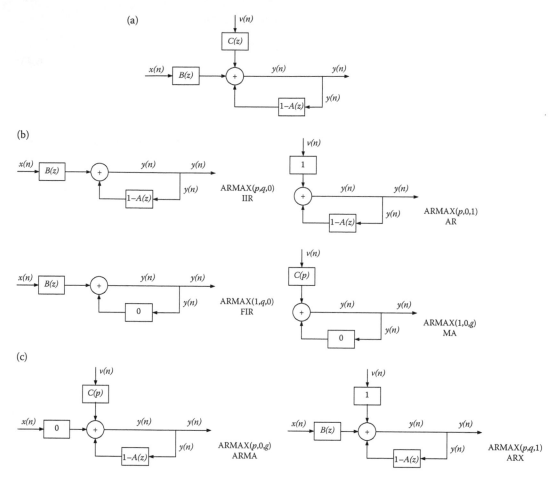

FIGURE 6.7

Taking the inverse z-transform of the previous equation and remembering that z^{-1} indicates shift of one unit in the time domain (see Chapter 3), we obtain

$$y(n) = -0.5y(n-1) + v(n)$$

The following Book m-file produces Figure 6.8.

BOOK M-FILE: EX6_5_1

```
%Book m-file: ex6_5_1
v=randn(1,256);
y(1)=0;
for i=1:256
    y(i+1)=-0.5*y(i)+v(i);
end
ry=xcorr(y)/256;
fys=fft(ry(1,257:290),256);
subplot(1,3,1)
plot(y(1,1:60),'k');
axis([0 70 -3 3]);
xlabel('n');ylabel('y(n)');
```

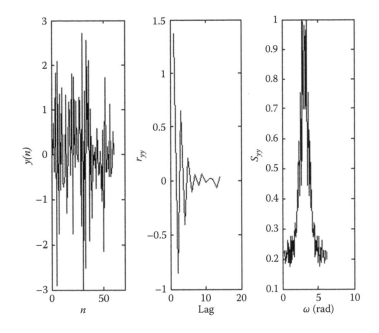

FIGURE 6.8

```
subplot(1,3,2)
plot(ry(1,257:270),'k');
xlabel('lag');ylabel('r_{yy}');
w=0:2*pi/256:2*pi-(2*pi/256);
subplot(1,3,3)
plot(w,abs(fys)/max(abs(fys)),'k')
xlabel('\omega rad');ylabel('S_{yy}');
```

Since the ARMAX model is essentially an IIR system, we can write

$$\frac{Y(z)}{V(z)} = H(z) = \sum_{i=0}^{\infty} h(i)z^{-i} \tag{6.44}$$

The covariance of the ARMAX model, with zero mean and white $\{v(n)\}$, is given by

$$r_{yy}(k) = r_{vv}(0)\sum_{i=0}^{\infty} h(i)h(i+k) \quad k \geq 0 \tag{6.45}$$

with corresponding variance

$$r_{yy}(0) = r_{vv}(0)\sum_{i=0}^{\infty} h^2(i) \tag{6.46}$$

Let's investigate the ARMA model (ARMAX(p, 0, g)), which is shown graphically in Figure 6.7. Therefore, in the figure we substitute the output $y(n)$ with the corresponding, in this case, with $h(n)$. Hence,

$$h(n) = h(n) + (1 - A(z))h(n) + C(z)\delta(n) \text{ or } h(n) = -A(z)h(n) + C(z)\delta(n) \tag{6.47}$$

In expanded form, remembering that z^{-1} means time shift by one unit, we have

$$h(n) = -\sum_{i=1}^{p} a(i)h(n-i) + \sum_{i=0}^{g} c(i)r_{w}(0)\delta(i) \quad c(0) = 1 \tag{6.48}$$

Example 6.5.2

Let's assume an AR model with $A(z) = 1 + 0.5z^{-1}$ and $r_{w}(0) = 1$. It is desired to find the variance of the output $\{y(n)\}$.

Solution: From Equation 6.48, we obtain

$$h(n) = -0.5h(n-1) + \delta(n)$$

$$h(0) = 1$$

$$h(1) = -0.5h(0) = -0.5$$

$$h(2) = -0.5h(1) = (-0.5)^2$$

$$\vdots \qquad \vdots$$

$$h(n) = (-0.5)^n$$

From Equation 6.26, we obtain the variance of y, $r_{yy}(0)$, to be equal to 1.3333. We can find this value by using MATLAB:

```
>>n=0:15;
>>h=(-0.5).^n;
>>ry0=sum(h.^n));
```

Figure 6.9 shows the desired results. ∎

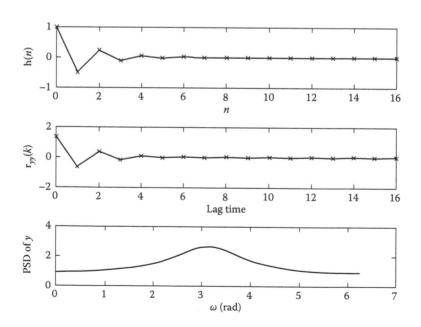

FIGURE 6.9

Exercise 6.5.1

Simulate the ARMAX (2, 1, 1) model described by the difference equation

$$A(z)Y(z) = B(z)X(z) + C(z)V(z) \quad \text{or}$$

$$\left(1 + \frac{3}{4}z^{-1} + \frac{1}{8}z^{-2}\right)y(n) = \left(1 + \frac{1}{8}z^{-1}\right)x(n) + \left(1 + \frac{1}{16}z^{-1}\right)v(n) \quad \text{or}$$

$$y(n) + \frac{3}{4}y(n-1) + \frac{1}{8}y(n-2) = x(n) + \frac{1}{8}x(n-1) + v(n) + \frac{1}{16}v(n-1)$$

The input is $x(n) = \sin 0.5\pi 0.1n$ and the exogenous source is $v = N(1, 0.02)$. Plot $y(n)$, $h(n)$, $m_y(n)$, $r_{yy}(k)$, $y(n) - m_y(n)$, and $S_{yy}(\omega)$. ▲

We summarize these results in Table 6.3 and in the two more tables, Tables 6.4 and 6.5.

TABLE 6.3
ARMAX Representation

Output Propagation

$$y(n) = (1 - A(z))y(n) + B(z)x(n) + C(z)v(n)$$

Mean Propagation

$$m_y(n) = (1 - A(z))m_y(n) + B(z)x(n) + C(z)m_v(n)$$

Impulse Propagation

$$h(n) = (1 - A(z))h(n) + C(z)\delta(n)$$

Variance/Covariance Propagation

$$R_{yy}(k) = r_{vv}(0) \sum_{i=0}^{\infty} h(i)h(i+k) \quad k \ge 0$$

where

y = the output or measurement sequence

x = the input sequence

v = the process (white) noise sequence with variance $r_{vv}(0)$

h = the impulse response sequence

δ = the impulse input of amplitude $\sqrt{r_{vv}(0)}$

m_y = the mean output or measurement sequence

m_v = the mean process noise sequence

r_{yy} = the stationary output covariance at lag k

A = the pth order system; characteristic (poles) polynomial

B = the qth order input (zeros) polynomial

C = the gth order noise (zeros) polynomial

TABLE 6.4

MA Representation

Output Propagation

$$y(n) = C(z)v(n)$$

Mean Propagation

$$m_y(n) = C(z)m_v(n)$$

Variance/Covariance Propagation

$$r_{yy}(k) = r_{vv}(0)\sum_{i=0}^{g} c(i)c(i+k) \quad k \geq 0$$

where

y = the output or measurement sequence

v = the process (white) noise sequence with variance $r_{vv}(0)$

m_y = the mean output or measurement noise sequence

m_v = the mean process noise sequence

r_{yy} = the stationary output covariance at lag k

C = the gth order noise (zeros) polynomial

TABLE 6.5

AR Representation

Output Propagation

$$y(n) = (1 - A(z))y(n) + v(n)$$

Mean Propagation

$$m_y(n) = (1 - A(z))m_y(n) + m_v(n)$$

Variance/Covariance Propagation

$$r_{yy}(k) = (1 - A(z))r_{yy}(k) + r_{vv}(0)\delta(k) \quad k \geq 0$$

where

y = the output or measurement sequence

v = the process (white) noise sequence with variance $r_{vv}(0)$

m_y = the mean output or measurement sequence

m_v = the mean process noise sequence

r_{yy} = the stationary output covariance at lag k

A = the pth order system characteristic (poles) polynomial

Δ = the unit impulse (Kronecker delta)

HINTS–SUGGESTIONS–SOLUTIONS FOR THE EXERCISES

6.1.1

The following Book MATLAB program was used to produce Figure 6.10.

```
>>n=1:256;
>>for i=1:200
>>    x(i,:)=sin(0.1*2*pi*n)+0.5*(0.3*2*pi*n)+5*randn((1,256);
>>end;
>>xn=sum(x,1)/200;xF5=fft(x(5,:),512);xnF=fft(xn,512);
>>w=0:2*pi/512:2*pi-(2*pi/512);
>>subplot(2,1,1);plot(w,abs((xF5)/max(abs(xF5))),'k');
>>subplot(2,1,2);plot(w,plot(abs(xnF)/max(abs(xnF))),'k');
```

6.1.2

First, we must find the mean:

$$m_x(n) = E\{A\sin(\omega_0 n + \varphi)\} = A\sin(\omega_0 n)E\{\cos(\varphi)\} + A\sin(\omega_0 n)E\{\sin(\varphi)\} = 0$$

where

$$E\{\cos(\varphi)\} = \int_0^{2\pi} \cos\varphi \, \mathrm{prob}(\varphi)\,d\varphi = \frac{1}{2\pi}\int_0^{2\pi} \cos\varphi \, d\varphi = 0, \text{ similarly } E\{\sin\varphi\} = 0$$

For a zero-mean process, the correlation is given by

$$r_{xx}(k) = E\{x(n)x(n+m)\} = E\{A\sin(\omega_0 n + \varphi)A\sin(\omega_0(n+m)+\varphi)\}$$

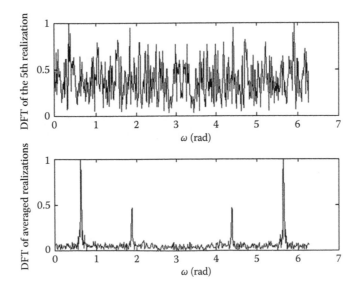

FIGURE 6.10

If we let $y = \omega_0 n + \varphi$, and $z = \omega_0(n + m) + \varphi$, then using the trigonometric identity

$$\sin y \sin z = \frac{1}{2}\cos(y - z) + \frac{1}{2}\sin(y + z)$$

we obtain

$$r_{xx}(k) = \frac{A^2}{2}E\{\cos\omega_0 k - \cos(2(\omega_0 n + \varphi) + \omega_0 k)\}$$

$$= \frac{A^2}{2}\cos\omega_0 k - \frac{A^2}{2}E\{\cos(2(\omega_0 n + \varphi) + \omega_0 k)\}$$

Using the fact that φ is uniform and calculating the expected value shows that the last term is zero. We then have

$$r_{xx}(k) = \frac{A^2}{2}\cos\omega_0 |k|$$

From the theorem, we obtain

$$S_{xx}(e^{j\omega}) = \text{DTFT}\{r_{xx}(k)\} = \text{DTFT}\left\{\frac{A^2}{2}\cos\omega_0 k\right\} = \pi A^2[\delta(\omega - \omega_0) + \delta(\omega + \omega_0)]$$

This development indicates that a sinusoidal signal with random phase can be characterized by a sinusoidal correlation and impulsive power spectrum at the specific frequency ω_0.

6.2.1

$$y(n) = x(n) * h(n) = \sum_{k=-\infty}^{\infty} h(k)x(n - k) \tag{1}$$

$$E\{y(n)\} = \sum_{k=-\infty}^{\infty} h(k)E\{x(n - k)\} = m_x \sum_{k=\infty}^{\infty} h(k) \tag{2}$$

The ensemble operates only on random variables (RVs). Next, we shift (1) by k, multiply by $x(n)$, and then take the ensemble average of both sides of the equation. Hence, we find

$$E\{y(n+k)x(n)\} = r_{yx}(n+k,n) = E\left\{\sum_{l=-\infty}^{\infty} h(l)x(n+k-l)x(n)\right\}$$

$$= \sum_{l=-\infty}^{\infty} h(l)E\{x(n+k-l)x(n)\} = \sum_{l=-\infty}^{\infty} h(l)r_{xx}(k-l) \tag{3}$$

(remember that $x(n)$ is stationary).

Since in (3) the summation is a function only of k, it indicates that the cross-correlation function is also only a function of k. Hence we write:

$$r_{yx}(k) = r_{xx}(k) * h(k) \tag{4}$$

The autocorrelation of $y(n)$ is found from (1) by shifting the time by k, $y(n+k)$, multiply next the new expression by $y(n)$ and then take the ensemble average. Therefore, we find

$$r_y(n+k,n) = E\{y(n+k)y(n)\} = E\left\{ y(n+k)\sum_{l=-\infty}^{\infty} x(l)h(n-l) \right\} = \sum_{l=-\infty}^{\infty} E\{y(n+k)x(l)\}h(n-l)$$

$$= \sum_{l=-\infty}^{\infty} h(n-l)r_{yx}(n+k-l) \tag{5}$$

Here, the input is a WSS process and, therefore, the output is also. Next, we set $m = n - l$ in (5) to obtain

$$r_y(k) = \sum_{m=-\infty}^{\infty} h(m)r_{yx}(m+k) = r_{yx}(k) * h(-k) \tag{6}$$

Remember that the convolution of two sequences, one been reversed, is equal to the correlation between the original sequences. Combine (6) and (4) to obtain

$$r_y(k) = r_{xx}(k) * h(k) * h(-k) = \sum_{l=-\infty}^{\infty}\sum_{m=-\infty}^{\infty} h(l)r_{xx}(m-l+k)h(m) = r_{xx}(k) * r_{hh}(k)$$

6.2.2

The power spectrum of $y(n)$ is $S_{yy}(z) = S_{vv}(z)H(z)H\left(z^{-1}\right) = \dfrac{1}{1-0.25z^{-1}}\dfrac{1}{1-0.25z}$. Here we substituted $S_{vv}(z) = \sigma_v^2 = 1$. Therefore, the power spectrum has a pair of poles, one at $z = 0.25$ and the other in the reciprocal location $z = 4$. The autocorrelation of $\{y(n)\}$ may be found from $S_{yy}(z)$ as follows. First we perform a partial expansion of $S_{yy}(z)$

$$S_{yy}(z) = \frac{z^{-1}}{\left(1-0.25z^{-1}\right)\left(z^{-1}-0.25\right)} = \frac{16/15}{1-0.25z^{-1}} + \frac{4/15}{z^{-1}-0.25} = \frac{16/15}{1-0.25z^{-1}} - \frac{16/15}{1-4z^{-1}}$$

From the tables in Chapter 3, the inverse z-transform yields the desired result:

$$r_{yy}(k) = \frac{16}{15}\left(\frac{1}{4}\right)^k u(k) + \frac{16}{15}4^k u(-k-1) = \frac{16}{15}\left(\frac{1}{4}\right)^{|k|}$$

6.2.3

From the figures, we conclude that by increasing the length of the random process, the spectrum does not becomes smoother. However, by increasing the length, the frequency resolution increases.

6.2.4

The following Book m-file was used to create Figure 6.11.

```
%exerc6_2_4
[b,a]=butter(4,0.1*pi);%b and a are two vectors that
    %identify the coefficients of a rational
    %transfer function, b are the numerator coefficient
    %and a the denominator coefficient;
x=randn(1024,50);%x=1024 by 50 matrix of normal rvs;
yx=filter(b,a,x);%yx=1024 by 50 matrix that identifies
    %the output of the butterworth filter, MATLAB
    %operates on the columns;
fyx=abs(fft(yx,1024));%fyx=1024 by 50 FFT results, the
    %fft function operates on the columns;
subplot(2,1,1);
plot(20*log10((fyx(:,1))/sqrt(1024)),'.k');
xlabel('Frequency bins');ylabel('Amplitude, dB');
subplot(2,1,2);
plot(20*log10((sum(fyx,2)/50)/sqrt(1024)),'.k');
xlabel('Frequency bins');ylabel('Amplitude, dB');
```

6.3.1

It is apparent that the closer the poles are to the unit circle, the sharper the peak becomes. This indicates that a better resolution of sinusoidal signals may be accomplished and are shown in Figure 6.12.

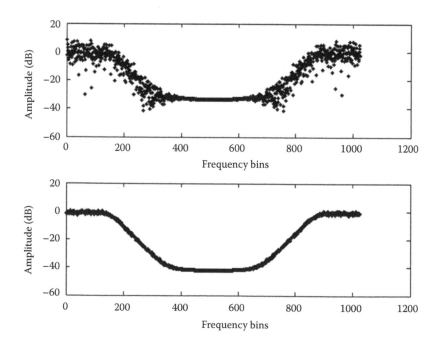

FIGURE 6.11

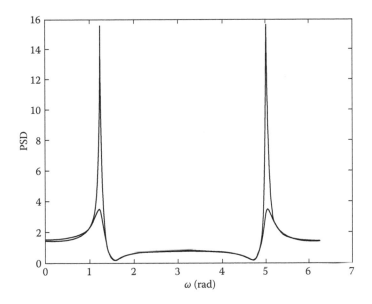

FIGURE 6.12

6.4.1

The division gives: $H(z) = b(0) - b(0)a(1)z^{-1} \dots$, which indicates that $h(0) = b(0)$.

6.4.2

The following Book m-file gives the desired results, which are shown in Figure 6.13.

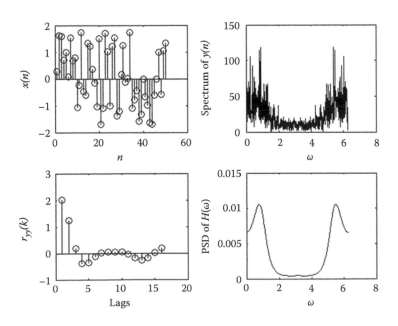

FIGURE 6.13

```
%Book m-file: exerc6_4_2;
x=(rand(1,1000)-0.5)*3.5;%uniformly
    %distributed WGN of zero mean
    %and unit variance; we have set x=v
    %as the input to the filter;
y=filter(1,[1  -0.8  0.4],x);% y=observed
    %signal;
[ry1,lags]=xcorr(y,y,15,'biased');
    %ry1=autocorrelation of y;
ry=ry1(1,16:31);%one sided;
Ry=toeplitz(ry(1,1:6));%Ry=autocorrelation
    %matrix of the output;
Rya=Ry(2:6,2:6);%creates a 5x5 matrix from Ry matrix;
a=inv(Rya)*(Ry(2:6,1)*(-1));%solves for a's from
    %the lower part of the
    %system which is given in the example;
b02=[Ry(1,1:6)]*[1  a']';
H=b02./(fft([1 a'],512));%spectrum of the
    %system function;
SH=H.*conj(H)/512;%PSD of the transfer function;
nn=0:511;w=2*pi*nn/512;
subplot(2,2,1);stem(x(1,1:50),'k');xlabel('n');
ylabel('x(n)');
subplot(2,2,2);plot(w,abs(fft(y(1,1:512),512)),'k');
xlabel('\omega');ylabel('Spectrum of y(n)')
subplot(2,2,3);stem(ry,'k');xlabel('Lags');
ylabel('r_{yy}(k)');
subplot(2,2,4);plot(w,SH,'k');xlabel('\omega');
ylabel('PSD of H(\omega)');
```

The estimated a's are as follows: -0.8897, 0.4239, 0.0646, -0.0895, and 0.0360. The last three are close to zero as they should be.

7 Least Squares-Optimum Filtering

7.1 INTRODUCTION

The method of **least squares** was proposed by Carl Friedrich Gauss back in 1795. The salient feature of this method is that no probabilistic assumptions are made about the data, but only a signal model is assumed. Even no assumption can be made about specific probabilistic structures of the data; nonetheless, the least-squares estimator (LSE) is widely used in practice due to its ease of implementation, amounting to the minimization of the least-squares error optimization. The data model and the least-squares error are shown in Figure 7.1.

7.2 THE LEAST-SQUARES APPROACH

The LSE chooses q such that makes the signal $s(n)$ to be closest to the received (observed) data $x(n)$. Closeness is measured by the least-squared error criterion:

$$J(q) = \sum_{n=0}^{N-1}[x(n)-s(n)]^2 \tag{7.1}$$

where the observational interval is assumed to be N, $n=0,1,\ldots,N-1$, and the dependence on q is via $s(n)$. The value of q that minimizes $J(q)$ is the LSE. Because no probabilistic assumptions were made, LSEs are applied in situations where precise statistical characteristics of the data is unknown or where an optimal estimator cannot be found or is difficult to find.

Example 7.2.1

Let $s(n) = A + Bn$ for $n=0,1,\ldots,N-1$, where the unknown parameters are A and B. Given the received data $x(n)$, we apply the least-squares method to find the unknowns A and B.

Solution: Specifically, we choose A and B such that they minimize the error criterion:

$$J(A,B) = \sum_{n=0}^{N-1}(x(n)-(A+Bn))^2 \tag{7.2}$$

Since $x(n)$ is known, we must minimize J with respect to A and B. Expanding the square in Equation 7.2 and taking the derivatives with respect to two unknowns and setting the results equal to zero, we obtain

$$\sum_{n=0}^{N-1}(x(n)-A-Bn)=0$$

$$\sum_{n=0}^{N-1}(x(n)-A-Bn)n=0 \tag{7.3}$$

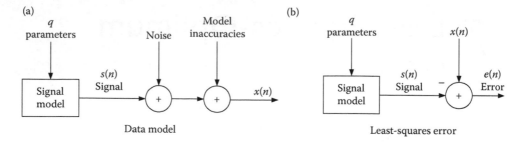

FIGURE 7.1

or equivalently,

$$\left(\sum_{n=0}^{N-1} 1\right) A + \left(\sum_{n=0}^{N-1} n\right) B = \sum_{n=0}^{N-1} x(n)$$

$$\left(\sum_{n=0}^{N-1} n\right) A + \left(\sum_{n=0}^{N-1} n^2\right) B = \sum_{n=0}^{N-1} nx(n)$$

(7.4)

or in matrix/vector form,

$$\begin{bmatrix} \sum_{n=0}^{N-1} 1 & \sum_{n=0}^{N-1} n \\ \sum_{n=0}^{N-1} n & \sum_{n=0}^{N-1} n^2 \end{bmatrix} \begin{bmatrix} A \\ B \end{bmatrix} = \begin{bmatrix} \sum_{n=0}^{N-1} x(n) \\ \sum_{n=0}^{N-1} nx(n) \end{bmatrix}$$

(7.5)

The preceding equations can also be written as

$$\mathbf{H}^T \mathbf{H} \mathbf{q} = \mathbf{H}^T \mathbf{x}$$

(7.6)

where

$$\mathbf{x} = \begin{bmatrix} x(0) \\ x(1) \\ \vdots \\ x(N-1) \end{bmatrix} \quad \mathbf{H} = \begin{bmatrix} 1 & 0 \\ 1 & 1 \\ 1 & 2 \\ \vdots & \vdots \\ 1 & N-1 \end{bmatrix} \quad \mathbf{q} = \begin{bmatrix} A \\ B \end{bmatrix}$$

(7.7)

The matrix **H** has dimensions $N \times 2$ and T denotes a transpose. The LSE for the unknown parameters becomes

$$\hat{\mathbf{q}} = \begin{bmatrix} \hat{A} \\ \hat{B} \end{bmatrix} = \left(\mathbf{H}^T \mathbf{H}\right)^{-1} \mathbf{H}^T \mathbf{x}$$

(7.8)

The −1 exponent means inverse of the 2×2 matrix $\mathbf{H}^T\mathbf{H}$. ■

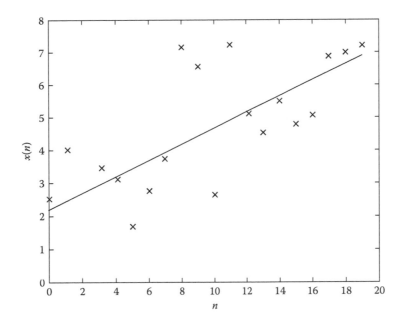

FIGURE 7.2

Exercise 7.2.1

If $N = 3$, obtain $\mathbf{H}^T\mathbf{H}$ and $\mathbf{H}^T\mathbf{x}$.   ■

Exercise 7.2.2

Assume that $s(n) = A$, a constant, and $x(n)$ has been observed for $n = 0, 1,\ldots, N - 1$. Find the estimate of A.   ■

Example 7.2.2

Find a and b of the straight line $a + bn$ if the received signal is given by $x(n) = a + bn + v(n)$, where $v(n)$ is a zero-mean normally distributed random variable (RV; see Figure 7.2).

Solution: The following Book MATLAB programs can be used to find a and b:

```
>>n=0:19;x=2+0.2*n+randn(1,20);
>>h=[20   sum(n);sum(n)   sum(n.^2)];
>>c=[sum(x)   sum(n.*x)];
>>ab=inv(h)*c; ab=[2.2068;  0.2482];
```

We can also use Equation 7.8 as follows:

```
>>H=[ones(1,20)'   [0  n(1,2:20)]'];
>>AB=inv(H'*H)*H'*x;AB=[2.2068; 0.2482];
>>%Note that the accent in MATLAB means transpose;
```

Observe that both ab and AB vectors are identical as it should be.   ■

A signal model that is **linear in the unknown parameters** is said to generate a **linear least squares** (LLS) problem. We will delve into this type of problem in this text. Otherwise, the problem is a nonlinear least-squares problem.

7.3 LINEAR LEAST SQUARES

Using the important LLS property for a scalar parameter, we set the signal

$$s(n) = qh(n) \tag{7.9}$$

where $h(n)$ is a known sequence. Therefore, the least-squares error criterion becomes

$$J(q) = \sum_{n=0}^{N-1} (x(n) - qh(n))^2 \tag{7.10}$$

Expanding the square and taking the derivative of $J(q)$ with respect to q and setting the resulting expression equal to zero, we obtain the LSE minimization of q (see Exercise 7.3.1):

$$\hat{q} = \frac{\sum_{n=0}^{N-1} x(n)h(n)}{\sum_{n=0}^{N-1} h^2(n)} \tag{7.11}$$

The minimum least-squares error is given by

$$J_{\min} = J(\hat{q}) = \sum_{n=0}^{N-1} x^2(n) - \hat{q} \sum_{n=0}^{N-1} x(n)h(n) \tag{7.12}$$

Exercise 7.3.1

Verify Equation 7.12.   ▲

Using Equation 7.11 in Equation 7.12, we obtain an alternative formulation for the minimum:

$$J_{\min} = \sum_{n=0}^{N-1} x^2(n) - \frac{\left(\sum_{n=0}^{N-1} x(n)h(n) \right)^2}{\sum_{n=0}^{N-1} h^2(n)} \tag{7.13}$$

Example 7.3.1

Obtain the estimates of a, b, and c if the output is given by $x(n) = a + bn + cn^2 + v(n)$. The signal $v(n)$ is normal with zero mean and unit variance and $a = 2$, $b = 0.1$, and $c = 0.1$.

Solution: The following Book MATLAB program was used and the results are shown in Figure 7.3.

```
>>n=0:15;
>>x=2+0.1*n+0.1*n.^2+randn(1,16);
>>H=[ones(1,16)'    n'    (n.^2)'];
>>q=inv(H'*H)*H'*x';
>>xa=3.1470-0.1941*n+0.1156*n.^2;
>>plot(n,x,'xk',n,xa,'k')
```
                                                                                      ■

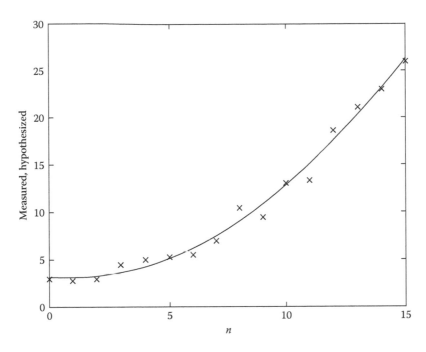

FIGURE 7.3

Exercise 7.3.2

Find the amplitude constant of the signal:

$$s(n) = A\sin(0.1\pi n) + B\sin(0.4\pi n)$$

if the received signal is $x(n) = s(n) + \text{randn}(1, N)$ and their exact values are $A = 0.2$ and $B = 5.2$. ▲

*7.3.1 Matrix Formulation of Linear Least Squares (LLS)

The extension to matrix formulation we set

$$\underset{N\times 1}{s} = \underset{N\times p}{H}\ \underset{p\times 1}{q}\quad N > p \tag{7.14}$$

Matrix H has full rank p and is known as the **observation matrix**. The LSE is found by minimizing

$$J(q) = \sum_{n=0}^{N-1}[x(n) - s(n)]^2 = (x - Hq)^T(x - Hq)$$

$$= x^T x - x^T Hq - q^T H^T x + q^T H^T Hq \tag{7.15}$$

$$= x^T x - 2x^T Hq + q^T H^T Hq, x^T Hq = q^T H^T x$$

Referring to Appendix 3, we find the following relations:

$$(x^T Hq)^T = (Hq)^T(x^T)^T = q^T H^T x = \text{scalar};\quad \frac{\partial x^T q}{\partial q} = x;\frac{\partial q^T Hq}{\partial q} = 2Hq$$

---

* Means section may be skipped.

Example 7.3.1.1

Let us verify the relation $\partial x^T q / \partial q = x$.

Solution: We proceed with a two-element vectors as follows:

$$\frac{\partial x^T q}{\partial q} = \begin{bmatrix} \dfrac{\partial x^T q}{q_1} \\[2ex] \dfrac{\partial x^T q}{q_1} \end{bmatrix} = \begin{bmatrix} \dfrac{\partial \begin{bmatrix} x_1 & x_2 \end{bmatrix} \begin{bmatrix} q_1 \\ q_2 \end{bmatrix}}{q_1} \\[4ex] \dfrac{\partial \begin{bmatrix} x_1 & x_2 \end{bmatrix} \begin{bmatrix} q_1 \\ q_2 \end{bmatrix}}{q_2} \end{bmatrix} = \begin{bmatrix} \dfrac{\partial}{q_1}(x_1 q_1 + x_2 q_2) \\[2ex] \dfrac{\partial}{q_2}(x_1 q_1 + x_2 q_2) \end{bmatrix} = \begin{bmatrix} x_1 \\ x_2 \end{bmatrix} = x$$

Similarly, we can prove the rest.   ∎

Therefore, we set the gradient equal to zero:

$$\frac{\partial J(q)}{\partial q} = -2H^T x + 2H^T H q = 0$$

Therefore, the estimated value of q is

$$\hat{q} = (H^T H)^{-1} H^T x \tag{7.16}$$

The minimum least-squares error is found from Equations 7.15 and 7.16 as

$$\begin{aligned} J_{\min} = J(\hat{q}) &= (x - H\hat{q})^T (x - H\hat{q}) \\ &= x^T (I - H(H^T H)^{-1} H^T)(I - H(H^T H)^{-1} H^T) x \\ &= x^T (I - H(H^T H)^{-1} H^T) x \\ &= x^T (x - H\hat{q}) \end{aligned} \tag{7.17}$$

The matrix $I - H(H^T H)^{-1} H^T$ is **idempotent** which means that $A^2 = A$.

7.4   POINT ESTIMATION

In a typical case, we are faced with the problem of extracting **parameter** values from a discrete-time waveform or a data set. For example, let us have an N-point set $\{x(0), x(1), \dots, x(N-1)\}$ which depends on an unknown parameter q. Our aim is to determine q based on the data or define an **estimator**:

$$\hat{q} = g(x(0), x(1), \dots, x(N-1)) \tag{7.18}$$

where g is some function (**statistic**) of the data and its numerical value is called an **estimate** of q.

To determine a good estimator, we must model the data mathematically. Because the data are inherently random, we describe them by the probability density function (PDF) or

$p(x(0), x(1), \ldots, x(N-1); q)$ which is **parameterized** by the unknown parameter q. This type of dependence is denoted by semicolon.

As an example, let $x(0)$ be an RV from a normal (Gaussian) population and with mean value $q = \mu$. Hence, the PDF is

$$p(x(0); q) = \frac{1}{\sqrt{2\pi\sigma^2}} e^{-\left(1/2\sigma^2\right)\left(x(0)-q\right)^2} \tag{7.19}$$

The plots of $p(x(0); q)$ for different values of q are shown in Figure 7.4. From the figure, we can **infer** that, if the value of $x(0)$ is positive, it is doubtful that $q = q_2$ or $q = q_3$ and, hence, the value $q = q_1$ is more probable. In the area of point estimation, the specification of the appropriate PDF is critical in determining a good estimator. In the case when the PDF is not given, which is more often than not, we must try to choose one that is consistent with the constraints of any prior knowledge of the problem, and, furthermore, that is mathematically tractable.

After the PDF has been selected, it is our concern to determine an optimal estimator (a function of data). The estimator may be thought of as rule that assigns a **value** to q for each realization of the sequence $\{x\}_N$. The estimate of q is the value of q obtained at a particular realization $\{x\}_N$.

7.4.1 ESTIMATOR PERFORMANCE

Let the set $\{x(0), x(1), \ldots, x(N-1)\}$ of random data be the sum of a constant c and a zero-mean white noise $v(n)$:

$$x(n) = c + v(n) \tag{7.20}$$

Intuitively, we may set as an estimate of c the sample mean of the data:

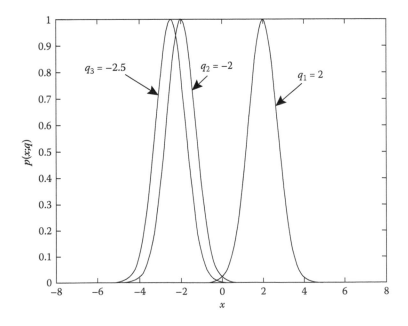

FIGURE 7.4

$$\hat{c} = \frac{1}{N} \sum_{n=0}^{N-1} x(n) \tag{7.21}$$

From Figure 7.5, we find that $x(0) = 1.0812$ and may accept the random variable $x(0)$ as another estimate of the mean:

$$\tilde{c} = x(0) \tag{7.22}$$

The basic question is, which of these two estimators will produce the more accurate mean value? Instead of repeating the experiment a large number of times, we proceed to prove that the sample mean is a better estimator than $x(0)$. To do this, we first look at their mean value (expectation):

$$E\{\hat{c}\} = E\left\{\frac{1}{N} \sum_{n=0}^{N-1} x(n)\right\} = \frac{1}{N} \sum_{n=0}^{N-1} E\{x(n)\} = \frac{1}{N} E\{x(0) + x(1) + \cdots + x(N-1)\}$$

$$= \frac{N\mu}{N} = \mu \tag{7.23}$$

$$E\{\tilde{c}\} = E\{x(0)\} = \mu \tag{7.24}$$

which indicates that on the average, both estimators produce the true mean value of the population. Next, we investigate their variations, which are (see Exercise 7.4.1.1)

$$\text{var}\{\hat{c}\} = \text{var}\left\{\frac{1}{N} \sum_{n=0}^{N-1} x(n)\right\} = \frac{\sigma^2}{N}$$

$$\text{var}\{\tilde{c}\} = \text{var}\{x(0)\} = \sigma^2 \tag{7.25}$$

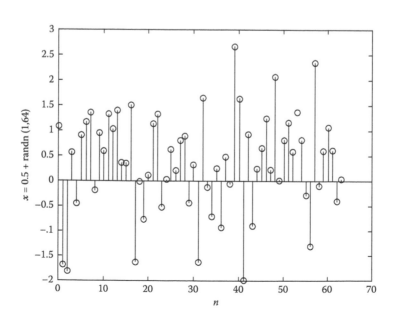

FIGURE 7.5

Exercise 7.4.1.1

Verify Equation 7.25.   ▲

The preceding results show that $\text{var}\{\tilde{c}\} > \text{var}\{\hat{c}\}$, and the $\text{var}\{\hat{c}\}$ approaches 0 as $N \to \infty$. To prove Equation 7.25, we assume that $v(n)$s are independent and identically distributed (IID) and have the same variance σ^2.

7.4.2 Biased and Unbiased Estimators

An estimator which on the average yields the true value, the population one, of the unknown parameter is known as **unbiased estimator** and, mathematically, is given by

$$E\{\hat{q}\} = q \qquad a < q < b \tag{7.26}$$

where (a, b) denotes the range of possible values of q (see Exercise 7.4.2.1).

Exercise 7.4.2.1

Let $v(n)$ be a white Gaussian noise (WGN) with zero mean and equal variance. If the observation is $x(n) = c + v(n)$ for $n = 0, 1,..., N - 1$, find the estimate of the parameter c and verify if it is biased or unbiased.   ▲

A biased estimator is given mathematically by the relation (see Exercise 7.4.2.2):

$$E\{\hat{q}\} = q + b(q) \tag{7.27}$$

where the bias of the estimator is given by

$$b(q) = E\{\hat{q}\} - q \tag{7.28}$$

Exercise 7.4.2.2

Show that $\hat{c} = \dfrac{1}{4N} \sum_{n=0}^{N-1} x(n)$, with $x(n) = c + v(n)$ is a biased estimator (see also Exercise 7.4.2.1).   ▲

7.4.3 Cramer–Rao Lower Bound (CRLB)

It is helpful to place a lower bound on the variance of any unbiased estimator. The Cramer–Rao lower bound (CRLB) is the appropriate measure. It assures us if the estimator is the minimum variance unbiased estimator (MVUE). It also provide us with a benchmark to compare the performance of the estimator.

Theorem 7.4.31 (CRLB) It is assumed that the PDF, $p(X:q)$ (the capital letter designates RV), satisfies the regularity condition:

$$E\left\{\frac{\partial \ln p(X;q)}{\partial q}\right\} = 0 \quad \text{for all } q \tag{7.29}$$

where the expectation is taken with respect to $p(X;q)$. Then, the variance of any unbiased estimator $\hat{q}$ must satisfy

$$\text{var}\{\hat{q}\} \geq \frac{1}{-E\left\{\dfrac{\partial^2 \ln p(X;q)}{\partial q^2}\right\}} \tag{7.30}$$

and

$$E\left\{\frac{\partial^2 \ln p(X;q)}{\partial q^2}\right\} = \int \frac{\partial^2 \ln p(x;q)}{\partial q^2} p(x;q)dx \tag{7.31}$$

where the derivative is evaluated at the true value q and the expectation is taken with respect to $p(X;q)$. Furthermore, an unbiased estimator may be found that contains the bound for all q if and only if

$$\frac{\partial \ln p(X;q)}{\partial q} = I(q)(g(x) - q) \tag{7.32}$$

for some function of g and I. That estimator, which is the MVUE, is $\hat{q} = g(x)$. And the minimum variance (MV) is $1/I(q)$. ∎

Example 7.4.3.1

Let the observed data be

$$x(0) = A + v(0) \tag{7.33}$$

where $v(0) \sim N(0, \sigma^2)$, and it is desired to estimate the constant A and to find the CRLB for A.

Solution: Because $E\{x(0)\} = E\{A\} + E\{v(0)\} = A + 0 = A$. This implies that the estimate is $\hat{A} = x(0)$. From Equation 7.33, we obtain the relation $x(0) - A = v(0)$. Therefore, the PDF of $x(0)$ is

$$p(x(0); A) = \frac{1}{\sqrt{2\pi\sigma^2}} e^{-\frac{1}{2\sigma^2}(x(0)-A)^2} \tag{7.34}$$

By plotting Equation 7.34 for, say, $\sigma = 3.5$ and $\sigma = 1/4$, we will observe that the PDF curve in the second variance will produce a sharper curve, which indicates that the estimator accuracy improves with the decrease of the variance of the additive noise.

When the PDF is viewed as a function of the unknown fixed parameter x, it is termed the **likelihood function**. Taking the natural logarithm of Equation 7.34, we obtain

$$\ln p(x(0); A) = -\ln \sqrt{2\pi\sigma^2} - \frac{1}{2\sigma^2}(x(0) - A)^2 \tag{7.35}$$

Then, the first derivative of the preceding equation is

$$\frac{\partial \ln p(x(0); A)}{\partial A} = \underbrace{\frac{1}{\sigma^2}}_{I(q)} \underbrace{(x(0) - A)}_{(g(x)-q)} \tag{7.36}$$

The negative of the second derivative becomes

$$-\frac{\partial^2 \ln p(x(0); A)}{\partial A^2} = \frac{1}{\sigma^2} \tag{7.37}$$

Since it was shown previously that the estimator $\hat{A} = x(0)$ has variance σ^2, for this case

$$\text{var}(\hat{A}) \geq \frac{1}{-\dfrac{\partial^2 \ln p(x(0); A)}{\partial A^2}} = \sigma^2 \quad \text{for all } A \tag{7.38}$$

∎

Example 7.4.3.2

We consider the multiple observations

$$x(n) = A + v(n) \quad n = 0, 1, 2, \ldots, N-1 \tag{7.39}$$

where $v(n)$ is WGN with variance σ^2.

Solution: To determine the CRLB, we start with the PDF:

$$p(\mathbf{x}; A) = p(x(0); A)p(x(1); A) \cdots p(x(N-1); A) = \frac{1}{\sqrt{2\pi\sigma^2}} e^{-\frac{1}{2\sigma^2}(x(0)-A)^2}$$

$$\frac{1}{\sqrt{2\pi\sigma^2}} e^{-\frac{1}{2\sigma^2}(x(1)-A)^2} \cdots \frac{1}{\sqrt{2\pi\sigma^2}} e^{-\frac{1}{2\sigma^2}(x(N-1)-A)^2} \tag{7.40}$$

$$p(\mathbf{x}; A) = \prod_{n=0}^{N-1} \frac{1}{\sqrt{2\pi\sigma^2}} e^{-\frac{1}{2\sigma^2}(x(n)-A)^2} = \frac{1}{(2\pi\sigma^2)^{N/2}} e^{-\frac{1}{2\sigma^2}\sum_{n=0}^{N-1}(x(n)-A)^2}$$

Taking the first derivative of the preceding equation, we obtain

$$\frac{\partial \ln p(\mathbf{x}; A)}{\partial A} = \frac{\partial}{\partial A}\left[-\ln\left[(2\pi\sigma^2)^{N/2}\right] - \frac{1}{2\sigma^2} \sum_{n=0}^{N-1}(x(n)-A)^2 \right]$$

$$= \frac{1}{\sigma^2} \sum_{n=0}^{N-1}(x(n)-A) = \frac{N}{\sigma^2}(m_x - A) \quad m_x = \frac{1}{N}\sum_{n=0}^{N-1} x(n) \tag{7.41}$$

where m_x is the sample mean. Differentiating the preceding equation once again, we find

$$\frac{\partial^2 \ln p(\mathbf{x}; A)}{\partial A^2} = -\frac{N}{\sigma^2} \tag{7.42}$$

From Equation 7.30, we find the CRLB to be

$$\text{var}(\hat{A}) \geq \frac{\sigma^2}{N} \tag{7.43}$$

It is apparent that the longer the data, the smaller the variance becomes and, thus, the more accurate the constant A is found. ∎

Exercise 7.4.3.1

Let us assume that a deterministic signal with an unknown parameter $q (=f_0)$ is observed in a WGN $v(n)$, as $x(n) = s(n; f_0) + v(n) = A\cos(2\pi f_0 n + \phi) + v(n) \quad 0 < f_0 < 1/2$, ϕ and A are known. Find the CRLB. ▲

Exercise 7.4.3.2

Find the CRLB, using Exercise 7.4.3.1, if we have the following data: $f_0 = 0.15, n = 0:8, \sigma = 3, A = 0.1$. ▲

7.4.4 Mean Square Error Criterion

When we seek the minimum mean square estimator, we can use a simplified form of the estimator where the statistic is a linear combination of the random data set $\{x(0), x(1), \ldots, x(N-1)\}$. Hence, we need to determine $a_0, a_1, \ldots, a_{N-1}$ such that

$$\text{mse}(\hat{q}) = E\{(q-\hat{q})^2\} = E\left\{[q - a_0 x(0) + a_1 x(1) + \cdots a_{N-1} x(N-1)]^2\right\} \tag{7.44}$$

is minimized. Since expectation is linear operation, the preceding equation becomes

$$\text{mse}(\hat{q}) = E\left\{q^2\right\} - 2\sum_{n=0}^{N-1} a_n E\{qx(n)\} + \sum_{m=0}^{N-1}\sum_{n=0}^{N-1} a_m a_n E\{x(m)x(n)\} \tag{7.45}$$

The preceding equation is quadratic in a_ns. Setting the derivatives equal to 0, we obtain the following set of equations (see Exercise 7.4.4.1):

$$r_{00}a_0 + r_{01}a_1 + \cdots + r_{0N-1}a_{N-1} = r(q, x(0)) \triangleq r_{qx}(0)$$
$$r_{10}a_0 + r_{11}a_1 + \cdots + r_{1N-1}a_{N-1} = r(q, x(1)) \triangleq r_{qx}(1)$$
$$\vdots$$
$$r_{N-1,0}a_0 + r_{N-1,1}a_1 + \cdots + r_{N-1,N-1}a_{N-1} = r(q, x(N-1)) \triangleq r_{qx}(N-1)$$

or

$$\boldsymbol{Ra = r_{\theta x}}$$
$$r_{mn} = E\{x(m)x(n)\} \tag{7.46}$$
$$r_{qx} = E\{qx(m)\}$$

From Equation 7.45, we can also proceed as follows:

$$\boldsymbol{mse(\hat{q})} = E\{[(\hat{q}-E\{\hat{q}\})+(E\{\hat{q}\}-q)]^2\} = E\{[(\hat{q}-E\{\hat{q}\}]^2\}+E\{[E\{\hat{q}\}-q]^2\}$$
$$+2E\{[\hat{q}-E\{\hat{q}\}][E\{\hat{q}\}-q]\} = \text{var}(\hat{q})+[E\{\hat{q}\}-q]^2 = \text{var}(\hat{q})+b^2(q) \tag{7.47}$$

which shows that the mean square error (MSE) is the sum of the error due to the variance of the estimator as well as its bias. Constraining the bias to zero, we can find the estimator, which minimizes its variance. Such an estimator is known as the **minimum variance unbiased estimator** (MVUE).

Exercise 7.4.4.1

Verify Equation 7.46.   ▲

7.4.5 Maximum Likelihood Estimator

Very often, MVUEs may be difficult or impossible to determine. For this reason, in practice, many estimators are found using the **maximum likelihood estimation** (MLE) principle. Besides being easy to implement, its performance is optimal for large number of data. The basic idea is to find a statistic

$$\hat{q} = g(x(0), x(1), \ldots, x(N-1)) \tag{7.48}$$

so that if the random variables $x(m)$'s take the observed experimental values $x(m)$'s, the number $\hat{q} = g(x(0), x(1), \ldots, x(N-1))$ will be a good estimate of q. Here, we use the same lower case letter for both the RV and its value. The reader should noticed that we have specifically defined their identity so that no confusion exists.

Definition 7.4.5.1 Let $x = [x(0), x(1), \ldots, x(N-1)]^T$ be a random vector with density function:

$$p(x(0), x(1), \ldots, x(N-1); q) \quad q \in Q$$

The function

$$l(q; x(0), x(1), \ldots, x(N-1)) = p(x(0), x(1), \ldots, x(N-1); q) \tag{7.49}$$

is considered as a function of the parameter $q = [q_0, q_1, \ldots, q_{N-1}]^T$ and is called the **likelihood function**. ∎

The random variables $x(0), x(1), \ldots, x(N-1)$ are IID with a density function $p(X; q)$; the likelihood function is

$$L(q; x(0), x(1), \ldots, x(N-1)) = \prod_{n=0}^{N-1} p(x(n); q) \tag{7.50}$$

Example 7.4.5.1

Let $x(0), x(1), \ldots, x(N-1)$ be random sample from the normal distribution $N(q, 1)$, $-\infty < q < \infty$ ($q \triangleq m_x$ = mean of the population) and one stands for the variance. Using Equation 7.50, we write

$$L(q; x(0), x(1), \ldots, x(N-1)) = \frac{1}{\sqrt{2\pi}} \exp\left(-(x(0)-q)^2/2\right) \frac{1}{\sqrt{2\pi}} \exp\left(-(x(1)-q)^2/2\right) \cdots$$

$$\frac{1}{\sqrt{2\pi}} \exp\left(-(x(N-1)-q)^2/2\right) = \left(\frac{1}{\sqrt{2\pi}}\right)^N \exp\left(-\sum_{n=0}^{N-1}(x(n)-q)^2/2\right) \tag{7.51}$$

Since the likelihood function $L(q)$ and its logarithm, $\ln\{L(q)\}$, are maximized for the same value of the parameter q, we can use either $L(q)$ or $\ln\{L(q)\}$. Therefore, we have

$$\frac{\partial \ln\{L(q; x(0), x(1), \ldots, x(N-1))\}}{\partial q} = \frac{\partial}{\partial q}\left(N \ln\{1/\sqrt{2\pi}\} - \sum_{n=0}^{N-1}(x(n)-q)^2/2\right) \tag{7.52}$$

$$= -\sum_{n=0}^{N-1}(x(n)-q) = -\sum_{n=0}^{N-1}x(n) + Nq = 0$$

The solution of Equation 7.52 for q is

$$\hat{q} = g(x(0), x(1), \ldots, x(N-1)) = \frac{1}{N}\sum_{n=0}^{N-1}x(n) \tag{7.53}$$

This equation shows that the estimator maximizes $L(q)$. Therefore, the statistic $g(.)$ (the sample mean value of the data) is the maximum likelihood estimator of the mean, $y = (x - 2.5)/2$. Since $E\{\hat{q}\} = \dfrac{1}{N}\displaystyle\sum_{n=0}^{N-1} E\{x(n)\} = \dfrac{Nq}{N} = q$, the estimator is an **unbiased** one. ∎

Definition 7.4.5.2 If we choose a function $g(x) = g(x(0), x(1), \ldots, x(N-1))$ such that q is replaced by $g(x)$, the likelihood function L is maximum; that is, $L(g(x); x(0), \ldots, x(N-1))$ is at least as great as $L(q; x(0), x(1), \ldots, x(N-1))$ for all $q \in Q$, or in mathematical form

$$L(\hat{q}; x(0), x(1), \ldots, x(N-1)) = \sup_{q \in Q} L(q; x(0), x(1), \ldots, x(N-1)) \qquad (7.54)$$ ∎

Definition 7.4.5.3 Any statistic whose expectation is equal to a parameter q is called an unbiased estimator of the parameter q. Otherwise, the statistic is said to be biased.

$$E\{\hat{q}\} = q \quad \text{unbiased estimator} \qquad (7.55)$$ ∎

Definition 7.4.5.4 Any statistic that converges stochastically to a parameter q is called a **consistent estimator** of that parameter. Mathematically, we write $\lim_{N \to \infty} \text{pr}\{|\hat{q} - q| > \varepsilon\} = 0$. ∎

If as $N \to \infty$, the relation

$$\hat{q} \to q$$

holds, the estimator $\hat{q}$ is said to be a **consistent estimator**. If, in addition, as $N \to \infty$, the relation

$$E\{\hat{q}\} = q$$

holds, it is said that the estimator $\hat{q}$ is said to be **asymptotically unbiased**. Furthermore, if as $N \to \infty$, the relation

$$\text{var}\{\hat{q}\} \to \text{lowest value}, \quad \text{for all } q$$

holds, it is said that $\hat{q}$ is **asymptotically efficient**.

Example 7.4.5.2

Let the observed the random data $x(n)$ be the set

$$x(n) = c + v(n) \quad n = 0,1,\ldots,N-1, c > 0 \qquad (7.56)$$

The signal $v(n)$ is WGN with zero mean and with unknown variance c, which is desired to obtain. The constant c is reflected in the mean and variance of the RV.

Solution: The PDF is (see also Equation 7.51)

$$p(x;c) = \frac{1}{(2\pi c)^{N/2}} \exp\left[-\frac{1}{2c}\sum_{n=0}^{N-1}(x(n)-c)^2\right] \qquad (7.57)$$

Considering the preceding equation as a function of c, it becomes a likelihood function $L(c;x)$. Differentiating using the natural logarithm, we obtain

$$\frac{\partial \ln[(2\pi)^{N/2} p(x;c)]}{\partial c} = \frac{\partial}{\partial c}\left[-\frac{N}{2}\ln c - \frac{1}{2c}\sum_{n=0}^{N-1}(x(n)-c)^2\right]$$

$$= -\frac{N}{2}\frac{1}{c} + \frac{1}{c}\sum_{n=0}^{N-1}(x(n)-c) + \frac{1}{2c^2}\sum_{n=0}^{N-1}(x(n)-c)^2 = 0 \tag{7.58}$$

where multiplying the PDF by a constant does not change the maximum point. From Equation 7.58, we obtain

$$\hat{c}^2 + \hat{c} - \frac{1}{N}\sum_{n=0}^{N-1}x^2(n) = 0$$

Solving for $\hat{c}$ and keeping the positive sign of the quadratic root, we find

$$\hat{c} = -\frac{1}{2} + \sqrt{\frac{1}{N}\sum_{n=0}^{N-1}x^2(n) + \frac{1}{4}} \tag{7.59}$$

Note that $\hat{c} > 0$ for all values of the summation under the square root. Since

$$E\{\hat{c}\} = E\left\{-\frac{1}{2} + \sqrt{\frac{1}{N}\sum_{n=0}^{N-1}x^2(n) + \frac{1}{4}}\right\} \neq -\frac{1}{2} + \sqrt{E\left\{\frac{1}{N}\sum_{n=0}^{N-1}x^2(n) + \frac{1}{4}\right\}} \tag{7.60}$$

it implies that the estimator is biased. From the law of the large numbers as $N \to \infty$,

$$\frac{1}{N}E\left\{\sum_{n=0}^{N-1}x^2(n)\right\} = \frac{NE\{x^2(n)\}}{N} = E\{x^2(n)\} = \text{var}\{x(n)\} + E^2\{x(n)\} = c + c^2 \tag{7.61}$$

$$\left(E\{(x-\bar{x})^2\}\right) \triangleq \text{var}(x) = E\{x^2\} - 2E\{x\}\bar{x} + \bar{x}^2 = E\{x^2\} - \bar{x}^2, \bar{x} = \text{mean value})$$

Therefore, Equation 7.59 gives

$$\left(\hat{c}+\frac{1}{2}\right)^2 = \left(c+c^2+\frac{1}{4}\right) \quad \text{or} \quad \hat{c}^2 + \frac{1}{2} + \hat{c} = c + c^2 + \frac{1}{4}$$

which indicates that

$$\bar{c} \to c \tag{7.62}$$

And, hence, the estimator is a consistent estimator. ∎

Example 7.4.5.3

It is required to find the maximum likelihood estimator for the mean m and variance σ^2 of a set of data $\{x(n)\}$ provided by a Gaussian random generator.

Solution: The Gaussian PDF pr($x;m, \sigma^2$) for one RV is

$$pr(x; m, \sigma^2) = \frac{1}{\sqrt{2\pi\sigma^2}} \exp\left[-\frac{1}{2}\left(\frac{x-m}{\sigma}\right)^2\right] \tag{7.63}$$

Its natural algorithm is

$$\ln\left(pr(x; m, \sigma^2)\right) = -\frac{1}{2}\ln(2\pi\sigma^2) - \frac{1}{2}\left(\frac{x-m}{\sigma}\right)^2 \tag{7.64}$$

The likelihood function for the data, IID, is given by

$$L(m, \sigma^2) = pr(x(0); m, \sigma^2) pr(x(1); m, \sigma^2) \cdots pr(x(N-1); m, \sigma^2) \tag{7.65}$$

and its logarithm is

$$\ln L(m, \sigma^2) = \sum_{n=0}^{N-1} \ln\left(pr(x(n); m, \sigma^2)\right) \tag{7.66}$$

Substituting Equation 7.64 in Equation 7.66, we obtain

$$\begin{aligned}
\ln L(m, \sigma^2) &= \sum_{n=0}^{N-1}\left[-\frac{1}{2}\ln(2\pi\sigma^2) - \frac{1}{2}\left(\frac{x(n)-m}{\sigma}\right)^2\right] \\
&= -\frac{N}{2}\ln(2\pi\sigma^2) - \frac{1}{2\sigma^2}\sum_{n=0}^{N-1}(x(n)-m)^2
\end{aligned} \tag{7.67}$$

There are two unknowns in the ln of the likelihood function. Differentiating Equation 7.67 with respect to the mean and variance, we find

$$\frac{\partial \ln L}{\partial m} = \frac{1}{\sigma^2}\sum_{n=0}^{N-1}(x(n)-m) \tag{7.68}$$

$$\frac{\partial \ln L}{\partial(\sigma^2)} = -\frac{N}{2}\frac{1}{\sigma^2} + \frac{1}{2\sigma^4}\sum_{n=0}^{N-1}(x(n)-m)^2$$

Equating the partial derivatives to 0, we obtain

$$\frac{1}{\hat{\sigma}^2}\sum_{n=0}^{N-1}(x(n)-\hat{m}) = 0 \tag{7.69}$$

$$-\frac{N}{2}\frac{1}{\hat{\sigma}^2} + \frac{1}{2\hat{\sigma}^4}\sum_{n=0}^{N-1}(x(n)-\hat{m})^2 = 0 \tag{7.70}$$

From Equation 7.69, we obtain the relation

$$\sum_{n=0}^{N-1}(x(n)-\hat{m}) = 0 \text{ or } \hat{m} = \frac{1}{N}\sum_{n=0}^{N-1}x(n) = \text{sample mean} \tag{7.71}$$

Therefore, the MLE of the mean of a Gaussian population is equal to the sample mean, which indicates that the sample mean is an optimal estimator.

Multiplying Equation 7.70 by $4\sigma^4$ leads to

$$-N\hat{\sigma}^2 + \sum_{n=0}^{N-1}(x(n)-\hat{m})^2 = 0 \text{ or } \hat{\sigma}^2 = \frac{1}{N}\sum_{n=0}^{N-1}(x(n)-\hat{m})^2 \tag{7.72}$$

Therefore, the MLE of the variance of a Gaussian population is equal to the sample variance. ■

Let the PDF of the population be the Laplacian (a is a positive constant):

$$\text{pr}(x;q) = \frac{a}{2}\exp(-a|x-q|) \tag{7.73}$$

Then, the likelihood function corresponding to the preceding PDF for a set of data {x(n)}, IID, is

$$L(q) = \ln\left\{\left(\frac{a}{2}\right)^N \prod_{n=0}^{N-1}\exp(-a|x(n)-q|)\right\}$$

$$= -a\sum_{n=0}^{N-1}|x(n)-q| + N\ln\{a\} - N\ln\{2\} \tag{7.74}$$

Before we proceed further, we must define the term **median** for a set of random variables. Median is the value of an RV of the set when half of the RVs of the set have values less than the median and half have higher values (odd number of terms). Hence, the mcd{1,4,2,3,5,9,8,7,11} = med{1,2,3,4,5,7,8,9,11} = 5.

Definition 7.4.5.5 Let us consider an RV whose cumulative density function (distribution) is F_x. The point x_{med} is the median of x if

$$F_x(x_{\text{med}}) = \frac{1}{2} \tag{7.75}$$

■

From Equation 7.74, we observe that the derivative of $-\sum_{n=0}^{N-1}|x(n)-q|$ with respect to q is negative if q is larger than the sample median and positive if it is less than the sample median (remember that $|x(n)-q| = [(x(n)-q)(x(n)-q)]^{1/2}$ for real values). Therefore, the estimator is (see Exercise 7.4.5)

$$\hat{q} = \text{med}\{x(0),x(1),\ldots,x(N-1)\} \tag{7.76}$$

which maximizes $L(q)$ for the Laplacian PDF likelihood function.

Exercise 7.4.5.1

Verify Equation 7.76. ▲

Note: (1) From the previous discussion, we have observed that the minimization of $L(q)$ created from a Gaussian distribution is equivalent in minimizing $\sum_{n=0}^{N-1}(x(n)-q)^2$. The minimization results in finding the estimator $\hat{q} = \frac{1}{N}\sum_{n=0}^{N-1}x(n)$, which is the sample mean of the data $x(n)$.

(2) Similarly, the minimization of the likelihood function $L(q)$ created from a Laplacian distribution is equivalent in minimizing $\sum_{n=0}^{N-1}|x(n)-q|$. This minimization results in finding the estimator $\hat{q}=\text{med}\{x(0),x(1),\ldots,x(N-1)\}$.

Exercise 7.4.5.2

Generate realizations of the RV (see Equation 7.59)

$$\hat{c}=-\frac{1}{2}+\sqrt{\frac{1}{N}\sum_{n=0}^{N-1}x^2(n)+\frac{1}{4}}$$

where $x(n)=c+v(n)$ for $v(n)$ WGN with variance $c>0$. Produce M realizations of N RVs and determine (1) the mean of $\hat{c}$, (2) the variance of $\hat{c}$, and (3) its PDF. ▲

7.5   MEAN SQUARE ERROR (MSE)

We proceed to develop a class of linear optimum discrete-time filters known as the **Wiener filters**. These filters are optimum in the sense of minimizing an appropriate function of the error, known as the **cost function**. The cost function that is commonly used in filter design optimization is the MSE. Minimizing MSE involves only second-order statistics (correlation functions) and leads to a theory of linear filtering that is useful in many practical applications.

In searching for optimal estimators, we need to adopt some optimality criterion. A natural one is the MSE, defined by

$$\text{mse}(\hat{q})=E\left\{(\hat{q}-q)^2\right\}=E\left\{(q-\hat{q})^2\right\}=E\left\{e^2\right\} \tag{7.77}$$

is minimized. Conceptually, the error is shown in Figure 7.6. This measures the average mean squared deviation of the estimator from the true value. To understand the MSE problem, we write

$$\text{mse}(\hat{q})=E\left\{\left[(\hat{q}-E\{\hat{q}\})+(E\{\hat{q}\}-q)\right]^2\right\}=E\left\{\left[\hat{q}-E\{\hat{q}\}\right]^2\right\}+E\left\{(E\{\hat{q}\}-q)^2\right\}$$

$$+E\left\{2[\hat{q}-E\{\hat{q}\}][E\{\hat{q}\}-q]\right\},\{[E\{\hat{q}\}-E\{\hat{q}\}][E\{\hat{q}\}-q]\}=0 \text{ or}$$

$$\text{mse}(\hat{q})=\text{var}(\hat{q})+(E\{\hat{q}\}-q)^2=\text{var}(\hat{q})+b^2(q) \tag{7.78}$$

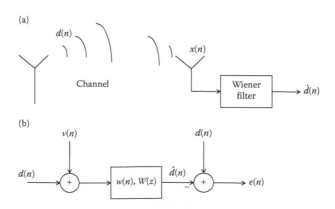

FIGURE 7.6

The preceding equation shows that the MSE is composed of errors due to the variance of the estimator as well as the bias.

Example 7.5.1

Let the estimator be constrained in the form

$$\hat{y} = ax + b$$

Find the values for a and b that minimize the MSE:

$$J = E\{(y-\hat{y})^2\} = E\{(y-ax-b)^2\} = E\{e^2\} \tag{7.79}$$

Solution: Solving the linear mean square estimation problem may be accomplished by differentiating J with respect to a and b and setting the derivatives equal to zero:

$$\frac{\partial J}{\partial a} = -2E\{(y-ax-b)x\} = -2E\{xy\} + 2aE\{x^2\} + 2bm_x = 0 \tag{7.80}$$

$$\frac{\partial J}{\partial b} = -2E\{(y-ax-b)\} = -2m_y + 2am_x + 2b = 0 \tag{7.81}$$

From Equation 7.80, we observe that

$$E\{(y-\hat{y})x\} = E\{ex\} = 0 \tag{7.82}$$

The error $e = y - \hat{y}$ is the **estimation error**. This relation is known as the **orthogonality principle** and shows that the **estimation error is orthogonal to the data** x. Figure 7.7 shows graphically the orthogonality principle.
 Solving Equations 7.80 and 7.81 for a and b, we obtain

$$a = \frac{E\{xy\} - m_x m_y}{\sigma_x^2} \tag{7.83}$$

$$b = \frac{E\{x^2\}m_y - E\{xy\}m_x}{\sigma_x^2} \tag{7.84}$$

From Equation 7.83, it follows that

$$E\{xy\} = a\sigma_x^2 + m_x m_y \tag{7.85}$$

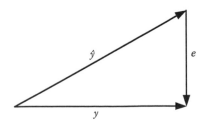

FIGURE 7.7

which, when substituted into Equation 7.84, gives the following expression for b:

$$b = \frac{E\{x^2\}m_y - \left[a\sigma_x^2 + m_x m_y\right]m_x}{\sigma_x^2} = m_y - am_x \qquad (7.86)$$

where we have used the relation $\sigma_x^2 = E\{x^2\} - m_x^2$. As a result, the estimate for y may be written as

$$\hat{y} = ax + (m_y - am_x) = a(x - m_x) + m_y \qquad (7.87)$$

where

$$a = \frac{E\{xy\} - m_x m_y}{\sigma_x^2} = \frac{\rho_{xy}\sigma_x\sigma_y}{\sigma_x^2} = \rho_{xy}\frac{\sigma_y}{\sigma_x} \qquad (7.88)$$

Substituting Equation 7.88 into Equation 7.87, we obtain the optimum linear estimator of y, which is

$$\hat{y} = \rho_{xy}\frac{\sigma_y}{\sigma_x}(x - m_x) + m_y, \quad \rho_{xy} = \text{correlation coefficient} = \frac{r_x r_y}{\sigma_x \sigma_y} \qquad (7.89)$$

∎

This approach is common to all optimum filter designs. Figure 7.6 shows the block diagram of the optimum filter problem. The basic idea is to recover a desired signal $d(n)$ given a noisy observation $x(n) = d(n) + v(n)$, where both $d(n)$ and $v(n)$ are assumed to be wide-sense stationary processes. Therefore, the problem can be stated as follows: Design a filter that produces an estimate $\hat{d}(n)$ using a linear combination of the data $x(n)$ such that the MSE function (cost function)

$$J = E\{(d(n) - \hat{d}(n))^2\} = E\{e^2(n)\} \qquad (7.90)$$

is minimized.

Depending on how the data $x(n)$ and the desired signal $d(n)$ are related, there are four basic problems that need to be solved. These are filtering, smoothing, prediction, and deconvolution.

Exercise 7.5.1

Let the desired signal be a constant A and the received signal be the set $\{x(n)\}$, $n = 0,..., N - 1$. Estimate the constant A. ▲

7.6   FINITE IMPULSE RESPONSE (FIR) WIENER FILTER

Let the sample response (filter coefficients) of the desired filter be denoted by the vector $\mathbf{w}$. This filter will process the real-valued stationary process $\{x(n)\}$ to produce an estimate $\hat{d}(n)$ of the desired real-valued signal $d(n)$. Without loss of generality we will assume, unless otherwise stated, that the process $\{x(n)\}$, $\{d(n)\}$ and so on have zero-mean values. Furthermore, assuming that the filter coefficients do not change with time, the output of the filter is equal to the convolution of the input signal $\{x(n)\}$ and the filter coefficients $\mathbf{w}$s. Hence, we obtain

$$\hat{d}(n) = \{w(n)\} * \{x(n)\} = \sum_{m=0}^{M-1} w(m)x(n-m) = \mathbf{w}^T \mathbf{x}(n) = \mathbf{x}^T(n)\mathbf{w} \qquad (7.91)$$

where

$$\mathbf{w} = \begin{bmatrix} w(0) & w(1) & w(2) & \cdots & w(M-1)]^T \end{bmatrix}$$

$$\mathbf{x}(n) = [x(n) \quad x(n-1) \quad \cdots \quad x(n-M)]^T$$

The MSE is given by (see Equation 7.77):

$$J(\mathbf{w}) = E\{e^2(n)\} = E\{[d(n) - \hat{d}(n)]^2\} = E\{[d(n) - \mathbf{w}^T\mathbf{x}(n)]^2\}$$

$$= E\{[d(n) - \mathbf{w}^T\mathbf{x}(n)][d(n) - \mathbf{w}^T\mathbf{x}(n)]^T\}$$

$$= E\{d^2(n) - \mathbf{w}^T\mathbf{x}(n)d(n) - d(n)\mathbf{x}^T(n)\mathbf{w} + \mathbf{w}^T\mathbf{x}(n)\mathbf{x}^T(n)\mathbf{w}\}$$

$$= E\{d^2(n)\} - 2\mathbf{w}^T E\{d(n)\mathbf{x}(n)\} + \mathbf{w}^T E\{\mathbf{x}(n)\mathbf{x}^T(n)\}\mathbf{w}$$

$$= \sigma_d^2 - 2\mathbf{w}^T \mathbf{p}_{dx} + \mathbf{w}^T \mathbf{R}_x \mathbf{w} \tag{7.92}$$

where

$$\mathbf{w}^T\mathbf{x}(n) = \mathbf{x}^T(n)\mathbf{w} = \text{number}$$

$\sigma_d^2 = $ variance of the desired signal, $d(n)$

$$\mathbf{p}_{dx} = \begin{bmatrix} p_{dx}(0) & p_{dx}(1) & \cdots & p_{dx}(M-1) \end{bmatrix}^T = \text{cross-correlation vector}$$

$$p_{dx}(0) \triangleq r_{dx}(0), p_{dx}(1) \triangleq r_{dx}(1), \ldots, p_{dx}(M-1) \triangleq r_{dx}(M-1)(n) = \mathbf{x}^T(n)\mathbf{w} = \text{number}$$

$\sigma_d^2 = $ variance of the desired signal, $d(n)$

$$\mathbf{p}_{dx} = \begin{bmatrix} p_{dx}(0) & p_{dx}(1) & \cdots & p_{dx}(M-1) \end{bmatrix}^T = \text{cross-correlation vector}$$

$$p_{dx}(0) \triangleq r_{dx}(0), p_{dx}(1) \triangleq r_{dx}(1), \ldots, p_{dx}(M-1) \triangleq r_{dx}(M-1)$$

The autocorrelation matrix is

$$\mathbf{R}_x = E\left\{ \begin{bmatrix} x(n) \\ x(n-1) \\ \vdots \\ x(n-M+1) \end{bmatrix} [x(n) \ x(n-1) \ \cdots \ x(n-M+1)] \right\}$$

$$= \begin{bmatrix} E\{x(n)x(n)\} & E\{x(n)x(n-1)\} & \cdots & E\{x(n)x(n-M+1)\} \\ E\{x(n-1)x(n)\} & E\{x(n-1)x(n-1)\} & \cdots & E\{x(n-1)x(n-M+1)\} \\ \vdots & & & \vdots \\ E\{x(n-M+1)x(n)\} & E\{x(n-M+1)x(n-1)\} & \cdots & E(x(n-M+1)x(n-M+1)\} \end{bmatrix}$$

$$= \begin{bmatrix} r_x(0) & r_x(1) & \cdots & r_x(M-1) \\ r_x(-1) & r_x(0) & \cdots & r_x(M-2) \\ \vdots & & \vdots & \\ r_x(-M+1) & r_x(-M+2) & \cdots & r_x(0) \end{bmatrix} \tag{7.93}$$

For simplicity, we have set $r_{xx}(i) = r_x(i)$ and we should have in mind these two identical representations.

The preceding matrix is the autocorrelation matrix of the input data to the filter. The matrix is symmetric because the random process is assumed to be stationary and, hence, we have the equality, $r_x(k) = r_x(-k)$. Since in practical cases we have only one realization, we will assume that the signal is ergodic. Therefore, we will use the sample autocorrelation function, which is an estimate. However, in this text, we will not differentiate the estimate with an over-bar since all of our simulations will be based only on estimate quantities and not on ensemble averages.

Example 7.6.1

Let's assume that we have found the sample autocorrelation coefficients ($r_x(0) = 1.0$, $r_x(1) = 0$) from given data $x(n)$, which, in addition to noise, contain the desired signal. Furthermore, let the variance of the desired signal $\sigma_d^2 = 24$ and the cross-correlation vector be $\boldsymbol{p}_{dx} = [2, 4.5]^T$. We want to find the surface defined by the mean-square function $J(\boldsymbol{w})$.

Solution: Introducing the preceding values in Equation 7.92, we obtain

$$J(\boldsymbol{w}) = 24 - 2\begin{bmatrix} w_0 & w_1 \end{bmatrix}\begin{bmatrix} 2 \\ 4.5 \end{bmatrix} + \begin{bmatrix} w_0 & w_1 \end{bmatrix}\begin{bmatrix} 1 & 0 \\ 0 & 1 \end{bmatrix}\begin{bmatrix} w_0 \\ w_1 \end{bmatrix} = 24 - 4w_0 - 9w_1 + w_0^2 + w_1^2$$

Note that the equation is quadratic with respect to filter coefficients and it is true for any number of filter coefficients. In the general case, products $w_i w_j$ will also be present. The data are the sum of the desired signal and noise. From the data, we find the correlation matrix and the cross-correlation between the desired signal and the data. Note that to find the optimum Wiener filter coefficients, the desired signal is needed. Figure 7.8 shows the **mean-square error surface**. This surface is found by inserting different values of w_0 and w_1 in the function $J(\boldsymbol{w})$. The values of the coefficients that correspond to the bottom of the surface are the **optimum** Wiener coefficients. The vertical distance from the $w_0 - w_1$ plane to the bottom of the surface is known as the **minimum error**, J_{min}, and corresponds to the optimum Wiener coefficients. We observe that the minimum height of the surface corresponds to about $w_0 = 2$ and $w_1 = 4.5$, which are the optimum coefficients as we will learn how to find them in the next section. The following Book m-File was used:

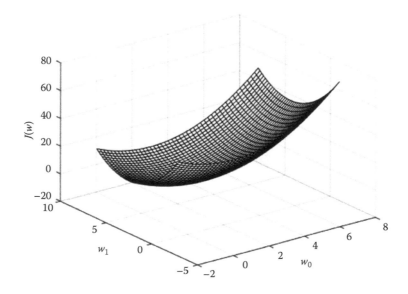

FIGURE 7.8

BOOK M-FILE: EX7_6_1

```
%Book m-file:ex7_6_1
w0=-1:0.2:8;
w1=w0;
[W0,W1]=meshgrid(w0,w1);%meshgrid(.) MATLAB
    %function that produces a pair
    %of matrices representing a rectangular
    %grid;
j=24-4*W0-9*W1+W0.^2+W1.^2;
colormap(gray)% a MATLAB function producing a gray figure;
mesh(w0,w1,j);% a MATLAB function producing  a three dimensional
    %graph;
xlabel('w_0');ylabel('w_1');zlabel('J(w)');
```
■

Figure 7.9 shows a schematic representation of the Wiener filter and Figure 7.10 shows an adaptive finite impulse response (FIR) filter.

Example 7.6.2

The input signal to a FIR filter, with filter coefficients w = [0.37905, 1.02546, 1.71687, 1.19739], is given by $x(n) = \sin(0.1\pi n) + 0.1\text{randn}(n)$. Find the MSE.

Solution: The following Book m-File finds the MSE and plots the filter.

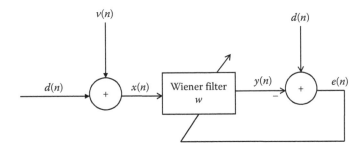

FIGURE 7.9

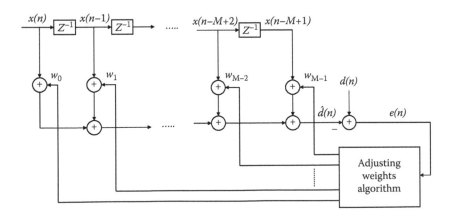

FIGURE 7.10

BOOK MATLAB FUNCTION: EX7_6_2

```
%Book m-File:ex7_6_2
n=0:3;
w=[0.37905 1.02546 1.71687 1.19739];%low pass filter;
x=sin(0.2*pi*n)+0.1*randn(1,4);
rx1=xcorr(x);
Tm=toeplitz(rx1(1,4:7));
dap=conv(w,x);
d=sin(0.2*pi*n);
pdx=xcorr(x,d);
J=var(d(1,1:4))-2*w*pdx(1,4:7)'+w*Tm*w';
wf=0:0.01:pi;
sw=0.37905*exp(-j*wf)+1.02546*exp(-j*2*wf)...
    +1.71687*exp(-j*3*wf)+1.19739*exp(-j*4*wf);
plot(wf,abs(sw));
```

Observe that we use the number of signal values equal to the filter number of coefficients. ∎

Example 7.6.3

Assume that the desired signal is represented by the relation $d(n) = qh(n)$, where $h(n)$ is a known sequence. If the received signal is $x(n)$, find the constant q that minimizes in the least-squares sense.

Solution: The least-squares error criterion becomes

$$J(q) = \sum_{n=0}^{N-1} [x(n) - qh(n)]^2 \qquad (7.94)$$

Taking the derivative with respect to q and equating the result to zero, we obtain the estimate

$$\hat{q} = \frac{\displaystyle\sum_{n=0}^{N-1} x(n)h(n)}{\displaystyle\sum_{n=0}^{N-1} h^2(n)} \qquad (7.95)$$

∎

7.7 WIENER SOLUTION—ORTHOGONAL PRINCIPLE

From Figure 7.8, we observe that there exists a plane touching the parabolic surface at its minimum point and is parallel to the w-plane. Furthermore, we observe that the surface is concave upwards and, therefore from calculus, the first derivatives of the MSE with respect to w_0 and w_1 must be zero at the minimum point and the second derivative must be positive. Hence, we write

$$\frac{\partial J(w_0, w_1)}{\partial w_0} = 0 \qquad \frac{\partial J(w_0, w_1)}{\partial w_1} = 0 \qquad (7.96a)$$

$$\frac{\partial^2 J(w_0, w_1)}{\partial w_0^2} > 0 \qquad \frac{\partial^2 J(w_0, w_1)}{\partial w_1^2} > 0 \qquad (7.96b)$$

For the two-coefficient filter, Equation 7.92 becomes

$$J(w_0, w_1) = w_0^2 r_{xx}(0) + 2w_0 w_1 r_{xx}(1) + w_1^2 r_{xx}(0) - 2w_0 r_{dx}(0) - 2w_1 r_{dx}(1) + \sigma_d^2 \qquad (7.97)$$

Introducing Equation 7.97 in Equation 7.96a produces the following set of equations:

$$2w_0^o r_{xx}(0) + 2w_1^o r_{xx}(1) - 2r_{dx}(0) = 0 \qquad (7.98a)$$

$$2w_1^o r_{xx}(0) + 2w_0^o r_{xx}(1) - 2r_{dx}(1) = 0 \qquad (7.98b)$$

The preceding system of equations can be written in the form known as the discrete form of the Wiener–Hopf equation, which is

$$\begin{bmatrix} r_{xx}(0) & r_{xx}(1) \\ r_{xx}(1) & r_{xx}(0) \end{bmatrix} \begin{bmatrix} w_0^o \\ w_1^o \end{bmatrix} = \begin{bmatrix} r_{dx}(0) \\ r_{dx}(1) \end{bmatrix}$$

or in matrix form

$$\boldsymbol{R}_{xx}\boldsymbol{w}^o = \boldsymbol{p}_{dx} \qquad (7.99)$$

The superscript "*o*" indicates the optimum Wiener solution for the filter. Note that to find the correlation matrix, we must know the second-order statistics (autocorrelation). If, in addition, the matrix is invertible, the optimum filter is given by

$$\boldsymbol{w}^o = \boldsymbol{R}_{xx}^{-1}\boldsymbol{p}_{dx} \qquad (7.100)$$

It turns out that in most practical signal processing applications, the matrix $\boldsymbol{R}_{xx}$ is invertible. For an M-order filter, $\boldsymbol{R}_{xx}$ is a $M \times M$ matrix, $\boldsymbol{w}^o$ is a $M \times 1$ vector, and $\boldsymbol{p}$ is an $M \times 1$ vector.

If we differentiate once more (Equation 7.98) with respect to w_0^o and w_1^o, which is equivalent in differentiating $J(\boldsymbol{w})$ twice, we find that it is equal to $2r_{xx}(0)$. But $r_{xx}(0) = E\{x(m)x(m)\} = \sigma_x^2 > 0$, the surface is concave upwards. Therefore, the extremum is the minimum point of the surface. If we, next, introduce Equation 7.100 in Equation 7.92, we obtain the minimum error in the mean-$\boldsymbol{w}^o$ square sense (see Exercise 7.7.1):

$$J_{\min} = \sigma_d^2 - \boldsymbol{p}_{dx}^T \boldsymbol{w}^o = \sigma_d^2 - \boldsymbol{p}_{dx}^T \boldsymbol{R}_{xx}^{-1} \boldsymbol{p}_{dx} \qquad (7.101)$$

which indicates that the minimum point of the error surface is at a distance $J_{\min}$ above the $\boldsymbol{w}$-plane. The preceding equation shows that if no correlation exists between the desired signal and the data, or equivalently $\boldsymbol{p}_{dx} = 0$, the error is equal to the variance of the desired signal.

The problem we are facing is how to choose the length M of the filter. In the absence of **a priori** information, we compute the optimum coefficients, starting from a small reasonable number. As we increase the number, we check the MMSE and if its value is small enough, for example, MMSE < 0.01, we accept the corresponding number of the coefficients.

Exercise 7.7.1

Obtain the minimum error in the mean square sense.   ▲

Exercise 7.7.2

Let $\boldsymbol{R} = \begin{bmatrix} 1.1 & 0.5 \; ; & 0.5 & 1.1 \end{bmatrix}$, $\boldsymbol{p} = \begin{bmatrix} 0.53 & -0.45 \end{bmatrix}^T$, and $\sigma_d^2 = 0.95$, find J, J_m, and $\boldsymbol{w}^o$.   ◼

Example 7.7.1

We would like to find the optimum filter coefficients w_0 and w_1 of the Wiener filter, which approximates (models) the unknown system with coefficients $b_0 = 1$ and $b_1 = 0.38$ (see Figure 7.11).

Solution: The following Book m-file was used:

```
%Book m-file: ex7_7_1
v=0.5*(rand(1,20)-0.5);%v=noise vector(20 uniformly distributed rv's
    %with mean zero);
x=randn(1,20);% x=data vector entering the system and the Wiener
    %filter (20 normal distributed rv's with mean zero;
sysout=filter([1 0.38],1,x);% sysout=system output with x as input;
    %filter(b,a,x) is a
    %MATLAB function, where b is the vector of the
    %coefficients of the ARMA numerator,
    % a is the vector of the coefficients of the
    %ARMA denominator;
dn=sysout+v;
rx1=xcorr(x,2,'biased');
rx=rx1(1,3:4);
Rx=toeplitz(rx);%toeplitz() is a MATLAB function that
               % gives the symmetric
               % autocorrelation matrix;
pdx=xcorr(x,dn,2,'biased');%xcorr() a MATLAB function that
               %gives a symmetric biased crosscorrelation;
p=pdx(1,2:3);
w=inv(Rx)*p';
dnsig1=xcorr(dn,1,'biased');
dnsig=dnsig1(2);
jmin=dnsig-p*w;
```

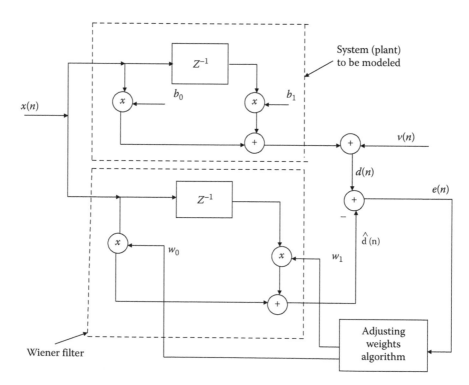

FIGURE 7.11

From the workspace, we read the following outputs from a typical run; these values change from run to run:

$$R_{xx} = \begin{bmatrix} 1.6084 & 0.4470 \\ 0.4470 & 1.6084 \end{bmatrix}, \quad p = \begin{bmatrix} 0.9642 & 1.7385 \end{bmatrix}$$

$$w = \begin{bmatrix} 0.3241 & 0.9908 \end{bmatrix}, \quad J_{min} = 0.0291 \qquad \blacksquare$$

The reader should run this m-file with different strengths of noise and observe the results, specifically, the minimum error.

7.7.1 ORTHOGONALITY CONDITION

In order for the set of filter coefficients to minimize the cost function $J(w)$, it is necessary and sufficient that the derivatives of $J(w)$ with respect to w_k be equal to zero for $k = 0, 1, 2,..., M - 1$ (see Equation 7.92):

$$\frac{\partial J}{\partial w_k} = \frac{\partial}{\partial w_k} E\{e(n)e(n)\} = 2E\left\{e(n)\frac{\partial e(n)}{\partial w_k}\right\} = 0 \qquad (7.102)$$

But

$$e(n) = d(n) - \sum_{m=0}^{M-1} w_m x(n-m) \qquad (7.103)$$

and, hence, it follows (note that the derivatives is for one variable w_k only and, therefore, the rest of the factors in the summation become zero besides one):

$$\frac{\partial e(n)}{\partial w_k} = -x(n-k) \qquad (7.104)$$

Therefore, Equation 7.102 becomes

$$E\{e^o(n)x(n-k)\} = 0 \quad k = 0,1,2,...,M-1 \qquad (7.105)$$

where the superscript "o" denotes that the corresponding w_k's used to find the estimation error $e^o(n)$ are the optimal ones. Figure 7.12 illustrates the orthogonality principle, where the error $e^o(n)$ is orthogonal (perpendicular) to the data set $\{x(n)\}$ when the estimator employs the optimum set of filter coefficients.

7.8 WIENER FILTERING EXAMPLES

The following examples elucidate the use of the Wiener filter.

Example 7.8.1 (Filtering)

Filtering of noisy signals (noise reduction) is extremely important and the method has been used in many applications such as speech in a noisy environment, reception of data across a noisy

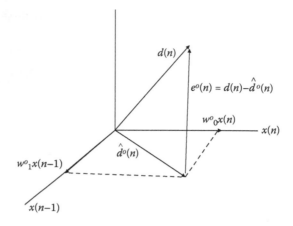

FIGURE 7.12

channel, enhancement of images, and so on. Let the received signal be $x(n) = d(n) + v(n)$, where $v(n)$ is IID noise with zero mean, variance σ_v^2, and it is uncorrelated with the desired signal, $d(n)$. Hence,

$$p_{dx}(m) = E\{d(n)x(n-m)\} = E\{d(n)d(n-m)\} + E\{d(n)\}E\{v(n-m)\}$$

$$= r_{dd}(m) + E\{d(n)\}0 = r_{dd}(m) \tag{7.106}$$

Similarly, we obtain

$$r_{xx}(m) = E\{x(n)x(n-m)\} = r_{dd}(m) + r_{vv}(m) \tag{7.107}$$

where we used the assumption that $d(n)$ and $v(n)$ are uncorrelated and $v(n)$ has zero-mean value. Therefore, the Wiener–Hopf equation (7.99) becomes

$$(\boldsymbol{R}_{dd} + \boldsymbol{R}_{vv})\boldsymbol{w}^o = \boldsymbol{p}_{dx} \tag{7.108}$$

The following Book m-file was used to produce the results shown in Figure 7.13.

BOOK M-FILE: EX7_8_1

```
%m-file:ex7_8_1
n=0:511;
d=sin(.2*pi*n);%desired signal
v=0.5*randn(1,512);%white Gaussian noise;
x=d+v;  %input signal to Wiener filter;
rd1=xcorr(d,20,'biased');
rd=rd1(1,20+1:39);      %rdx=rd=biased autocorrelation
                        %function of the desired signal;
rv1=xcorr(v,20,'biased');
rv=rv1(1,20+1:39);%rv=biased autoc. function of the noise;
R=toeplitz(rd(1,1:15))+toeplitz(rv(1,1:15));
pdx=rd(1,1:15);
w=inv(R)*pdx';
y=filter(w',1,x);%output of the filter;
plot(x(1,1:100));hold on;plot(y(1,1:100),'g');
```

But

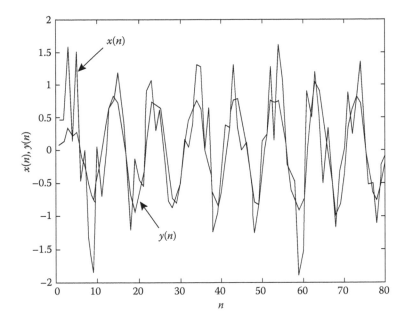

FIGURE 7.13

$$\sigma_x^2 = \sigma_d^2 + \sigma_v^2, \sigma_x^2 = r_{xx}(0), \sigma_d^2 = r_{dd}(0), \sigma_v^2 = r_{vv}(0) \tag{7.109}$$

and, hence, from the MATLAB workspace, we obtain var(d) = var(x) − var(v) = 0.4686 and $J_{\min} = 0.4686 - p_{dx}w^o = 0.0166$. ▲

Example 7.8.2 (Filtering)

Say you want to find a two-coefficient Wiener filter for the communication channel shown in Figure 7.14. Let $v_1(n)$ and $v_2(n)$ be white noses with zero mean, uncorrelated with each other and with $d(n)$, and have the following variances: $\sigma_{v1}^2 = 0.31$, $\sigma_{v2}^2 = 0.12$. The desired signal produced by the first filter shown in Figure 7.14 is

$$d(n) = -0.796d(n-1) + v_1(n) \tag{7.110}$$

Therefore, the autocorrelation function of the desired signal becomes

$$E\{d(n)d(n)\} = 0.796^2 E\left\{d^2(n-1)\right\} - 2 \times 0.796 E\left\{d(n-1)\right\} E\left\{v_1(n)\right\} + E\left\{v_1^2(n)\right\} \tag{7.111}$$

$$\sigma_d^2 = 0.796^2 \sigma_d^2 + \sigma_{v1}^2 \text{ or } \sigma_d^2 = 0.31/(1 - 0.796^2) = 0.8461$$

From the second filter, we obtain the relation

$$d(n) = u(n) - 0.931u(n-1) \tag{7.112}$$

Introducing the preceding equation, for the expression of $d(n)$, into Equation 7.110, we obtain

$$u(n) - 0.135u(n-1) - 0.741u(n-2) = v_1(n) \tag{7.113}$$

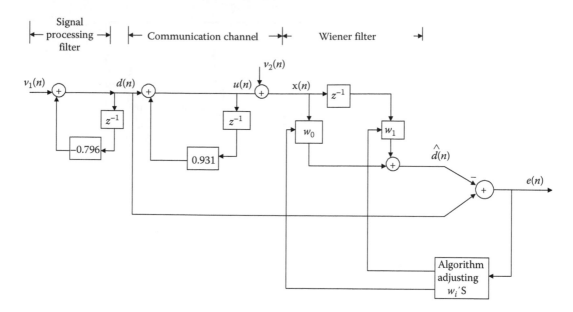

FIGURE 7.14

But $x(n) = u(n) + v_2(n)$, and hence, the column vector form of the set becomes $x(n) = u(n) + v_2(n)$. Therefore, the autocorrelation matrix R_{xx} becomes

$$E\{x(n)x^T(n)\} = R_{xx} = E\{[u(n) + v_2(n)][u^T(n) + v_2^T(n)]\} = R_{uu} + R_{v_2v_2} \qquad (7.114)$$

We used the assumption that $u(n)$ and $v_2(n)$ are uncorrelated zero-mean random sequences, which implies that $E\{v_2(n)u^T(n)\} = E\{u(n)\}E\{u^T(n)\} = 0$. Next, we multiply Equation 7.113 by $u(n - m)$ and take the ensemble average of both sides, which results in the following expression:

$$r_{uu}(m) - 0.135r_{uu}(m-1) - 0.7411r_{uu}(m-2) = r_{v_1u}(m) = E\{v_1(n)u(n-m)\} \qquad (7.115)$$

Setting $m = 1$ and then $m = 2$ in the preceding equation, we obtain

$$\begin{bmatrix} r_{uu}(0) & r_{uu}(-1) \\ r_{uu}(1) & r_{uu}(0) \end{bmatrix} \begin{bmatrix} -0.1350 \\ -0.7411 \end{bmatrix} = \begin{bmatrix} -r_{uu}(1) \\ -r_{uu}(2) \end{bmatrix} = \text{Yule} - \text{Walker equation} \qquad (7.116)$$

because $v_1(n)$ and $v(n - m)$ are uncorrelated. If we set $m = 0$ in Equation 7.115, it becomes

$$r_{uu}(0) - 0.135r_{uu}(-1) - 0.741r_{uu}(-2) = E\{v_1(n)u(n)\} \qquad (7.117)$$

Next, we substitute the value of $u(n)$ from Equation 7.113 in the preceding equation, taking into consideration that v and u are independent RVs. We obtain

$$r_{uu}(1) = \frac{0.135}{1 - 0.7411} r_{uu}(0) = \frac{0.135}{0.2589} \sigma_u^2 \qquad (7.118)$$

Substitute the preceding equation in the second equation of the set of Equation 7.116, we find

$$r_{uu}(2) = 0.135 \frac{0.135}{0.2589} \sigma_u^2 + 0.7411\sigma_u^2 \qquad (7.119)$$

Therefore, the last three equations give the variance of u:

$$\sigma_u^2 = \frac{\sigma_{v_1}^2}{0.3282} = \frac{0.31}{0.3282} = 0.9445 \tag{7.120}$$

Using Equations 7.120, 7.118, and the given value $\sigma_{v_2}^2 = 0.12$, we obtain the correlation matrix:

$$R_{xx} = R_{uu} + R_{v_2v_2} = \begin{bmatrix} 0.9445 & 0.4925 \\ 0.4925 & 0.9445 \end{bmatrix} + \begin{bmatrix} 0.12 & 0 \\ 0 & 0.12 \end{bmatrix} = \begin{bmatrix} 1.0645 & 0.4925 \\ 0.4925 & 1.0645 \end{bmatrix} \tag{7.121}$$

We multiply Equation 7.112 by $u(n)$ first and next by $u(n-1)$, and taking the ensemble average of the results, we obtain the vector p equal to

$$p = \begin{bmatrix} 0.4860 & -0.3868 \end{bmatrix}^T \tag{7.122}$$

∎

Exercise 7.8.1

Find the spectrum of the optimum filter. ▲

Minimum MSE

Introducing the previous results in Equation 7.92, we obtain the MSE surface (cost function) as a function of the filter coefficients, w's. Hence, we write

$$J(w) = 0.8461 - 2\begin{bmatrix} w_0 & w_1 \end{bmatrix}\begin{bmatrix} 0.4860 \\ -0.3868 \end{bmatrix} + \begin{bmatrix} w_0 & w_1 \end{bmatrix}\begin{bmatrix} 1.0645 & 0.4925 \\ 0.4925 & 1.0645 \end{bmatrix}$$

$$= 0.8461 - 0.972w_0 + 0.7736w_1 + 1.0645w_0^2 + 1.0645w_1^2 + 0.985w_0w_1 \tag{7.123}$$

The MSE surface and its contour plots are shown in Figure 7.15 by using the following Book MATLAB program:

Book MATLAB Program

```
>>%fig7_8_3
>>w0=-5:0.2:5;
>>w1=w0;
>>[W0,W1]=meshgrid(w0,w1);
>>j=0.8461-0.972*W0+0.7736*W1...
   >>+1.0645*W0.^2+1.0645*W1.^2+...
   >>0.985*W0.*W1;
>>colormap(gray);
>>subplot(1,2,1);
>>mesh(w0,w1,j);
>>xlabel('w_0');ylabel('w_1');
>>zlabel('J(w)');
>>subplot(1,2,2)
>>contour(W0,W1,j,10);
>>xlabel('w_0');ylabel('w_1');
```

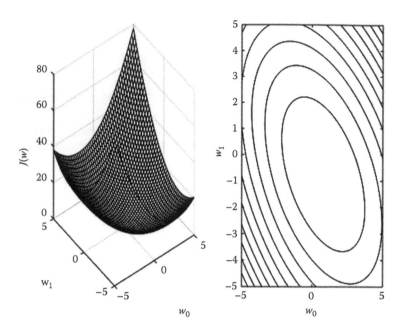

FIGURE 7.15

Observing Figure 7.15, we can estimate that the optimum filter coefficients are about 0.7 and −0.7. From Exercise 7.8.1, we obtain the exact values to be 0.7948 and −0.7311 (see also Equation 7.124).

Optimum Filter (w^o)

The optimum filter was found previously to be $R_{xx}w^o = p_{dx}$, and, in this case, it takes the following form:

$$w^o = R_{xx}^{-1}p_{dx} = \begin{bmatrix} 1.1953 & -0.5531 \\ -0.5531 & 1.1953 \end{bmatrix}\begin{bmatrix} 0.4860 \\ -0.3868 \end{bmatrix} = \begin{bmatrix} 0.7948 \\ -0.7311 \end{bmatrix} = \begin{bmatrix} w_0^o \\ w_2^o \end{bmatrix} \qquad (7.124)$$

The MMSE was found previously, and for this case, we find

$$J_{min} = \sigma_d^2 - p_{dx}^T R_{xx}^{-1}p_{dx} = 0.8461$$

$$-\begin{bmatrix} 0.4860 & -0.3868 \end{bmatrix}\begin{bmatrix} 1.1953 & -0.5531 \\ -0.5531 & 1.1953 \end{bmatrix}\begin{bmatrix} 0.4860 \\ -0.3868 \end{bmatrix}$$

$$= 0.1770 \qquad (7.125)$$

Example 7.8.3 (System Identification)

It is desired, using a Wiener filter, to estimate the unknown impulse response coefficients h_i's of an unknown FIR system (see Figure 7.16). The input $\{x(n)\}$ is a zero-mean IID RVs with variance σ_x^2. Let the impulse response h of the filter be: $h = \begin{bmatrix} 0.9 & 0.6 & 0.2 \end{bmatrix}^T$. Since the input sequence $\{x(n)\}$ is zero mean and IID RVs, the correlation matrix R_{xx} is a diagonal matrix with elements

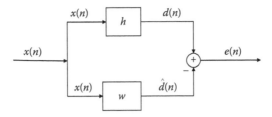

FIGURE 7.16

having values σ_x^2. The desired signal $d(n)$ is the output of the unknown filter and it is given by $d(n) = 0.9x(n) + 0.6x(n-1) + 0.2x(n-2)$ (see Exercise 7.8.2).

Solution: Therefore, the cross-correlation output is given by

$$p_{dx}(i) = E\{d(n)x(n-i)\} = E\{[0.9x(n)+0.6x(n-1)+0.2x(n-2)]x(n-i)\}$$

$$= 0.9E\{x(n)x(n-i)\} + 0.6E\{x(n-1)x(n-i)\} + 0.2E\{x(n-2)x(n-i)\}$$

$$= 0.9r_{xx}(i) + 0.6r_{xx}(i-1) + 0.2r_{xx}(i-2) \qquad (7.126)$$

Hence, we obtain $[r_{xx}(m) = 0$ for $m \neq 0]$: $p_{dx}(0) = 0.9\sigma_x^2$, $p_{dx}(1) = 0.6\sigma_x^2$. The optimum filter is

$$w^o = R_{xx}^{-1}p_{dx} = \frac{1}{\sigma_x^2}\begin{bmatrix} 1 & 0 \\ 0 & 1 \end{bmatrix}\begin{bmatrix} 0.9 \\ 0.6 \end{bmatrix} = \frac{1}{\sigma_x^2}\begin{bmatrix} 0.9 \\ 0.6 \end{bmatrix}$$

and the MMSE is (assuming $\sigma_x^2 = 1$)

$$J_{\min} = \sigma_d^2 - \begin{bmatrix} 0.9 & 0.6 \end{bmatrix}\begin{bmatrix} 1 & 0 \\ 0 & 1 \end{bmatrix}\begin{bmatrix} 0.9 \\ 0.6 \end{bmatrix}$$

But

$$\sigma_d^2 = E\{d(n)d(n)\} = E\{[0.9x(n)+0.6x(n-1)+0.2x(n-2)]^2\} = 0.81 + 0.36 + 0.04 = 1.21 \text{ and,}$$
hence, $J_{\min} = 1.21 - (0.9^2 + 0.6^2) = 0.04$.

BOOK M-FUNCTION FOR SYSTEM IDENTIFICATION (WIENER FILTER): [W,JM]=LMS_WIENER_FILTER(X,H,M)

```
function[w,jm]=lms_wiener_filter(x,h,M)
    %[w,jm]=lms_wiener_fir_filter(x,h,M);
    %x=data entering both the unknown filter(system)
    %and the Wiener filter;
    %d=the desired signal=output of the unknown
    %system; length(d)=length(x);
    %M=number of coefficients of the Wiener filter;
    %w=Wiener filter coefficients;
    %jm=minimum mean square error;
d1=conv(x,h);%h=vector with coefficients of the
    %unknown system;
d=d1(1,1:length(x));
pdx=xcorr(d,x,'biased');
```

```
p=pdx(1,(length(pdx)+1)/2:((length(pdx)+1)/2)+M-1);
rx=sample_biased_autoc(x,M);%M plays the role of
     %the lag number in correlation functions;
R=toeplitz(rx);
w=inv(R)*p';
jm=var(d)-p*w;% var() is a MATLAB function;
```

We set for the unknown system coefficients the vector $h = \begin{bmatrix} 0.9 & 0.6 & 0.2 \end{bmatrix}$. We assume $M = 6$ for the unknown filter number of coefficients. The input signal was random normally distributed with zero mean and variance one, $N(0,1)$, or $x = $ rand(1,512). The results, for a typical run, were $w = \begin{bmatrix} 0.8997 & 0.5993 & 0.1958 & -0.0054 & -0.0039 & -0.0039 \end{bmatrix}$, $jm = 0.0034$ as we were expecting. ■

Exercise 7.8.2

Draw the FIR filter given by $d(n) = 0.9x(n) + 0.6x(n - 1) + 0.2x(n - 2)$, where $d(n)$ is the output of the filter and $x(n)$ is its input. ▲

Example 7.8.4 Wiener Filtering (Noise Canceling)

In many practical applications, there exists a need to cancel the noise added to a signal. For example, we are talking to the cell phone inside the car and the noise of the car, radio, and so on, is added to the message we are trying to transmit. Similar circumstance appears when pilots in planes and helicopters try to communicate or tank drivers try to do the same. Figure 7.17 shows pictorially the noise contamination situations. Observe that the noise added to the signal and the other component entering the Wiener filter emanate from the same source but follow different paths in the same environment. This indicates that there is some difference in value. The output of the Wiener filter will approximate the noise added to the desired signal and, thus, the error will be close to the desired signal. The Wiener filter in this case is

$$R_{v_2} w^o = p_{v_1 v_2} \tag{7.127}$$

because the signal from the Wiener filter $\{\hat{v}_1\}$ should be approximately equal to $\{v_1\}$ so that the desired signal is recaptured, in this case, the signal $e(n)$. The individual components of the vector $p_{v_1 v_2}$ are

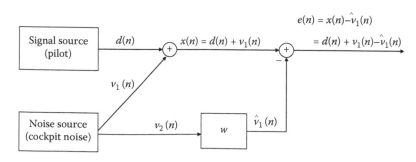

FIGURE 7.17

$$p_{v_1v_2}(m) = E\{v_1(n)v_2(n-m)\} = E\{[x(n)-d(n)]v_2(n-m)\}$$

$$= E\{x(n)v_2(n-m)\} - E\{d(n)v_2(n-m)\}$$

$$= E\{x(n)v_2(n-m)\} - E\{d(n)\}E\{v_2(n-m)\}$$

$$= E\{x(n)v_2(n-m)\} - E\{d(n)\}0 = p_{xv_2}(m) \tag{7.128}$$

since $d(n)$ and v_2 are uncorrelated, and it was assumed that the noise had zero-mean value. Therefore, Equation 7.127 becomes

$$\boldsymbol{R}_{v_2v_2}\boldsymbol{w}^o = \boldsymbol{p}_{xv_2} \tag{7.129}$$

To demonstrate the effect of the Wiener filter, let $d(n) = 2\left[0.99^n \sin(0.1n\pi + 0.2\pi)\right]$, $v_1(n) = 0.8v_1(n-1) + v(n)$ and $v_2(n) = 0.2v_1$, where $v(n)$ is white noise with zero-mean value. The correlation matrix $\boldsymbol{R}_{v2v2}$ and cross-correlation vector $\boldsymbol{p}_{xv2}$ are found using the sample biased correlation equation:

$$\hat{r}_{v_2v_2}(k) = \frac{1}{N}\sum_{n=0}^{N-1} v_2(n)v_2(n-k) \qquad k = 0,1,\dots,K-1,\ K \ll N$$

$$\hat{p}_{xv_2}(k) = \frac{1}{N}\sum_{n=0}^{N-1} x(n)v_2(n-k) \qquad k = 0,1,\dots,K-1,\ K \ll N \tag{7.130}$$

Figure 7.18 shows simultaneous results for a sixth-order Wiener filter using the Book m-function given below. We set a1 = 0.9.

BOOK M-FUNCTION FOR NOISE CANCELING

```
function[d,w,xn]=wiener_noise_canceling(dn,a1,M,N)
   %[d,w,xn]=wiener_noise_canceling(dn,a1,M,N);
   %dn=desired signal;
   %a1=first order IIR coefficient,a2=first order
   %IIR coefficient;
   %v=noise;M=number of Wiener filter coefficients;
```

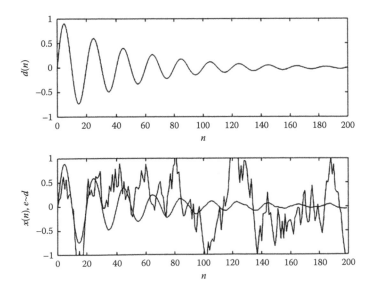

FIGURE 7.18

```
        %N=number of sequence
        %elements of dn(desired signal) and v(noise);
        %d=output desired signal;
        %w=Wiener filter coefficients;xn=corrupted
        %signal;en=xn-v1=d;
v1(1)=0;v2(1)=0;
for n=2:N
        v1(n)=a1*v1(n-1)+(rand-0.5);
        v2(n)=0.2*v1(n);
end;
v2autoc=sample_biased_autoc(v2,M);%Book MATLAB function;
xn=dn+v1;
Rv2=toeplitz(v2autoc);%MATLAB function;
p1=xcorr(xn,v2,'biased');%MATLAB function;
if M>N
        disp(['error:M must be less than N']);
end;
R=Rv2(1:M,1:M);
p=p1(1,(length(p1)+1)/2:(length(p1)+1)/2+M-1);
w=inv(R)*p';
yw=filter(w,1,v2);%MATLAB function;
d=xn-yw(:,1:N);
```

To produce the figure, the following program was used:

```
N=0:511;dn=0.89.^n.*sin(0.1*pi*n);
subplot(2,1,1);plot(n(1,1:200),dn(1,1:200),'k');
xlabel('n');ylabel('d(n)');
subplpot(2,1,2);plot(n(1,1:200),xn(1,1:200),'k');
hold on
plot(n(1,1:200),d(1,1:200),'k');
xalbel('n');ylabel('x(n), e~d');
```

∎

Example 7.8.5 Wiener Filter [Book Proposed Self-Correcting Wiener Filter (SCWF)]

We can also arrange the standard single Wiener filter in a series form as shown in Figure 7.19. This book proposed configuration permits us to process the signal using filters with fewer coefficients, thus saving in computation. Figure 7.20a shows the desired signal, which is a decaying cosine wave. Figure 7.20b shows the signal with a white zero-mean additive noise. Figure 7.20c shows the output of the first stage of a SCWF, $\{x_1(n)\}$. Each stage Wiener filter has two coefficients. Figure 7.20d shows the output of the second stage of the SCWF.   ∎

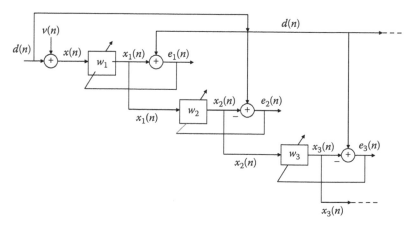

FIGURE 7.19

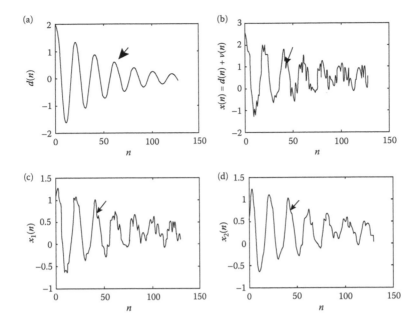

FIGURE 7.20

It is proposed in this text to extend a given input signal to a filter so that better results are found. This idea will also be used in the chapter of spectra estimation of signals having small number of elements with impressive results.

Example 7.8.6

Use the Book proposed **extended method** of the input signal to improve the desired results.

Solution: It is desired, by using the extended proposed method, to improve the detection of the Wiener filter coefficients. The following Book MATLAB program was used (this method has been used in spectra detection and adaptive filtering in the coming chapters):

```
>>x=randn(1,9);
>>[w,jm]=lms_wiener_filter(x,[0.9  0.6  0.2],5)%Book
    %MATLAB function;
```

Typical values obtained are:

```
w=[-1.0093   -0.6602  -1.6387   2.2865   1.9447];
jm=0.3677;
>>x1=[x  0.2*x  0.1*x];
>>[w1,jm1]=lms_wiener_filter(x1,[0.9  0.6  0.2],5);
w1=[0.9030  0.6061  0.1.952  -0.0017  -0.0130];
```

jm1=-0.1025;   ■

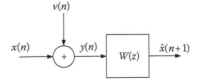

FIGURE 7.21

7.8.1 LINEAR PREDICTION

Let's investigate the prediction problem as shown in Figure 7.21. Say you want to predict the signal in the presence of noise. The input to Wiener filter is

$$y(n) = x(n) + v(n) \tag{7.131}$$

The goal is to design a filter such that the value of the signal $x(n+1)$ is predicted based on p linear combination of previous values of the input signal $y(n)$. Hence, we write

$$\hat{x}(n+1) = \sum_{k=0}^{p-1} w(k)y(n-1) = \sum_{k=0}^{p-1} w(k)[x(n-k) + v(n-k)] \tag{7.132}$$

The Wiener equation is

$$\mathbf{R}_{yy}\mathbf{w} = \mathbf{r}_{dy} \tag{7.133}$$

If the noise $v(n)$ is uncorrelated with the signal $x(n)$, the autocorrelation of the input signal $y(n)$ to the Wiener filter is

$$r_{yy}(k) = E\{y(n)y(n-k)\} = E\{(x(n) + v(n))(x(n-k) + v(n-k)\}$$
$$= E\{x(n)x(n-)\} + E\{v(n)x(n-k)\} + E\{x(n)v(n-k)\} + E\{v(n)v(n-k)\} \tag{7.134}$$
$$= r_{xx}(k) + 0 + 0 + r_{vv}(k) = r_{xx}(k) + r_{vv}(k)$$

The vector cross-correlation between $d(n)$, in this case $x(n+1)$, and $y(n)$ is

$$r_{dy}(k) = E\{d(n)y(n-k)\} = E\{x(n+1)y(n-k)\}$$
$$= E\{x(n+1)[x(n-k) + v(n-k)]\} = E\{x(n+1)x(n-k)\} = r_{xx}(k+1) \tag{7.135}$$

Equation 7.134, with $k = 0, 1, 2, \ldots, p-1$, becomes $\mathbf{R}_{yy} = \mathbf{R}_{xx} + \mathbf{R}_{vv}$. Therefore, the MSE is

$$J_{\min} = \sigma_d^2 - \mathbf{p}_{dy}^T\mathbf{w}^o = \sigma_d^2 - \mathbf{p}_{dy}^T\mathbf{R}_{yy}^{-1}\mathbf{p}_{dy} \tag{7.136}$$

gives the minimum error between the desired signal and the output of the Wiener filter.

Example 7.8.7

Let the noise be a white noise with variance σ_v^2. The noise is assumed to be uncorrelated with the signal $\{x(n)\}$. Assume that the autocorrelation function of $\{x(n)\}$ is $r_{xx}(k) = a^{|k|}$. We also assume a two-coefficient predictor filter, $[w_0 \ w_1]$.

Solution: The Wiener equation is

$$\left(\mathbf{R}_{xx} + \sigma_v^2\mathbf{I}\right)\mathbf{w} = \mathbf{r}_{dy} \tag{7.137}$$

The preceding equation becomes

$$\begin{bmatrix} r_{xx}(0)+\sigma_v^2 & r_{xx}(1) \\ r_{xx}(1) & r_{xx}(0)+\sigma_v^2 \end{bmatrix}\begin{bmatrix} w_0 \\ w_1 \end{bmatrix}=\begin{bmatrix} r_{dy}(0) \\ r_{dy}(1) \end{bmatrix}, \quad r_{dy}(k)=r_{xx}(k+1)$$

or

$$\begin{bmatrix} 1+\sigma_v^2 & a \\ a & 1+\sigma_v^2 \end{bmatrix}\begin{bmatrix} w_0 \\ w_1 \end{bmatrix}=\begin{bmatrix} a \\ a^2 \end{bmatrix} \; or \; \begin{bmatrix} w_0 \\ w_1 \end{bmatrix}=\frac{a}{(1+\sigma_v^2)^2-a^2}\begin{bmatrix} 1+\sigma_v^2-a^2 \\ a\sigma_v^2 \end{bmatrix}$$

Let $a = 0.8$, $\sigma_v^2 = 1.2$, $x(n) = 2\sin(0.1n\pi)$, and $y(n) = x(n)+0.5(\text{rand}-0.5)$, and then the Wiener filter coefficients are $w_0 = 0.2971$, $w_1 = 0.1829$. Hence, $J_{\min} = [1 \quad 0.8]^2 [0.2971 \quad 0.1829]^T = 0.5858$. ■

Exercise 7.8.3

Verify Equation 7.137.   ▲

HINTS, SUGGESTIONS, AND SOLUTIONS OF THE EXERCISES

7.2.1

$$\mathbf{H}^T\mathbf{H}=\begin{bmatrix} 3 & 3 \\ 3 & 5 \end{bmatrix} \quad \mathbf{H}^T\mathbf{x}=\begin{bmatrix} x(0)+x(1)+x(2) \\ x(1)+2x(2) \end{bmatrix}$$

7.2.2

$$J(A)=\sum_{n=0}^{N-1}(x(n)-A)^2 \Rightarrow \hat{A}=\frac{1}{N}\sum_{n=0}^{N-1}x(n)=\bar{x}=\text{mean value}$$

7.3.1

$$J_{\min}=\sum_{n=0}^{N-1}[x(n)-\hat{q}h(n)][x(n)-\hat{q}h(n)]=\sum_{n=0}^{N-1}x(n)[x(n)-\hat{q}h(n)]-\hat{q}\underbrace{\sum_{n=0}^{N-1}h(n)[x(n)-\hat{q}h(n)]}_{S}$$

$$=\sum_{n=0}^{N-1}x^2(n)-\hat{q}\sum_{n=0}^{N-1}x(n)h(n)$$

The last summation S becomes zero by setting $\hat{q}=\left[\sum_{n=0}^{N-1}h(n)x(n)\right]/\sum_{n=0}^{N-1}h^2(n)$ in the expression S and, thus, we obtain the desired results.

7.3.2

For this case, and for $n = 0, 1, 2, 3$ and $N = 4$, we obtain

$$J(A, B) = \sum_{n=0}^{3} (x(n) - h_1(n)A - h_2(n)B)^2$$

$$h_1(n) = \sin(0.1\pi n) \quad h_2(n) = \sin(0.4\pi n)$$

If we differentiate $J(A, B)$ first with respect to A and next with respect to B, and then equate the results to zero, we obtain

$$\begin{bmatrix} A \\ B \end{bmatrix} = \begin{bmatrix} \sum_{n=0}^{3} h_1^2(n) & \sum_{n=0}^{3} h_1(n)h_2(n) \\ \sum_{n=0}^{3} h_1(n)h_2(n) & \sum_{n=0}^{3} h_2^2(n) \end{bmatrix}^{-1} \begin{bmatrix} \sum_{n=0}^{3} x(n)h_1(n) \\ \sum_{n=0}^{3} x(n)h_2(n) \end{bmatrix}$$

The following Book MATLAB program was used to find $\hat{A} = 0.5372$, $\hat{B} = 4.7868$ from the vector ab:

```
>> n = 0 : 3; x = 0.2 * sin(0.1 * pi * n) + 5.2 * sin(0.4 * pi * n)...
>> +0.5 * randn(1, 4);
>> h1 = sin(0.1 * pi * n); h2 = sin(0.4 * pi * n);
>> ab = inv([sum(h1.^ 2) sum(h1.* h2); sum(h1.* h2)...
>> sum(h2.^ 2)]) * [sum(x.* h1); sum(x.* h2)];
```

7.4.1.1

$$\text{var}\{\hat{c}\} = \text{var}\left\{ \frac{1}{N} \sum_{n=0}^{N-1} x(n) \right\} = E\{[\hat{c} - E\{\hat{c}\}]^2\} = E\left\{ \left[\frac{1}{N} \sum_{n=0}^{N-1} x(n) - c \right]^2 \right\}$$

$$= E\left\{ \left(\frac{1}{N} \sum_{n=0}^{N-1} x(n) \right)^2 + c^2 - 2c \frac{1}{N} \sum_{n=0}^{N-1} x(n) \right\} = E\left\{ \left(\frac{1}{N} \sum_{n=0}^{N-1} x(n) \right)^2 \right\} + c^2 - 2cE\left\{ \frac{1}{N} \sum_{n=0}^{N-1} x(n) \right\}$$

$$= E\left\{ \left(\frac{1}{N} \sum_{n=0}^{N-1} x(n) \right)^2 \right\} - c^2, \left(\text{since } E\left\{ \frac{1}{N} \sum_{n=0}^{N-1} x(n) \right\} = \frac{1}{N} \sum_{n=0}^{N-1} E\{x(n)\} = \frac{Nc}{N} = c \right)$$

$$= E\left\{ \frac{1}{N^2} \sum_{n=0}^{N-1} x^2(n) + \frac{1}{N} \sum_{n \neq m} x(n)x(m) \right\} - c^2 = \frac{1}{N^2} \sum_{n=0}^{N-1} E\{x^2(n)\} + \frac{1}{N} \sum_{n=0}^{N-1} c^2 - c^2$$

$$= \frac{1}{N^2} \sum_{n=0}^{N-1} \text{var}\{x(n)\} + \frac{Nc^2}{N} - c^2 = \frac{N\sigma^2}{N^2} = \frac{\sigma^2}{N}; x(n), x(m) \text{ are independent}$$

7.4.2.1

If we set $\hat{c} = \dfrac{1}{N} \sum\limits_{n=0}^{N-1} x(n) = g(x(0), x(1), \dots, x(N-1))$ (1) then $E\{\hat{c}\} = \dfrac{1}{N} \sum\limits_{n=0}^{N-1} E\{x(n)\}$

$= \dfrac{1}{N} \sum\limits_{n=0}^{N-1} E\{c + v(n)\} = \dfrac{1}{N} \sum\limits_{n=0}^{N-1} (c + E\{v(n)\}) = \dfrac{Nc}{N} = c$ for all c. This implies that the estimator (1) is unbiased.

7.4.2.2

$E\{\hat{c}\} = \dfrac{1}{4N} \sum\limits_{n=0}^{N-1} E\{x(n)\} = \dfrac{1}{4} c$ which indicates that if $c = 0$, then $E\{\hat{c}\} = c$, and if $c \neq 0$, then $E\{\hat{c}\} \neq c$, and, thus, $\hat{c}$ is a biased estimator.

7.4.3.1

$$s(n; f_0) = A\cos(2\pi f_0 n + \varphi);\ p(x(n); f_0) = \frac{1}{(2\pi\sigma^2)^{N/2}} \exp\left(-\frac{1}{2\sigma^2} \sum_{n=0}^{N-1} (x(n) - s(n; f_0))\right) \quad (1)$$

Differentiate the logarithm of (1) once to find

$$\frac{\partial \ln p(x; f_0)}{\partial f_0} = \frac{1}{\sigma^2} \sum_{n=0}^{N-1} (x(n) - s(n; f_0)) \frac{\partial s(n; f_0)}{\partial f_0} \quad (2)$$

The second derivative of (2) gives

$$\frac{\partial^2 \ln p(x; f_0)}{\partial f_0^2} = \frac{1}{\sigma^2} \sum_{n=0}^{N-1} \left[(x(n) - s(n; f_0)) \frac{\partial^2 s(n; f_0)}{\partial f_0^2} - \left(\frac{\partial s(n; f_0)}{\partial f_0}\right)^2 \right] \quad (3)$$

Taking the expectation of (3), we obtain: $E\left\{\dfrac{\partial^2 \ln p(x; f_0)}{\partial f_0^2}\right\} = -\dfrac{1}{\sigma^2} \sum\limits_{n=0}^{N-1} \left(\dfrac{\partial s(n; f_0)}{\partial f_0}\right)^2$. Therefore, the CRLB is given by

$$\mathrm{var}(f_0) \geq \frac{\sigma^2}{\sum\limits_{n=0}^{N-1} \left(\dfrac{\partial s(n; f_0)}{\partial f_0}\right)^2} = \frac{\sigma^2}{A^2 \sum\limits_{n=0}^{N-1} [2\pi n \sin(2\pi f_0 n + \varphi)]^2}.$$

7.4.3.2
var(0.15) ≥ 0.2060

7.4.4.1
Taking the partial derivative of Equation 7.45 with respect to a_m, we find

$$\partial \mathrm{mse}(\hat{q}) / \partial a_m = -2E\{qx(m)\} + \frac{\partial}{\partial a_m}\left[\sum_{m=0}^{N-1}\sum_{n=0}^{N-1} a_m a_n E\{x(m)x(n)\} \right] = -2r_{qx}(m) + 2\sum_{n=0}^{N-1} a_n r_{mn} = 0 \quad (1)$$

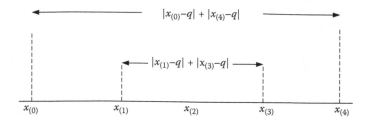

FIGURE 7.22

since $a_m a_n$ appears twice in the double summation expansion. Therefore, (1) is the mth equation of Equation 7.46.

7.4.5.1

Arrange a set of RVs $\{x(0), x(1), x(2), x(3), x(4)\}$ in assenting order of their values $\{x_{(0)}, x_{(1)}, x_{(2)}, x_{(3)}, x_{(4)}\}$ that have a Laplacian PDF, and it is desired to find the estimator.

Referring to Figure 7.22, the sum $|x_{(0)} - q| + |x_{(4)} - q|$ becomes minimum in the range $x_{(1)} \le q \le x_{(4)}$ and remains constant in this range. The sum $|x_{(1)} - q| + |x_{(3)} - q|$ becomes minimum in the range $x_{(2)} \le q \le x_{(4)}$ and stays constant. The last factor of the summation $\sum_{n=0}^{4-1} |x(n) - q|$ is $|x_{(2)} - q|$, and this factor becomes minimum if $q = x_{(2)}$, where $x_{(2)}$ is the median of the set.

7.4.5.2

The following Book MATLAB program was used to find the desired values:

```
>>N=50;
>>for m=1:2000
>>     c(m)=-(1/2)+sqrt((1/N)*sum((1+randn(1,N)).^2)+(1/4));
>>end
```

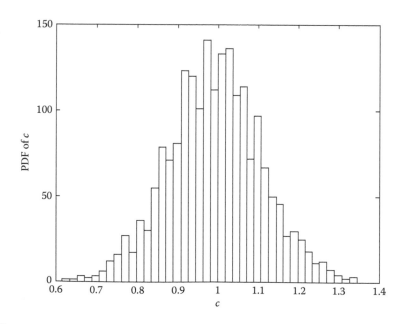

FIGURE 7.23

```
>>hist(c,50,'k');
>>colormap([1  1  1]);
```

Results: for $N = 5$, mean(c) = 0.9513 and variance(c) = 0.1276.
Results: for $N = 50$, mean(c) = 0.9927 and variance(c) = 0.0130.
Figure 7.23 was created using $N = 50$ and 2,000 realizations.

7.5.1

$$y(n) = A \text{ and } x(n) \ n = 0 \ \ldots \ N - 1 \Rightarrow J(A) = \sum_{n=0}^{N-1}(x(n)-A)^2 \Rightarrow \frac{\partial J(A)}{\partial A} = -2\sum_{n=0}^{N-1}(x(n)-A) \text{ or}$$

$$NA = \sum_{n=0}^{N-1} x(n) \Rightarrow A = \frac{1}{N}\sum_{n=0}^{N-1} x(n) = \text{mean of } x(n). \text{ This cannot be accepted as the optimal; just tell}$$

us that it minimizes the least-squares error.

7.7.1

$$J(w) = J_{\min} = \sigma_d^2 - 2w^{oT}R_xw^o = \sigma_d^2 - w^{oT}R_xw^o = \sigma_d^2 - (R_x^Tw^o)^T w^o = \sigma_d^2 - (R_xw^o)^T w^o$$

$$= \sigma_d^2 - p_{dx}^T w^o = \sigma_d^2 - p_{dx}^T R_x^{-1} p_{dx} \quad (R_x \text{ is symmetric and } w\text{'s and } p\text{'s are vectors})$$

7.7.2

$$J_m = 0.1477, \quad w^o = [0.8417; -0.7917]$$

7.8.1

$$w = 0:0.1:\text{pi}; \ w^o = \text{inv}([1.0645\ 0.4925; \ 0.4925\ 1.0645]) * [0.4860; -0.3868]$$

$$w^o = 0.7948$$

$$-0.7311$$

plot (w, abs([0.7948−0.7311*exp(−j*w)]));

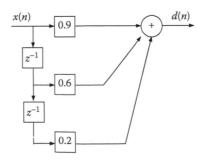

FIGURE 7.24

7.8.2

See Figure 7.24.

7.8.3

$$r_{yy}(k) = r_{xx}(k) + r_{vv}(k) \Rightarrow r_{yy}(0) = r_{xx}(0) + r_{vv}(0); r_{yy}(1) = r_{xx}(1) + r_{vv}(1);$$

$$r_{yy}(-1) = r_{xx}(-1) + r_{vv}(-1) \text{ but } r_{yy}(1) = r_{yy}(-1), r_{xx}(1) = r_{xx}(-1), r_{vv}(1) = r_{vv}(-1) \Rightarrow$$

$$\begin{bmatrix} r_{yy}(0) & r_{yy}(1) \\ r_{yy}(-1) & r_{yy}(0) \end{bmatrix} = \begin{bmatrix} r_{xx}(0) + r_{vv}(0) & r_{xx}(1) + r_{vv}(1) \\ r_{xx}(-1) + r_{vv}(-1) & r_{xx}(0) + r_{vv}(0) \end{bmatrix} = \begin{bmatrix} r_{xx}(0) + \sigma_v^2 & r_{xx}(1) + 0 \\ r_{xx}(1) + 0 & r_{xx}(0) + \sigma_v^2 \end{bmatrix}$$

$$= \begin{bmatrix} r_{xx}(0) & r_{xx}(1) \\ r_{xx}(1) & r_{xx}(0) \end{bmatrix} + \begin{bmatrix} \sigma_v^2 & 0 \\ 0 & \sigma_v^2 \end{bmatrix} = R_x + \sigma_v^2 \begin{bmatrix} 1 & 0 \\ 0 & 1 \end{bmatrix}$$

8 Nonparametric (Classical) Spectra Estimation

8.1 PERIODOGRAM AND CORRELOGRAM SPECTRA ESTIMATION

Estimating the power spectrum, given a long random sequence $\{x(n)\}$, can be accomplished by, for example, creating its autocorrelation function and then taking the Fourier transform (FT) of it. However, several problems appear in establishing power spectra densities. First, the sequence may not be long enough and some of the times can be very short. Second, the spectra characteristics may change with time. Third, data very often are corrupted with noise. Therefore, the spectrum estimation is a problem that involves estimating $S_x\left(e^{j\omega}\right)$ from a finite noisy measurements of a sequence, $\{x(n)\}$.

8.1.1 DETERMINISTIC SIGNALS (SEE ALSO CHAPTER 2)

If the sequence $\{x(n)\}$ is a deterministic real data sequence and has finite energy, for example,

$$\sum_{n=-\infty}^{\infty} x^2(n) < \infty \tag{8.1}$$

the sequence possesses a discrete-time Fourier transform (DTFT),

$$X\left(e^{j\omega}\right) = \sum_{n=-\infty}^{\infty} x(n)e^{-j\omega n} \tag{8.2}$$

where it is assumed that the sampling time is $T=1$ and ω has units of radians.

The inverse DTFT (IDTFT) is

$$x(n) = \frac{1}{2\pi} \int_{-\pi}^{\pi} X\left(e^{j\omega}\right) e^{j\omega n} \, d\omega \tag{8.3}$$

Note: *The time function is discrete and the transformed one is continuous. The reader should remember that, if the sequence is the result of sampling of a continuous signal, the discrete frequency can be easily converted to physical frequency rad/s by dividing it by the sampling time T_s. Hence, the frequency range becomes $-\pi/T \le \omega \le \pi/T$. In addition, we make the following changes in Equation 8.2: $X(e^{j\omega T})$, $Tx(nT)e^{-jn\omega T}$.*

We can define the **energy spectral density** as follows:

$$S_x\left(e^{j\omega}\right) = \left|X\left(e^{j\omega}\right)\right|^2 \tag{8.4}$$

The Parseval's theorem for real sequence states the following

$$\sum_{n=-\infty}^{\infty} x^2(n) = \frac{1}{2\pi} \int_{-\pi}^{\pi} S_x\left(e^{j\omega}\right) d\omega \tag{8.5}$$

Note: *This formula states that the total energy in the time domain is equal to the total power in the frequency domain. It does not tell us what part of the time domain energy is distributed in a specific frequency range.*

Exercise 8.1.1

Verify Equations 8.2, 8.5, and 8.7.   ▲

The autocorrelation of a sequence is defined as follows:

$$r_{xx}(k) = \sum_{n=-\infty}^{\infty} x(n)x(n-k) \tag{8.6}$$

Its DTFT is equal to its power density spectrum:

$$\sum_{k=-\infty}^{\infty} r_{xx}(k)e^{-j\omega k} = S_x\left(e^{j\omega}\right) \tag{8.7}$$

8.1.2 THE PERIODOGRAM-RANDOM SIGNALS

Stationary random signals, which we will study in this text, have usually finite average power and therefore can be characterized by an average power spectral density (PSD). We shall call such a quantity the PSD. Without loss of generality, the discrete real random sequences, which we will study in this text, will assume that they have zero mean value, for example,

$$E\{x(n)\} = 0 \quad \text{for all } n \tag{8.8}$$

The **periodogram spectral density** is defined as follows:

$$S_x\left(e^{j\omega}\right) = \lim_{N\to\infty} E\left\{ \frac{1}{2N+1} \left| \sum_{n=-N}^{N} x(n)e^{-j\omega n} \right|^2 \right\} \tag{8.9}$$

The periodogram spectral density estimator of the data segment $\{x(0), x(1), ..., x(N-1)\}$, finite number of data, is based on the following formula:

$$\hat{S}_x\left(e^{j\omega}\right) = \frac{1}{N} \left| \sum_{n=0}^{N-1} x_w(n)e^{-j\omega n} \right|^2 = \frac{1}{N}\left| X\left(e^{j\omega}\right)\right|^2 \tag{8.10}$$

where $X\left(e^{j\omega}\right)$ is the DTFT of the windowed signal:

$$x_w(n) = x(n)w(n) \quad 0 \le n \le N-1 \tag{8.11}$$

In Equation 8.10, the signal can be multiplied by a unit value square window $\{1, 1, 1, ..., 1\}$. The PSD of the signal, $\hat{S}_x\left(e^{j\omega}\right)$, is periodic with period 2π, $-\pi \le \omega \le \pi$. The spectrum $X\left(e^{j\omega}\right)$ is the DTFT of the windowed signal $\{x_w(n)\}$. The periodicity is easily shown by introducing $\omega + 2\pi$ in place of ω in Equation 8.10 and remembering that $\exp(j2\pi) = 1$. The values of the periodogram at discrete points in the frequency domain are located at $\{\omega_k = 2\pi k / N\}$, $k = 0, 1, 2, ..., N-1$ and are found by

$$\hat{S}_x(k) \triangleq \hat{S}_x\left(e^{j2\pi k/N}\right) = \frac{1}{N}\left|X(\omega_k)\right|^2 \quad k = 0,1,\ldots,N-1 \quad \omega_k = \frac{2\pi}{N}k \tag{8.12}$$

where $X(k)$ is the N-point DFT of the windowed signal $x_w(n)$. It is recommended that we multiply the random signal with an appropriate window for a smoother PSD. The following example and the Book m-file explain the use of the windows. Some of the MATLAB window functions are given in the upcoming sections. See also Appendix 1 at the end of the book for more types of windows.

Example 8.1.1

Plot the PSD of the signal $y(n)$ that is produced by the filter $y(n) - 0.9y(n-1) + 0.2y(n-2) = v(n)$, where $v(n) = 0.8 * \text{randn}(n)$. Use a rectangular and a Hamming window with $N = 128$.

Solution: The results are shown in Figure 8.1 using the following Book m-file:

BOOK M-FILE: EX 8_1_1

```
%Book m-file:ex8_1_1
y(1)=0;y(2)=0;
for n=3:1024
    y(n)=0.9*y(n-1)-0.2*y(n-2)+0.8*randn;
end;
N=128;
xw=y(1,50:50+N-1).*rectwin(N)';%the window
    %is given as a column vector, here is
    %a rectangular window, rectwin() is a MATLAB
    %function;
psd=(abs(fft(xw,N))).^2/N;
xw1=y(1,1:N).*hamming(N)';%hamming() stands for
    %hamming window and is a MATLAB function;
psd1=(abs(fft(xw1,N))).^2/N;
sx128=psd(1,1:N);
sx1128=psd1(1,1:N);
om=0:2*pi/N:2*pi-(2*pi/N);
subplot(2,1,1);stem(om,sx128,'k');
xlabel('\omega rad');ylabel('PSD');
subplot(2,1,2);stem(om,sx1128,'k');
xlabel('\omega rad');ylabel('PSD');
```

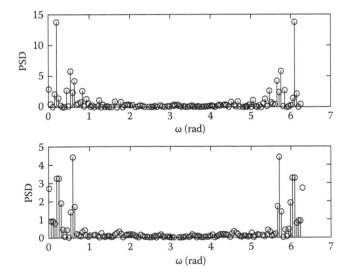

FIGURE 8.1

The reader will observe that the Hamming window (and any other one) produces a smoother PSD than the rectangular one. We must point out that the results showing in the figure is for only one realization. This means that any time we run the program, we will observe different PSDs. We can also proceed to create a number of these realizations and then average them. ∎

The periodogram is an **asymptotic estimator** since as $N \to \infty$ the estimate periodogram converges to the power spectrum of the process. However, the periodogram is **not a consistent estimate** of the power spectrum. This means that the variance of the periodogram does not decrease by increasing the number of terms.

The Book m-Function for producing the periodogram of a sequence $\{x(n)\}$ is as follows:

Book m-Function: [s,as,ps]=periodogram(x,L)

```
function[s,as,ps]=periodogram(x,L)
%[s,as,ps]=periodogram(x,L);
%L=desired number of points (bins) of the spectrum;
%x=data in row form; s=complex form of the DFT;
for m=1:L
    s(m)=sum(x.*exp(-j*(m-1)*(2*pi/L)*(1:length(x))));
end;
as=((abs(s)).^2/length(x));%as=amplitude spectral density;
ps=(atan(imag(s)./real(s))/length(x));%ps=phase spectrum;

%To plot as or ps we can use the command:
%plot(0:2*pi/L:2*pi-(2*pi/L),as);for the phase spectrum
%we change as to ps;
```

To create an ensemble of a signal, we use the following Book m-function:

```
function[s]=periodogram1(N,M,L)
%L=desired number of points (bins) of the spectrum;
%N=number of realizations of the PSD;
%x=data in row form; s=complex form of the DFT;
%M is the length of data vector x;
for n=1:N
    x=randn(1,M);
    for m=1:L
    s(m,n)=sum(x.*exp(-j*(m-1)*(2*pi/L)*(1:length(x))));
    end;
end;


%To plot PSD we can use the command:
%plot(0:2*pi/L:2*pi-(2*pi/L),abs(sum(s,2)));
%Note, line 8 can be changed and asigned
%different types of signals;
%Note, if we like to plot only the sum of 3
%columns you write: abs(sum(s(:,4:6),2)/M);
```

8.1.3 Correlogram

The **correlogram spectra estimator** is based on the formula:

$$\hat{S}_x(e^{j\omega}) = \sum_{m=-(N-1)}^{N-1} \hat{r}_{xx}(m)e^{-j\omega m} \quad \text{(correlogram)} \tag{8.13}$$

The **biased** autocorrelation function is given by

$$\hat{r}_{xx}(k) = \frac{1}{N} \sum_{n=k}^{N-1} x(n)x(n-k) \quad 0 \le k \le N-1 \tag{8.14}$$

The unbiased one is given by the same formula, with a difference: Instead of dividing by N, the expression is divided by $N-k$. In practice, the biased autocorrelation is used since it produces positive definite matrices and, thus, they can be inverted.

To produce the biased correlation function, the following Book MATLAB function may be used:

BOOK M-FUNCTION FOR BIASED AUTOCORRELATION FUNCTION:

[R]=SAMPLE _ BIASED _ AUTOC(X,LG)

```
function[r]=urdsp_sample_biased_autoc(x,lg)
    %Book MATLAB function:[r]=sample_biased_autoc(x,lg);
    %this function finds the biased autocorrelation function
    %with lag from 0 to lg;it is recommended that lg is 20-30%
    %of N;
N=length(x);%x=data;lg=lag;
for m=1:lg
    for n=1:N+1-m
        xs(m,n)=x(n-1+m);
    end;
end;
r1=xs*x';
r=r1'./N;
```

The reader can also use the Book m-function: [r]=sample _ biased _ autoc(x,lg).

To plot the PSD after we find the autocorrelation r, we write, for example, the following:

```
>>rf=fft(r,128);w=0:2*pi/128:2*pi-(2*pi/128); plot(w,abs(rf));
```

We can also use MATLAB functions to obtain the sample biased and unbiased autocorrelation and cross-correlation functions. The functions are

```
r_xy=xcorr(x,y); % x,y are N length vectors; rxy= a 2N-1
    % symmetric cross-correlation vector, in case the vectors %do not
    have the same length, the shorter
    %one will be padded with zero; Note: The correlation %sequence is not
    divided by N or N-1;
r_xy=xcorr(x,y,'biased'); %will give the biased
    %cross-correlation function and the
    %sequence is divided by N;
r_xy=xcorr(x,y,'unbiased');%will give the unbiased
    %autocorrelation function;
```

Note that the MATLAB functions give both sides of the autocorrelation function since $r(n) = r(-n)$.

8.1.4 COMPUTATION OF PERIODOGRAM AND CORRELOGRAM USING FFT

Since both functions are continuous with respect to frequency (DTFT), we can sample the frequency as follows (see also Chapter 2):

$$\omega_k = \frac{2\pi}{N}k \qquad k = 0,1,2,\dots,N-1 \tag{8.15}$$

Introducing the above values of the discrete frequency in Equations 8.10 and 8.13, we obtain

$$\hat{S}_x\left(e^{j\omega_k}\right) = \frac{1}{N}\left|X(e^{j\omega_k})\right|^2;$$

$$X\left(e^{j\omega_k}\right) = \sum_{n=0}^{N-1} x(n)e^{-j\frac{2\pi}{N}kn} = \sum_{n=0}^{N-1} x(n)W^{kn} \quad W = e^{-j2\pi/N} \quad 0 \le k \le N-1 \qquad (8.16)$$

$$\hat{S}_r\left(e^{j\omega_k}\right) = \sum_{m=-(N-1)}^{N-1} \hat{r}_{xx}(m)e^{-j\frac{2\pi}{N}km} \quad 0 \le k \le N-1$$

The most efficient way to obtain the DFT using FFT is to set $N=2^r$ for some integer r. The following two Book m-functions give windowed periodogram and correlogram.

BOOK M-FUNCTION FOR WINDOWED PERIODOGRAM:
[s,as,phs]=WINDOWED _ PERIODOGRAM(x,w,L)

```
%Book function: [s,as,phs]=windowed_periodogram(x,w,L)
function[s,as,phs]=windowed_periodogram(x,w,L)
    %w=name(length),w is in column form
    %(see also Appendix 1
    %name=hamming,kaiser,hann,rectwin,
    %bartlett,tukeywin,blackman,gausswin,nattallwin,
    %triang,blackmanharris);
    %L=desired number of points (bins) of the spectrum;
    %x=data in row form;s=complex form of its DFT;
xw=x.*w';
for m=1:L
    n=1:length(x);
    s(m)=sum(xw.*exp(-j*(m-1)*(2*pi/L)*n));
end;%the for loop finds the set of complex numbers
    %of the summation for each m;
as=((abs(s)).^2/length(x))/norm(w);
    %as=amplitude value of the periodogram;
phs=(atan(imag(s)./real(s))/length(x))/norm(w);
    %phase of the periodogram;
```

To plot the PSD, we write plot(as).

BOOK M-FUNCTION FOR WINDOWED CORRELOGRAM:
[s,as,ps]=WINDOWED _ CORRELOGRAM(x,w,L)

```
function[s,as,ps]=windowed_correlogram(x,w,lg,L)
%Book m-file function: function [s,as,phs]=windowed_correlogram(x,w,lg,L);
    %x=data w),
    %L=desired
    %number of spectral points;
    %lg=lag number<<N best 20-30% of N;rc=symmetric
    %autocorrelation function;
r=sample_biased_autoc(x,lg);
rc=[fliplr(r(1,2:lg)) r 0];
rcw=rc.*w';
```

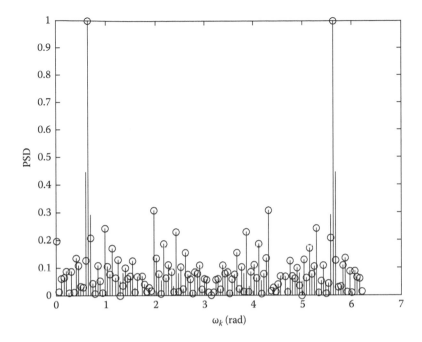

FIGURE 8.2

```
for m=1:L
    n=-lg+1:lg;
    s(m)=sum(rcw.*exp(-j*(m-1)*(2*pi/L)*n));
end;
as=(abs(s).^2)/norm(w)/L;%amplitude spectrum;
ps=(atan(imag(s))/real(s))/norm(w);%phase spectrum;
```

We used the following Book MATLAB program to produce Figure 8.2. The circles correspond to periodogram and the dots to correlogram spectra.

```
>>n=1:256;x=sin(0.2*pi*n)+2*randn(1,256);
>>lg=64;L=128;w=hamming(256);w1=hamming(2*lg);
>>[s,as,ps]=windowed_periodogram(x,w,L);
>>[s1,as1,ps1]=windowed_correlogram(x,w1,lg,L);
>>wb=0:2*pi/L:2*pi-(2*pi/L);
>>stem(wb,as/max(as),'k');hold on;
>>stem(wb,as1/max(as1),'.k');
>>xlabel('\omega_k rad');ylabel('PSD');
```

General Remarks

1. The variance of the periodogram does not tend to zero as $N \to \infty$. This indicates that the periodogram is not a **consistent** estimator; that is, its distribution does not tend to cluster more closely around the true spectrum as N increases.
2. To reduce the variance and thus produce smoother spectral estimator we must (1) average contiguous values of the periodogram or (2) average periodogram obtained from multiple data segments.
3. The effect of the side-lobes of the windows on the estimated spectrum consists of transferring power from strong bands to less strong bands or bands with no power. This process is known as the **leakage** problem.

4. The correlation sequence produces a positive semi-definite sequence and, as a consequence, the correlation matrix is also positive semi-definite. This property guarantees that the PSD is positive for all frequencies.
5. Show that $S_{px}\left(e^{j\omega}\right) = S_{cx}\left(e^{j\omega}\right)$ (see Exercise 8.1.2).
6. The periodogram is **asymptotically unbiased** (see Exercise 8.1.3).

Exercise 8.1.2

Prove that the PSD due to periodogram is equal to the one produced by the correlogram.   ▲

Exercise 8.1.3

Show that the periodogram is asymptotically unbiased.   ▲

Exercise 8.1.4

Create random sequences and find the PSD using the periodogam1 function and average for different values of M (10, 100, 300). Compare the average PSDs with any single PSD. The results are shown in Figure 8.3.   ▲

Exercise 8.1.5

Use the periodogram and correlogram to plot the PSDs on the same figure. The signal is $x(n) = \sin(0.05\pi n) + 2\text{randn}, \quad n = 1, 2, \ldots, 256$.   ▲

Example 8.1.2

Find the periodogram for a white Gaussian (WG) signal with $N = 128$ and 1024 and observe the variance (the variability) of the signal.

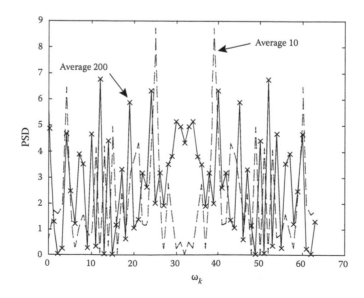

FIGURE 8.3

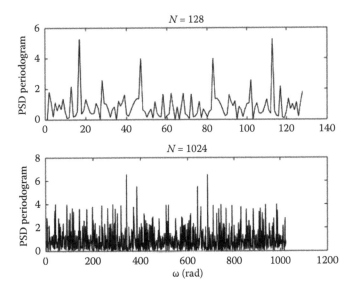

FIGURE 8.4

Solution: Figure 8.4 shows the two periodograms by simulation. It is apparent that the variance does not decreases and it is shown analytically in advanced texts. For this particular case, the two variances were variance for the signal with $N = 128$ was 0.7815, and variance for the signal with $N = 1024$ was 0.8910. ∎

Example 8.1.3

Demonstrate what effects the spectrum resolution using the periodogram.

Solution: Figure 8.5 shows two plots of the PSD of a sequences made up of two sinusoids plus a White gaussian (WG) noise. The signals were $s = \sin(0.3\pi n) + \sin(0.35\pi n) + 0.5\mathrm{randn}(1,64)$ and $s1 = \sin(0.3\pi n) + \sin(0.35\pi n) + 0.5\mathrm{randn}(1,256)$. The two frequencies were at 0.9425 rad and at 1.0996 rad. The difference between these two frequencies is 0.0571 rad. The resolution for $N = 64$ is $2\pi / 64 = 0.0982$ and for $N = 256$ is $2\pi / 256 = 0.0245$. From these values and the figures, it is apparent how the resolution property of the periodogram depends on. The effect of the number of samples, N, is also apparent from the figure. ∎

Example 8.1.4

It is instructive to obtain the average of a number of realizations of a periodogram.

Solution: The signal is the sum of two sinusoids and a white noise of unit variance. Using the following Book m-file, the results are shown in Figure 8.6. The top plot shows 50 realizations of the signal, the second shows 50 realizations of the periodograms, the third shows the average PSD of all 50 PSDs, and the last one shows just the PSD of one of the realizations.

BOOK M-FILE: EX8_1_4

```
%Book m-file: ex8_1_4
for m=1:60
    n=0:127;
    x(m,:)=sin(0.35*pi*n)+sin(0.37*pi*n)...
    +3.5*(rand(1,128)-0.5);
    subplot(4,1,1);plot(x(m,1:64),'k');xlabel('n');
```

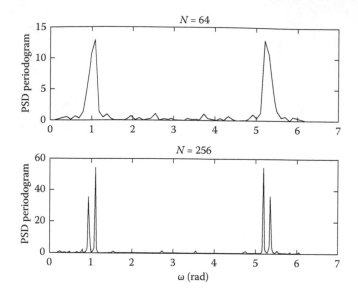

FIGURE 8.5

```
    ylabel('x(n)')
    hold on;
    x(m+1,:)=sin(0.35*pi*n)+sin(0.37*pi*n) ...
    +3.5*(rand(1,128)-0.5);
    ax(m,:)=abs(fft(x(m,:),128)).^2/128;
    ax(m+1,:)=abs(fft(x(m+1,:),128)).^2/128;
    ax(m+2,:)=(ax(m,:)+ax(m+1,:))/2;
    subplot(4,1,2);plot(ax(m,1:64),'k');xlabel('\omega_k');
    ylabel('PSD');
    hold on;
end;
    subplot(4,1,3);plot(ax(m,1:64),'k');xlabel('\omega_k');
    ylabel('Ave. PSD');
    subplot(4,1,4);plot(ax(13,1:64),'k');xlabel('\omega_k');
    ylabel('PSD 1 real.');
```

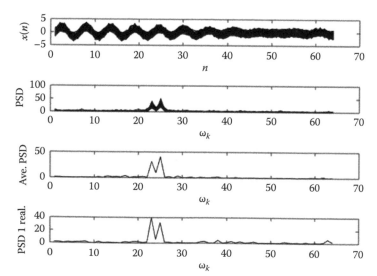

FIGURE 8.6

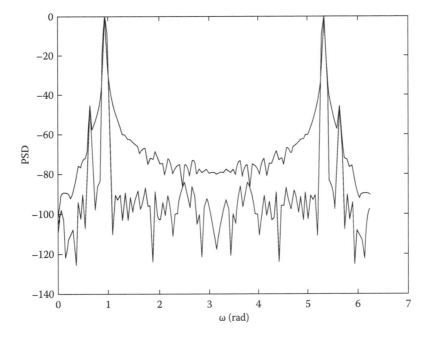

FIGURE 8.7

WINDOWED PERIODOGRAM

We can define a window periodogram as the DTFT of the time function $\{x(n)\}$ multiplied by a window, where $\{w(n)\}$ is any desired window function. Hence, the periodogram with the temporal window is given by

$$\hat{S}_x\left(e^{j\omega}\right) = \frac{1}{NP_w}\left|\sum_{n=0}^{N-1} x(n)w(n)e^{-j\omega n}\right|^2$$

$$P_w = \frac{1}{N}\sum_{n=0}^{N-1}|w(n)|^2 = \text{average power of window.}$$

(8.17)

Example 8.1.5

Apply the rectangular and Hamming window to the following signal and compare the resulting spectra.

$$x(n) = [0.06\sin(0.2\pi n) + \sin(0.3\pi n) + 0.05(\text{rand}(1,128) - 0.5)]w(n) \quad 0 \le n \le 127$$

Solution: The window provides a trade-off between spectra resolution (main lobe) and spectral masking (side-lobe amplitude). In this case, the Hamming window has side-lobe at about −45 dB, compared to the rectangular window having a side-lobe at about −13 dB. Figure 8.7 shows that although the side-lobe of the rectangular window just about obscures the 0.2π frequency, the Hamming window clearly resolves it. ∎

It can be shown that that the amount of smoothing in the periodogram is determined by the type of the window that is applied to the data. Selecting the type of window, we must weight our intention between resolution and smoothing.

The following Book m-function calculates the temporal-windowed data with the ability to introduce any one of the several windows that are given in the function without the obligation to introduce it in the command window. This helps if you wants to repeat the simulation using different windows.

BOOK M-FUNCTION FOR WINDOWED PERIODOGRAM

```
function[s,as,ps]=general_win_periodogram(x,win,L)
    %[s,as,ps]=general_win_periodogram(x,win,L);
    %window names=hamming,kaiser,hann,rectwin,
    %bartlett,tukeywin,blackman,gausswin,nattallwin,
    %triang,blackmanharris;
    %L=desired number of points (bins)
    %of the spectrum;
    %x=data in row form;s=complex
    %form of the DFT;
if (win==2) w=rectwin(length(x));
elseif (win==3) w=hamming(length(x));
elseif (win==4) w=bartlett(length(x));
elseif (win==5) w=tukeywin(length(x));
elseif (win==6) w=blackman(length(x));
elseif (win==7) w=triang(length(x));
elseif (win==8) w=blackmanharris(length(x));
end;
xw=x.*w';
for m=1:L
    n=1:length(x);
    s(m)=sum(xw.*exp(-j*(m-1)*(2*pi/L)*n));
end;
as=((abs(s)).^2/length(x))/norm(w);
    %as=amplitude spectral density;
ps=(atan(imag(s)./real(s))/length(x))/norm(w);
    %ps=phase spectrum;
    %To plot as or ps we can use the command:
    %plot(0:2*pi/L:2*pi-(2*pi/L),as);
```

Figure 8.7 was produced using the following plotting expression:

```
plot(0:2*pi/L:2*pi-(2*pi/L),20*log10(as/max(as)),'k').
```

Appendix A8.1 presents some important windows and their corresponding spectrums.

8.2   BOOK PROPOSED METHOD FOR BETTER RESOLUTION USING TRANSFORMATION OF THE RANDOM VARIABLES

One of the difficulties is that, if the sequence is short (N small number), we may not be able to sufficiently resolve frequencies being close together. Since we have at hand one realization and we can't extract another one from the population in the probability space, we propose to create a pseudo-sequence from the data in hand by linear transformation of the RVs. Next, we overlap (by about 25%) these two series. These series can also be multiplied by a window and then processed.

Note: *The proposed modified method is based on the linear transformation of RVs.*

Example 8.2.1

Compare the spectrum based on the periodogram and proposed modified one.

Solution: Figure 8.8 shows the resolution capabilities of the sequence that is made up from the original sequence as follows: y=0.2*[x zeros(1,48)]+0.2*[zeros(1,48) x];. The original

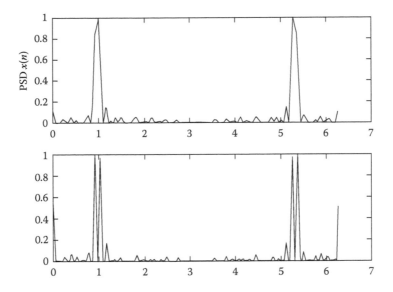

FIGURE 8.8

signal, a 64-term series, was `x=sin(0.3*pi*n)+sin(0.32*pi*n)+0.5*randn(1,64)`, `n=1:64`. The following Book m-function produced the figure.

BOOK M-FUNCTION FOR LINEARLY MODIFIED PERIODOGRAM

```
function[ax,ay,w,y]=linear_modified_periodogram(x)
    %Book MATLAB function:
    %[ax,ay,w,y]:linear_modified_periodogram(x);
    %x must be produced in the command window;
[sx,ax,px]=general_win_periodogram(x,2,512);
y1=[x zeros(1,48)]+0.2*rand(1,112);
y2=[zeros(1,48) x]+0.2*rand(1,112);
y=(0.2*y1+(y2)*0.2);
[sy,ay,py]=general_win_periodogram(y,2,512);
w=0:2*pi/512:2*pi-(2*pi/512);
    %2 implies rectangular window, see Book
    %MATLAB function general_win_periodogram
    %for other windows; to plot we must write
    %in the command window:
    %plot(w,20*log10(ax/max(ax)),'k')
    %and similar for the ay;.
```

Exercise 8.2.1

Use the proposed linear modified method and the help of correlogram to obtain the PSD of small-length sequences. Use the functions given in Example 8.2.1. ▲

8.3 DANIEL PERIODOGRAM

Daniel suggested that, for a discrete Fourier transform (DFT) periodogram, the sampled PSD $\hat{S}_{xD}\left(e^{j\omega_i}\right)$ at a particular frequency is found by averaging K points on either side of this frequency. The Daniel formula is

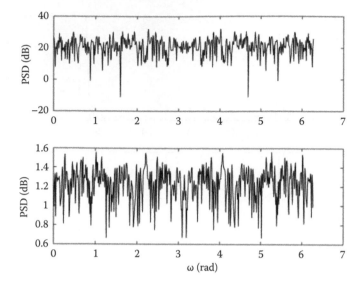

FIGURE 8.9

$$\hat{S}_{xD}\left(e^{j\omega_i}\right) = \frac{1}{2K+1}\sum_{k=i-K}^{i+K}\hat{S}_{xD}\left(e^{j\omega_i}\right) \tag{8.18}$$

This procedure is equivalent to passing the periodogram through a low-pass filter.

8.4   BARTLETT PERIODOGRAM

The variance of the periodogram, as given in Equation 8.9, presupposes an ensemble averaging for variance decrease. However, for a single realization, the variance does not decrease as $N \to \infty$ (inconsistent estimator) and hence there is a need for other approaches to reduce the variance. We can improve the statistical properties of the periodogram by replacing the expectation operator with averaging a set of periodograms. Figure 8.9 shows the ensemble averaging effect on the variance using ten realizations of the PSD only. The variance of one realization was found as follows: var(abs(fft(x(10,:))))=48.2135. The variance of the ensemble of thirty realizations was found as follows: var(abs(fft(sum(x,1)/30)))=2.0479. It is understood that any time the file is run will give different variance values but around the values which were found in one run. Compare this figure with Figure 8.4. The following Book MATLAB script file produces Figure 8.9.

BOOK M-FILE: fig8_4_1

```
%m-file:fig8_4_1
N=256;
for m=1:30
    x(m,1:N)=randn(1,N);%x=matrix 30xN;
end;
w=0:2*pi/512:2*pi-(2*pi/512);
subplot(2,1,1);
plot(w,20*log10(abs(fft(x(10,:),512))),'k');
ylabel('PSD (dB)');
    %FFT of row 10 of the matrix x;
subplot(2,1,2);
```

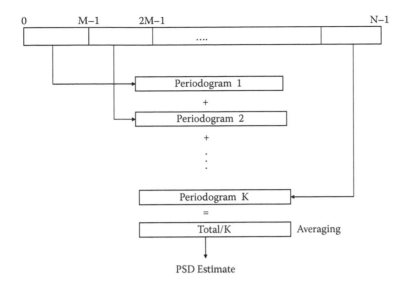

FIGURE 8.10

```
plot(w,20*log10(abs(fft(sum(x,1),512)))/30,'k');
    %sums first the 30 rows of the matrix x;
xlabel('\omega rad');ylabel('PSD (dB)');
```

Bartlett proposed to split the data in K nonoverlapping segments, which are assumed to be statistically independent. This condition presupposes that the autocorrelation of each segment decays much faster than the length of the segment. The N samples of the data signal is divided in K nonoverlapping segments of M samples each such that $KM \leq N$. The ith segment will then consists of the samples

$$x^{(i)}(n) = x(iM + n) \quad n = 0,1,\ldots,M-1$$

$$i = \text{segment number} = 0,1,\ldots,K-1$$

(8.19)

Each of the segments has the spectrum

$$S_x^{(i)}\left(e^{j\omega}\right) = \frac{1}{M}\left|\sum_{n=0}^{M-1} x(iM+n)e^{-j\omega n}\right|^2$$

(8.20)

and, hence, the Bartlett average periodogram is given by

$$S_{xB} = \frac{1}{K}\sum_{i=0}^{K-1} S_x^{(i)}\left(e^{j\omega}\right)$$

(8.21)

See Figure 8.10 for a diagrammatic representation of the above equations defining the Bartlett approach. Since the power spectrum is due to the reduced number of terms, $M \ll N$, comparing to the total available data, the resolution reduces from the order of $1/N$ to the order of $1/M$. This implies that the resolution is reduced by $N/M = K$ comparing to the resolution of the original data set. It can be shown that the variance is reduced by the same amount.

The following Book MATLAB function produces the Bartlett periodogram with a rectangular window.

BOOK M-FUNCTION FOR BARTLETT PERIODOGRAM: URDSP_BARTLETT_PSD(X,WIN,K,L)

```
function[s,as,ps]=urdsp_bartlett_psd(x,win,k,L)
    %x=data;k=number of sections;
    %L=number of points desired in the FT domain;
    %M=number of points in each section;
    %kM<=N=length(x);
    %the number 2 in the
    %function means rectangular window;
    %win=2 means rectangular window and
    %for other windows see general_win_periodogram
    %book m-function;
M=floor(length(x)/k);
s=0;
ns=1;
for m=1:k
    s=s+general_win_periodogram(x(1,ns:ns+M-1),2);
    ns=ns+M;
end;
as=((abs(s)/k).^2)/length(x);
ps=(atan(imag(s/k)./real(s/k)))/length(x);
```

8.4.1   BOOK-MODIFIED METHOD

Figure 8.11 is produced to demonstrate the ability of the proposed modified method that uses a linear transformation of the signal at hand to produce a longer pseudo-realization of the data and thus improve the resolution. The data for this figure were the following: $n = 0:127$; $x = \sin(0.3*n*pi) + \sin(0.315*n*pi) + 0.5*randn(1,128)$, as indicated above. Figure 8.11a shows the PSD of the 128-long sequence using the Bartlett approach with two sections of 64 elements each. The window for this case was the rectangular window. Figure 8.11b is the same PSD with the difference that a Hamming window is used. The reader will observe that the PSD is smoother as it was expected. The last three rows of same figure present the three proposed modified approaches, with the first column presenting the use of the rectangular window and the second column the use of the Hamming window. The three different Book m-functions for producing the modified spectrums are given later. The figure shows the average PSD of ten realizations. However, since the amplitude spectrum *as*, for example, is an Rx512 matrix, we can plot only one realization by invoking the command: plot(as(6,:)). Hence, in this case the sixth realization, out of R such realizations, is plotted. In the modified cases, the sequences were split in to two equal sequences, as was done in Figure 8.11a and b.

BOOK M-FUNCTIONS FOR THE PROPOSED MODIFIED BARTLETT SPECTRUMS

Modified No. 1

```
function[s,as,ps,avpsd]=urdsp_aver_modif_bartlett_psd1(x,win,k,L,R)
    %x=data; k=number of sections; L=number of points
    %desired in the FT domain; M=number of points in each
    %section;kM<=N=length(x);R=number of realizations;
    %win=2 for rectwin, 3 for Hamming etc, see
    %general_win_periodogram function;
    %s, as and ps are RxL amtrices;apsd is the average
    %spectrum of the R realizations;
```

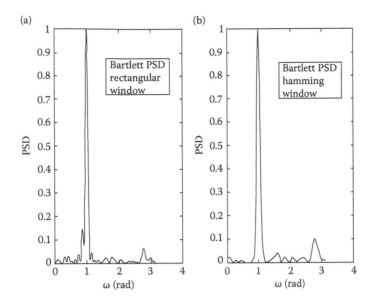

```
%if desired we can plot the amplitude spectrum
%of one realization:plot(as(5,1:512));
for m=1:R
xm(m,:)=([x   zeros(1,floor(0.8*length(x)))]+...
[zeros(1,floor(0.8*length(x)))   x+0.1*rand(1,length(x))])*.2;
[s(m,:),as(m,:),ps(m,:)]=urdsp_bartlett_psd(xm,win,k,L);
end;
avpsd=sum(as,1)/R;
```

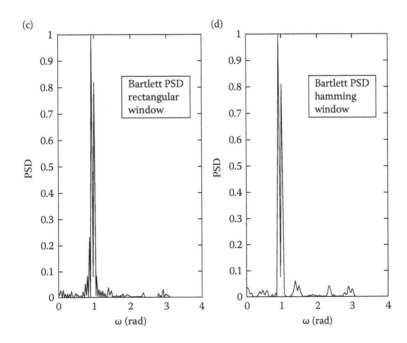

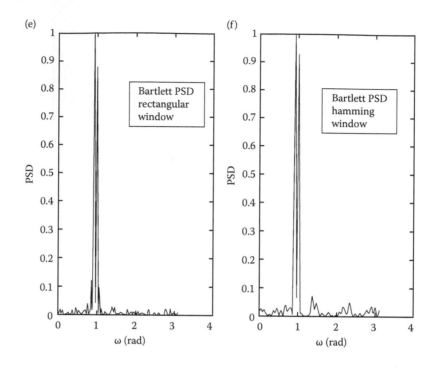

Modified No. 2

```
function[s,as,ps,apsd]=urdsp_aver_modif_bartlett_psd2(x,win,k,L,R)
    %x=data; k=number of sections; L=number of points
    %desired in the FT domain; M=number of points in each
    %section;kM<=N=length(x);R=number of realizations;
for m=1:R
x1(m,:)=[ x*0.5+0.05*randn(1,length(x))  x*0.5+0.05*randn(1,length(x))];
```

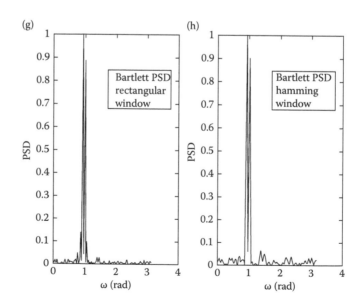

FIGURE 8.11

```
[s(m,:),as(m,:),ps(m,:)]=urdsp_bartlett_psd(x1,win,k,L);
end;
apsd=sum(as,1)/R;
```

Modified No. 3

```
function[s,as,ps,apsd]=urdsp_aver_modif_bartlett_psd3(x,win,k,L,R)
    %x=data; k=number of sections; L=number of points
    %desired in the FT domain; M=number of points in each
    %section;kM<=N=length(x);R=number of realizations;
for m=1:R
x1(m,:)=[x   0.2*x+0.1*randn(1,length(x))];
[s(m,:),as(m,:),ps(m,:)]=urdsp_bartlett_psd(x1,win,k,L);
end;
apsd=sum(as,1);
```

Exercise 8.4.1

Do the following and observe the results: (1) Create a signal of two close in frequency sinusoids plus white noise with $N=512$ values, (2) Overlay the plots of 50 periodograms with rectangular window and then average, (3) Overlay the plots of 50 Bartlett estimates with $l=4$ (four sections, $M=128$) and then average, and (4) Overlay the plots of 50 Bartlett estimates with $l=8$ (eight sections, $M=64$) and then average. ▲

Exercise 8.4.2

Repeat the modified Bartlett method as it was proposed and vary the constants in the transformation of the RVs, length of data, and window type, and observe the results. ▲

8.5 BLACKMAN–TUKEY (BT) METHOD

Because the correlation function at its extreme lag values is not reliable due to the small overlapping of the correlation process, it is recommended to use lag values of about 30%–40% of the total length of the data. The Blackman–Tukey estimator is a windowed correlogram and is given by

$$\hat{S}_{BT}\left(e^{j\omega}\right) = \sum_{m=-(M-1)}^{M-1} w(m)\hat{r}(m)e^{-j\omega m} \tag{8.22}$$

where $w(m)$ is the window with zero values for $|m| > M - 1$ and $M \ll N$. This window is the **lag window**. The above equation can also be written in the form

$$\hat{S}_{BT}\left(e^{j\omega}\right) = \sum_{m=-\infty}^{\infty} w(m)\hat{r}(m)e^{-j\omega m}$$

$$= \frac{1}{2\pi}\hat{S}_c(e^{j\omega}) * W(e^{j\omega}) = \frac{1}{2\pi}\int_{-\pi}^{\pi}\hat{S}_c(e^{j\tau})W(e^{j(\omega-\tau)})d\tau \tag{8.23}$$

where we applied the DTFT frequency convolution property (the DTFT of the multiplication of two functions is equal to the convolution of their Fourier transforms). Since windows have a dominant and relatively strong main lob, the BT estimator corresponds to a "locally" weighting average of the periodogram. Although the convolution produces smoother periodogram, it reduces resolution in the same time. It is expected that the smaller the M, the larger the reduction in variance and the lower the resolution. It turns out that the resolution of this spectral estimator is on the order of $1/M$, whereas its variance is on the order of M/N. Observe that the trade-off between resolution (smoothing) and variance depends on the total number of lags (M) we retain.

BOOK M-FUNCTION FOR BT PERIODOGRAM:[S,AS,PS]=URDSP_BTPERIODOGRAM(X,WIN,PER,L)

```
function[s,as,ps]=urdsp_BTperiodogram(x,win,per,L)
    %[s,as,ps]=urdsp_BTperiodogram(x,win,per,L)
    %window names=hamming,kaiser,hann,rectwin,
    %bartlett,tukeywin,blackman,gausswin,
    %nattallwin,triang,blackmanharris;
    %L=desired number of points (bins) of the spectrum;
    %x=data in row form;s=complex form
    %of the DFT;NOTE:per=the percentage
    %of points (length(rxt/2))deleted from
    %the correlation function
    %to decrease the edge effect;
rxt=xcorr(x,'biased');
wn=[zeros(1,floor((length(rxt)*per/2)))   ones(1,length(rxt)-(...
    2*floor((length(rxt)*per/2))))   zeros(1,floor((length(rxt)...
    *per/2)))];
rx=rxt.*wn;
if (win==2) w=rectwin(length(rx));
elseif (win==3) w=hamming(length(rx));
elseif (win==4) w=bartlett(length(rx));
elseif (win==5) w=tukeywin(length(rx));
elseif (win==6) w=blackman(length(rx));
elseif (win==7) w=triang(length(rx));
elseif (win==8) w=blackmanharris(length(rx));
end;
rxw=rx.*w';
for m=1:L
    n=1:length(rx);
    s(m)=sum(rxw.*exp(-j*(m-1)*(2*pi/L)*n));
end;
as=((abs(s)).^2/length(x))/norm(w);%as=amplitude
%spectral density;
ps=atan((imag(s)./(real(s)+eps))/((length(x))/(norm(w)+eps)));
    %ps=phase spectrum;
    %To plot as or ps we can use the command:
    %plot(0:2*pi/L:2*pi-(2*pi/L),as);
```

The top drawing of Figure 8.12 was produced using the following: 50% of the correlation function which was produced from a 128 values of a time function and a rectangular window. The middle drawing was produced using 80% of its correlation function and a rectangular window. The last figure used 80% of its correlation function but with a Hamming window, which produces a spectrum with smaller resolution but a smoother one. The signal used for all three results was $x = \sin(0.3\pi n) + \sin(0.32\pi n) + 0.5\mathrm{randn}(1,128)$.

A second proposed Book-modified method is put forward for the BT method by using linear transformation to RVs. The upper left part of Figure 8.13 shows the original data:

$$x = \sin(0.3\pi n) + \sin(0.32\pi n) + 0.8\mathrm{randn}(1,64).$$

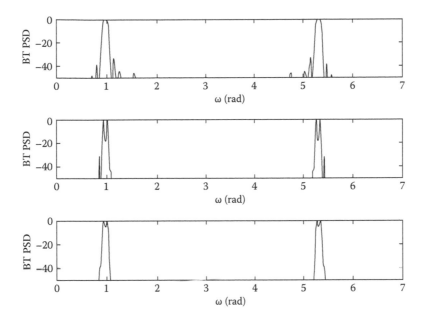

FIGURE 8.12

The second signal is equal to the original plus a linear transformation of the original: $z = ([x * 0.5 \text{ zeros}(1,5)] + [\text{zeros}(1,5) \, 0.5 * x]) * 0.2$. This signal is plotted in the upper right side of Figure 8.13. The lower section and left part of Figure 8.13 shows the BT PSD of the original sequence $\{x(n)\}$ on a linear scale. The right side of the lower section depicts the BT PSD of the modified sequence on a linear scale. For both cases, the percentage of eliminating the trailing correlations terms was 30%.

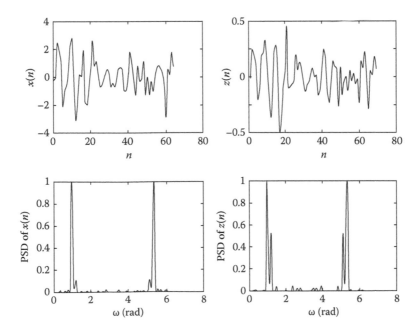

FIGURE 8.13

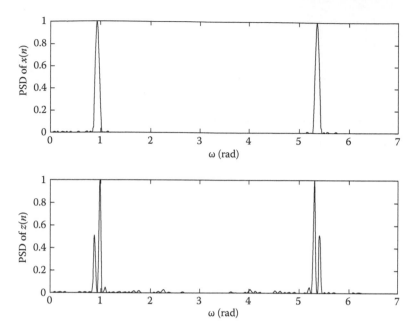

FIGURE 8.14

A different modified approach, without overlapping but using window, is given by the following Book MATLAB program. The spectrums are shown in Figure 8.14.

Book MATLAB Program for Modified BT Using Linear RVs Transformation and Window

```
>>n=0:63;
>>x=sin(0.30*pi*n)+sin(0.32*pi*n)+0.8*randn(1,64);
>>z=[x*2.*hamming(64)'     x.*hamming(64)'].*hamming(128)';
>>[sx,asx,apx]=urdsp_BTperiodogram(x,2,0.4,512);
>>[sz,asz,apz]=general_win_periodogram(z,2,0.4,512);
```

It is interesting to see, by creating an ensemble of modified sequences, whether the resolution increases or not. Figure 8.15a shows the BT PSD of the sequence: $x = \sin(0.3*pi*n) + \sin(0.325*pi*n) + \mathrm{randn}(1,64)$, and Figure8.15b shows the average (20 sequences) modified BT PSD. Figure 8.15b was created by the following Book m-function given. The following constants were used: win=2=rectangular, per=0.5, avn=20, L=512.

Book m-Function for Modified BT Method Using RV Transformation and Averaging

```
function[sz,az,pz]=aver_modified_BTpsd(x,win,per,avn,L)
    %win=2 implies rectangular, see the
    %function general_win_periodogram
    %to identify the appropriate numbers
    %for other windows;per=
    %percentage of deleted one sided
    %aucorralation function;
    %avn=number of ensemble sequences
    %to be averaged;L=number
    %of frequency bins;
```

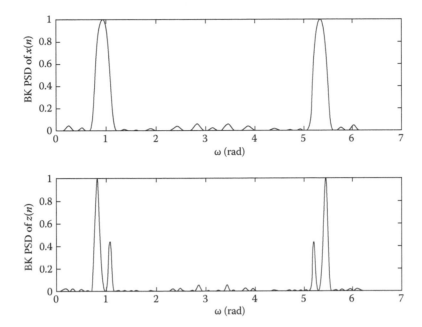

FIGURE 8.15

```
for m=1:avn
    z1(m,:)=[x*(rand)*0.5.*hamming(64)'...
        x*(rand)*0.2].*hamming(2*length(x))';
end;
z=sum(z1,1);
[sz,az,pz]=urdsp_BTperiodogram(z,win,per,L);
```

Exercise 8.5.1

Use the BT periodogram function and observe the spectrums for different windows, different variances of the white additive noise, and different sinusoids having close frequencies or not.   ▲

8.6   WELCH METHOD

Welch proposed modifications to the Bartlett method, as follows: Data segments are allowed to overlap and each segment is windowed prior to computing the periodogram. Since in most practical applications only a single realization is available, we create smaller sections as follows:

$$x_i(n) = x(iD+n)w(n) \quad 0 \le n \le M-1 \; 0 \le i \le K-1 \tag{8.24}$$

where

$w(n)$ is the window of length M
D is an offset distance
K is the number of sections that the sequence $\{x(n)\}$ is divided into

Pictorially, the Welch method is shown in Figure 8.16.
    The ith periodogram is given by

$$S_i\left(e^{j\omega}\right) = \frac{1}{M}\left|\sum_{n=0}^{M-1} x_i\left(e^{-j\omega n}\right)\right|^2 \tag{8.25}$$

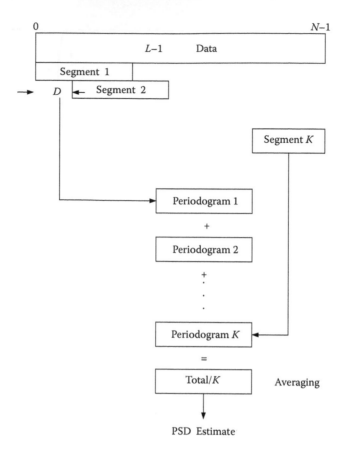

FIGURE 8.16

and the average Welch periodogram is given by

$$S\left(e^{j\omega}\right) = \frac{1}{K}\sum_{i=0}^{K-1} S_i\left(e^{j\omega}\right) \tag{8.26}$$

BOOK M-FUNCTION FOR WELCH PERIODOGRAM

```
function[s,ast,ps,K,M]=welch_periodogram(x,win,frac,frac1,L)
    %function[as,ps,s,K,M]=welch_periodogram(x,win,frac,frac1,L);
    %x=data; M=section length;
    %L=number of samples desired in the frequency domain;
    %win=2 means rectwin, number 3 means hamming window;
    %see general_win_periodogram function to add
    %more windows if you desire;
    %frac defines the number of data for each
    %section, depending on the data
    %length it is recommended the number
    %to vary between 0.2 and 1;
    %frac1 defines the overlapping of the sections, it is
    %recommended the frac1 to vary from 1
    %(no overlap) to 0.5
    %which means a 50% overlap; M<<N=length(x);
```

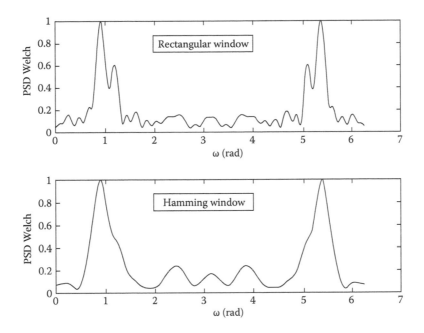

FIGURE 8.17

```
if (win==2) w=rectwin(floor(frac*length(x)));
elseif (win==3) w=hamming(frac*length(x));
end;
N=length(x);
M=floor(frac*length(x));
K=floor(floor((N-M+floor(frac1*M)))/floor(frac1*M));
    %K=number of processings;
s=0;as=0;
for i=1:K
    s=s+fft(x(1,(i-1)*(floor(frac1*M))...
        +1:(i-1)*floor(frac1*M)+M).*w',L);
    as=as+abs(s);
end;
ast=as/(M*K);%as=amplitude spectral density;
ps=atan(imag(s)./real(s))/(M*K);
    %phase spectral density;
```

Figure 8.17 shows the result using the above Book m-function with the following constants and functions: $n = [0,1,2,...,63]; x = \sin(0.3*pi*n) + \sin(0.37*pi*n) + 0.2*randn(1,64);$ frac $= 0.5$, frac1 $= 0.25$, L $= 256$, win $= 2$ (rectangular window), win $= 3$ (Hamming window).

8.6.1 Proposed Modified Methods for Welch Method

Modified Method Using Different Types of Overlapping

It is evident from Figure 8.16 that, if the lengths of the sections are not long enough, frequencies close together can't be differentiated. Therefore, we propose a procedure, defined as **symmetric modified Welch method**. Its implementation is shown in Figure 8.18. Windowing of the segments can also be incorporated. This approach and the rest of the proposed schemes have the advantage of progressively incorporating longer and longer segments of the data and thus introducing better and better resolution. In addition, due to the averaging process, the variance decreases and smoother

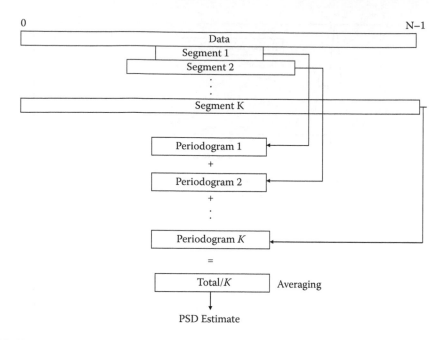

FIGURE 8.18

periodograms are obtained but not as smooth as the Welch method. It is up to the reader to decide between smoothness of the periodogram and resolution of frequencies. Figure 8.19 shows another proposed method that is defined as the **asymmetric modified Welch method**. Figure 8.20 shows another suggested approach for better resolution and reduced variance. The procedure is based on the method of prediction and averaging. This proposed method is defined as the **symmetric prediction modified Welch method**. This procedure can be used in all the other forms, for example, nonsymmetric. The abovementioned methods can also be used for spectral estimation if we substitute the word "periodogram" with the word "correlogram."

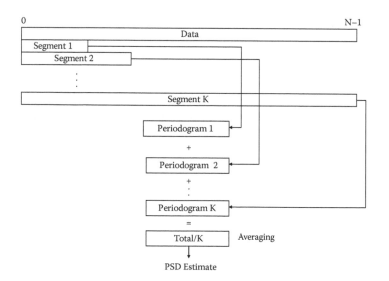

FIGURE 8.19

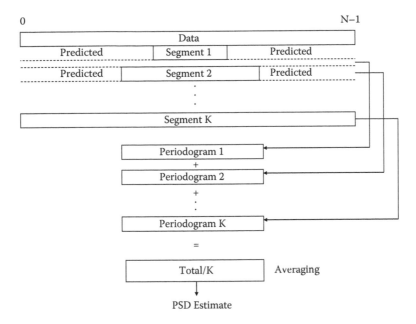

FIGURE 8.20

The drawing at the top of Figure 8.21 shows data created by the equation

$$x(n) = \sin(0.3\pi n) + \sin(0.33\pi n) + 0.5\text{randn}(1,128) \tag{8.27}$$

and for 128 sample values. The middle drawing is a Welch periodogram with two sections of the 128 points, 50% overlap, and a Hamming window. For the bottom drawing, the symmetric proposed method was used, shown in Figure 8.20, and Hamming window was used. It is apparent that the method produces a better resolution and with smaller variance.

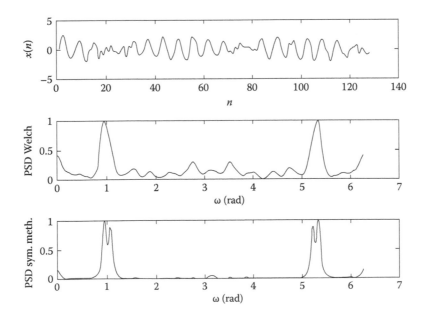

FIGURE 8.21

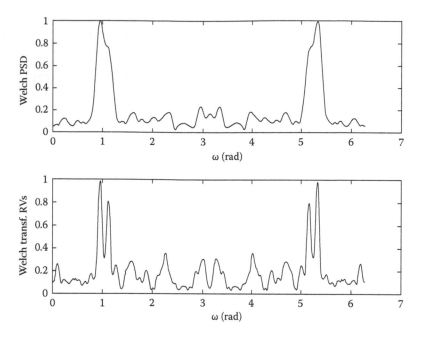

FIGURE 8.22

Modified Welch Method Using RV Transformation

Top drawing in Figure 8.22 shows the result using the Welch method for the signal $x = \sin(0.3\pi n) + \sin(0.36\pi n) + 0.5\mathrm{randn}(1,128)$, $n = 0,1,\ldots,127$. The signal was split into two sections with 50% overlap. For the second drawing of Figure 8.22, the following transformed signal was used by incorporating a linear transformation of the RVs: $z = [x + 0.1 * x + 0.05 * \mathrm{randn}(1,128)]$. For this signal, the Welch method was used, splitting the series into two parts and using 50% overlapping. The figure was produced using the following Book MATLAB program:

BOOK M-FILE USING RV TRANSFORMATION

```
%welch_transf_rvs
n=0:127;
x=sin(0.3*pi*n)+sin(0.36*pi*n)+0.5*randn(1,128);
 [sx,as ps,K,M]=welch_periodogram(x,3,0.5,0.5,512);%
     %50% overlap;
w=0:2*pi/512:2*pi-(2*pi/512);
subplot(2,1,1)
plot(w,as/max(as),'k')
xlabel('\omega rad');ylabel('welch PSD');
z=[x    0.1*x+0.05*randn(1,128)];
 [sz,asz psz,Kz,Mz]=welch_periodogram(z,3,0.5,0.5,512);
subplot(2,1,2)
plot(w,asz/max(asz),'k')
xlabel('\omega rad');ylabel('Welch transf. rvs');
```

Exercise 8.6.1

Use the Welch approach with different variances of the noise, different distance between frequencies of sinusoids, different windows, and different lengths of segments.   ▲

HINTS, SUGGESTIONS, AND SOLUTIONS OF THE EXERCISES

8.1.1

Multiply Equation 8.2 by $e^{j\omega m}$ and integrate in the range $-\pi \le \omega \le \pi$. Hence,

$$\int_{-\pi}^{\pi} X\left(e^{j\omega}\right) e^{j\omega m} \, d\omega = \sum_{n=-\infty}^{\infty} x(n) \int_{-\pi}^{\pi} e^{-j\omega(n-m)} \, d\omega = 2\pi x(n) \text{ since the integral is zero for } n \ne m \text{ (an inte-}$$

ger) and equal to 2π for $n=m$.

(8.5)

$$\frac{1}{2\pi} \int_{-\pi}^{\pi} S_x\left(e^{j\omega}\right) d\omega = \frac{1}{2\pi} \int_{-\pi}^{\pi} \sum_{n=-\infty}^{\infty} \sum_{m=-\infty}^{\infty} x(n)x(m) e^{-j\omega(n-m)} d\omega = \sum_{n=-\infty}^{\infty} \sum_{m=-\infty}^{\infty} x(n)x(m)$$

$$\left[\frac{1}{2\pi} \int_{-\pi}^{\pi} e^{-j\omega(n-m)} \, d\omega\right] = \sum_{n=-\infty}^{\infty} x^2(n); (1/2\pi) \int_{-\pi}^{\pi} e^{-j\omega(n-m)} \, d\omega = \delta(n-m) \text{(Kronecker delta)}$$

which is equal to 1 if $m=n$ and zero if $m \ne n$. Hence, in the expansion of the summations only the terms with square elements of $x(n)$'s remain.

(8.7)

$$\sum_{k=-\infty}^{\infty} r_x(k) e^{-j\omega k} = \sum_{k=-\infty}^{\infty} \sum_{n=-\infty}^{\infty} x(n)x(n-k) e^{-j\omega n} e^{j\omega(n-k)} = \left[\sum_{n=-\infty}^{\infty} x(n) e^{-j\omega n}\right]\left[\sum_{s=-\infty}^{\infty} x(s) e^{-j(-\omega)s}\right]$$

$$= X\left(e^{j\omega}\right) X\left(e^{-j\omega}\right) = \left|X(e^{j\omega})\right|^2 = S_x\left(e^{j\omega}\right)$$

8.1.2

$$y(n) = \frac{1}{\sqrt{N}} \sum_{m=0}^{N-1} h(m)v(n-m), h() = \text{constant}, v() = \text{white noise with variance one} \qquad (1)$$

$$Y\left(e^{j\omega}\right) = \frac{1}{\sqrt{N}} \sum_{m=0}^{N-1} y(m) e^{-j\omega m} \qquad (2)$$

$$E\{v(n)v(m)\} = r_v(k) = 1\delta(k), \quad k = m - n;$$

$$V\left(e^{j\omega}\right) = \sum_{k=0}^{N-1} r_v(k) e^{-j\omega k} = \sum_{k=0}^{N-1} \delta(k) e^{-j\omega k} = 1 \qquad (3)$$

from (4.26),

$$S_y\left(e^{j\omega}\right) = \left|H\left(e^{j\omega}\right)\right|^2 = \hat{S}_{py}\left(e^{j\omega}\right) \qquad (4)$$

$$r_y(k) = E\{y(n)y(n-k)\} = \frac{1}{N} \sum_{p=0}^{N-1} \sum_{s=0}^{N-1} h(p)h(s)E\{v(n-p)v(n-k-s)\}$$

$$= \frac{1}{N} \sum_{p=0}^{N-1} \sum_{s=0}^{N-1} h(p)h(s)\delta(k+s-p) = \frac{1}{N} \sum_{p=0}^{N-1} h(p)h(p-k) = r_h(k) \quad \text{for } 0 \le k \le N-1 \quad (5)$$

Using the DTFT of (5), we find $S_y\left(e^{j\omega}\right) = S_h\left(e^{j\omega}\right) \triangleq \left|H\left(e^{j\omega}\right)\right|^2$ and, hence, from (4), we obtain the desired results.

8.1.3

To show that the periodogram is asymptotically unbiased, we must show that the mean value of the correlation tend to zero as $N \to \infty$. Hence, from the correlation formula, we have

$$E\{\hat{r}_x(k)\} = \frac{1}{N}\sum_{n=k}^{N-1} E\{x(n)x(n-k)\} = \frac{1}{N}\sum_{n=k}^{N-1}\hat{r}_x(k) = \hat{r}_x(k)\frac{1}{N}(N-k) = \left(1 - \frac{k}{N}\right)\hat{r}_x(k) \qquad (1)$$

Taking into consideration the correlation symmetry, we write

$$E\{\hat{r}_x(k)\} = \begin{cases} \left(1 - \dfrac{|k|}{N}\right)\hat{r}_x(k) & |k| \le N \\ 0 & \text{elsewhere} \end{cases} \quad ; \quad w_B = \begin{cases} \left(1 - \dfrac{|k|}{N}\right) & |k| \le N \\ 0 & \text{elsewhere} \end{cases} = \text{Bartlett window.}$$

Taking the ensemble average of the periodogram, we obtain

$$E\left\{\hat{S}_{px}\left(e^{j\omega}\right)\right\} = E\left\{\sum_{k=-N+1}^{N-1}\hat{r}_x(k)e^{-j\omega k}\right\} = \sum_{k=-N+1}^{N-1}E\{\hat{r}_x(k)\}e^{-j\omega k} = \sum_{k=-\infty}^{\infty}r_x(k)w_B e^{-j\omega k}$$

Since the ensemble average of the periodogram (DTFT) is equal to DTFT of the product $r_x(k)w_B(k)$, the product property of DTFT gives $E\left\{S_{px}\left(e^{j\omega}\right)\right\} = \dfrac{1}{2\pi}S_x\left(e^{j\omega}\right) * W_B\left(e^{j\omega}\right)$ where $W_B\left(e^{j\omega}\right) = [\sin(N\omega/2)/\sin(\omega/2)]^2/N$ is the DTFT of the Bartlett window. Since the DTFT of the Bartlett window approaches a delta function as N approaches infinity, the convolution becomes $\lim\limits_{N\to\infty} E\left\{S_{px}\left(e^{j\omega}\right)\right\} = S_x\left(e^{j\omega}\right)$.

8.1.5

Correlogram is more accurate with smaller variations of the PSD.

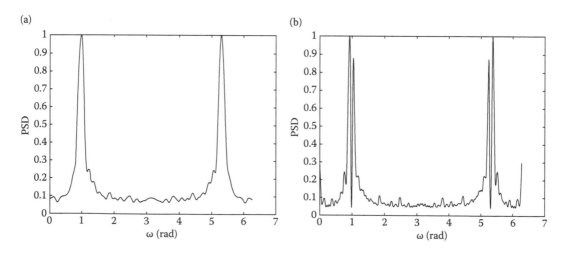

FIGURE 8.23

8.2.1

Figure 8.23a shows the PSD of the signal x and Figure 8.23b shows the PSD modified signal by linear extension, as is done in Example 8.2.1.

APPENDIX A8.1

IMPORTANT WINDOWS AND THEIR SPECTRA

1. Rectangular:

$$w_R(n) = \begin{cases} 1 & 0 \leq n \leq N-1 \\ 0 & \text{elsewhere} \end{cases}$$

Figure A8.1a presents both the window and its PSD in log scale.

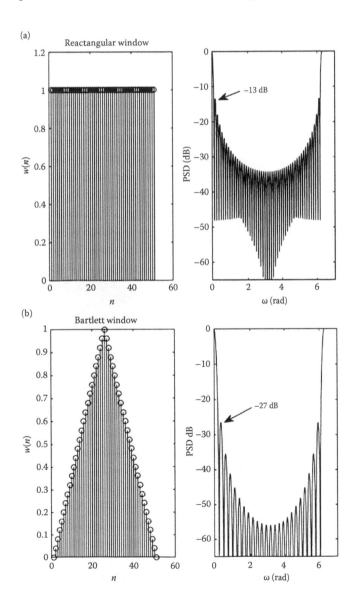

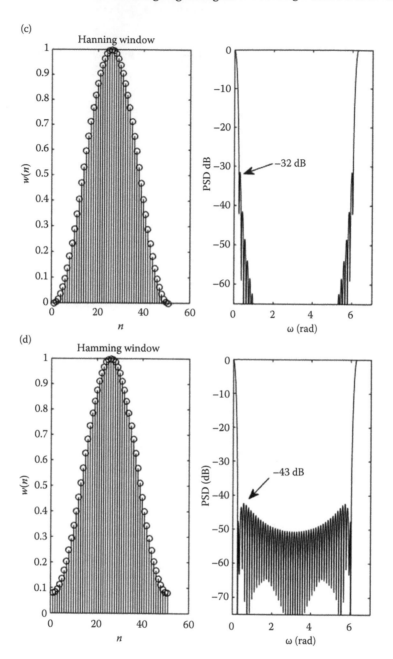

2. Bartlett:

$$
w_B(n) = \begin{cases} 2n/(N-1) & 0 \le n \le (N-1)/2 \\ 2 - 2n/(N-1) & (N-1)/2 \le n \le N-1 \\ 0 & \text{elsewhere} \end{cases}
$$

Figure A8.1b presents both the window and its PSD in log scale.

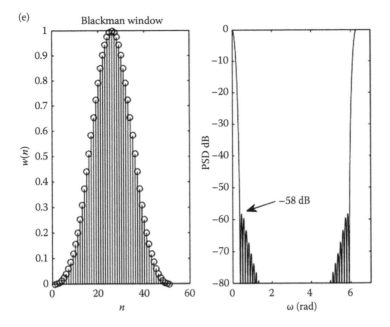

FIGURE A8.1

3. Hanning:

$$w_{\text{Han}}(n) = \begin{cases} (1 - \cos[2\pi n / (N-1)]) / 2 & 0 \le n \le N-1 \\ 0 & \text{elsewhere} \end{cases}$$

Figure A8.1c presents both the window and its PSD in log scale.

4. Hamming:

$$w_{\text{Ham}} = \begin{cases} 0.54 - 0.46\cos[2\pi n / (N-1)] & 0 \le n \le N-1 \\ 0 & \text{elsewhere} \end{cases}$$

Figure A8.1d shows both the window and its PSD in log scale.

5. Blackman:

$$w_{\text{Bl}}(n) = \begin{cases} 0.42 - 0.5\cos[2\pi n / (N-1)] + 0.08\cos[4\pi n / (N-1)] & 0 \le n \le N-1 \\ 0 & \text{elsewhere} \end{cases}$$

Figure A8.1e shows both the window and its PSD in log scale.

9 Parametric and Other Methods for Spectral Estimation

9.1 INTRODUCTION

In the previous chapter, we presented the classical (nonparametric) spectra estimation. The procedure was to find the autocorrelation function (ACF) of the data and then take the discrete-time Fourier transform (DTFT) of the ACF. In this chapter, we will consider the **parametric** approach that is based on model parameters rather than on ACF. These models include the autoregressive (AR) model, the moving average (MA) model, and the autoregressive-moving average (ARMA) model. It is assumed that the data are the result of the output of one of these systems with input white noise having finite variance. The parameters and the noise variance will be determined as described in this chapter.

The parametric approach has been devised to produce better spectral resolution and better spectra estimation. However, we must be careful in using this method since the degree of improvement in resolution and spectral fidelity is solely determined by the appropriateness of the selected model.

Primarily, we will consider **discrete spectra** (sinusoidal signals) embedded in white noise. The motivation for studying parametric models for spectrums estimation is based on the ability to achieve better power spectral density (PSD) estimation, assuming that we incorporate the appropriate model. Furthermore, in the nonparametric case, the PSD was obtained from a windowed set of data or autocorrelation sequence (ACS) estimates. The unavailable data, in both the sequences and ACS, imposed the unrealistic assumption that the data outside the windows have zero values. If some knowledge about the underlined process that produces the sequence is present, then the extrapolation to unavailable data outside the window is a more realistic process than in the nonparametric case. This brings up the idea that the window is not needed and, as a consequence, less distortion may occur to the spectrum.

9.2 AR, MA, AND ARMA MODELS

The difference equation that describes the general model is (see also Chapter 3)

$$y(n) = -\sum_{m=1}^{p} a(m)y(n-m) + \sum_{m=0}^{q} b(m)v(n-m) \tag{9.1}$$

For simplicity and without loss of generality, we will assume that $b(0)$ is equal to 1. In the preceding equation, $y(n)$ represents the output of the system and $v(n)$ is the input white noise to the system. This is an innate-type of input to the system. Any external noise that is present must be separately added to the system. If we would like to present the system with its impulse response, $\{h(n)\}$, the preceding equation takes the form

$$y(n) = \sum_{m=0}^{\infty} h(m)v(n-m) \tag{9.2}$$

For causal systems, we have the relation $h(m)=0$ for $m < 0$. If we take the z-transform of both sides of the preceding equations, we obtain the following (remember that the z-transform of the convolution of two sequences is equal to the multiplication of their z-transforms):

$$\frac{Y(z)}{V(z)} = \frac{B(z)}{A(z)}, \quad \frac{Y(z)}{V(z)} \triangleq H(z) \text{ or } H(z) = \frac{B(z)}{A(z)} \tag{9.3}$$

We will assume that all the roots and zeros of the transfer function $H(z)$ are within the unit circle so that the systems are **stable**, **causal**, and **minimum phase**. The expanded forms of the preceding polynomials are

$$A(z)=1+\sum_{m=1}^{p}a(m)z^{-m} \quad B(z)=1+\sum_{m=1}^{q}b(m)z^{-m}$$

$$H(z)=1+\sum_{m=1}^{\infty}h(m)z^{-m}$$

(9.4)

The power spectrum of the output is given by (see Section 6.2 and $z=e^{j\omega}$)

$$S_y(z)=S_v(z)H(z)H(z^{-1})=S_v(z)\frac{B(z)B(z^{-1})}{A(z)A(z^{-1})}$$

(9.5)

This equation represents the ARMA model for the time series $y(n)$ when the input is zero mean white noise. The ARMA PSD is obtained by setting $z=e^{j\omega}$ in Equation 9.5, yielding

$$S_y(e^{j\omega})=\sigma_v^2\frac{B(e^{j\omega})B(e^{-j\omega})}{A(e^{j\omega})A(e^{-j\omega})}=\sigma_v^2\frac{\left|B(e^{j\omega})\right|^2}{\left|A(e^{j\omega})\right|^2}$$

(9.6)

where σ_v^2 is the variance of the white noise that entirely characterizes the PSD of the process $y(n)$. The polynomials in the preceding equation are

$$A(e^{j\omega})=1+\sum_{m=1}^{p}a(m)e^{-j\omega m} \quad B(e^{j\omega})=1+\sum_{m=1}^{q}b(m)e^{-j\omega m}$$

(9.7)

When the $a(m)$'s are zero, the system is the MA one, and when the $b(m)$'s are zero, the system is an AR. In this text, we concentrate on the AR-type systems since these systems provide the sharpest peaks in the spectrum at the corresponding sinusoidal frequencies and will provide us with the fundamentals of dealing with the processing of random discrete signals. The AR system is characterized in the time and frequency domain by the following two equations, respectively:

$$y(n)=-\sum_{m=1}^{p}a(m)y(n-m)+v(n)$$

(9.8)

$$S_y(e^{j\omega})=\frac{\sigma_v^2}{\left|1+\sum_{m=1}^{p}a(m)e^{j\omega m}\right|^2}$$

(9.9)

For causal system, the ARMA and AR correlation equations are

$$r_{yy}(k)+\sum_{m=1}^{p}a(m)r_{yy}(k-m)=\sigma_v^2\sum_{m=0}^{q}b(m)h(m-k)$$

(9.10)

Exercise 9.2.1

Verify Equation 9.10.   ▲

Since, in general, $h(k)$ is a nonlinear function of the $a(m)$'s and $b(m)$'s coefficients and we impose the causality condition, $h(s)=0$ for $s < 0$, Equation 9.10 for $k \geq q+1$ becomes

$$r_{yy}(k) + \sum_{m=1}^{p} a(m)r_{yy}(k-m) = 0 \qquad k > q \tag{9.11}$$

which is the basis for many estimators of the AR coefficients of the AR process. For the values $k \geq q+1$, and the restriction of $m \leq q$, the impulse response of a causal system is zero.

9.3   YULE–WALKER (YW) EQUATIONS

Equations 9.10 and 9.11 indicate a linear relationship between the correlation coefficients and the AR parameters. From Equation 9.10 and for $k=0$, we obtain the relation

$$r_{yy}(0) + \sum_{m=1}^{p} a(m)r_{yy}(0-m) = \sigma_v^2 + \sigma_v^2 \sum_{m=1}^{p} b(m)h(m-0) = \sigma_v^2 \tag{9.12}$$

because the system is an AR system and by definition all the $b(m)$'s are zero for $m > 0$.

The last equality of the preceding equation is true since $b(0)=h(0)=1$ (see Equation 9.4) and any other $b(m)$ is zero for an AR model to be true. In the expansion of Equation 9.11 (for AR model $q=0$) in connection with Equation 9.12, we obtain the following set of equations:

$$\begin{bmatrix} r_{yy}(0) & r_{yy}(-1) & \cdots & r_{yy}(-p) \\ r_{yy}(1) & r_{yy}(0) & \cdots & r_{yy}(-p+1) \\ \vdots & \vdots & & \vdots \\ r_{yy}(p) & r_{yy}(p-1) & \cdots & r_{yy}(0) \end{bmatrix} \begin{bmatrix} 1 \\ a(1) \\ \vdots \\ a(p) \end{bmatrix} = \begin{bmatrix} \sigma_v^2 \\ 0 \\ \vdots \\ 0 \end{bmatrix} \tag{9.13}$$

The preceding equations are known as the **Yule–Walker** (YW) or the **normal ones** (see also Section 6.4). For additional information about vectors and matrices operations, see Appendix 2 at the end of the book. These equations form the basis of many AR estimate methods. The preceding matrix is Toeplitz because $r_{yy}(k) = r_{yy}(-k)$. If the correlation coefficients are known, the lower part of the preceding equation can be written as follows:

$$\begin{bmatrix} r_{yy}(0) & \cdots & r_{yy}(-p+1) \\ \vdots & & \vdots \\ r_{yy}(p-1) & \cdots & r_{yy}(0) \end{bmatrix} \begin{bmatrix} a(1) \\ a(2) \\ \vdots \\ a(p) \end{bmatrix} = \begin{bmatrix} r_{yy}(1) \\ \vdots \\ r_{yy}(p) \end{bmatrix} \tag{9.14}$$

In matrix form, the preceding equation can be written as

$$R_{yy(p)}a = -r_{yy(p)} \text{ and the solution is } a = -R_{yy(p)}^{-1}r_{yy(p)} \tag{9.15}$$

To find σ_v^2, we introduce the $a(n)$'s, just found from Equation 9.15, in the first row of Equation 9.13. It can be shown that the correlation matrix in Equation 9.13 is positive definite and, hence, it has a unique solution.

Note: *We must keep in mind that the true correlation elements $\{r(k)\}$ are replaced with the sample correlation $\{\hat{r}(k)\}$.*

Exercise 9.3.1

Produce the autocorrelation sequences of the output of a first-order AR filter (system) and use $N=100$ and $N=10000$ to produce them. Compare the results and draw your conclusion. In our case, we used $a=0.7$. The reader should also try the value $a=-0.7$. ▲

Example 9.3.1

Find the spectrum using the YW equations for the sequence produced by the infinite impulse response (IIR) filter:

$$y(n) = 1.3847y(n-1) - 1.5602y(n-2)$$

$$+0.8883y(n-3) - 0.4266y(n-4) + v(n)$$

$v(n)$ is white Gaussian noise with zero mean and unit variance values. Use 30th and 60th order AR estimating models. Produce a sequence with 256 elements and add the following two different sinusoids:

$$s = 2\sin(0.3\pi n) + 2\sin(0.32\pi n)$$

$$s = 3\sin(0.3\pi n) + 3\sin(0.32\pi n)$$

Solution: To produce Figure 9.1, we used the AR model given earlier where $v(n)$ is a white Gaussian noise with variance one and mean value zero. To the preceding data, the following signal was added: $s(n) = 2\sin(0.3n\pi) + 2\sin(0.32n\pi)$. In the figure, we observe that as the order of the AR estimating model increases, 30 and 60 respectively, the resolution increases. Figure 9.2 was produced with the same data and the same orders respectively, but with the following signal: $s(n) = 3\sin(0.3n\pi) + 3\sin(0.32n\pi)$. For this case, the separation of the two frequencies is clearly defined. Comparing Figures 9.1 and 9.2, we observe that as the signal-to-noise ratio increases, identification of the line spectrum improves considerably. The total number of data in the sequence

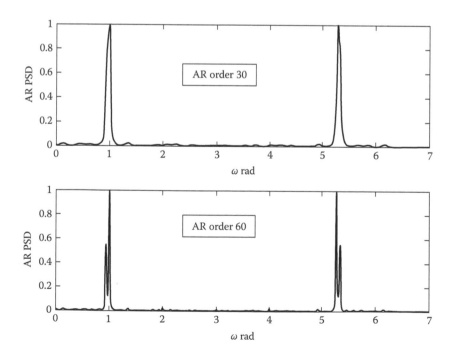

FIGURE 9.1

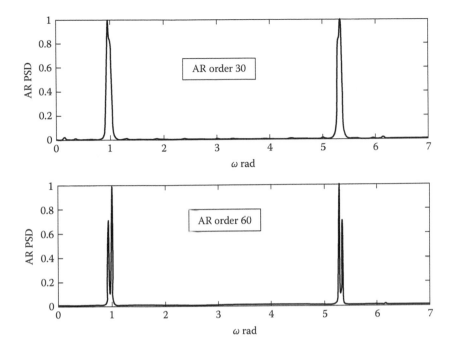

FIGURE 9.2

was 256 and the correlation lag was 128 units. The results were found using the following Book m-function:

BOOK M-FUNCTION: [R]=SAMPLE_BIASED_AUTOC(X,LG)

```
function[r]=sample_biased_autoc(x,lg)
    %the function finds the biased autocorrelation
    %with lag from 0 to lg,it is recommended that lg
    %is 20-30% of N;N=total number of elements of
    %the observed vector x (data);N=length(x);lg
    %stands for lag number;
N=length(x);%length(x)=MATLAB function returning
            %the number of elements of the vector x
for m=1:lg
    for n=1:N+1-m
        x1(n)=x(n-1+m)*x(n);
    end;
    r(m)=sum(x1)/N;
end;                                                    ■
```

Figure 9.3 was produced using the following Book m-function. The left side of Figure 9.3 gives one of the realizations of the spectrum, and the right side of Figure 9.3 is the result of averaging 15 realizations. The order of the AR model estimator was 60, the number of data was 256, and the lag number for the autocorrelation function was 128. The averaging reduces the variance of the spectrum.

BOOK M-FUNCTION: [W,R,ARFT]=AR_PSD_REALIZ(LG,ORD,K,L,N)

```
function[w,r,arft]=AR_psd_realiz(lg,ord,K,L,N)
    %x=output of an IIR with input white Gaussian
    %noise of variance 1 and mean zero;
    %a(ord)=coefficients
```

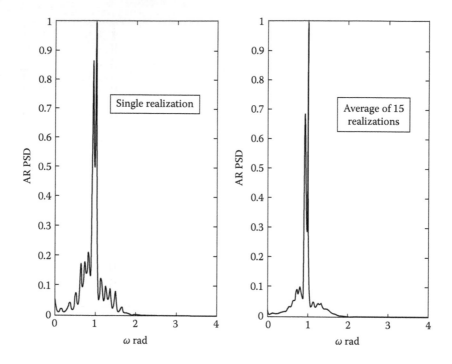

FIGURE 9.3

```
%with order equal to ord;lg=number of
%lags for the autocorrelation
%function and MUST be at least
%one larger than the ord;
%ord=order of the AR model;
%L=desired bins in the spectrum power of 2^n;
%on the command window the
%plot(w,arft) will plot the PSD;
%K=number of realizations; N=length
%of IIR output sequence (vector x);
%arft=a LxK matrix with each column
%representing of the PSD for each
%realization, to plot the average you
%write: plot(w,sum(arft,2)), the
%number 2 indicates the summation of columns;
%the reader can easily change
%the signal and the IIR filter to
%fit the desired outputs;
for k=0:K-1
    for n=0:N-1
    x(4)=0;x(3)=0;x(2)=0;x(1)=0;
    x(n+5)=1.3847*x(n+4)-1.5602*x(n+3)+0.8883*x(n+2)-...
    0.4266*x(n+1)+2*randn;
end;
q=0:N-1;
s=4*sin(0.3*pi*q)+4*sin(0.32*pi*q);
x1=x(1,5:N+4)+s;
r=sample_biased_autoc(x1,lg);
m=1:ord;
R=toeplitz(r(1,1:ord));
a=-inv(R)*r(1,2:ord+1)';
w=0:2*pi/L:2*pi-(2*pi/L);
arft(:,k+1)=(1./abs((1+(exp(-j*w'*m)*a)).^2))/lg;
end;
```

9.4 LEAST-SQUARES (LS) METHOD AND LINEAR PREDICTION

An alternative approach would be to perform a least squares (LS) minimization with respect to the linear prediction coefficients. There exist two types of LS estimates, the forward and backward linear prediction estimates, and the combination of forward and backward linear prediction.

Assume the N-data sequence $y(1), y(2), \ldots, y(N)$ to be used to estimate the pth order AR parameters. The forward **linear prediction** data $\hat{y}(n)$, based on the previously given data, $\{y(n-1) \quad y(n-2) \quad \cdots \quad y(n-p)\}$ is

$$\hat{y}(n) = -\sum_{i=1}^{p} a(i)y(n-i) = -y^T(n)a,$$

$$y(n) = [\quad y(n-1) \quad y(n-2) \quad \cdots \quad y(n-p) \quad]^T \tag{9.16}$$

$$a = [\quad a(1) \quad a(2) \quad \cdots \quad a(p) \quad]^T$$

Therefore, the **prediction error** is

$$e_p(n) = y(n) - \hat{y}(n) = y(n) + \sum_{i=1}^{p} a(i)y(n-i) = y(n) + \hat{y}^T(n)a \tag{9.17}$$

The vector **a**, which minimizes the prediction error variance $\sigma_p^2 \triangleq E\{|e(n)|^2\}$, is the desired AR coefficients. From Equation 9.17, we obtain (see Exercise 9.4.1)

$$\sigma_p^2 \triangleq E\{|e(n)|^2\} = E\{[y(n) + a^T y(n)][y(n) + y^T(n)a]\}$$

$$= r_y(0) + r_{y(p)}^T a + a^T r_{y(p)} + a^T R_{y(p)} a \tag{9.18}$$

where $r_{y(p)}$ and $R_{y(p)}$ are defined in Equation 9.15. Without any confusion, we use the one subscript. If, however, we had a cross-correlation, then we would have written $r_{yx(p)}$ and $R_{yx(p)}$. The vector **a** that minimizes Equation 9.18 (see Exercise 9.4.2) is given by

$$a = -R_{y(p)}^{-1} r_{y(p)} \tag{9.19}$$

which is identical to Equation 9.15 and was derived from the YW equations.

Exercise 9.4.1

Verify Equation 9.18.   ▲

Exercise 9.4.2

Verify Equation 9.19.   ▲

The minimum prediction error is found to be (see Exercise 9.4.3)

$$\sigma_p^2 \triangleq E\{|e(n)|^2\} = y(0) - r_{y(p)}^T R_{y(p)}^{-1} r_{y(p)} \tag{9.20}$$

Exercise 9.4.3

Verify Equation 9.20.   ▲

The LS AR estimation method approximates the AR coefficients found by Equation 9.19, using a finite sample instead of an ensemble, by minimizing the finite-sample cost function:

$$J(a) = \sum_{n=N_1}^{N_2} |e(n)|^2 = \sum_{n=N_1}^{N_2} \left| y(n) + \sum_{i=1}^{p} a(i)y(n-i) \right|^2 \tag{9.21}$$

From the complex number properties, we know that the absolute value square of a set of complex numbers is equal to the square values of the real and imaginary parts of each number (in our case here is only the real factors present) and, therefore, the summing of these factors from N_1 to N_2 is equivalent to the Frobenius norm of the following matrix (see Appendix 2)

$$J(a) = \left\| \begin{bmatrix} y(N_1) \\ y(N_1+1) \\ \vdots \\ y(N_2) \end{bmatrix} + \begin{bmatrix} y(N_1-1) & \cdots & y(N_1-p) \\ y(N_1) & \cdots & y(N_1+1-p) \\ \vdots & & \vdots \\ y(N_2-1) & \cdots & y(N_2-p) \end{bmatrix} \begin{bmatrix} a(1) \\ a(1) \\ \vdots \\ a(p) \end{bmatrix} \right\|^2 = \|y + Ya\|^2 \tag{9.22}$$

where we assumed that $y(0)=0$ for $n < N_1$ and $n > N_2$.

The vector a that minimizes the cost function $J(a)$ is given by (see Exercise 9.4.4)

$$\hat{a} = -\left(Y^T Y\right)^{-1}\left(Y^T y\right) \tag{9.23}$$

Exercise 9.4.4

Verify Equation 9.23. See also Appendix 2.   ▲

If we set $N_1=p+1$ and $N_2=N$, the preceding vectors take the form

$$y = \begin{bmatrix} y(p+1) \\ y(p+2) \\ \vdots \\ y(N) \end{bmatrix} \quad Y = \begin{bmatrix} y(p) & y(p-1) & y(1) \\ y(p+1) & y(p) & y(2) \\ \vdots & \vdots & \vdots \\ y(N-1) & y(N-2) & y(N-p) \end{bmatrix} \tag{9.24}$$

Example 9.4.1

Use the following AR model to find 64 term output:

$$y(n) = 1.3847y(n-1) - 1.5602y(n-2) + 0.8883y(n-3) - 0.4266y(n-4) + v(n)$$

The noise v is a zero mean and unit variance Gaussian process. In addition, the following signal was added: $s(n) = 3\sin(0.2n\pi) + 3\sin(0.5n\pi) + 2\sin(0.6n\pi)$.

Solution: The following Book m-function was used to produce Figure 9.4. The reader can easily change the coefficients of the IIR filter or the signal.

BOOK M-FUNCTION: [Y,A,PSD,OM]=COVARIANCE_METHOD_PSD(P,N)

```
function[y,a,psd,om]=covariance_method_psd(p,N)
    %p=order of the AR filter for the PSD estimate;
    %N=the number of IIR output and signal length;
```

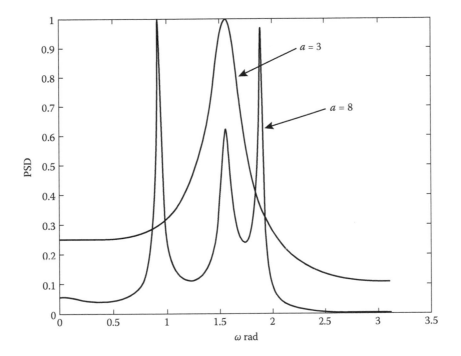

FIGURE 9.4

```
    %y=signal and IIR output combined;a is a vector
    %of the AR coefficients;
for m=0:N-1
y1(1)=0;y1(2)=0;y1(3)=0;y1(4)=0;
y1(m+5)=1.3847*y1(m+4)-1.5602*y1(m+3)+0.8883*y1(m+2)...
    -0.4266*y1(m+1)+2*randn;
s(m+1)=3*sin(0.3*pi*m)+3*sin(0.4*pi*m)+3*sin(0.6*pi*m);
end;
y=y1(1,5:length(y1))+s;
for t=0:-p+N-1
    yv(t+1)=y(t+p+1);
end;
for r=0:N-p-1
    for q=0:p-1
    Y(r+1,q+1)=y(1,p+r-q);
end;
end;
a=-(inv(Y'*Y))*(Y'*yv');

m=1:length(a);
for w=1:N-1
X(w+1)=sum(1+exp(-j*w*2*pi*m/512)*a);
end;
psd=1./abs(X);
om=0:2*pi/N:2*pi-(2*pi/N);
```

To plot the results given by the preceding Book m-file, we write the following in the command window: plot(om,psd/max(psd),'k'). Figure 9.4 shows the spectrum of all-pole filter AR. The plots are for three coefficients and for eight coefficients. It is obvious that the resolution increases with the number of coefficients. The three frequencies that were used were as follows: 0.942 rad, 1.57 rad, and 1.884 rad and the noise was white Gaussian noise (WGN) 0.5*randn. Try to vary the number of coefficients and the strength of the noise to get an understanding of these effects. ∎

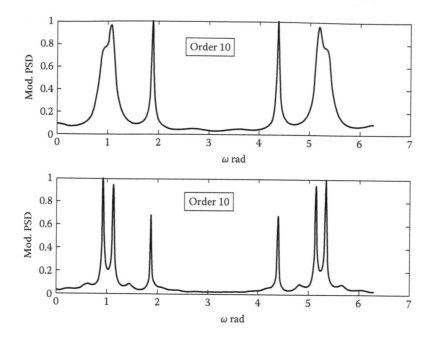

FIGURE 9.5

Figure 9.5 shows clearly the effect of the variance increase on the resolution.

9.5 MINIMUM VARIANCE METHOD

The minimum variance (MV) method produces a spectrum estimate by filtering the signal through a bank of narrowband band-pass filter. Let us have a band-pass filter that has unit gain at ω_i. This constraint dictates that

$$H_i(e^{j\omega_i}) = \sum_{n=0}^{p-1} h_i(n)e^{-j\omega_i n} = 1; \quad h_i = [h_i(0) \quad h_i(1) \quad \cdots \quad h_i(p-1)]^T;$$

$$e_i = [1 \quad e^{j\omega_i} \quad \cdots \quad e^{j\omega_i(p-1)}]^T$$

(9.25)

Therefore, Equation 9.25 becomes

$$h_i^H e_i = e_i^H h_i = 1 \tag{9.26}$$

The superscript H stands for conjugating each vector and then transposing the vector. For real data, the superscript H indicates transpose T. The spectrum of the output of the preceding band-pass filter (system) is given by Equation 6.13 to be equal

$$E\left\{|y_i(n)|^2\right\} = h_i^H R_{xx} h_i \tag{9.27}$$

It can be shown that the filter design problem becomes one of minimizing Equation 9.27 subject to the linear constraint given by Equation 9.26. Hence, we write

$$h_i = \frac{R_{xx}^{-1} e_i}{e_i^H R_{xx}^{-1} e_i} \tag{9.28}$$

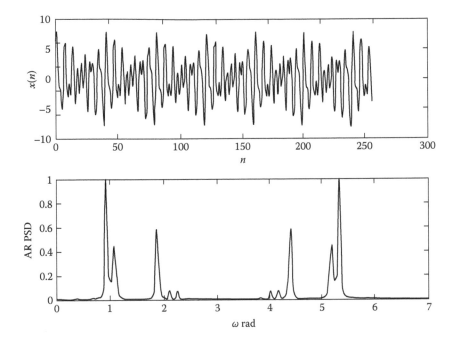

FIGURE 9.6

for one of the filters in the bank and for frequency ω_i. Because this is any frequency, the preceding equation is true for all frequencies. Hence, we write

$$h = \frac{R_{xx}^{-1}e}{e^H R_{xx}^{-1}e} \qquad (9.29)$$

and the power estimate is

$$\hat{\sigma}_x^2\left(e^{j\omega}\right) \triangleq E\left\{|y(n)|^2\right\} = h^T R_{xx}h = \frac{e^H R_{xx}^{-1}}{e^H R_{xx}^{-1}e} R_{xx} \frac{R_{xx}^{-1}e}{e^H R_{xx}^{-1}e}$$

$$= \frac{e^H R_{xx}^{-1}e}{e^H R_{xx}^{-1}e} \frac{1}{e^H R_{xx}^{-1}e} = \frac{1}{e^H R_{xx}^{-1}e} \qquad (9.30)$$

where $e = \begin{bmatrix} 1 & e^{j\omega} & e^{j\omega 2} & \cdots & e^{j\omega(p-1)} \end{bmatrix}^T$. Remember that R is symmetric and therefore $R = R^H, R^{-H} = R^{-1}$.

Figure 9.6 shows the PSD using the MV method and averaging 20 spectrums with increasing the correlation matrix from 5×5 to 20×20 in increments of 5. The top figure of Figure 9.6 shows the signal $x = 3 * \sin(0.3 * pi * [1:256]) + 3 * \sin(0.35 * pi * [1:256]) + 3 * \sin(0.6 * pi * [1:256]) + 0.2 * randn(1,256)$ and the bottom Figure 9.6 shows the PSD. The following Book m-function produces Figure 9.6. The frequencies were 0.9425, 1.0995, and 1.8850 radians. The value of P was 50.

BOOK M-FUNCTION: [PSD]=MINIMUM_VARIANCE_PSD1(X,P,L)

```
function [psd]=minimum_variance_psd1(x,P,L)
%x=data;P=number divided exactly by 5;
%L=number of frequency bins;
%psd=averaged power spectra P/5 times;
```

```
r=sample_biased_autoc(x,floor(length(x)*0.8));
psd1=0;
for p=5:5:P;
    R=toeplitz(r(1,1:p));
for m=0:L-1;
    n=0:p-1;
e=exp(-j*m*(2*pi/L)*(n-1));
ps(m+1)=abs(1/(e*inv(R)*(conj(e))'));
end;
psd1=psd1+abs(ps)/max(abs(ps));
end;
psd=psd1/max(psd1);
```

Figure 9.6 was created using the following Book MATLAB program:

```
>> subplot(2,1,1)
>> plot(x,'k')
>> xlabel('n');ylabel('x(n)')
>> subplot(2,1,2)
>> om=0:2*pi/128:2*pi-(2*pi/128);
>> plot(om,psd,'k')
>> xlabel('\omega   rad');ylabel('MV  PSD')
```

Exercise 9.5.1

Use the data used to produce Figure 9.6 to find the MV PSD with $N=128$. Next, use the MV using the book proposed linear extended method. Compare the PSD for two results. ▲

Example 9.5.1

Find the power spectrum estimate using the MV approach. Assume a white noise signal $\{x(n)\}$ having zero mean value and variance σ_x^2. The autocorrelation matrix is a 4×4 matrix.

Solution: The autocorrelation of white noise signal is given by $r_x(n,m)=\sigma_x^2\delta(n-m)$. Therefore, the autocorrelation matrix takes the form

$$
R_{xx} = \begin{bmatrix} \sigma_x^2 & 0 & 0 & 0 \\ 0 & \sigma_x^2 & 0 & 0 \\ 0 & 0 & \sigma_x^2 & 0 \\ 0 & 0 & 0 & \sigma_x^2 \end{bmatrix} = \sigma_x^2 \begin{bmatrix} 1 & 0 & 0 & 0 \\ 0 & 1 & 0 & 0 \\ 0 & 0 & 1 & 0 \\ 0 & 0 & 0 & 1 \end{bmatrix} = \sigma_x^2 I
$$

Since we have the following relations (see Appendix 2 for matrix inversion)

$$
R_{xx}^{-1} = \frac{1}{\sigma_x^2} I; e^H = \begin{bmatrix} 1 & e^{-j\omega} & e^{-j2\omega} & e^{-j3\omega} \end{bmatrix}; e = \begin{bmatrix} 1 & e^{j\omega} & e^{j2\omega} & e^{j3\omega} \end{bmatrix}^T
$$

Equation 9.30 becomes

$$
\hat{\sigma}_x^2\left(e^{j\omega}\right) \triangleq E\left\{|y(n)|^2\right\} = \frac{1}{e^H R_{xx}^{-1} e} = \frac{\sigma_x^2}{4}
$$

■

9.6   MODEL ORDER

It is important that we are able to approximately estimate the order of the system from which the data were produced. To estimate the order of the filter, Akaike has developed two criteria based on

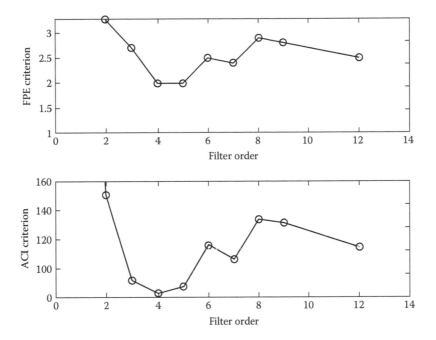

FIGURE 9.7

concepts in mathematical statistics. These are the **final prediction error** (FPE) and the **Akaike information criterion** (AIC). Their corresponding equations are

$$\text{FPE} = s_p^2 \frac{N+p}{N-p} \tag{9.31a}$$

$$\text{AIC} = N \ln(s_p^2) + 2p \tag{9.31b}$$

Studies of the FPE criterion show that it tends to have a minimum at values of the order p are less than the model order. The shortcoming of the AIC criterion is that it tends to overestimate model order. Despite these shortcomings, both of these criteria are often used in practical applications. Figure 9.7 shows typical plots for the two criteria. It is obvious to conclude, from both criteria, that the order is about four, which is the correct value. For the present case, we used 128 data points and 30 lag times for the autocorrelation function.

9.7   LEVINSON–DURBIN ALGORITHM

The Levinson–Durbin algorithm is a recursive algorithm for solving the YW equations to estimate the model coefficients. This scheme is based upon the concept of estimating the parameters of the model of order p from the parameters of a model of order $p-1$. This is possible because of the Toeplitz form of the matrix (see Appendix 2). From Equation 9.10, with $m=0$ and $k=1$, and taking into consideration that the impulse response is zero for negative times, we obtain (the a's are estimates)

$$r_{yy}(1) = -a_1(1)r_{yy}(0) \text{ or } a_1(1) = -\frac{r_{yy}(1)}{r_{yy}(0)} \tag{9.32}$$

From Equation 9.12, and for first-order system, we find the relation:

$$\sigma_{v,1}^2 = r_{yy}(0)\left(1 - a_1(1)^2\right) \tag{9.33}$$

After finding the preceding two initial relations, the recursive scheme starts from the second-order model and is found from Equation 9.14 to be equal:

$$\begin{bmatrix} r_{yy}(0) & r_{yy}(1) \\ r_{yy}(1) & r_{yy}(0) \end{bmatrix}\begin{bmatrix} a_2(1) \\ a_2(2) \end{bmatrix} = \begin{bmatrix} -r_{yy}(1) \\ -r_{yy}(2) \end{bmatrix} \tag{9.34}$$

Observe that we used the equality of the correlation coefficients $r_{yy}(k) = r_{yy}(-k)$ and we specifically identified the order of approximation of the a's coefficients with a subscript. Using the first equation of Equation 9.34 and solving for $a_2(1)$, we obtain

$$r_{yy}(0)a_2(1) = -r_{yy}(1) - r_{yy}(1)a_2(2)$$

or $\tag{9.35}$

$$a_2(1) = -\frac{r_{yy}(1)}{r_{yy}(0)} - a_2(2)\frac{r_{yy}(1)}{r_{yy}(0)} = a_1(1) + a_2(2)a_1(1)$$

To obtain $a_2(2)$, we must use the augmented YW equations (Equation 9.13) of order 2:

$$\begin{bmatrix} r_{yy}(0) & r_{yy}(1) & r_{yy}(2) \\ r_{yy}(1) & r_{yy}(0) & r_{yy}(1) \\ r_{yy}(p) & r_{yy}(1) & r_{yy}(0) \end{bmatrix}\begin{bmatrix} 1 \\ a_2(1) \\ a_2(2) \end{bmatrix} = \begin{bmatrix} \sigma_{v,2}^2 \\ 0 \\ 0 \end{bmatrix} \tag{9.36}$$

The crucial point is to express the left side of Equation 9.36 in terms of $a_2(2)$. Using Equation 9.35, the column vector of a's becomes

$$\begin{bmatrix} 1 \\ a_2(1) \\ a_2(2) \end{bmatrix} = \begin{bmatrix} 1 \\ a_1(1) + a_1(1)a_2(2) \\ a_2(2) \end{bmatrix} = \begin{bmatrix} 1 \\ a_1(1) \\ 0 \end{bmatrix} + a_2(2)\begin{bmatrix} 0 \\ a_1(1) \\ 1 \end{bmatrix} \tag{9.37}$$

Introducing the preceding results in Equation 9.36 and applying the matrix properties (see Appendix 2), we obtain

$$\begin{bmatrix} r_{yy}(0) & r_{yy}(1) & r_{yy}(2) \\ r_{yy}(1) & r_{yy}(0) & r_{yy}(1) \\ r_{yy}(2) & r_{yy}(1) & r_{yy}(0) \end{bmatrix}\begin{bmatrix} 1 \\ a_1(1) \\ 0 \end{bmatrix} + a_2(2)\begin{bmatrix} r_{yy}(0) & r_{yy}(1) & r_{yy}(2) \\ r_{yy}(1) & r_{yy}(0) & r_{yy}(1) \\ r_{yy}(2) & r_{yy}(1) & r_{yy}(0) \end{bmatrix}\begin{bmatrix} 1 \\ a_1(1) \\ 0 \end{bmatrix}$$

$$= \begin{bmatrix} r_{yy}(0) + r_{yy}(1)a_1(1) \\ r_{yy}(1) + r_{yy}(0)a_1(1) \\ r_{yy}(2) + r_{yy}(1)a_1(1) \end{bmatrix} + a_2(2)\begin{bmatrix} r_{yy}(1)a_1(1) + r_{yy}(2) \\ r_{yy}(0)a_1(1) + r_{yy}(1) \\ r_{yy}(1)a_1(1) + r_{yy}(0) \end{bmatrix} = \begin{bmatrix} \sigma_{v,2}^2 \\ 0 \\ 0 \end{bmatrix} \tag{9.38}$$

The preceding equation can be written in the form

$$t_1 + t_2 = \begin{bmatrix} r_{yy}(0)[1 - a_1(1)^2] \\ r_{yy}(1) + r_{yy}(0)a_1(1) \\ r_{yy}(2) + r_{yy}(1)a_1(1) \end{bmatrix} + a_2(2) \begin{bmatrix} r_{yy}(1)a_1(1) + r_{yy}(2) \\ r_{yy}(0)a_1(1) + r_{yy}(1) \\ r_{yy}(0)[1 - a_1(1)^2] \end{bmatrix} = \begin{bmatrix} \sigma_{v,1}^2 \\ \Delta_2 \\ \Delta_3 \end{bmatrix} + a_2(2) \begin{bmatrix} \Delta_3 \\ \Delta_2 \\ \sigma_{v,1}^2 \end{bmatrix}$$

$$= \begin{bmatrix} \sigma_{v,2}^2 \\ 0 \\ 0 \end{bmatrix} \tag{9.39}$$

From the third equation of the preceding system, we obtain the unknown $a_2(2)$ as follows:

$$\Delta_3 + a_2(2)\sigma_{v,1}^2 = 0 \text{ or } a_2(2) = -\frac{\Delta_3}{\sigma_{v,1}^2} = -\frac{r_{yy}(1)a_1(1) + r_{yy}(2)}{\sigma_{v,1}^2} \tag{9.40}$$

Using Equation 9.40 and the first equation of Equation 9.39, we obtain the variance for step two of first order:

$$\sigma_{v,2}^2 = \sigma_{v,1}^2 + a_2(2)\Delta_3 = \left(1 - a_2(2)^2\right)\sigma_{v,1}^2 \tag{9.41}$$

Example 9.7.1

The autocorrelation matrix of a second-order IIR system output, with an input of zero mean Gaussian noise, is given by

$$R = \begin{bmatrix} 5.5611 & 1.3291 & 0.8333 \\ 1.3291 & 5.5611 & 1.3291 \\ 0.8333 & 1.3291 & 5.5611 \end{bmatrix}$$

Find the variances and filter coefficients.

Solution: From $a_1(1) = -[r_{yy}(1)/r_{yy}(0)] = -(1.3291/5.5611) = -0.2390$ and from Equation 9.33, we obtain $\sigma_{v,1}^2 = r_{yy}(0)\left(1 - a_1(1)^2\right) = 5.5611\left(1 - 0.2390^2\right) = 5.2434$. From Equation 9.40, we obtain

$$a_2(2) = -\frac{\Delta_3}{\sigma_{v,1}^2} = -\frac{r_{yy}(1)a_1(1) + r_{yy}(2)}{\sigma_{v,1}^2} = -\frac{-5.5611 \times 0.2390 + 1.3291}{5.2434} = -0.1929, \text{ and from Equation 9.35,}$$

we obtain $a_2(1) = a_1(1) + a_2(2)a_1(1) = -0.2390 - 0.1929(-0.2390) = -0.1929$. The noise variance is $\sigma_{v,2}^2 = \sigma_{v,1}^2 + a_2(2)\Delta_3 = \left(1 - a_2(2)^2\right)\sigma_{v,1}^2 = \left(1 - (-0.1929)^2\right)5.2434 = 5.0483.$ ∎

LEVINSON–DURBIN ALGORITHM (REAL-VALUED CASE)

1. Initialize the recursion with the zero-order model

  a. $a_0(0) = 1$
  b. $\sigma_{v,0}^2 = r_{yy}(0)$

2. For $j = 0, 1, 2, \cdots, p-1$

  a. $k_j = r_{yy}(j+1) + \sum_{i=1}^{j} a_j(i)r_{yy}(j-i+1)$

  b. $K_{j+1} = -k_j / \sigma_{v,j}^2$

c.   For $i = 1, 2, \cdots, j$

$$a_{j+1}(i) = a_j(i) + K_{j+1} a_j(j - i + 1),$$

d. $a_{j+1}(j+1) = K_{j+1}$

e. $\sigma_{v,j+1}^2 = \sigma_{v,j}^2 \left\lfloor 1 - K_{j+1}^2 \right\rfloor$

3. $b(0) = \sqrt{\sigma_{v,p}^2}$

BOOK M-FUNCTION: [A,VAR]=LEVINSON(R,P)

```
function[a,var]=levinson(r,p)
r=r(:);%p=number of a's+1;to find r we can use
        %the Book MATLAB function
        %r=sample_biased_autoc(x,lg), p<lg,
        %lg=the lag number of the autocorrelation
        %function r;
a=1;
var=r(1);
for j=2:p
    g=r(2:j)'*flipud(a);
    gamma=-g/var;
    a=[a;0]+gamma*[0;(flipud(a))];
    var=var*(1-abs(gamma)^2);
end;
```

The following two additional functions were used:

BOOK M-FUNCTION: [Y,R,S]=URDSP_OUTPUTSIGN_5TERM_ARSYST(N,LG)

```
function[y,r,s]=urdsp_outputsign_5term_ARsyst(N,lg)
    %5 term AR system;sig=multisine signal;N=length
    %of the signal, for example 256, 2^n;
    %lg the lag for the correlation that
    %p<lg, p is number of AR coefficients;
for m=0:N-1
    y1(1)=0;y1(2)=0;y1(3)=0;y1(4)=0;
    y1(m+5)=1.38*y1(m+4)-1.56*y1(m+3)+0.89*y1(m+2)...
        -0.43*y1(m+1)+0.5*randn;
    s(m+1)=6*sin(0.3*pi*m)+6*sin(0.34*pi*m)+6*sin(0.6*pi*m);
end;
y=y1(1,5:length(y1))+s;
r=sample_biased_autoc(y,lg);
```

BOOK M-FUNCTION: [PSD,OM]=URDSP_OUTPUTSIGN_ARSYSTEM(A,N)

```
function[psd,om]=urdsp_psd_ARsystem(a,N)
    %N=number of frequency bins;
    %a=AR coefficients;
m=1:length(a);
for w=0:N-1
    X(w+1)=sum(1+exp(-j*w*2*pi*m/512)*a);
end;
psd=1./abs(X);
om=0:2*pi/N:pi-(2*pi/N);
```

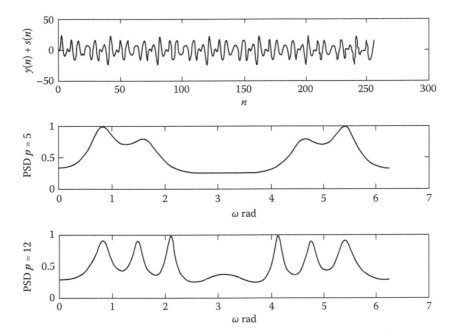

FIGURE 9.8

To produce Figure 9.8, we used an AR system described by the difference equation:

$$y(n) = 1.38y(n-1) - 1.56y(n-2) + 0.89y(n-3) - 0.43y(n-4) + 0.1\text{randn}$$

To the preceding signal we added the signal

$$s(n) = 4\sin(0.3\pi n) + 4\sin(0.50\pi n) + \sin(0.7\pi n)$$

The number of signal was 256 and the lag number for the correlation was 64. To plot the power spectra density, we used the following Book MATLAB program for $p=5$. The reader can produce the other subfigures by setting $p=12$.

```
>> [y,r]=urdsp_outputsign_5term_ARsyst(256,64);
>> [a5,var]=levinson(r,5);
>> [psd5,om]=urdsp_psd_ARsystem(a5,256);
>>plot(om,psd5,'k');
```

Similarly, we can also use the YW equations to obtain the desired coefficients.

BOOK YULE–WALKER M-FUNCTION: [A,VAR]=YULE_WALKER_LG(X,LG,P)

```
function[a,var]=yule_walker_lg(x,lg,p)
    %x=signal; lg=lag number for the
    %autocorrelation function r;
    %p=order of AR model=number of a's<lg;
r=sample_biased_autoc(x,lg);
R=toeplitz(r(1,1:p));%toeplitz() is a MATLAB
    %function;
a=inv(R)*r(1,2:p+1)';%inv() stands for inverse
    %of a matrix, a MATLAB function;
var=r(1)+sum(r(1,p+1).*a');
```

Exercise 9.7.1

Use the AR system described by the difference equation $x(n)=1.38x(n-1)-1.56\ x(n-2)+0.89$ $x(n-3)-0.43\ x(n-4)+0.5$randn and add the signal $s(n)=6\sin(0.3\pi n)+6\sin(0.34\pi n)$ to its output. Use the Book Yule–Walker function to produce the spectrum for $p=10$ and $p=40$. ▲

9.8 MAXIMUM ENTROPY METHOD

Because we are limited to small number of lags when finding the autocorrelation function, it is natural to ask how we can expand the autocorrelation function so that the extension be as accurate as possible. One such method that was suggested is the **maximum entropy method** (MEM). It can be shown that the estimate of the PSD using the MEM approach is given by

$$S_{\text{MEM}}\left(e^{j\omega}\right)=\frac{b(0)^2}{\left|e^H a\right|^2} \tag{9.42}$$

where

$$\begin{bmatrix} r_{xx}(0) & r_{xx}(-1) & \cdots & r_{xx}(-p) \\ r_{xx}(1) & r_{xx}(0) & \cdots & r_{xx}(p-1) \\ & \vdots & & \vdots \\ r_{xx}(p) & r_{xx}(p-1) & \cdots & r_{xx}(0) \end{bmatrix} \begin{bmatrix} 1 \\ a(1) \\ \vdots \\ a(p) \end{bmatrix} = \begin{bmatrix} b(0)^2 \\ 0 \\ \vdots \\ 0 \end{bmatrix} \tag{9.43}$$

$$b(0)^2 = r_{xx}(0) + \sum_{k=1}^{p} a(k)r_{xx}(k) \tag{9.44}$$

$$e^H = \begin{bmatrix} 1 & e^{-j\omega} & e^{-j\omega 2} & \cdots & e^{-j\omega p} \end{bmatrix} \tag{9.45}$$

The superscript H stands for the conjugate transpose of a matrix or vector. For a real situation the H is equivalent to transpose T. The vector a is the solution of Equation 9.43 and $r_{xx}(k)$ is the correlation of the data. First, the autocorrelation normal equations (Equation 9.43) are solved for the a's and $b(0)^2$. The following Book m-Functions produces the PSD using the MEM method.

BOOK M-FUNCTION: FUNCTION[PSD,A,VAR]=MAX_ENTROPY_METH(X,LG,P)

```
function[psd,a,var]=max_entropy_meth(y,lg,p)
%y=output of a %-term IIR system
%when the input to the system is s;
%p=order of AR system;p<lg=lag number of the
%autocorrelation function;
[a,var]=yule_walker_lg(y,lg,p);
psd=var./abs(fft([1;-a],512));
w=0:2*pi/512:2*pi-(2*pi/512);
plot(w,psd/max(psd),'k');
function[y]=output_5term_ARsyst(N)
    %5 term AR system;s=multisine signal;N=length
    %of the signal, for example 256, 2^n;
    %lg the lag for the correlation that
    %p<lg, p is number of AR coefficients;
for m=0:N-1
```

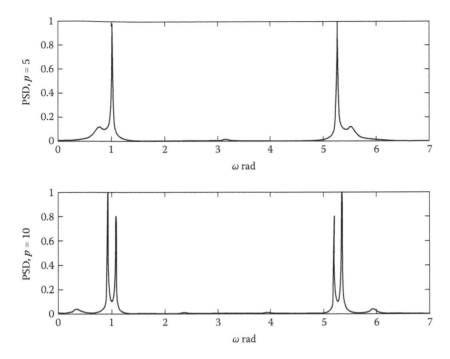

FIGURE 9.9

```
        y1(1)=0;y1(2)=0;y1(3)=0;y1(4)=0;
        y1(m+5)=1.38*y1(m+4)-1.56*y1(m+3)+0.89*y1(m+2)...
            -0.43*y1(m+1)+0.5*randn+...
        6*sin(0.3*pi*m)+6*sin(0.34*pi*m);
end;
y=y1(1,5:length(y1));
```

To produce the spectrum, we used the following Book MATLAB program:

```
>> [y]=output_of_5term_ARsys(256);
>> [psd,a,var]=max_entropy_meth(y,64,5);
>> om=0:2*pi/512:2*pi-(2*pi/512);
>> plot(om,psd/max(psd));
```

Figure 9.9 was produced using the following signal $s = 0.5\text{randn} + 6\sin(0.3\pi n) + 6\sin(0.34\pi n)$ as input to an AR system: $y(n) = 1.38y(n-1) - 1.56y(n-2) + 0.89y(n-3) - 0.43y(n-4)$. The lag number was 64, and the number of AR coefficients were $p=5$ and $p=10$.

Exercise 9.8.1

Use the same signal and AR system with the difference of multiplying the normal noise by six: 6randn. Plot the signal, and the psd for $p=5$ and $p=10$.   ▲

9.9   SPECTRUMS OF SEGMENTED SIGNALS

Many times, we are faced with the situation where some data are missing, for example, the breakdown of a detector or receiving data from space that is interrupted by the rotation of the earth.

The sequence to be used is

$$x(n) = \sin(0.3\pi n) + \sin(0.36\pi n) + v(n) \tag{9.46}$$

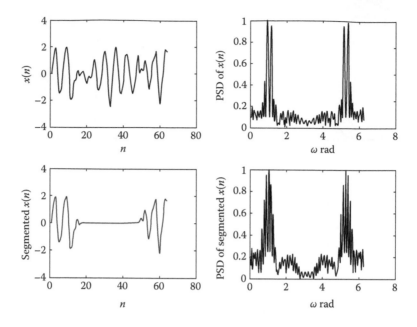

FIGURE 9.10

The sequence $\{x(n)\}$ has 64 elements of which $x(17) = x(18) = \cdots = x(48) = 0$. (are missing). The noise $v(n)$ is white Gaussian (WG) of zero mean and variance σ_v^2. Figure 9.10 shows the sequence, its segmented form, and their corresponding spectra.

9.9.1 Method 1: The Average Method

In this method, we obtain the spectrum, S_1, of the first segment. Next we obtain the spectrum, S_2, of the last segment and then average them: $S = (S_1 + S_2) / 2$. Figure 9.11 shows (1) the spectrum of the

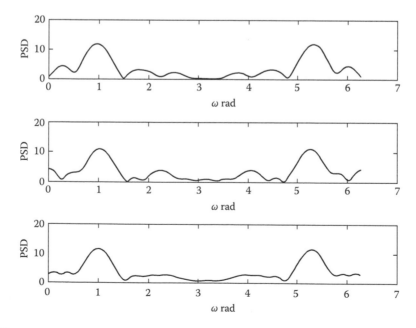

FIGURE 9.11

first segment, (2) the spectrum of the third segment, and (3) the average spectrum of the first two spectra. The signal is the same given in the previous paragraph.

9.9.2 Method 2: Extrapolation Method

One of the basic ideas is to assume that a signal is the result of an AR process and then find extra values of the signal using the extrapolation method. Let the estimate $x(n)$ be given by the linear combination of the previous values of the system. Hence,

$$\hat{x}(n) = -a_1 x(n-1) - a_2 x(n-2) - \cdots - a_p x(n-p) \tag{9.47}$$

then, the error in the estimate is given by the difference

$$\varepsilon(n) = x(n) - \hat{x}(n) \tag{9.48}$$

or

$$\varepsilon(n) = x(n) + a_1 x(n-1) + a_2 x(n-2) + \cdots + a_p x(n-p) \tag{9.49}$$

The normal equations, whose solution provides the optimal coefficients, can be derived directly in matrix form by first defining the vectors:

$$\boldsymbol{x}(n) = \begin{bmatrix} x(n-p) \\ x(n-p+1) \\ \vdots \\ x(n) \end{bmatrix}; \quad \boldsymbol{a} = \begin{bmatrix} 1 \\ a_1 \\ \vdots \\ a_p \end{bmatrix}; \quad \tilde{\boldsymbol{x}}(n) = \begin{bmatrix} x(n) \\ x(n-1) \\ \vdots \\ x(n-p) \end{bmatrix}; \quad \boldsymbol{l} = \begin{bmatrix} 1 \\ 0 \\ \vdots \\ 0 \end{bmatrix} \tag{9.50}$$

The vector $\boldsymbol{a}$, of the coefficients and the prediction error variance $\sigma_\varepsilon^2 = E\{\varepsilon^2(n)\}$, constitute the **linear prediction parameters**. Based on Equation 9.50, we write (see Equation 9.49)

$$\varepsilon(n) = \boldsymbol{a}^T \tilde{\boldsymbol{x}}(n) \tag{9.51}$$

To find the optimal filter coefficients, we apply the orthogonality principle, which states that $E\{x(n-i)\varepsilon(n)\} = 0, i = 1,2,\ldots,p$ and that $\sigma_\varepsilon^2 = E\{x(n)\varepsilon(n)\}$. These assertions can be stated in compact matrix form:

$$E\{\tilde{\boldsymbol{x}}(n)\varepsilon(n)\} = \begin{bmatrix} \sigma_\varepsilon^2 \\ 0 \\ \vdots \\ 0 \end{bmatrix} = \sigma_\varepsilon^2 \boldsymbol{l} \tag{9.52}$$

Drop the argument n for simplicity and substitute Equation 9.51 in 9.52. We obtain

$$E\{\tilde{\boldsymbol{x}}(\boldsymbol{a}^T \tilde{\boldsymbol{x}})\} = E\{\tilde{\boldsymbol{x}}(\tilde{\boldsymbol{x}}^T \boldsymbol{a})\} = E\{\tilde{\boldsymbol{x}}\tilde{\boldsymbol{x}}^T\}\boldsymbol{a} = \sigma_\varepsilon^2 \boldsymbol{l} \text{ or } \boldsymbol{R}\boldsymbol{a} = \sigma_\varepsilon^2 \boldsymbol{l} \tag{9.53}$$

The preceding equation is the well-known normal equation.

Example 9.9.1.1

The following Book MATLAB function finds the spectrum of a segmented sequence with 64 elements, of which the 28 central ones are missing. The results are shown in Figures 9.12 and 9.13. The signal $x(n) = \sin(0.3\pi n) + \sin(0.35\pi n) + 0.5\mathrm{randn}(1,64)$ with 64 values is segmented as

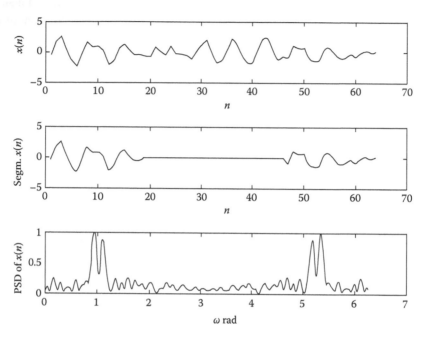

FIGURE 9.12

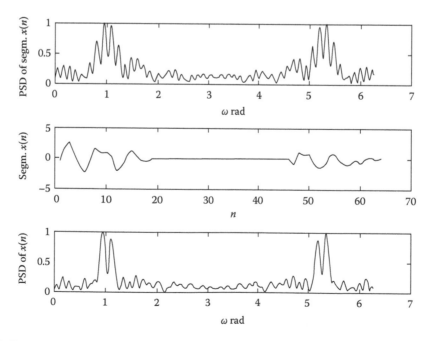

FIGURE 9.13

follows: $xs(n) = [x(1,1:18) \quad zeros(1,28) \quad x(1,47:64)]$. In Figure 9.12 presents the noisy signal, the segmented signal and the PSD of the noisy signal. Figure 9.13 presents the PSD of the segmented signal, the segmented signal and the PSD of the backward extension of the second segment. It is apparent that the extrapolation gives good results.

BOOK M-FUNCTION: []=EXTRAP_PSD_METH_1(X,SEG1,SEG2,P)

```
function[x1,a1,xe1,x2,a2,xe2,xe1f,xe2f]=extrap_psd_meth_1(x,seg1,seg2,p)
    %r=autocorrelation of x;seg1=length of first segment;seg2=length
    %of the second segment (see fig 2);length(x)=2^n;p=filter order;
subplot(3,1,1);plot(x,'k');xlabel('n');ylabel('x(n)');
xx12=[x(1,1:seg1) zeros(1,(length(x)-seg1-seg2))...
x(1,(length(x)-seg2+1):length(x))];
subplot(3,1,2);plot(xx12,'k');xlabel('n');ylabel('Segm. x(n)');
%----------------------
    %Spectrum of the complete sample and segment samples;
w=0:2*pi/256:2*pi-(2*pi/256);
fx=fft(x,256);
subplot(3,1,3);plot(w,abs(fx)/max(abs(fx)),'k');
xlabel('\omega rad');ylabel('PSD of x(n)');
fxs=fft(xx12,256);
figure(2);
subplot(3,1,1);plot(w,abs(fxs)/max(abs(fxs)),'k');
xlabel('\omega rad');ylabel('PSD of segm. x(n)');
%----------------------
    %Extrapolation
r1=xcorr(x(1,1:seg1),'biased');
rs1=r1(1,seg1:2*seg1-1);%p<length(rs1);
[a1,var1]=ssp_yule_walker_alg_rp(rs1,p);
Ext1=length(x)-seg1-seg2;
for m=1:Ext1
    x1(seg1+m)=sum(a1'.*fliplr(x(1,seg1-p+m:seg1-1+m)));
end;
xe1=[x(1,1:length(x)-seg1-seg2) x1(length(x)-seg1-seg2+1:length(x)...
    -seg2) x(1,length(x)-seg2+1:length(x))];
r2=xcorr(x(1,length(x)-seg2:length(x)),'biased');
rs2=r2(1,seg2:2*seg2-1);
[a2,var2]=ssp_yule_walker_alg_rp(rs2,p);
    %the above p can change to p1 if different number of a's
    %is desired; p<min{rs1,rs2};
Ext2=length(x)-seg1-seg2;
for k=1:Ext2
    x2(length(x)-seg2+1-k)=-sum((a2/max(a2))'.*x(1,length(x)-seg2...
        +2-k:length(x)-seg2+1+p-k));
end;
xe2=[x(1,1:length(x)-seg1-seg2) x2(length(x)-seg1-seg2+1:length(x)...
    -seg2) x(1,length(x)-seg2+1:length(x))];
xe1f=(abs(fft(xe1,256))/max(abs(fft(xe1,256))));
subplot(3,1,2)
plot(w,xe1f,'k');xlabel('\omega rad/unit');ylabel('PSD of ext1');
subplot(3,1,3);
xe2f=(abs(fft(xe2,256))/max(abs(fft(xe2,256))));
plot(w,xe2f,'k');xlabel('\omega rad/unit');ylabel('PSD ext2');    ■
```

We can further introduce the following approaches for the extrapolation method: (1) from the forward extrapolated signal, find the autocorrelation function and then the PSD; (2) from the backward extrapolated signal, find the autocorrelation signal and its PSD; (3) find the average of the previous two spectrums; and (4) using Monte Carlo method by repeating the process and then averaging.

9.10   EIGENVALUES AND EIGENVECTORS OF MATRICES (SEE ALSO APPENDIX 2)

Let us create a relationship between a vector v and a $n \times n$ matrix A and a constant λ as follows:

$$Aq = \lambda q \text{ or } (A - \lambda q) = 0 \tag{9.54}$$

where I is the identity matrix having ones along the diagonal and zeros for the rest of the elements. Since $q = 0$ is always a solution, we must find the nonzero solution if it exists. The number λ is called the **eigenvalue** and the vector q belonging to that particular eigenvalue is called **eigenvector**. To find the eigenvalues, the determinant $|A - \lambda I|$ must be equal to zero. If, for example,

$$A = \begin{bmatrix} 4 & -5 \\ 2 & -3 \end{bmatrix} \text{then} |A - \lambda I| = \begin{vmatrix} \begin{bmatrix} 4 & -5 \\ 2 & -3 \end{bmatrix} - \begin{bmatrix} \lambda & 0 \\ 0 & \lambda \end{bmatrix} \end{vmatrix} = \begin{vmatrix} 4 - \lambda & -5 \\ 2 & -3 - \lambda \end{vmatrix} = 0 \tag{9.55}$$

then

$$(4 - \lambda)(-3 - \lambda) + 10 = 0 \text{ or } \lambda^2 - \lambda - 2 = 0 \text{ or } \lambda_1 = -1, \lambda_2 = 2 \tag{9.56}$$

MATLAB has the following function to find the eigenvalues and the corresponding eigenvectors:

$$[Q, D] = \text{eig}(R) \tag{9.57}$$

where Q is a matrix containing the eigenvectors (the columns) and D is a diagonal matrix containing the eigenvalues. For this present case, and referring to Equation 9.54, we have the following relations:

$$\lambda_1 = -1, (A - \lambda I)q = \begin{bmatrix} 5 & -5 \\ 2 & -2 \end{bmatrix} \begin{bmatrix} q_1 \\ q_2 \end{bmatrix} = \begin{bmatrix} 0 \\ 0 \end{bmatrix}, \quad \begin{cases} 5q_1 - 5q_2 = 0 \\ 2q_1 - 2q_2 = 0 \end{cases}$$

It is apparent that the solution to the system is any eigenvector that is the multiple of the vector $q_1 = [\ 1 \quad 1\ ]^T$. Similarly, the second eigenvalue corresponds to any multiple of the vector $q_2 = [\ 5 \quad 2\ ]^T$. If we use the MATLAB function, Equation 9.57, we obtain

$$[Q, D] = \text{eig}(A), \quad Q = \begin{bmatrix} 0.9285 & 0.7071 \\ 0.3714 & 0.7071 \end{bmatrix}, \quad D = \begin{bmatrix} 2 & 0 \\ 0 & -1 \end{bmatrix}$$

In the preceding equation, the columns of Q are the eigenvectors and the diagonal values are the corresponding eigenvalues. Although the eigenvector matrix, given by MATLAB, seems that gives different results, the values are proportional. If we divide the first column by 0.3714 and multiply by 2, we obtain the vector $[\ 5 \quad 2\ ]^T$. Similarly, if we divide the second by 0.7071, we obtain the vector $[\ 1 \quad 1\ ]^T$.

After finding the eigenvectors and eigenvalues, the following matrix operations are true

$$AQ = QD \text{ or } Q^{-1}AQ = D \text{ or } A = Q^{-1}DQ \qquad (9.58)$$

It is known in matrix theory that

If the eigenvectors of a matrix correspond to different eigenvalues, then those eigenvectors are linear independent. Therefore, if the eigenvectors of a matrix are linear independent, then the eigenvalues are distinct. If, in addition, the matrix is symmetric, the eigenvectors corresponding to distinct eigenvalues are orthogonal.

BOOK MATLAB PROGRAM

```
>>x=randn(1,64);
>>rx=sample_based_autoc(x,3);%book m-function giving 3-term
                %autocorrelation;
>>R=toeplitz(rx);%a 3x3 autoc. matrix, toeplitz()=MATLAB
                % function;
>>[Q,D]=eig(R);
>>R
R=
    0.8076          0.0106          -0.1027
    0.0106          0.8076           0.0106
   -0.1027          0.0106           0.8076
>>Q
Q=
   -0.7000         -0.0998          -0.7071
    0.1411         -0.9900          -0.0000
   -0.7000         -0.0998           0.7071
>>D
D=
            0.7030              0              0
                 0         0.8100              0
                 0              0         0.9106
>>Q(:,1)'*Q(:,2)%the product of all rows and first column
                %with all the rows of the second column;
ans=
        2.3592e-16%which is approximately zero as it should;
```

The eigenvalue properties are given in Table 9.1.

9.10.1   EIGENDECOMPOSITION OF THE AUTOCORRELATION MATRIX

Let us consider a sinusoid having random phase

$$x(n) = A\sin(n\omega_0 + \phi) \qquad (9.59)$$

where A and ω_0 are fixed constants and the phase is a RV that is uniformly distributed over the interval $-\pi \le \phi \le \pi$. The mean value and the autocorrelation of $x(n)$ is given by (see Exercise 9.10.1.1)

$$E\{x(n)\} \triangleq m_x(n) = 0, \quad E\{x(k)x(l)\} \triangleq r_{xx}(k,l) = \frac{1}{2}A^2\cos[(k-l)\omega_0] \qquad (9.60)$$

TABLE 9.1
Eigenvalue Properties

$x(n) = [x(n)\ x(n-1)\cdots x(n-M+1)]$	Wide-sense stationary stochastic process
$R_x = E\{x(n)x^T(n)\}$	Correlation matrix
λ_i	The eigenvalues of R_x are real and positive
	The eigenvalues of a Hermitian ($R_x^H = R_x^{*T}$) matrix are real; a Hermitian matrix is positive definite, $A > 0$, if and only if its eigenvalues are positive, $\lambda_i > 0$;
$q_i^T q_j = 0$	Two eigenvectors belonging to two different eigenvalues are orthogonal; the eigenvectors of a Hermitian matrix corresponding to distinct eigenvalues are orthogonal
$Q^T Q = I$	$Q = [\ \ q_0 \ \ \ q_1 \ \ \cdots \ \ q_{M-1} \ \ \ ];\ Q$ is a unitary matrix
$Q_x^T = Q_x^{-1}\ \left(Q_x^H = Q_x^{-1} \right)$	
$R_x = Q\Lambda Q^T = \displaystyle\sum_{i=0}^{M-1} \lambda_i q_i\, q_i^T$	$\Lambda = \mathrm{diag}\ [\ \ \lambda_0 \ \ \ \lambda_1 \ \ \ \lambda_2 \ \ \cdots \ \ \lambda_{M-1} \ \ ]$
$R_x = Q\Lambda Q^H = \displaystyle\sum_{i=0}^{M-1} \lambda_i q_i\, q_i^H$	R_x is a Hermitian matrix
$R_x^{-1} = \left(Q\Lambda Q^T \right)^{-1} = \displaystyle\sum_{i=0}^{M-1} \frac{1}{\lambda_i} q_i\, q_i^T$	
$R_x^{-1} = \left(Q\Lambda Q^H \right)^{-1} = \displaystyle\sum_{i=0}^{M-1} \frac{1}{\lambda_i} q_i\, q_i^H$	R_x is nonsingular and Hermitian
$\mathrm{tr}\{R_x\} = \displaystyle\sum_{i=0}^{M-1} \lambda_i$	$\mathrm{tr}\{R\}=$ trace of $R=$ sum of the diagonal elements of R
$A = B + aI$	A and B have the same eigenvectors and the eigenvalues A are $\lambda_i + a$
$\lambda_{\max} \le S_x^{\max} = \displaystyle\max_{-\pi \le \omega \le \pi} S_x(e^{j\omega})$	
$\lambda_{\min} \ge S_x^{\min} = \displaystyle\min_{-\pi \le \omega \le \pi} S_x(e^{j\omega})$	

Exercise 9.10.1.1

Verify Equation 9.60.   ■

Since the mean is a constant and the autocorrelation depends only on the difference (lag time), the process is wide sense stationary (WSS).

If a signal is made up of sinusoids with additive white noise

$$x(n) = \sum_{m=1}^{M} A_m \sin(n\omega_m + \phi_m) + v(n) \tag{9.61}$$

its autocorrelation function is

$$r_{xx}(k,l) = \frac{1}{2} \sum_{m=1}^{M} A_m^2 \cos[(k-l)\omega_m] + r_{vv}(k,l) \tag{9.62}$$

where the amplitudes and frequencies are fixed constants and the phases are RVs that are uniformly distributed.

We can consider the complex form of the sinusoids as follows:

$$x(n) = Ae^{j(n\omega_0 + \phi)} \tag{9.63}$$

The phase is a random variable uniformly distributed in the interval $-\pi \leq \phi \leq \pi$. Therefore, the mean is (expand the exponent in Euler's format):

$$m_x(n) = E\left\{Ae^{j(n\omega_0 + \phi)}\right\} = 0 \tag{9.64}$$

and the autocorrelation function is

$$r_{xx}(k,l) = E\{x(k)x*(l)\} = E\{Ae^{j(n\omega_0 + \phi)}A * e^{-j(n\omega_0 + \phi)}\} = |A|^2 E\{e^{j(k-l)\omega_0}\}$$

$$= Pe^{j(k-l)\omega_0} = r_{xx}(k-l) \tag{9.65}$$

where $k-l$ is the lag of the autocorrelation function and can be substituted simply by k (setting $l=0$). P is the power of the sinusoidal (complex) signal. Note that the mean is constant and the autocorrelation is independent of the time origin. Therefore, the process is a WSS.

If the noise is white with mean value zero, then the autocorrelation function is $\sigma_v^2 \delta(n)$ and, therefore, the autocorrelation of p sinusoids embedded in noise is

$$r_{xx}(k) = \sum_{i=1}^{p} P_i e^{j\omega_i k} + \sigma_v^2 \delta(k) \tag{9.66}$$

For $p=2$, we obtain

$$r_{xx}(0) = P_1 + P_2 + \sigma_v^2, \quad (\delta(0) = 1); \quad r_{xx}(1) = P_2 e^{j\omega_1} + P_2 e^{j\omega_2}, \quad (\delta(1) = 0) \tag{9.67}$$

Therefore, the correlation matrix is

$$R = \begin{bmatrix} r_{xx}(0) & r_{xx}(1) \\ r_{xx}(1) & r_{xx}(0) \end{bmatrix}$$

If we define (the exponent H stands for Hermitian or, equivalently, stands for transposed conjugate of a vector or matrix and just conjugation for a complex quantity)

$$e_1 = \begin{bmatrix} 1 & e^{j\omega_1} \end{bmatrix}^T; e_1^H = \begin{bmatrix} 1 & e^{-j\omega_1} \end{bmatrix}; e_2 = \begin{bmatrix} 1 & e^{j\omega_2} \end{bmatrix}^T; e_2^H = \begin{bmatrix} 1 & e^{j\omega_2} \end{bmatrix}$$

then (see Appendix 2)

$$R_{xx} = P_1 e_1 e_1^H + P_2 e_2 e_2^H + \sigma_v^2 I = R_s + R_n$$

The preceding equation can be put in the compact form:

$$R_{xx} = EPE^H + \sigma_v^2 I, E = \begin{bmatrix} e_1 & e_2 \end{bmatrix}, E^H = \begin{bmatrix} e_1^H & e_2^H \end{bmatrix}^T, P = \begin{bmatrix} P_1 & 0 \\ 0 & P_2 \end{bmatrix} \tag{9.68}$$

Therefore, for p sinusoids, the general autocorrelation matrix is

$$R_{xx} = R_s + R_n = \sum_{i=1}^{p} P_i e_i e_i^H + \sigma_v^2 I$$

(9.69)

$$e_i = \begin{bmatrix} 1 & e^{j\omega_i} & e^{j2\omega_i} & \cdots & e^{j(M-1)\omega_i} \end{bmatrix}, i = 1, 2, \ldots, p$$

Example 9.10.1.1

Find the eigenvalues and eigenvectors of a signal having the following correlation matrices:

$$R_s = \begin{bmatrix} 0.5 & 0.25 \\ 0.25 & 1.00 \end{bmatrix}, R_v = \begin{bmatrix} 0.1 & 0 \\ 0 & 0.1 \end{bmatrix}, R_{xx} = R_s + R_v = \begin{bmatrix} 0.6 & 0.25 \\ 0.25 & 1.10 \end{bmatrix}$$

Solution: Using MATLAB, we obtain

1. Signal eigenvectors and eigenvalues

$$Q_s = \begin{bmatrix} q_1 & q_2 \end{bmatrix} = \begin{bmatrix} -0.9239 & 0.3827 \\ 0.3827 & -0.9239 \end{bmatrix}, D_s = \begin{bmatrix} 0.3964 & 0 \\ 0 & 1.1036 \end{bmatrix}$$

$$q_1^H * q_1 = 1, \quad q_2^H * q_2 = 1, \quad q_1^H * q_2 = 0$$

2. Noise eigenvectors and eigenvalues

$$Q_v = \begin{bmatrix} q_1 & q_2 \end{bmatrix} = \begin{bmatrix} 1 & 0 \\ 0 & 1 \end{bmatrix}, D_v = \begin{bmatrix} 0.1 & 0 \\ 0 & 0.1 \end{bmatrix}$$

$$q_1^H * q_1 = 1, \quad q_2^H * q_2 = 1, \quad q_1^H * q_2 = 0$$

3. Data eigenvectors and eigenvalues

$$R_x v_i = (R_s + \sigma_v^2 I)v_i = \lambda_i^s v_i + \sigma_v^i v_i = (\lambda_i^s + \sigma_v^i)v_i$$

Therefore, the eigenvectors of R_x are the same as those of R_s, and the eigenvalues of R_x, are

$$\lambda_i = \lambda_i^s + \sigma_v^i$$

Therefore, the largest eigenvalue of R_x is

$$\lambda_{max} = MP_1 + \sigma_v^2$$

And the remaining $M-1$ eigenvalues are equal to σ_v^2.

Note: Parameter extraction for one frequency in the data

1. Perform an eigendecomposition of the autocorrelation matrix R_x. The largest eigenvalue is equal to $\lambda_{max} = MP_1 + \sigma_v^2$ and the remaining eigenvalues are equal to σ_v^2.
2. Use the eigenvalues of R_x to solve for the power P_1 and the noise variance

$$\sigma_v^2 = \lambda_{min} \quad P_1 = \frac{1}{M}(\lambda_{max} - \lambda_{min})$$

3. Since R_x is the result of noisy data $\{x(n)\}$, we consider weighted averages as follows. Let v_i be a noise eigenvector of R_x, for example, one that has an eigenvalue σ_v^2, and let $v_i(k)$ be the kth component of v_i. If we compute the DTFT of the coefficients in v_i

$$V_i(e^{j\omega}) = \sum_{k=0}^{M-1} v_i(k)e^{-jk\omega} = e^H v_i \tag{9.70}$$

then the orthogonality condition (see Equation 9.68) implies that at $\omega = \omega_i$, the value of $V_i(e^{j\omega})$ will be equal to zero and, hence, the **PSD estimation function**

$$\hat{S}(e^{j\omega}) = \frac{1}{\left| \sum_{k=0}^{M-1} v_i(k)e^{-jk\omega} \right|^2} = \frac{1}{\left| e^H v_i \right|^2} \tag{9.71}$$

will be extremely large at $\omega - \omega_i$. This is an effective way to estimate the frequency. To avoid errors in estimating the frequency, using only one eigenvector, it is recommended that a weighted average of all the noise eigenvectors are used. Hence, we write

$$\hat{S}(e^{j\omega}) = \frac{1}{\sum_{k=1}^{M-1} a_i \left| e^H v_i \right|^2} \tag{9.72}$$

where a_i's are some appropriate chosen constants. ∎

9.10.2 HARMONIC MODEL

Consider the signal model that consists of p complex exponentials in white noise:

$$x(n) = \sum_{i=1}^{p} A_i e^{j\omega_i n} + v(n) \tag{9.73}$$

where ω_i is the discrete-time frequency (rad), and A_i is a complex number of the form

$$A_i = |A_i| e^{j\phi_i} \tag{9.74}$$

As discussed in Chapter 2, the spectrum of sinusoidal functions are impulses in the frequency domain. The additive noise produces a constant background level at the power level of the white noise

$$\sigma_v^2 = E\left\{ |v(n)|^2 \right\} \tag{9.75}$$

Since we will be dealing with discrete signals, it is advantageous to form a vector of the signal over a time window of length M. Therefore, the data take the form

$$x(n) = \begin{bmatrix} x(n) & x(n+1) & x(n+2) & \cdots & x(n+M-1) \end{bmatrix}^T \tag{9.76}$$

The signal can now be written in the form

$$x(n) = \sum_{i=1}^{p} A_i e(\omega_i) e^{j\omega_i n} + v(n) = s(n) + v(n)$$

$$s(n) = \sum_{i=1}^{p} A_i e(\omega_i) e^{j\omega_i n} = \text{signal}, v(n) = \begin{bmatrix} v(n) & v(n+1) & \cdots & v(n+M-1) \end{bmatrix}^T \tag{9.77}$$

$$e(\omega_i) = \begin{bmatrix} 1 & e^{j\omega_i} & e^{j\omega_i 2} & \cdots & e^{j\omega_i (M-1)} \end{bmatrix}^T = \text{time-window frequency vector}$$

Note that we differentiate the signal made up of sinusoidal functions (complex) and the white noise signal. We observe that for the data $x(n)=s(n)+v(n)$, the autocorrelation function of $x(n)$ is

$$r_{xx}(k) = r_{ss}(k)+\sigma_v^2\delta(k) \quad k = 0, \pm 1, \pm 2, ..., \pm(M-1) \tag{9.78}$$

Therefore, the $M \times M$ autocorrelation matrix of $x(n)$ may be expressed as

$$\boldsymbol{R}_{xx} = \boldsymbol{R}_s + \boldsymbol{R}_v = \sum_{i=1}^{p}|A_i|^2 e(\omega_i)e^H(\omega_i)+\sigma_v^2\boldsymbol{I} = \boldsymbol{EAE}^H +\sigma_v^2\boldsymbol{I} \tag{9.79}$$

$$\boldsymbol{E} = \begin{bmatrix} e(\omega_1) & e(\omega_2) & \cdots & e(\omega_p) \end{bmatrix}$$

the matrix where $\boldsymbol{E}$ is a $M \times p$ matrix whose columns are the time-window frequency vectors of length M (see Equation 9.77) at frequencies ω_i and $\boldsymbol{A}$ stands for

$$\boldsymbol{A} = \begin{bmatrix} |A_1|^2 & 0 & \cdots & 0 \\ 0 & |A_2|^2 & \cdots & 0 \\ \vdots & \vdots & & \vdots \\ 0 & 0 & \cdots & |A_p|^2 \end{bmatrix} \tag{9.80}$$

We must always take the time-window length M to be greater than the number of sinusoids p.

We can write the $M \times M$ autocorrelation matrix of the data in the form (see Appendix 2)

$$\boldsymbol{R}_x = \boldsymbol{QDQ}^H = \sum_{i=1}^{M}\lambda_i\boldsymbol{q}_i\boldsymbol{q}_i^H, \quad \boldsymbol{Q} = \begin{bmatrix} \boldsymbol{q}_1 & \boldsymbol{q}_2 & \cdots & \boldsymbol{q}_M \end{bmatrix}, \quad \boldsymbol{D} = \begin{bmatrix} \lambda_1 & 0 & \cdots & 0 \\ 0 & \lambda_2 & \cdots & 0 \\ \vdots & \vdots & & \vdots \\ 0 & 0 & & \lambda_M \end{bmatrix} \tag{9.81}$$

Correspondingly, we can also write the preceding equation as follows:

$$\boldsymbol{R}_{xx} = \sum_{i=1}^{p}\lambda_i\boldsymbol{q}_i\boldsymbol{q}_i^H + \sum_{i=p+1}^{M}\sigma_v^2\boldsymbol{q}_i\boldsymbol{q}_i^H = \boldsymbol{Q}_s\boldsymbol{D}_s\boldsymbol{Q}_s^H +\sigma_v^2\boldsymbol{Q}_v\boldsymbol{Q}_v^H \tag{9.82}$$

$$\boldsymbol{Q}_s = \begin{bmatrix} \boldsymbol{q}_1 & \boldsymbol{q}_2 & \cdots & \boldsymbol{q}_p \end{bmatrix} \quad \boldsymbol{Q}_v = \begin{bmatrix} \boldsymbol{q}_{p+1} & \boldsymbol{q}_{p+2} & \cdots & \boldsymbol{q}_M \end{bmatrix}$$

Thus, the M-dimensional subspace that contains the observations of the time-window signal vector can be split into two subspaces spanned by the signal and noise eigenvectors respectively. These two subspaces are known as the **signal subspace** and **noise subspace**. These subspaces are orthogonal to each other since the correlation matrix is Hermitian symmetric (the eigenvectors of a Hermitian symmetric matrix are orthogonal $\boldsymbol{q}_j^H\boldsymbol{q}_i$).

The eigendecomposition separates the eigenvectors into two sets. The set $\{\boldsymbol{q}_1 \ \boldsymbol{q}_2 \cdots \boldsymbol{q}_p\}$, which are principal eigenvectors, span the signal subspace. The set $\{\boldsymbol{q}_{p+1} \ \boldsymbol{q}_{p+2} \cdots \boldsymbol{q}_M\}$, which are orthogonal to the principal eigenvectors belong to the noise subspace. Since the signal vectors $\boldsymbol{e}_i$ are in the signal subspace, it simply follows that they are a linear combination of the principal eigenvectors and, hence, they are orthogonal to the vectors in the noise subspace (see Figure 9.14).

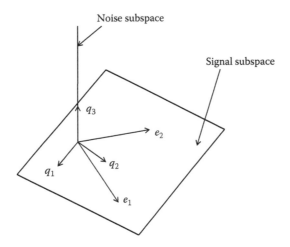

Noise subspace

Signal subspace

FIGURE 9.14

Example 9.10.2.1

Find the eigenvalues of a signal having the following correlation matrices:

$$R_s = \begin{bmatrix} 0.5 & 0.25 \\ 0.25 & 1.0 \end{bmatrix}, \quad R_v = \begin{bmatrix} 0.1 & 0 \\ 0 & 0.1 \end{bmatrix}, \quad R_{xx} = R_s + R_v = \begin{bmatrix} 0.6 & 0.25 \\ 0.25 & 1.1 \end{bmatrix}$$

Solution: Using MATLAB, we obtain

1. Signal eigenvectors and eigenvalues:

$$Q_s = [\ q_1 \quad q_2\ ] = \begin{bmatrix} -0.9239 & 0.3827 \\ 0.3827 & 0.9239 \end{bmatrix}, \quad D_s = \begin{bmatrix} 0.3964 & 0 \\ 0 & 1.1036 \end{bmatrix}$$

$$q_1^H * q_1 = 1, \quad q_2^H * q_2 = 1, \quad q_1^H * q_2 = 0$$

2. Noise eigenvectors and eigenvalues:

$$Q_v = [\ q_1 \quad q_2\ ] = \begin{bmatrix} 1 & 0 \\ 0 & 1 \end{bmatrix}, \quad D_v = \begin{bmatrix} 0.1 & 0 \\ 0 & 0.1 \end{bmatrix}$$

$$q_1^H * q_1 = 1, \quad q_2^H * q_2 = 1, \quad q_1^H * q_2 = 0$$

3. Data eigenvectors and eigenvalues:

$$Q_x = [\ q_1 \quad q_2\ ] = \begin{bmatrix} -0.9239 & 0.3827 \\ 0.3827 & 0.9239 \end{bmatrix}, \quad D_x = \begin{bmatrix} 0.4964 & 0 \\ 0 & 1.2036 \end{bmatrix}$$

$$q_1^H * q_1 = 1, \quad q_2^H * q_2 = 1, \quad q_1^H * q_2 = 0$$

4. Expansion of the autocorrelation matrix (the exponent H (Hermitian) becomes T (transpose) for real matrices:

$$Q_s D_s Q_s^H + \sigma_v^2 Q_v Q_v^H = \begin{bmatrix} -0.9239 & 0.3827 \\ 0.3827 & 0.9239 \end{bmatrix} \begin{bmatrix} 0.3964 & 0 \\ 0 & 1.1036 \end{bmatrix} \begin{bmatrix} -0.9239 & 0.3827 \\ 0.3827 & 0.9239 \end{bmatrix}$$

$$+0.1 \begin{bmatrix} 1 & 0 \\ 0 & 1 \end{bmatrix} \begin{bmatrix} 1 & 0 \\ 0 & 1 \end{bmatrix} = \begin{bmatrix} 0.6000 & 0.2500 \\ 0.2500 & 1.1001 \end{bmatrix}$$

Based on the preceding results, we conclude as before that the following relations hold:

$$R_x = Q_s \sigma_v^2 D_s Q_s^H + Q_v Q_v^H$$

$$Q_s = [\ q_1 \quad q_2 \quad \cdots \quad q_p\ ] \quad Q_v = [\ q_{p+1} \quad \cdots \quad q_M\ ]$$

(9.83)

Note: (1) the eigenvalues of the data (signal plus noise) is equal to sum of the eigenvalues of the signal plus the variance of the noise; (2) the eigenvectors of the data are identical to the eigenvectors of the signal; and (3) Equation 9.83 relation holds. ∎

If there exist p sinusoids in the data, then the first p eigenvalues in descending order correspond to the first part of Equation 9.83 (signal) and the remaining correspond to the second part (noise). These columns of these matrices consist of the signal and noise eigenvectors. This first part of Equation 9.83 is the signal subspace and the second part is the noise subspace.

The following book MATLAB function produces the autocorrelation matrix if the data vector (sequence) $\{x(n)\}$ is given. The autocorrelation is given by

$$\hat{R}_{xx} = \frac{1}{N} X^H X$$

$$X = \begin{bmatrix} x^T(1) \\ x^T(2) \\ \vdots \\ x^T(n) \\ \vdots \\ x^T(N-1) \\ x^T(N) \end{bmatrix} = \begin{bmatrix} x(1) & x(2) & \cdots & x(M) \\ x(2) & x(3) & \cdots & x(M+1) \\ & \vdots & & \\ x(n) & x(n+1) & \cdots & x(n+M-1) \\ & \vdots & & \\ x(N-1) & x(N) & \cdots & x(N+M-2) \\ x(N) & x(N+1) & \cdots & x(N+M-1) \end{bmatrix}$$

(9.84)

BOOK M-FUNCTION: [RX]=EST_AUTOCR_MATRIX(X)

```
function[rx]=ssp_est_autocor_matrix(x,M)
%rx=NxM matrix;N/M=integer;N=length(x)=2^n;
%M=time window along the vector x;x=row data;
N=length(x);
for n=1:N
    for m=1:M
        X(n,m)=x(1,m+n-1);
    end;
end;
rx=(conj(X'))*X/length(x);
```

9.10.3 Pisarenko Harmonic Decomposition

Pisarenko observed that in an ARMA (AR moving average) process consisting of p sinusoids in additive white noise, the noise variance corresponds to the minimum eigenvalue of R_x and, hence, the method proceeds as follows: (1) estimate R_x from the data; (2) find the minimum eigenvalue (the MATLAB function [Q, D] = eig(Rx) gives the eigenvalues in the descent order and to those values correspond the eigenvectors); (3) find the minimum eigenvector; and (4) use the equation

$$S(e^{j\omega}) = \frac{1}{\left|q^H e\right|^2} \tag{9.85}$$

Example 9.10.3.1

Find the spectrum (known as pseudospectrum) using the method of Pisarenko harmonic decomposition.

Solution: The top of Figure 9.15 was produced using the following signal:

$$x(n) = 2\sin(0.3\pi n) + 2\sin(0.4\pi n) + \sin(0.5\pi n)$$

and the bottom one was produced using the following signal:

$$x(n) = 2\sin(0.3\pi n) + 2\sin(0.4\pi n) + \sin(0.5\pi n) + 3\text{randn}$$

We observe that the noise energy produces shifting.

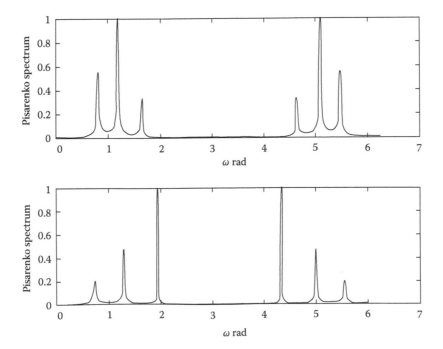

FIGURE 9.15

BOOK M-FUNCTION FOR PISARENKO HARMONIC DECOMPOSITION: [QMIN,SIGMA,Q,D]=PISAR_HARM_DECOMP(X,P)

```
function[qmin,sigma,q,d]=ssp_pisar_harm_decomp(x,p)
    %this function graphs the PSD using Pisarenko
    %harmonic decomposition; qmin=minimum eigenvector;
    %sigma=noise variance;q=matrix eigenvectors;
    %d=diagonal matrix of the eigenvalues by descending
    %order;p=number of harmonic present;
r=xcorr(x,'biased');
R=toeplitz(r(:,length(x):length(x)+2*p));
[q,d]=eig(R);
sigma=min(diag(d));
coln=find(diag(d)==sigma);
qmin=q(:,coln);
w=0:2*pi/256:2*pi-(2*pi/256);
plot(w,((abs(1./fft(qmin,256))/abs(1./...
fft(qmin,256)))),'k');
```

We observe that the strength of the noise plays a dominant effect on the location of the spectra lines. The Pisarenko method is important from the conceptual and analytical perspective. It lucks robustness to be used for most applications. Furthermore, the correlation matrix must be estimated and, therefore, the resulting noise eigenvectors are only estimated. Because the roots of the minimum estimator can be close to the unit circle, splitting of the line spectrum can also occur. ∎

9.10.4   MUSIC Algorithm

The MUSIC (**multiple signal classification**) algorithm is based on two disciplines: (1) the window M is not set equal to $p+1$ but larger than that; and (2) we average the noise spectra. Therefore, we write

$$S_{\mathrm{music}}(e^{j\omega}) = \frac{1}{\displaystyle\sum_{i=1}^{M-p}\left|e^H q_i\right|^2} = \frac{1}{\displaystyle\sum_{i=1}^{M-p}\left|Q_i(e^{j\omega})\right|^2}$$

(9.86)

$$e = [\ \ 1 \quad e^{j\omega} \quad e^{j\omega 2} \quad \cdots \quad e^{j\omega(M-1)}\ \ ], \quad Q_i(e^{j\omega}) = \text{FT of } i\text{th eigenvector}$$

It has also been suggested to multiply the ith factor of the summation by $1/\lambda_i$. The following Book m-function plots the MUSIC spectrum.

Book m-Function for MUSIC Algorithm: [psd,q,d,R]=music_alg(x,p,M,nfft)

```
function[psd,q,d,R]=music_alg(x,p,M,nfft)
    %length(x)+M<2*length(x); R=correlation matrix;
    %p=number of sinusoids; M=time window;nfft=the
    %desired number of bins in fft e.g. 256 or 512;
    %to observe the spectrum write plot(psd/max(psd));
r=xcorr(x,'biased');
R=toeplitz(r(:,length(x):length(x)+M));
[q,d]=eig(R);
qin=zeros(nfft,1);
for i=1:M-p-1
    qin=qin+abs(fft(q(:,i),nfft));
end;
psd=1./(qin/max(qin));
```

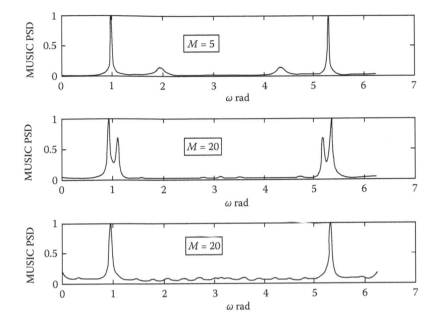

FIGURE 9.16

The top two spectra shown in Figure 9.16 were produced using the following signal:

$$x(n) = \sin(0.3\pi n) + \sin(0.35\pi n) + 0.5\text{randn}$$

The bottom spectrum of the figure was produced using the following signal:

$$x(n) = \sin(0.3\pi n) + \sin(0.35\pi n) + 1.5\text{randn}$$ ■

HINTS, SUGGESTIONS, AND SOLUTIONS OF THE EXERCISES

9.2.1

From Equation 9.1, we write

$$y(n) + \sum_{m=1}^{p} a(m)y(n-m) = \sum_{m=0}^{q} b(m)v(n-m) \quad (b(0)=1) \tag{1}$$

Multiply (1) by $y(n-k)$ and take expectations to obtain

$$E\{y(n)y(n-k)\} + \sum_{m=1}^{p} a(m)E\{y(n-m)y(n-k)\} = \sum_{m=0}^{q} b(m)E\{v(n-m)y(n-k)\} \text{ or}$$

$$r_{yy}(k) + \sum_{m=1}^{p} a(m)r_{yy}(k-m) = \sum_{m=0}^{q} b(m)E\{v(n-m)y(n-k)\} \tag{2}$$

but

$$y(n) = \sum_{s=0}^{\infty} h(s)v(n-s) \tag{3}$$

and hence the last expectation of (2) becomes $E\{v(n-m)y(n-k)\} = E\{v(n-m)\sum_{s=0}^{\infty} h(s)v(n-k-s)\}$

$$= \sum_{s=0}^{\infty} h(s)E\{v(n-m)v(n-k-s)\} = \sum_{s=0}^{\infty} h(s)\sigma_v^2\delta(k+s-m) = \sigma_v^2 h(m-k) \Rightarrow 2 \text{ becomes } r_{yy}(k)$$

$$+ \sum_{m=1}^{p} a(m)r_{yy}(k-m) = \sigma_v^2 \sum_{m=0}^{q} b(m)h(m-k).$$

9.3.1

Figure 9.17 shows the results. The accuracy depends on the number of data used.

9.4.1

$$\sigma_p^2 \triangleq E\{|e(n)|^2\} = E\{[y(n) + \mathbf{a}^T \mathbf{y}(n)][y(n) + \mathbf{y}^T(n)\mathbf{a}]\} = E\{y^2(n)\} + \mathbf{a}^T E\{\mathbf{y}(n)y(n)\}$$

$$+ E\{y(n)\mathbf{y}^T(n)\}\mathbf{a} + \mathbf{a}^T E\{\mathbf{y}(n)\mathbf{y}^T(n)\}\mathbf{a}$$

$$r(0) = E\{y^2(n)\}; \ \mathbf{a}^T \mathbf{r}_{y(p)} = \mathbf{a}^T E\{\mathbf{y}(n)y(n)\}; \mathbf{r}_{y(p)}^T \mathbf{a} = E\{y(n)\mathbf{y}^T(n)\}\mathbf{a}; \mathbf{a}^T \mathbf{R}_{y(p)}\mathbf{a}$$

$$= \mathbf{a}^T E\{\mathbf{y}(n)\mathbf{y}^T(n)\}\mathbf{a}$$

We obtain [for a 2×2 matrices, which can be extrapolated to any dimension (see also Appendix 2)]

$$\frac{\partial \sigma^2}{\partial a(1)} = \begin{bmatrix} r(1) & r(2) \end{bmatrix}\begin{bmatrix} 1 \\ 0 \end{bmatrix} + \begin{bmatrix} 1 & 0 \end{bmatrix}\begin{bmatrix} r(1) \\ r(2) \end{bmatrix} + \begin{bmatrix} 1 & 0 \end{bmatrix}\begin{bmatrix} r(1) & r(2) \\ r(2) & r(1) \end{bmatrix}\begin{bmatrix} a(1) \\ a(2) \end{bmatrix}$$

$$+ \begin{bmatrix} a(1) & a(2) \end{bmatrix}\begin{bmatrix} r(1) & r(2) \\ r(2) & r(1) \end{bmatrix}\begin{bmatrix} 1 \\ 0 \end{bmatrix} = 2r(1) + 2r(1)a(1) + 2r(2)a(2) = 0, r(1)$$

$$+ \begin{bmatrix} r(1) & r(2) \end{bmatrix}\begin{bmatrix} a(1) \\ a(2) \end{bmatrix} = 0 \tag{1}$$

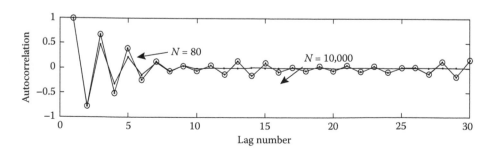

FIGURE 9.17

$$\frac{\partial \sigma^2}{\partial a(2)} = \begin{bmatrix} r(1) & r(2) \end{bmatrix}\begin{bmatrix} 0 \\ 1 \end{bmatrix} + \begin{bmatrix} 0 & 1 \end{bmatrix}\begin{bmatrix} r(1) \\ r(2) \end{bmatrix} + \begin{bmatrix} 0 & 1 \end{bmatrix}\begin{bmatrix} r(1) & r(2) \\ r(2) & r(1) \end{bmatrix}\begin{bmatrix} a(1) \\ a(2) \end{bmatrix}$$

$$+ \begin{bmatrix} a(1) & a(2) \end{bmatrix}\begin{bmatrix} r(1) & r(2) \\ r(2) & r(1) \end{bmatrix}\begin{bmatrix} 0 \\ 1 \end{bmatrix} = 2r(2) + 2r(2)a(1) + 2r(1)a(2)$$

$$= 0, r(1) + \begin{bmatrix} r(2) & r(1) \end{bmatrix}\begin{bmatrix} a(1) \\ a(2) \end{bmatrix} = 0 \qquad (2)$$

(1) and (2) above form the following matrix form: $\begin{bmatrix} r(1)a(1) + r(2)a(2) \\ r(2)a(1) + r(1)a(2) \end{bmatrix} = -\begin{bmatrix} r(1) \\ r(2) \end{bmatrix}$ or $\mathbf{Ra} = -\mathbf{r}$.

9.4.3

$J(a) = r_y(0) + r_{y(p)}^T a + a^T r_{y(p)} + a^T R_{y(p)}a$ (1), let $a = a_0 + \tilde{a}$, $a_0 = -R_{y(p)}^{-1}r_{y(p)}$ then (1) becomes

$$J(a) = r_y(0) + r_{y(p)}^T a + a^T r_{y(p)} + a^T R_{y(p)}a = r_y(0) + r_{y(p)}^T(a_0 + \tilde{a}) + (a_0^T + \tilde{a}^T)r_{y(p)}$$

$$+ (a_0^T + \tilde{a}^T)R_{y(p)}(a_0 + \tilde{a}) = r_y(0) - r_{y(p)}^T R_{y(p)}^{-1}r_{y(p)} + r_{y(p)}^T\tilde{a} - r_{y(p)}^T R_{y(p)}^{-1}r_{y(p)} + \tilde{a}^T r_{y(p)}$$

$$r_{y(p)}^T R_{y(p)}^{-1}r_{y(p)} - \tilde{a}^T r_{y(p)} - r_{y(p)}^T\tilde{a} + \tilde{a}^T R_{y(p)}\tilde{a} = \tilde{a}^T R_{y(p)}\tilde{a} + y(0) - r_{y(p)}^T R_{y(p)}^{-1}r_{y(p)};$$

where we used the following identities (see also Appendix 2 Matrices):

$$R_{y(p)}^{-1}R_{y(p)} = I,$$

$$\tilde{a}^T r_{y(p)} = r_{y(p)}^T\tilde{a}$$

because $J(a)$ is a constant, the first factor after the last equality is constant since matrix R is a positive definite matrix and, hence, the last two terms are the minimum mean square error.

9.4.4

$$J(a) = (y + Ya)(y^T + a^T Y^T) = yy^T + Yay^T + ya^T Y^T + Yaa^T Y^T = yy^T + 2ya^T Y^T + a^T Y^T Ya$$

$$\frac{\partial J(a)}{\partial a} = 0 + 2Y^T y + 2Y^T Ya = 0 \Rightarrow a = (Y^T Y)^{-1}(Y^T y)$$

The lowercase letters are vectors and the uppercase letters are matrices. Since J is a number, so it is every part of the expression.

9.5.1

The signal and the inputs to the function for obtaining the MV PSD were $x = 6 * \sin(0.3 * pi * [1:128]) + 6 * \sin(0.34 * pi * [1:128]) + 6 * \sin(0.6 * pi * [1:128]) + 0.1*randn$ (1,128) $P=40$ and $L=512$. The Linear Extended Book method signal was $xlt = [0.5 * x. * hamming(1,128) \quad 0.2 * x. * hamming(1,128)]$. The results are shown in Figure 9.18.

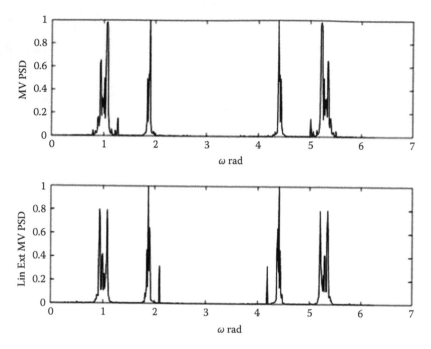

FIGURE 9.18

9.7.1

The following Book MATLAB program was used to produce Figure 9.19.

```
>>x(1)=0;x(2)=0;x(3)=0;x(4)=0;
>>for n=0:127
>>      x(n+5)=1.38*x(n+4)-1.56*x(n+3)+0.89*x(n+2)-0.43*x(n+1)+0.5*randn;
```

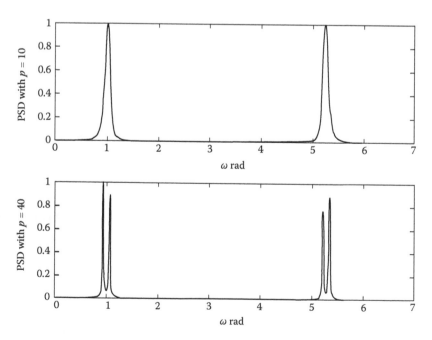

FIGURE 9.19

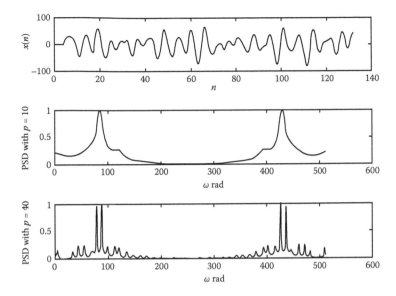

FIGURE 9.20

```
>>      s(n+1)=6*sin(0.3*pi*n)+6*sin(0.34*pi*n);
>>end
>>xs=s+x(1,1:128);
>>p=10;
>>[a10,var10]=yule_walker_lg(xs,64,p);
>>p=40;
>>[a40,var40]=yule_walker_lg(xs,64,p);
>>ps10=abs(sqrt(var10)./(exp(-…
>>   j*[0:0.01:6.2]'*[1:length(a10)+1]*[1;-a10]).^2);
>>ps40=abs(sqrt(var40)./(exp(-…
>>   j*[0:0.01:6.2]'*[1:length(a40)+1]*[1;-a40]).^2);
>>subplot(2,1,1);plot([0:0.01:6.2],ps10,'k');
>> subplot(2,1,2);plot([0:0.01:6.2],ps40,'k');
>>xlabel('\omega rad');ylabel('PSD with p=40');
```

9.8.1

The results are shown in Figure 9.20.

9.10.1.1

Mean value:

$$m_x(n) = E\{x(n)\} = E\{A\sin(n\omega_0 + \phi)\} = \int_{-\infty}^{\infty} A\sin(n\omega_0 + x)f_\phi(x)\,dx =$$

$$\int_{-\pi}^{\pi} A\sin(n\omega_0 + x)f_\phi(x)\,dx = 0; r_{xx}(k,l) = E\{x(k)x(l)\} = E\{A\sin(k\omega_0 + \phi)A\sin(l\omega_0 + \phi)\} \text{ but}$$

$$2\sin A\sin B = \cos(A-B) - \cos(A+B) \Rightarrow r_{xx}(k,l) = \frac{1}{2}A^2 E\{\cos[(k-l)\omega_0]\} - \frac{1}{2}A^2$$

$$E\{\cos[(k+l)\omega_0 + 2\phi]\} = \frac{1}{2}A^2\cos[(k-l)\omega_0]$$

10 Newton's and Steepest Descent Methods

10.1 GEOMETRIC PROPERTIES OF THE ERROR SURFACE

The performance function of a transversal [Finite Impulse Response (FIR)] Wiener filter with real-valued input sequence $\{x(n)\}$ and a desired output sequence $\{d(n)\}$ is (see Equation 7.92)

$$J = w^T R w - 2 p^T w + E\{d^2(n)\}, E\{d^2(n)\} = \text{variance}$$

$$w = \begin{bmatrix} w_0 & w_1 & \cdots & w_{M-1} \end{bmatrix}^T, R_{xx} = E\{x(n)x^T(n)\} = \text{autocorrelation matrix},$$

$$x = \begin{bmatrix} x(n) & x(n-1) & \cdots & x(n-M+1) \end{bmatrix}^T$$

$$p = E\{d(n)x(n)\} = \text{crosscorrelation}$$

(10.1)

The optimum Wiener filter coefficients, w^o, are given by the relation

$$R_{xx} w^o = p \tag{10.2}$$

Using the optimum solution of the Wiener filter, the preceding equation, and, since $w^T p = p^T w = $ constant and $R_{xx}^T = R_{xx}$, the performance function of Equation 10.1 becomes

$$J = w^T R_{xx} w - w^T R_{xx} w^o - w^{oT} R_{xx}^T w + w^{oT} R_{xx} w^o + E\{d^2(n)\} - w^{oT} R_{xx} w^o$$

$$J = (w - w^o)^T R_{xx}(w - w^o) + E\{d^2(n)\} - w^{oT} R_{xx} w^o$$

$$= (w - w^o)^T R_{xx}(w - w^o) + J_{\min}$$

$$J = J_{\min} + (w - w^o)^T R_{xx}(w - w^o)$$

(10.3)

The cost function (performance function) can be written in the form:

$$w^T R_{xx} w - 2 p^T w - (J - E\{d^2(n)\}) = 0 \text{ or } w^T R_{xx} w - 2 p^T w - \left(J - \sigma_d^2\right) = 0 \tag{10.4}$$

If we set values of $J > J_{\min}$, the w-plane will cut the second order surface, for a two-coefficient filter, along a line whose projection on the w-plane are ellipses arbitrarily oriented as shown in Figure 10.1. The contours were found using the following Book MATLAB program.

Book MATLAB Program

```
>>w0=-3:0.05:3; w1=-3:0.05:3;
>> [x,y]=meshgrid(w0,w1);%MATLAB function
>>j=0.8461+0.972*x-0.773*y+1.0647*x.^2+1.064*y.^2+0.985*x.*y;
>>contours(x,y,j,30);%MATLAB function,30 is the number of %contours;
```

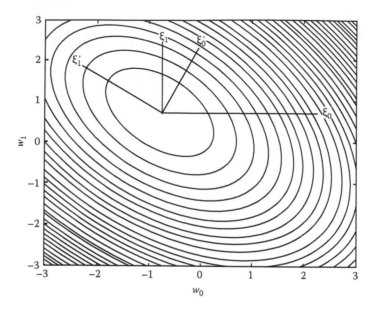

FIGURE 10.1

If we set `contour(j,[2.3   3.1   5])`, we will produce three contours at heights 2.3, 3.1, and 5.

The first simplification we can do is to shift the origin of the w-axes to another whose origin is on the center of the ellipses. This is accomplished using the transformation $\xi = w - w^o$. If we introduce this transformation in Equation 10.4 and setting $J=2J_{\min}$, we obtain the relationship (see Exercise 10.1.1)

$$\xi^T R_{xx} \xi - 2J_{\min} + \sigma_d^2 - p^T w^o = 0 \tag{10.5}$$

Exercise 10.1.1

Verify Equation 10.5.   ▲

But $\sigma_d^2 - p^T w^o = J_{\min}$, and thus, the preceding equation becomes

$$\xi^T R_{xx} \xi = J_{\min} \tag{10.6}$$

The matrix R_{xx} is **diagonal** with elements that are its eigenvalues.

Example 10.1.1

Let $\lambda_1 = 1$, $\lambda_2 = 0.5$, and $J_{\min} = 0.67$. The ellipse in the (ξ_0', ξ_0') plane is found by solving the system

$$\begin{bmatrix} \xi_0' & \xi_0' \end{bmatrix} \begin{bmatrix} 1 & 0 \\ 0 & 0.5 \end{bmatrix} \begin{bmatrix} \xi_0' \\ \xi_1' \end{bmatrix} = 0.67 \quad \text{or} \quad \left(\frac{\xi_0'}{\sqrt{0.67/1}} \right)^2 + \left(\frac{\xi_1'}{\sqrt{0.67/0.5}} \right)^2 = 1 \tag{10.7}$$

where 0.67/0.5 is the major axis and 0.67/1 is the minor axis. Hence, for the case $J=2J_{\min}$, the ellipse intersects the ξ_0' axis at 0.67 and ξ_1' at 2.68. To find, for example, the intersection between the ξ_1'-axis and the ellipse, we must set ξ_0' equal to zero because the projection on the ξ_0' axis of the intersection point is at $\xi_0' = 0$.

If we start with Equation 10.4 and apply the shift and rotation transformations, we obtain the relationships:

$$J = J_{\min} + (w - w^o)^T R_{xx} (w - w^o) = J_{\min} + \xi^T R_{xx} \xi$$

$$= J_{\min} + \xi^T (Q \Lambda Q^T) \xi = J_{\min} + \xi'^T \Lambda \xi' = J_{\min} + \sum_{i=0}^{M-1} \lambda_i \xi_i'^2 \qquad (10.8)$$

Note

- The contours intersect the ξ'-axes at values dependent upon the eigenvalues of R_{xx} and the specific mean square error (MSE) value chosen. The rotation and translation do not alter the shape of the MSE surface.
- If the successive contours for the values $2J_{\min}$, $3J_{\min}$, and so on are close to each other, the surface is steep, which in turn indicates that the mean-square estimation error is very sensitive to the choice of the filter coefficients.
- Choosing the filter values w is equivalent to choosing a point in the w-plane. The height of the MSE surface above the plane at that point is determined only by the signal correlation properties.

Let us rearrange Equation 10.8 in the form

$$J - J_{\min} = \xi'^T \Lambda \xi' = \begin{bmatrix} \xi_0' & \xi_0' \end{bmatrix} \begin{bmatrix} \lambda_0 & 0 \\ 0 & \lambda_1 \end{bmatrix} \begin{bmatrix} \xi_0' \\ \xi_1' \end{bmatrix} = \lambda_0 \xi_0'^2 + \lambda_1 \xi_1'^2 \qquad (10.9)$$

Next, we plot the contours of the performance surface (see Figure 10.2) for two different ratios of the eigenvalues. The contours with shapes closer to a circle were produced by the ratio $0.8/0.6 = 1.333$; for the most elongated ones, the ratio $1.2/0.2 = 6$ was used. Instead for ξ_i', we renamed them to ξ_i for plotting purposes. ∎

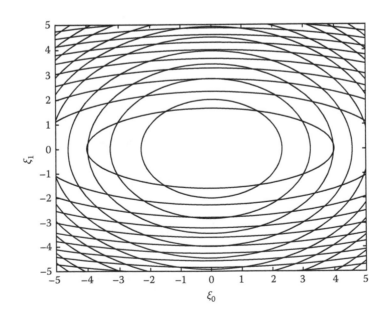

FIGURE 10.2

Exercise 10.1.2

Find the contours of the performance surface if the following data are given:

$$R_{xx} = \begin{bmatrix} 2 & 1 \\ 1 & 3 \end{bmatrix}, \quad p_{dx} = \begin{bmatrix} 6 \\ 7 \end{bmatrix}, \sigma_d^2 = 28 \qquad \blacktriangle$$

Exercise 10.1.3

A Wiener filter is characterized by the following parameters:

$$R = \begin{bmatrix} d & a \\ a & d \end{bmatrix}, \quad p = \begin{bmatrix} 1 \\ 1 \end{bmatrix}, \sigma_d^2 = 2$$

It is requested to explore the performance surface as the ratio λ_0/λ_1 varies. This is accomplished using different values of a and b. $\qquad \blacktriangle$

Exercise 10.1.4

Consider the performance function $J_{\min} = w_0^2 + w_1^2 + w_0 w_1 - w_0 + w_1 + 2$. First, convert this equation to its canonical form; next, plot the set of contour ellipses of the performance surface of J for values of $J=1, 2, 3,$ and 4. $\qquad \blacktriangle$

Exercise 10.1.5

If R is a correlation matrix, show (1) $R^n = Q\Lambda^n Q^T$. (2) If $R^{1/2}R^{1/2}=R$, show $R^{1/2} = Q\Lambda^{1/2}Q^T$. $\qquad \blacktriangle$

10.2   ONE-DIMENSIONAL GRADIENT SEARCH METHOD

In general, we can say that adaptive algorithms are nothing but iterative search algorithms derived from minimizing a cost function with the true statistics replaced by their estimates. To study the adaptive algorithms, it is necessary to have a thorough understanding of the iterative algorithms and their convergence properties. In this chapter, we discuss the **steepest descent** and the **Newton's method**.

The one-coefficient MSE surface (line in this case) is given by (see Equation 10.8)

$$J(w) = J_{\min} + r_{xx}(0)(w - w^o)^2; \quad J(w^o) \le J(w) \quad \text{for all } w \tag{10.10}$$

and it is pictorial shown in Figure 10.3. The first and second derivatives are

$$\frac{\partial J(w)}{\partial w} = 2r_{xx}(0)\left(w - w^o\right) \tag{10.11a}$$

$$\frac{\partial^2 J(w)}{\partial w^2} = 2r_{xx}(0) > 0 \tag{10.11b}$$

Because, at $w=w^o$, the first derivative is zero and the second derivative is positive and greater than zero, indicates that the surface has a global minimum and is concave upwards. To find the optimum value of w, we can use an iterative approach. We start first with an arbitrary value $w(0)$ and measure the slope of the curve $J(w)$ at $w(0)$. Next, we find $w(1)$ to be equal to $w(0)$ plus the negative of an

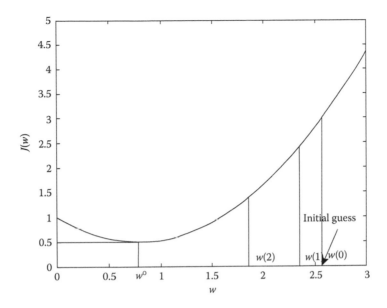

FIGURE 10.3

increment **proportional** to the slope of $J(w)$ at $w(0)$. Proceeding with the iteration procedure, we will eventually find the minimum value w^o. The values $w(0)$, $w(1)$, ..., are known as the **gradient estimates**.

10.2.1 GRADIENT SEARCH ALGORITHM

Based on the preceding development in Section 10.2, the filter coefficient at iteration n, $w(n)$, is found using the relation

$$w(n+1) = w(n) + \mu\left(-\frac{\partial J(w)}{\partial w}\Big|_{w=w(n)}\right)$$

$$= w(n) + \mu[-\nabla J(w(n))] = w(n) - 2\mu r_{xx}(0)\left(w(n) - w^o\right)$$

(10.12)

where
μ is a constant to be determined.
Rearranging Equation 10.12, we obtain

$$\underbrace{w(n+1)}_{\substack{\text{new}\\\text{coefficient}}} = \underbrace{[1 - 2\mu r_{xx}(0)]w(n)}_{\substack{\text{old}\\\text{coefficient}}} + \underbrace{2\mu r_{xx}(0)w^o}_{\substack{\text{adaptation}\\\text{gain}}}$$

(10.13)

$$w(n+1) = aw(n) + bw^o \qquad \text{(difference equation)}$$

The solution of the preceding equation, using the iteration approach (see Exercise 10.2.1.1), is

$$w(n) = w^o + [1 - 2\mu r_{xx}(0)]^n \left(w(0) - w^o\right)$$

(10.14)

The preceding equation gives $w(n)$ explicitly at any iteration n in the search procedure. This is the solution to the **gradient search algorithm**. Note that if we had initially guessed $w(0) = w^o$, which is the optimum value, we would have found $w(1) = w^o$ and that gives the optimum value in one step.

Exercise 10.2.1.1

Verify Equation 10.14.   ▲

Exercise 10.2.1.2

Plot Equation 10.14 for the following positive values: (1) $0 < \mu < \dfrac{1}{2r_{xx}(0)}$, (2) $\mu \cong \dfrac{1}{2r_{xx}(0)}$, and (3)

$\dfrac{1}{2r_{xx}(0)} < \mu < \dfrac{1}{r_{xx}(0)}$ ▲

To have convergence of $w(n)$ in Equation 10.14, we must impose the condition

$$\left|1 - 2\mu r_{xx}(0)\right| < 1 \tag{10.15}$$

The preceding inequality defines the range of the step-size constant μ so that the algorithm will converge. Hence, we obtain

$$-1 < 1 - 2\mu r_{xx}(0) < 1 \text{ or } 0 < 2\mu r_{xx}(0) < 2 \text{ or } 0 < \mu < \frac{1}{r_{xx}(0)} \tag{10.16}$$

Under the preceding condition, Equation 10.14 converges to the optimum value w^o as $n \to \infty$. If $\mu > 1/r_{xx}(0)$, the process is unstable and no convergence takes place.

When the filter coefficients has a value $w(n)$ (that is, at iteration n), then MSE surface (here a line) is (see Equation 10.1)

$$J(n) = J_{\min} + r_{xx}(0)\left(w(0) - w^o\right)^2 [1 - 2\mu r_{xx}(0)]^{2n} \tag{10.17}$$

Substituting the quantity $w(0) - w^o$ in Equation 10.17 from 10.14, we obtain

$$J(n) = J_{\min} + r_{xx}(0)\left(w(0) - w^o\right)^2 [1 - 2\mu \, r_{xx}(0)]^{2n} \tag{10.18}$$

The preceding two equations show that $w(n) = w^o$ as n increases to infinity, and the MSE undergoes a geometric progression towards $J_{\min}$. The plot of the performance surface $J(n)$ versus the iteration number n is known as the **learning curve**.

10.2.2 Newton's Method in Gradient Search

The Newton's method finds the solution (zeros) of the equation $f(w) = 0$. From Figure 10.4, we observe that the slope at $w(0)$ is

$$f'(w(0)) = \frac{df(w)}{dw}\bigg|_{w=w(0)} \cong \frac{f(w(0))}{w(0) - w(1)} \tag{10.19}$$

where $w(0)$ is the initial guessing. The preceding equation is the result of retaining the first two terms of Taylor's expansion and setting the rest of them equal to zero. This equation can be written in the form

$$w(1) = w(0) - \frac{f[w(0)]}{f'[w(0)]} \tag{10.20}$$

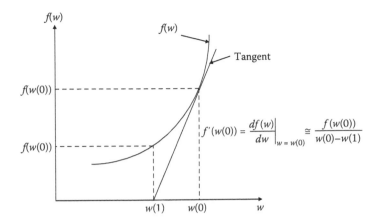

FIGURE 10.4

From Equation 10.20, it is apparent that if we know the value of the function and its derivative at the same point, in this case at $n=0$, $w(0)$ obtain $w(1)$. Hence the nth iteration of the preceding equation takes the form

$$w(n+1) = w(n) - \frac{f[w(n)]}{f'[w(n)]} \qquad n = 0, 1, 2, \ldots \qquad (10.21)$$

But we have $f'[w(n)] \cong [f[w(n)] - f[w(n-1)]] / [w(n) - w(n-1)]$, and hence, Equation 10.21 becomes

$$w(n+1) = w(n) - \frac{f[w(n)][w(n) - w(n-1)]}{f[w(n)] - f[w(n-1)]} \qquad (10.22)$$

As we have mentioned previously, Newton's method finds the roots of the function $f(w)$; that is, solving the polynomial $f(w)=0$. However, in our case, we need to find the minimum point of the performance surface (in the present case is a line). This is equivalent to setting $\partial J(w) / \partial w = 0$. Therefore, we substitute the derivative $\partial J(w) / \partial w$ for $f(w)$ in Equation 10.21; for $f'(w)$, we substitute the second order derivative $\partial^2 J(w) / \partial w^2$. Hence, Equation 10.22 becomes

$$w(n+1) = w(n) - \frac{\dfrac{\partial J[w(n)]}{\partial w(n)}}{\dfrac{\partial^2 [w(n)]}{\partial w^2(n)}} \qquad n = 0, 1, 2, \ldots \qquad (10.23)$$

Exercise 10.2.2.1

Using the one-dimension Newton's algorithm, find the third root of 8. Start with $x(1)=1.2$. ▲

10.3 STEEPEST DESCENT ALGORITHM

To find the minimum value of the MSE surface, $J_{\min}$, using the steepest descent algorithm, we proceed as follows: (1) we start with the initial value $w(0)$, usually using the null vector; (2) at the MSE surface point that corresponds to $w(0)$, we compute the **gradient vector**, $\nabla J[w(0)]$; (3) we compute the value $-\mu \nabla J[w(0)]$ and add it to $w(0)$ to obtain $w(1)$; (4) we go back to step (2) and continue the procedure until we find the optimum value of the vector coefficients, w^o.

If $w(n)$ is the filter-coefficient vector at step n (time), its updated value $w(n+1)$ is given by (see also Equation 10.12; it is the first order of Taylor's expansion)

$$w(n+1) = w(n) - \mu \nabla_w J[w(n)]$$

$$\nabla J_w[w(n)] = \begin{bmatrix} \dfrac{\partial J[w(n)]}{\partial w_0} \\[2mm] \dfrac{\partial J[w(n)]}{\partial w_1} \\[1mm] \vdots \\[1mm] \dfrac{\partial J[w(n)]}{\partial w_{M-1}} \end{bmatrix}; w(n) = \begin{bmatrix} w_0 & w_1 & w_2 & \cdots & w_{M-1} \end{bmatrix}^T \qquad (10.24)$$

10.3.1 STEEPEST DESCENT ALGORITHM APPLIED TO WIENER FILTER

The cost function $J[w(n)]$ at time n is quadratic function of the Wiener coefficients. Hence, we write

$$J[w(n)] = \sigma_d^2 - w^T(n)p - p^T w(n) + w^T(n) R_{xx} w(n) \qquad (10.25)$$

where

$$\sigma_d^2 = \text{variance of the desired response } \{d(n)\}$$

$$p = \text{cross-correlation vector between } \{x(n)\}$$

$$\text{and the desired response } \{d(n)\}$$

$$R_{xx} = \text{autocorrelation matrix of the coefficient vector } x(n)$$

Using Equation 10.24 and Section 2.6 of Appendix 2, we obtain

$$\nabla_w J[w(n)] = -2p + 2R_{xx} w(n) \qquad (10.26)$$

Example 10.3.1.1

Verify Equation 10.26 with a two-length vector.

Solution:

$$\nabla_w \{J[w(n)]\} = \nabla_w \{\sigma_d^2\} - \nabla_w \{w^T(n)p\} - \nabla_w \{p^T w(n)\} + \nabla_w \{w^T(n) R_{xx} w(n)\}$$

Hence, we have

$$\nabla_w(\sigma_d^2) = 0, \ -\nabla_w(w^T p) = -\begin{bmatrix} \dfrac{\partial(w_0 p_0 + w_1 p_1)}{\partial w_0} \\[2mm] \dfrac{\partial(w_0 p_0 + w_1 p_1)}{\partial w_1} \end{bmatrix} = -\begin{bmatrix} p_0 \\ p_1 \end{bmatrix} = -p,$$

$$-\nabla_w(p^T w) = -\begin{bmatrix} \dfrac{\partial(w_0 p_0 + w_1 p_1)}{\partial w_0} \\[2mm] \dfrac{\partial(w_0 p_0 + w_1 p_1)}{\partial w_1} \end{bmatrix} = -\begin{bmatrix} p_0 \\ p_1 \end{bmatrix} = -p,$$

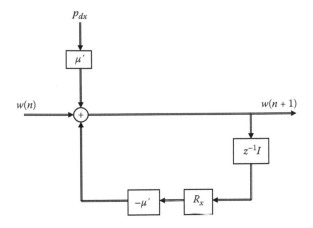

FIGURE 10.5

$$\nabla_w\left(w^T R_{xx} w\right) = \nabla_w\left([w_0 w_1]\begin{bmatrix} r_{xx}(0) & r_{xx}(1) \\ r_{xx}(1) & r_{xx(0)} \end{bmatrix}\begin{bmatrix} w_0 \\ w_1 \end{bmatrix}\right)$$

$$= \nabla_w\left[w_0^2 r_{xx}(0) + w_0 w_1 r_{xx}(1) + w_0 w_1 r_{xx}(1) + w_1^2 r_{xx}(1)\right]$$

$$= \begin{bmatrix} 2w_0 r_{xx}(0) + 2w_1 r_{xx}(1) \\ 2w_0 r_{xx}(1) + 2w_1 r_{xx}(0) \end{bmatrix} = 2\begin{bmatrix} r_{xx}(0) & r_{xx}(1) \\ r_{xx}(1) & r_{xx(0)} \end{bmatrix}\begin{bmatrix} w_0 \\ w_1 \end{bmatrix} = 2R_{xx} w$$

All vectors are taken at time n. ∎

Introducing the results in Example 10.3.1.1 in the first equation in Equation 10.24, we obtain

$$w(n+1) = w(n) + 2\mu[p_{dx} - R_{xx} w(n)] = w(n) + \mu'[p_{dx} - R_{xx} w(n)]$$

$$= [I - \mu' R_{xx}]w(n) + \mu' p_{dx}, \mu' = 2\mu, n = 0, 1, 2, \ldots \qquad (10.27)$$

where I is the identity matrix (ones along the diagonal and the rest zeros). The value of the primed step-size parameter, μ', must be much less than 1/2 for convergence. We may view Equation 10.27 as a **feedback model**, as illustrated by the **single-flow graph** in Figure 10.5.

To apply the method of steepest descent, we must find first the estimate of the autocorrelation matrix R_{xx} and the cross-correlation vector p_{dx} from the finite data. This is necessary because we do not have in practice an ensemble of data to obtain R_{xx} and p_{dx}.

Exercise 10.3.1.1

Let the MSE be given by the nonquadratic equation $J = 2 - (1/35)[(1-w^2)(4.5+3.5w)]$. Find and plot the recursive Newton's algorithm for the coefficients $\{w(n)\}$ for $n=0, 1, 2, \ldots$. Start with $w(0)=2$. ▲

10.3.2 STABILITY (CONVERGENCE) OF THE ALGORITHM

Let

$$\xi(n) = w(n) - w^o \tag{10.28}$$

be the difference between the filter coefficient vector and its optimum Wiener value w^o. Next we write the first part of Equation 10.27 in the form

$$w(n+1) - w^o = w(n) - w^o + \mu'[R_{xx}w^o - R_{xx}w(n)]$$

or

$$\xi(n+1) = [I - \mu' R_{xx}]\xi(n) \tag{10.29}$$

But because $R_{xx} = Q\Lambda Q^T$ (see Table 9.1) and $I = QQ^T$, Equation 10.29 becomes

$$\xi(n+1) = [I - \mu' Q\Lambda Q^T]\xi(n)$$

or

$$Q^T \xi(n+1) = [I - \mu'\Lambda]Q^T \xi(n)$$

or

$$\xi'(n+1) = [I - \mu'\Lambda]\xi'(n) \tag{10.30}$$

where the coordinate axis defined by ξ' is orthogonal to ellipsoids and is a diagonal matrix with its diagonal elements equal to the eigenvalues of R_{xx}. The kth row of Equation 10.30, which represents the kth **natural mode** of the steepest descent algorithm, is (see Exercise 10.3.2.1)

$$\xi_k'(n+1) = [1 - \mu'\lambda_k]\xi_k'(n) \tag{10.31}$$

Exercise 10.3.2.1

Verify Equation 10.31.   ▲

The preceding equation is a homogeneous difference equation, which has the following solution (see Exercise 10.2.1.1):

$$\xi_k'(n) = [1 - \mu'\lambda_k]^n \xi_k'(0) \qquad k = 1, 2, \ldots, M - 1, \quad n = 1, 2, 3, \ldots \tag{10.32}$$

For the preceding equation to converge (to be stable) as $n \to \infty$, we must set

$$-1 < 1 - \mu'\lambda_k < 1 \qquad k = 0, 1, 2, \ldots, M - 1 \tag{10.33}$$

First, subtract −1 from all the three elements of the inequality. Next, multiply by −1 and reverse the inequality. Hence, we obtain

$$0 < \mu' < \frac{2}{\lambda_k} \quad \text{or} \quad 0 < \mu < \frac{1}{\lambda_k} \qquad k = 0, 1, 2, \ldots, M - 1 \tag{10.34}$$

Under the preceding conditions and as $n \to \infty$, Equation 10.32 becomes $\lim\limits_{n \to \infty} \xi'_k(n) = 0$ or $\xi = 0$ since $Q^T \neq 0$ and, thus, $w(\infty) = w^o$. Because Equation 10.32 decays exponentially to zero, there exists a time constant that depends on the value of μ' and the eigenvalues of R_x. Furthermore, Equation 10.32 implies that immaterially of the initial value $\xi(0)$, $w(n)$ always converges to w^o provided, of course, that Equation 10.34 is satisfied. This is an important property of the steepest descent algorithm. Since each row of Equation 10.31 must decay as $n \to \infty$, it is necessary and sufficient that μ' obeys the following relationship:

$$0 < \mu' < \frac{2}{\lambda_{\max}} \tag{10.35}$$

Since $(1 - \mu'\lambda_k)^n$ decay exponentially, there exists an exponential function with time constant τ_k such that $\left(e^{-1/\tau_k}\right)^n = (1 - \mu'\lambda_k)^n$ or $(1 - \mu'\lambda_k) = e^{-1/\tau_k} = 1 - \dfrac{1}{\tau_k} + \dfrac{1}{2!\tau_k^2} - \cdots$. Therefore, for small μ' and λ_k (larger τ_k), we have

$$\tau_k \cong \frac{1}{\mu'\lambda_k} \qquad \mu'\lambda_k \ll 1 \tag{10.36}$$

In general, the kth time constant τ_k can be expressed in the form

$$\tau_k = -\frac{1}{\ln(1 - \mu'\lambda_k)} \tag{10.37}$$

10.3.3 Transient Behavior of MSE

Using Equation 10.8 at time n, we obtain the relation:

$$J(w(n)) = J_{\min} + \xi'^T \Lambda \xi' = J_{\min} + \sum_{k=0}^{M-1} \lambda_k \xi'^2_k(n) \tag{10.38}$$

Substituting the solution of $\xi'_k(n)$ from Equation 10.32 in 10.38, we obtain the relation:

$$J[w(n)] = J_{\min} + \sum_{k=0}^{M-1} \lambda_k (1 - \mu'\lambda_k)^{2n} \xi'^2(0) \tag{10.39}$$

It is obvious from the preceding equation (the factor in parentheses is less than one) that

$$\lim_{n \to \infty} J[w(n)] = J_{\min} \tag{10.40}$$

From Equation 10.38, we observe that the **learning curve** [the plot of $J[w(n)]$ versus n] consists of a sum of exponentials, each one corresponding to a natural mode of the algorithm.

Example 10.3.3.1

Consider the following data:

$$R_{xx} = \begin{bmatrix} 1 & 0.4 \\ 0.4 & 1 \end{bmatrix}, \; p_{dx} = \begin{bmatrix} 0 \\ 0.294 \end{bmatrix}, w_0(1) = -1.2, w_1(1) = -2.2, \mu = 0.2 \tag{10.41}$$

It is desired (1) to find the equation of the MSE and plot the contours, and (2) to plot the convergence path of the steepest descent algorithm in the MSE surface.

Solution: The MSE function is

$$J = E\{d^2(k)\} - 2w^T p_{dx} + w^T R_{xx} w$$

$$= \sigma_d^2 - 2\begin{bmatrix} w_0 & w_1 \end{bmatrix}\begin{bmatrix} 0 \\ 0.294 \end{bmatrix} + \begin{bmatrix} w_0 & w_1 \end{bmatrix}\begin{bmatrix} 1 & 0.4 \\ 0.4 & 1 \end{bmatrix}\begin{bmatrix} w_0 \\ w_1 \end{bmatrix}$$

$$= 0.48 + w_0^2 + w_1^2 + 0.8 w_0 w_1 - 0.588 w_1$$

The following Book MATLAB program was used to obtain the changes in the filter coefficients.

BOOK MATLAB PROGRAM

```
>>w0(1)=-1.2; w1(1)=-2.2; m=0.2;
>>for k=1:50
>>     w0(k+1)=w0(k)-2*m*(w0(k)+0.4*w1(k));
>>     w1(k+1)=w1(k)-2*m*(0.4*w0(k)+w1(k))+2*m*0.294;
>>end
>>x=-2:0.2:2; y=-2:0.2:2;
>>[X,Y]=meshgrid(x,y);%MATLAB function;
>>Z=0.48+X.^2+Y.^2+0.8*X*Y-0.588*Y;
>>contour(X,Y,Z,15,'k');%MATLAB function;
>>hold on; plot(w0,w1,'k');
```

Figure 10.6 shows the contours and the trajectory of the filter coefficients. The eigenvalue ratio for this case was 1.4/0.6 = 2.3333. ∎

As the eigenvalue ratio increases, the contours become elongated ellipses.

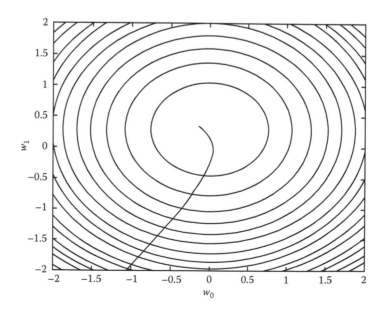

FIGURE 10.6

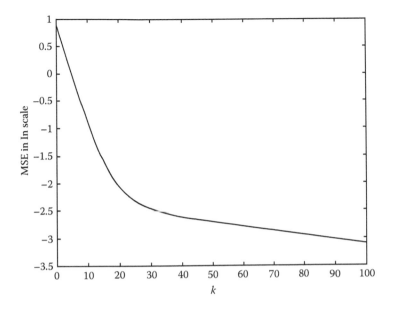

FIGURE 10.7

10.3.4 LEARNING CURVE

Accepting a small value of the constant μ, converging condition, the terms under the summation in Equation 10.39 converge to zero as k increases. As a result, the minimum MSE is achieved after a sufficient number of iterations.

The curve obtained by plotting J as a function of the iteration index k, is called the **learning curve**. A learning curve of the steepest descent algorithm, as can be seen from Equation 10.39, consists of the sum of M exponentially decaying terms, each of which corresponds to one of the modes of convergence of the algorithm. Assuming $\mu = 0.02$, $\xi'^2(0) = 1$ and the two eigenvalues, 0.1 and 2.4, we obtain the curve shown in Figure 10.7. We observe that the two straight sections in the graph indicate two different time constants (see Exercise 10.3.4.1).

Exercise 10.3.4.1

Find the time constant for one eigenvalue.   ▲

10.4 NEWTON'S METHOD

Using $p_{dx} = R_{xx}w^{o}$, Equation 10.27 becomes

$$w(n+1) = w(n) - 2\mu R_{xx}(w - w^{o}) \tag{10.42}$$

The presence of the correlation matrix R_{xx} in the preceding equation causes the eigenvalue-spread problem in the steepest descent algorithm. Newton's method overcomes this problem by replacing the scalar step-size parameter μ with the matrix step-size given by μR_{xx}^{-1}. Using Appendix 2, we obtain from Equation 10.25 the relation

$$\nabla_{w}J[w(n)] = -2p_{dx} + R_{xx}w(n) \tag{10.43}$$

Using Equation 10.43 in 10.42, we obtain the equation

$$w(n+1) = w(n) - 2\mu R_{xx}\left(w(n) - w^o\right) = w(n) - \mu \nabla_w J(w) \tag{10.44}$$

Replacing μ with Newton's relation μR_{xx}^{-1} overcomes the eigenvalue spread, which may produce large values of its elements, and this will produce difficulties in solving equations that involve inverse correlation matrices. In such cases, we say that the correlation matrices are **ill-conditioned**. This substitution has the effect of rotating the gradient vector to the direction pointing towards the minimum point of the MSE surface, as shown in Figure 10.8.

Substituting Equation 10.43 in 10.44 and the Newton's relation, we obtain

$$w(n+1) = w(n) - \mu R_{xx}^{-1}(2Rw(n) - 2p_{dx}) = (1 - 2\mu)w(n) + 2R_{xx}^{-1}p_{dx}$$

$$= (1 - 2\mu)w(n) + 2\mu w^o \tag{10.45}$$

Subtracting w^o from both sides of the preceding equation, we find

$$w(n+1) - w^o = (1 - 2\mu)(w(n) - w^o) \tag{10.46}$$

For $n=0$ and $\mu=1/2$, we obtain w^o in one step. However, in practice, $\nabla_w J(w)$ and R_{xx}^{-1} are estimated and, therefore, the value of the step-size parameter must be less than 0.5.

Introducing $w - w^o = \xi$ in Equation 10.46, we obtain the following relation:

$$\xi(n+1) = (1 - 2\mu)\xi(n) \tag{10.47}$$

Which has the solution (see Exercise 10.4.1)

$$\xi(n) = (1 - 2\mu)^n \xi(0) \quad \text{or} \quad w(n) - w^o = (1 - 2\mu)^n(w(0) - w^o) \tag{10.48}$$

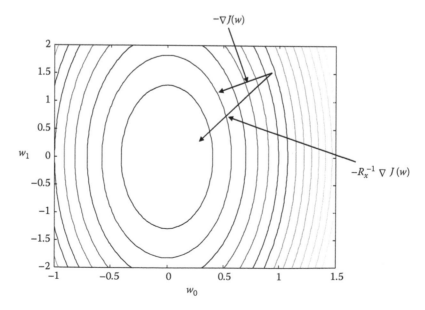

FIGURE 10.8

Exercise 10.4.1

Verify Equation 10.48.   ▲

Using Equation 10.8, we obtain

$$J = J_{\min} + \left(w - w^o\right)^T R_{xx}\left(w - w^o\right) = J_{\min} + \xi^T R_{xx}\xi$$

$$= J_{\min} + \xi^T (Q\Lambda Q^T)\xi = J_{\min} + \xi'^T \Lambda \xi'$$

In connection with Equation 10.48, we obtain the equation (see Exercise 10.4.2):

$$J(n) = J_{\min} + (1 - 2\mu)^{2n}(J(0) - J_{\min})$$ (10.49)

where $J(n)$ is the value of the performance function, J, when $w = w(n)$.

To obtain a decaying equivalent expression, we introduce the relation

$$(1 - 2\mu)^{2n} = e^{-(n/\tau)}$$ (10.50)

where τ is the **time constant**. Under the condition $2\mu \ll 1$, we can use the approximation $\ln(1 - 2\mu) \cong -2\mu$. Therefore, the time constant has the value

$$\tau \cong \frac{1}{4\mu}$$ (10.51)

The preceding equation shows that Newton's algorithm has only **one mode of convergence** (one time constant).

Exercise 10.4.2

Verify Equation 10.49.   ▲

*10.5 SOLUTION OF THE VECTOR DIFFERENCE EQUATION

If we set $n = 0$ in Equation 10.29, we obtain

$$w(1) = [I - \mu' R_{xx}]w(0) + \mu' p_{dx}$$ (10.52)

Similar, if we set $n = 1$, we find

$$w(2) = [I - \mu' R_{xx}]w(1) + \mu' p_{dx}$$ (10.53)

If we next substitute Equation 10.52 in 10.53, we obtain

$$w(2) = [I - \mu' R_{xx}]^2 w(0) + \mu' p_{dx} + [I - \mu' R_{xx}]\mu' p_{dx}$$

$$= [I - \mu' R_{xx}]^2 w(0) + \left(\sum_{j=0}^{1}[I - \mu' R_{xx}]^j\right)\mu' p_{dx}$$ (10.54)

---

* Means section may be skipped.

Therefore, the nth step is easily recognized to be equal to

$$w(n) = [I - \mu' R_{xx}]^n w(0) + \left(\sum_{j=0}^{n-1} [I - \mu' R_{xx}]^j \right) \mu' p_{dx} \tag{10.55}$$

The preceding equation does not provide us with a way to study the convergence of $w(n)$ to w^o as $n \to \infty$. We must decouple the equations and then we must describe them in a different coordinate system. To accomplish this task, we translate and then rotate the coordinate system.

Finding the eigenvalues and eigenvectors of R_{xx}, we create a diagonal matrix Λ consisting of the eigenvalues of R_{xx} and a matrix Q made up of the eigenvectors of R_{xx}. Since $Q^T Q = Q Q^T = I$, Equation 10.27 takes the form

$$w(n+1) = Q[I - \mu' \Lambda] Q^T w(n) + \mu' p_{dx} \tag{10.56}$$

To uncouple the weights (coefficients of the filter), we multiply both sides of the preceding equations by Q^T. Hence,

$$w'(n+1) = [I - \mu' \Lambda] w'(n) + \mu' p'_{dx} \tag{10.57}$$

where define the following quantities:

$$w'(n+1) = Q^T w(n+1), w'(n) = Q^T w(n),$$

$$p'_{dx} = Q^T p'_{dx}, w^{o'} = Q^T w^o \tag{10.58}$$

Next, we obtain the relation:

$$w'^o = Q^T w^o = Q^T R_{xx}^{-1} p_{dx} = Q^T \left(Q \Lambda Q^T \right)^{-1} p_{dx}$$

$$= Q^T Q \Lambda^{-1} Q^T p_{dx} = \Lambda^{-1} p'_{dx} \tag{10.59}$$

since

$$Q^{-1} = Q^T \text{ and } (Q \Lambda Q^T)^{-1} = (\Lambda Q^T)^{-1} Q^{-1} = (Q^T)^{-1} \Lambda^{-1} Q^T = Q \Lambda^{-1} Q^T$$

The ith equation of the system given in Equation 10.56 is

$$w'_i(n+1) = [1 - \mu' \lambda_i] w'_i(n) + \mu' p'_{idx} \qquad 0 \le i \le M - 1 \tag{10.60}$$

By iteration, the preceding equation has the following solution (see Exercise 10.5.1):

$$w'_i(n) = (1 - \mu' \lambda_i)^n w'_i(0) + \mu' p'_{idx} \left(\sum_{j=0}^{n-1} (1 - \mu' \lambda_i)^j \right) \tag{10.61}$$

If we set $\alpha_i = 1 - \mu' \lambda_i$, then Equation 10.61 becomes

$$w'_i(n) = \alpha_i^n w'_i(0) + \mu' p'_{idx} \sum_{j=0}^{n-1} \alpha_i^j = \alpha_i^n w'_i(0) + \mu' p'_{idx} \frac{1 - \alpha_i^n}{1 - \alpha_i}$$

$$= \alpha_i^n w'_i(0) + \frac{p'_{idx}}{\lambda_i} \left[1 - (1 - \mu' \lambda_i)^n \right] \tag{10.62}$$

Since the sum is a finite geometric series.

Exercise 10.5.1

Verify Equation 10.61.   ▲

Example 10.5.1

For the development of this example, we used the help of MATLAB. Hence, the data were found using `x=randn(1,10)` and the desired ones were found using `d=conv([1   0.2],x)` or `d=filter([1    0.2],1,x)`. The correlation matrix was found using the function `R=toeplitz(xc(1,10:11))`, where `xc  =xcorr(x,'biased')`. Similarly, we found the *pdx* cross-correlation. Hence,

$$R_x = \begin{bmatrix} 0.7346 & 0.0269 \\ 0.0269 & 0.7346 \end{bmatrix}; \quad [Q, \Lambda] = eig(R_x);$$

$$Q = \begin{bmatrix} -0.7071 & 0.7071 \\ 0.7071 & 0.7071 \end{bmatrix}; \quad \Lambda = \begin{bmatrix} 0.7077 & 0 \\ 0 & 0.7071 \end{bmatrix}$$

and $\mu' < 2/0.7071 = 2.8285$ for the solution to converge. We next choose $[\ w_0(0) \quad w_1(0)\ ]' = [\ 0 \quad 0\ ]'$ and hence $w'(0) = Q^T w(0) = 0$. Therefore, Equation 10.62 becomes

$$w_i'(n) = \mu' p_{idx}' \frac{1 - \alpha_i^n}{1 - \alpha_i} = \frac{p_{idx}'}{\lambda_i}\left[1 - (1 - \mu' \lambda_i)^n\right]$$

From Equation 10.58, we obtain

$$p_{dx}' = Q^T p_{dx} = \begin{bmatrix} -0.7071 & 0.7071 \\ 0.7071 & 0.7071 \end{bmatrix}\begin{bmatrix} 0.7399 \\ -0.0003 \end{bmatrix} = \begin{bmatrix} -0.5234 \\ 0.5230 \end{bmatrix}$$

Therefore, the system is (we set $\mu' = 3$ for convergence)

$$w_0'(n) = \frac{1}{0.7077}(-0.5234)\left[1 - (1 - 2 \times 0.7071)^n\right]$$

$$w_1'(n) = \frac{1}{0.7614} 0.5230\left[1 - (1 - 2 \times 0.7614)^n\right]$$

Since $w'(n) = Q^T w(n)$ and $Q^T = Q^{-T}$, we find $w(n) = Q^T w'(n)$ and at the limit value when n approaches infinity, the filter coefficients take the values

$$\begin{bmatrix} w_0^o \\ w_1^o \end{bmatrix} = Q^T w' = \begin{bmatrix} -0.7071 & 0.7071 \\ 0.7071 & 0.7071 \end{bmatrix}\begin{bmatrix} \dfrac{-0.5234}{0.7077} \\ \dfrac{0.5230}{0.7614} \end{bmatrix} = \begin{bmatrix} 1.0087 \\ -0.0373 \end{bmatrix}$$ ■

Exercise 10.5.2

Let

$$R_{xx} = \begin{bmatrix} 1 & 0.7 \\ 0.7 & 1 \end{bmatrix} \quad \text{and} \quad p_{dx} = \begin{bmatrix} 0.7 & 0.5 \end{bmatrix}^T$$

be the estimates derived from the data. Find $w(n)$.   ▲

ADDITIONAL EXERCISES

1. Let $R = \begin{bmatrix} 1 & 0.5 & ; & 0.25 & 1 \end{bmatrix}$ and $p = \begin{bmatrix} 0.5 & 0.2 \end{bmatrix}^T$. (1) Find a value of the step-size that ensures the convergence of the steepest descent method and (2) Find the recursion for computing the elements $w_1(n)$ and $w_2(n)$ of the $w(n)$.
2. Assuming one weight coefficient $w(n)$, find the following: (1) find the mean-squared error $J(n)$ as a function of $w(n)$, (2) find the Wiener solution and the minimum mean-squared error $J_{min}(n)$, and (3) sketch a plot of $J(n)$ versus $w(n)$.

HINTS, SUGGESTIONS, AND SOLUTIONS OF THE EXERCISES

10.1.1

$$(w - w^o + w^o)^T R(w - w^o + w^o) - 2p(w - w^o + w^o) - 2J_{min} + \sigma_d^2 = (\xi + w^o)^T R$$

$$(\xi + w^o) - 2p^T (\xi + w^o) - 2J_{min} + \sigma_d^2. \text{ With } Rw^o = p, w^{oT} p = p^T w^o = \text{number}$$

and $\sigma_d^2 - p^T w^o = J_{min}$ the equation is easily proved.

10.1.2

$$J = J_{min} + \xi'^T \Lambda \xi \text{ and } J_{min} = \sigma_d^2 - p^T w^o, |R - \lambda I| = \begin{vmatrix} 2 - \lambda & 1 \\ 1 & 3 - \lambda \end{vmatrix} = (2 - \lambda)(3 - \lambda) - 1 = 0$$

$$\lambda_1 = 1.382, \ \lambda_2 = 3.618 \begin{bmatrix} 2 & 1 \\ 1 & 3 \end{bmatrix} \begin{bmatrix} w_0^o \\ w_1^o \end{bmatrix} = \begin{bmatrix} 6 \\ 7 \end{bmatrix}, \begin{bmatrix} w_0^o \\ w_1^o \end{bmatrix} = \begin{bmatrix} 11/5 \\ 8/5 \end{bmatrix}, J_{min} = 28 - \begin{bmatrix} 6 & 7 \end{bmatrix} \begin{bmatrix} 11/5 \\ 8/5 \end{bmatrix} = 3.6$$

$$J = 3.6 + \begin{bmatrix} \xi_0' & \xi_1' \end{bmatrix} \begin{bmatrix} 1.328 & 0 \\ 0 & 3.618 \end{bmatrix} \begin{bmatrix} \xi_0^o \\ \xi_1^o \end{bmatrix} = 3.6 + 1.382\xi_0' + 3.618\xi_1'$$

10.1.3

The MSE surface is given by $J = 2 - 2 \begin{bmatrix} w_0 & w_1 \end{bmatrix} \begin{bmatrix} 1 \\ 1 \end{bmatrix} + \begin{bmatrix} w_0 & w_1 \end{bmatrix} \begin{bmatrix} d & a \\ a & d \end{bmatrix} \begin{bmatrix} w_0 \\ w_1 \end{bmatrix}$, and

the optimum tap weights are $\begin{bmatrix} w_0^o \\ w_1^o \end{bmatrix} = R^{-1} p = \begin{bmatrix} d & a \\ a & d \end{bmatrix}^{-1} \begin{bmatrix} 1 \\ 1 \end{bmatrix} = \begin{bmatrix} \dfrac{1}{d+a} \\ \dfrac{1}{d+a} \end{bmatrix}$. Therefore, the minimum

MSE surface is $J_{min} = \sigma_d^2 - w^{oT} p = 2 - \begin{bmatrix} \dfrac{1}{d+a} & \dfrac{1}{d+a} \end{bmatrix} \begin{bmatrix} 1 \\ 1 \end{bmatrix} = 2\dfrac{d+a-1}{d+a}$. We also have the

relation $J = J_{min} + (w - w^o)^T R(w - w^o) = 2\dfrac{d+a+1}{d+a} + \begin{bmatrix} \xi_0 & \xi_1 \end{bmatrix} \begin{bmatrix} d & a \\ a & d \end{bmatrix} \begin{bmatrix} \xi_0 \\ \xi_1 \end{bmatrix}$. From the relation $\begin{vmatrix} d-\lambda & a \\ a & d-\lambda \end{vmatrix} = 0$, we obtain the two eigenvalues $\lambda_0 = d+a$ and $\lambda_1 = d-a$. By incorpo-

rating different values for a and d, we can create different ratios of the eigenvalues. The larger the ratio, the more elongated the ellipses are.

10.1.4

$$J = \begin{bmatrix} w_0 & w_1 \end{bmatrix} \begin{bmatrix} 1 & 0.5 \\ 0.5 & 1 \end{bmatrix} \begin{bmatrix} w_0 \\ w_1 \end{bmatrix} - 2 \begin{bmatrix} \frac{1}{2} & \frac{1}{2} \end{bmatrix} \begin{bmatrix} w_0 \\ w_1 \end{bmatrix} + 2, E\{d^2(n)\} = 2, R = \begin{bmatrix} 1 & 0.5 \\ 0.5 & 1 \end{bmatrix}, p = \begin{bmatrix} \frac{1}{2} \\ -\frac{1}{2} \end{bmatrix}$$

$$w^o = R^{-1}p = \frac{1}{0.75} \begin{bmatrix} 1 & -0.5 \\ -0.5 & 1 \end{bmatrix} \begin{bmatrix} 1/2 \\ -1/2 \end{bmatrix} = \begin{bmatrix} 1 \\ -1 \end{bmatrix}, J_{min} = 2 - p^T w^o = 2 - 1 = 1,$$

$$J = 1 + \xi^T \begin{bmatrix} 1 & 0.5 \\ 0.5 & 1 \end{bmatrix} \xi, |\lambda I - R| = 0 \Rightarrow (\lambda - 1)^2 - 0.25 = 0 \Rightarrow \lambda_0 = 1.5, \lambda_1 = 0.5, \text{eigenvectors}$$

$$\begin{bmatrix} 1 & 0.5 \\ 0.5 & 1 \end{bmatrix} \begin{bmatrix} q_{00} \\ q_{01} \end{bmatrix} = 1.5 \begin{bmatrix} q_{00} \\ q_{01} \end{bmatrix} \Rightarrow q_{00} + 0.5q_{01} = 1.5q_{00} \Rightarrow q_{00} = q_{01} \text{ also } 0.5q_{00} + q_{01} = 1.5q_{01}$$

$$\Rightarrow q_{00} = q_{01} \Rightarrow \text{normalized } q_0 = \frac{1}{\sqrt{2}} \begin{bmatrix} 1 \\ 1 \end{bmatrix}; \begin{bmatrix} 1 & 0.5 \\ 0.5 & 1 \end{bmatrix} \begin{bmatrix} q_{10} \\ q_{11} \end{bmatrix} = 0.5 \begin{bmatrix} q_{10} \\ q_{11} \end{bmatrix} \Rightarrow q_{10} = -q_{11} \Rightarrow$$

$$\text{normalized } q_1 = \frac{1}{\sqrt{2}} \begin{bmatrix} 1 \\ -1 \end{bmatrix}; \text{eigenvalue and eigenvector matrices are as follows: } \Lambda = \begin{bmatrix} 1.5 & 0 \\ 0 & 0.5 \end{bmatrix}$$

$$Q = \frac{1}{\sqrt{2}} \begin{bmatrix} 1 & 1 \\ 1 & -1 \end{bmatrix}, J = 1 + \xi^T Q \Lambda Q^T \xi = 1 + \xi'^T \begin{bmatrix} 1.5 & 0 \\ 0 & 0.5 \end{bmatrix} \xi' = 1 + 1.5\xi_0'^2 + 0.5\xi_1'^2$$

$$= 1 + \left(\frac{\xi_0'}{1/\sqrt{1.5}} \right)^2 + \left(\frac{\xi_1'}{1/\sqrt{0.5}} \right)^2$$

We can use the following Book MATLAB program to produce four contours at different heights of the performance surface.

```
>>x=-5:0.1:5; y=-5:0.1:5;
>> [X,Y]=meshgrid(x,y);
>>Z=1+1.5*X.^2+0.5*Y.^2;
>>V=[1.2    2   3   4];% if we set 1 instead 1.2 the first
>>      % circle will be a dot since this indicates one point
>>      %of the bottom of the surface, as it should;
>>contour(X,Y,Z,V,'k');% the circles will be black;
```

10.1.5

1. $R^n = Q\Lambda Q^T Q\Lambda Q^T Q\Lambda Q^T \cdots Q\Lambda Q^T = Q\Lambda^n Q^T$, because $Q^T Q = I = QQ^T$

2. Let $P = Q^T R^{1/2} Q \Rightarrow R^{1/2} = QPQ^T \Rightarrow R^{1/2} R^{1/2} = R \Rightarrow QPQ^T QPQ^T = Q\Lambda Q^T \Rightarrow$

$$QP^2 Q^T = Q\Lambda Q^T \Rightarrow Q^T QQP^2 Q^T Q = Q^T Q\Lambda Q^T Q \Rightarrow P^2 = \Lambda \Rightarrow P = \Lambda^{1/2} \Rightarrow R^{1/2} = Q\Lambda^{1/2} Q^T$$

10.2.1.1

$$w(1) = aw(0) + bw^o \ (a = 1 - 2\mu r(0), b = 2\mu r(0)), w(2) = aw(1) + bw^o = a(aw(0) + bw^o) + bw^o$$

$$= a^2 w(0) + abw^o + bw^o \Rightarrow w(n) = a^n w(0) + (a^{n-1} + a^{n-2} + \cdots + 1)bw^o = a^n w(0) + \frac{1-a^n}{1-a} bw^o,$$

but $1 - a = b$

$$\Rightarrow w(n) = a^n w(0) + w^o - a^n w^o = w^o + (w(0) - w^o)a^n$$

10.2.1.2

(1) Overdamped case, (2) critically damped case, and (3) underdamped case.

10.2.2.1

Set $f(x) = x^3 - 8$ to obtain $x(n+1) = x(n) - [x(n)^3 - 8]/(3x(n)^2)$. For $x(1) = 1.2$, MATLAB gives us the results: $x(1) = 1.2000$, $x(2) = 2.6519$, $x(3) = 2.1471$, $x(4) = 2.0098$, $x(5) = 2.0000$, and $x(6) = 2.0000$. Book MATLAB program:

```
>>x(1)=1.2;
>>for n=1:10
>>x(n+1)=x(n)-(x(n)^3-8)/(3*x(n)^2);
>>end;
```

10.3.1.1

$$w(n+1) = w(n) - [(\partial J(w)/\partial w)/(\partial J^2(w)/\partial w^2)] = w(n) - \frac{10.5w^2(n) + 9w(n) - 3.5}{21w(n) + 9}$$

With the MATLAB help, we obtain 2.0000, 0.8922, 0.4275, 0.3014, 0.2905, and 0.2905.

10.3.2.1

$$\begin{bmatrix} \xi_0'(n+1) \\ \xi_1'(n+1) \\ \vdots \\ \xi_{M-1}'(n+1) \end{bmatrix} = \left(\begin{bmatrix} 1 & 0 & \cdots & 0 \\ 0 & 1 & \cdots & 0 \\ & & \vdots & \\ 0 & 0 & \cdots & 1 \end{bmatrix} - \mu \begin{bmatrix} \lambda_0 & 0 & \cdots & 0 \\ 0 & \lambda_1 & \cdots & 0 \\ & & \vdots & \\ 0 & 0 & \cdots & \lambda_{M-1} \end{bmatrix} \right) \begin{bmatrix} \xi_0'(n) \\ \xi_1'(n) \\ \vdots \\ \xi_{M-1}' \end{bmatrix} = \begin{bmatrix} (1 - \mu\lambda_0)\xi_0'(n) \\ (1 - \mu\lambda_1)\xi_1'(n) \\ \vdots \\ (1 - \mu\lambda_{M-1})\xi_{M-1}'(n) \end{bmatrix}$$

and the results are found since the equality of two vectors means equality of each corresponding element.

10.3.4.1

$$(1-2\mu\lambda_i)^{2k} = e^{-(k/\tau_i)} \Rightarrow \tau_i = \frac{-1}{2\ln(1-2\mu\lambda_i)} \text{ (1) for small } \mu \Rightarrow 2\mu\lambda_i \ll 1 \Rightarrow$$

$$\ln(1-2\mu\lambda_i) \approx 2\mu\lambda_i \text{, substituting in (1)} \Rightarrow \tau_i \approx \frac{1}{4\mu\lambda_i}$$

10.4.1

$$w(1) = aw(0) + bw^o (a = 1-2\mu r(0), b = 2\mu r(0)), w(2) = aw(1) + bw^o = a(aw(0) + bw^o) + bw^o$$

$$= a^2 w(0) + abw^o + bw^o \Rightarrow w(n) = a^n w(0) + (a^{n-1} + a^{n-2} + \cdots + 1)bw^o = a^n w(0) + \frac{1-a^n}{1-a} bw^o,$$

but $1-a=b$

$$\Rightarrow w(n) = a^n w(0) + w^o - a^n w^o = w^o + (w(0) - w^o)a^n$$

10.4.2

$$J(n) = J_{\min} + \xi^T R_{xx}\xi = J_{\min} + (1-2\mu)^n \xi^T(0)R_{xx}(1-2\mu)^n \xi(0) = J_{\min} + (1-2\mu)^{2n} \xi^T(0)R_{xx}\xi(0). \text{ But}$$
$$\xi^T(0)\mathbf{R_x}\xi(0) = J(0) - J_{\min} \text{ and Exercise 7.3.2 is proved.}$$

10.5.2

Using the MATLAB function $[v,d] = \text{eig}(R)$, we obtain the following eigenvalues and eigenvectors: $\lambda_1 = 0.3000, \lambda_2 = 1.7000, q_1 = \begin{bmatrix} -0.7071 & 0.7071 \end{bmatrix}^T, q_2 = \begin{bmatrix} 0.7071 & 0.7071 \end{bmatrix}^T$. The step-size factor must be set equal to $\mu' < 2/1.7 = 1.1765$, so that the solution converges. Choosing $w(0)=0$ implies that $w'(0) = Q^T w(0) = 0 \Rightarrow$

$$p'_{dx} = Q^T p_{dx} = \begin{bmatrix} -0.7071 & 0.7071 \\ 0.7071 & 0.7071 \end{bmatrix}\begin{bmatrix} 0.7 \\ 0.5 \end{bmatrix} = \begin{bmatrix} -0.1414 \\ 0.8485 \end{bmatrix}$$

$$w'_0(n) = \frac{1}{0.3}(-0.1414)[1-(1-\mu'0.3)^n]; w'_1(0) = \frac{1}{1.7}0.8485[1-(1-\mu'1.7)^n].$$

Since

$$w'(n) = Q^T w(n), Q^T = Q^{-1}, w(n) = Q^T w'(n)$$

$$\begin{bmatrix} w_0(n) \\ w_1(n) \end{bmatrix} = \begin{bmatrix} -0.7071 & 0.7071 \\ 0.7071 & 0.7071 \end{bmatrix}\begin{bmatrix} \frac{1}{0.3}(-0.1414)[1-(1-\mu'0.3)^n] \\ \frac{1}{1.7}0.8485[1-(1-\mu'1.7)^n] \end{bmatrix}$$

$$= \begin{bmatrix} \frac{-0.7071}{0.3}(-0.1414)[1-(1-\mu'0.3)^n] + \frac{0.7071}{1.7}0.8485[1-(1-\mu'1.7)^n] \\ \frac{0.7071}{0.3}(-0.1414)[1-(1-\mu'0.3)^n] + \frac{0.7071}{1.7}0.8485[1-(1-\mu'1.7)^n] \end{bmatrix}$$

ADDITIONAL PROBLEMS

1.

1. $[Q,L] = \text{eig}\left(\begin{bmatrix} 0 & 0.5 \; ; \; 0.25 & 1 \end{bmatrix}\right); Q = \begin{bmatrix} 0.8165 & 0.8165 \; ; \; 0.5774 & 0.5774 \end{bmatrix};$

$L = \begin{bmatrix} 1.3536 & 0.6464 \end{bmatrix}$. Therefore, $0 < \mu < 1/\lambda_{\max}$ and thus, $0 < \mu < 1/1.3536 = 0.7388$.

2. $w(n+1) = w(n) + \mu[p - Rw(n)], w(n+1) = w(n) + 1*\left([0.5;0.2] - \begin{bmatrix} 1 & 0.5 \; ; \; 0.25 & 1 \end{bmatrix}\right)$

$* w(n)$ or $w(n+1) = w(n) + \left(\begin{bmatrix} 1 & 0 \; ; \; 0 & 1 \end{bmatrix} - \begin{bmatrix} 1 & 0.5 \; ; \; 0.25 & 1 \end{bmatrix}\right) * w(n) - [0.5;0.2]$

$= \begin{bmatrix} 0 & -0.5 \; ; \; -0.25 & 0 \end{bmatrix} * w(n) - \begin{bmatrix} 0.5 & 0.2 \end{bmatrix}$.

Hence,

$\begin{bmatrix} w_1(n+1) \\ w_2(n+1) \end{bmatrix} = \begin{bmatrix} 0 & -0.5 \\ -0.25 & 0 \end{bmatrix} \begin{bmatrix} w_1(n) \\ w_2(n) \end{bmatrix} - \begin{bmatrix} 0.5 \\ 0.2 \end{bmatrix}$ or $w_1(n+1) = -0.5w_2(n) - 0.5; w_2(n+1)$

$= -0.25w_1(n) - 0.2$; for $n = 0$ $w_1(1) = -0.5$ and $w_2(1) = -0.2$ with $w_1(0) = 0$ and $w_2(0) = 0$;

$w_1(2) = -0.5(-0.2) - 0.5 = -0.4, w_2(2) = -0.25(-0.5) - 0.2 = -0.075$ etc.

2. $J(n) = \sigma_d^2 - 2p_{xd}w(n) + r_{xx}w^2(n); \dfrac{\partial J(n)}{\partial w(n)} = 0 - 2p_{xd} + 2r_{xx}w(n) = 0 \Rightarrow w^o = \dfrac{p_{dx}}{r_{xx}} \Rightarrow$

$J_{\min} = \sigma_d^2 - 2p_{xd}\dfrac{p_{dx}}{r_{xx}} + r_{xx}\left(\dfrac{p_{dx}}{r_{xx}}\right)^2 = \sigma_d^2 - \dfrac{p_{dx}^2}{r_{xx}}$

The curve is concave upward with minimum $J_{\min}$.

11 The Least Mean Square (LMS) Algorithm

11.1 INTRODUCTION

In this chapter, we present the celebrated least mean square (LMS) algorithm, developed by Widrow and Hoff in 1960. This algorithm is a member of stochastic gradient algorithms, and because of its robustness and low computational complexity, it has been used in a wide spectrum of applications.

The LMS algorithm has the following important properties:

1. It can be used to solve the Wiener–Hopf equation without finding matrix inversion. Furthermore, it does not require the availability of the autocorrelation matrix of the filter input and the cross-correlation between the filter input and its desired signal.
2. Its form as well as its implementation is simple, yet it is capable of delivering high performance during the adaptation process.
3. Its iterative procedure involves (1) computing the output of an FIR filter produced by a set of tap inputs (filter coefficients), (2) generation of an estimated error by computing the output of the filter to a desired response, and (3) adjusting the tap weights (filter coefficients) based on the estimation error.
4. The correlation term needed to find the values of the coefficients at the $n+1$ iteration contains the stochastic product $x(n)e(n)$ without the expectation operation that is present in the steepest-descent method.
5. Since the expectation operation is not present, each coefficient goes through sharp variations (noise) during the iteration process. Therefore, instead of terminating at the Wiener solution, the LMS algorithm suffers random variation around the minimum point (optimum value) of the error-performance surface.
6. It includes a step-size parameter, μ, that must be selected properly to control stability and convergence speed of the algorithm.
7. It is stable and robust for a variety of signal conditions.

11.2 THE LMS ALGORITHM

In Chapter 10, we developed the following relations using the steepest-descent method:

$$w(n+1) = w(n) - \mu \nabla_w J(w(n)) \tag{11.1a}$$

$$\nabla_w J(w(n)) = -2p_{dx} + 2R_{xx}w(n) \tag{11.1b}$$

The simplest choice of the estimators R_x and p_{dx} are the instantaneous estimates defined by

$$R_x \cong x(n)x^T(n), p_{dx} \cong d(n)x(n) \tag{11.2}$$

Substituting the preceding equation in Equation 11.1a and b, we obtain

$$\mathbf{w}(n+1) = \mathbf{w}(n) + 2\mu\mathbf{x}(n)[d(n) - \mathbf{x}^T(n)\mathbf{w}(n)]$$

$$= \mathbf{w}(n) + 2\mu\mathbf{x}(n)[d(n) - \mathbf{w}^T(n)\mathbf{x}(n)] \qquad (11.3)$$

$$= \mathbf{w}(n) + 2\mu e(n)\mathbf{x}(n)$$

where

$$y(n) = \mathbf{w}^T(n)\mathbf{x}(n) \quad \text{filter output}$$

$$e(n) = d(n) - y(n) \quad \text{error} \qquad (11.4)$$

$$\mathbf{w}(n) = [w_0 w_1 \cdots w_{M-1}]^T \quad \text{filter taps (coefficients) at time } n$$

$$\mathbf{x}(n) = [x(n)x(n-1)\cdots x(n-M+1)]^T \quad \text{input data} \qquad (11.5)$$

The algorithms defined by Equations 11.3 and 11.4 constitute the LMS algorithm. The algorithm at each iteration requires that $\mathbf{x}(n)$, $d(n)$ and $\mathbf{w}(n)$ are known. The LMS algorithm is a **stochastic gradient algorithm** if the input signal is a stochastic process. This results in varying the pointing direction of the coefficient vector during the iteration. An Finite Impulse Response (FIR) adaptive filter realization is shown in Figure 11.1. Figure 11.2 presents the block diagram representation of the LMS filter. Table 11.2.1 presents the steps for executing the LMS algorithm.

BOOK LMS M-FUNCTION

```
function[w,y,e,J,w1,Js]=lms1(x,dn,mu,M)
    %function[w,y,e,J,w1]=lms1(x,dn,mu,M);
    %all quantities are real-valued;
```

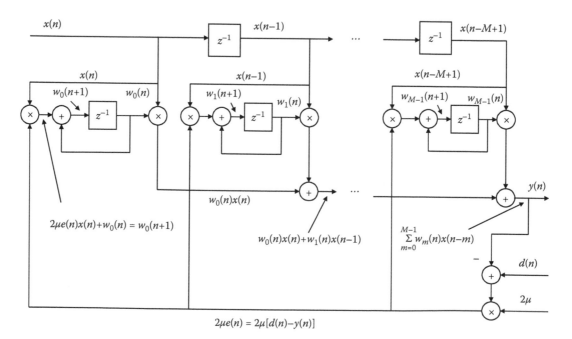

FIGURE 11.1

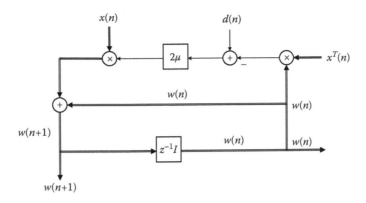

FIGURE 11.2

TABLE 11.2.1
The LMS Algorithm for an Mth-Order FIR Filter

Inputs	M = Filter length
	μ = Step-size factor
	$x(n)$ = Input data to the adaptive filter
	$w(0)$ = Initialization filter vector = 0
Outputs	$y(n)$ = Adaptive filter output = $w^T(n)x(n)$ approximate of the desired signal
	$e(n) = d(n) - y(n)$ = Error
	$w(n+1) = w(n) + 2\mu e(n)x(n)$

```
%x=input data to the filter; dn=desired signal;
%M=order of the filter;
%mu=step-size factor; x and dn must be
%of the same length;
%Js=smooths the learning curve;
%w1=is a matrix of dimensions: length(x)xM,
%each column represents the variation of
%each filter coefficient;
N=length(x);w=zeros(1,M);w1=zeros(1,M);
for n=M:N
    x1=x(n:-1:n-M+1); %for each n the vector x1 is
              %of length M with elements from x in
              %reverse order;
    y(n)=w*x1';
    e(n)=dn(n)-y(n);
    w=w+2*mu*e(n)*x1;
    w1(n-M+1,:)=w(1,:);
end;
J=e.^2;%J is the learning curve of the adaptation;
for n=1:length(x)-5
    Js(n)=(J(n)+J(n+1)+J(n+2))/3;
end;
```

Exercise 11.2.1

Plot the learning curve, using the LMS algorithm, with different variance of the noise. ▲

11.3   EXAMPLE USING THE LMS ALGORITHM

The following examples will elucidate the use of LMS algorithm to different areas of engineering and will bring forth the versatility of this very important algorithm.

Example 11.3.1 (Linear Prediction)

We can use an adaptive LMS filter as a predictor, as shown in Figure 11.3. The data $\{x(n)\}$ were created by passing a zero-mean white noise $\{v(n)\}$ through an autoregressive (AR) process described by the difference equation $x(n)=0.6010x(n-1)-0.7225x(n-2)+v(n)$. The LMS filter is used to predict the values of the AR filter parameters 0.6010 and −0.7225. A two-coefficient LMS filter predicts $x(n)$ by

$$\hat{x}(n) = \sum_{i=0}^{1} w_i(n)x(n-1-i) \equiv y(n) \tag{11.6}$$

Figure 11.4 shows the trajectory of w_0 and w_1 versus the number of iterations for two different values of step-size parameter ($\mu=0.02$ and $\mu=0.008$). The adaptive filter is a two-coefficient filter. The noise is white and Gaussian distributed. The figure shows fluctuations in the values of coefficients as they converge to a neighborhood of their optimum value, 0.6010 and −0.7225, respectively. As the step size μ becomes smaller, the fluctuations are not as large, but the convergence speed to the optimal values is slower.

BOOK ONE-STEP LMS PREDICTOR M-FUNCTION

```
function[w,y,e,J,w1,Js]=lms_one_step_predictor(x,mu,M)
    %function[w,y,e,J,w1,Js]=lms_one_step_predictor(x,mu,M);
    %x=data=signal plus noise;mu=step size factor;
    %M=number of filter
    %coefficients;w1 is a matrix and each column
    %is the history of each
    %filter coefficient versus time (iteration) n;
    %N=number of elements
    %of the vector x;Js=smoother learning curve;
```

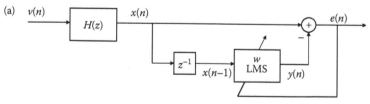

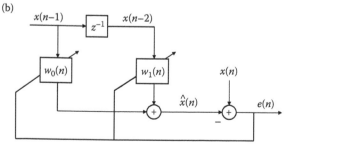

FIGURE 11.3

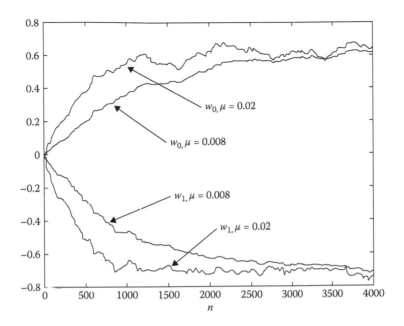

FIGURE 11.4

```
    %w1=is a matrix and each column depicts the
    %variation of each coefficient;
N=length(x);
y=zeros(1,N);
w=zeros(1,M);
for n=M:N-1
    x1=x(n:-1:n-M+1);
    y(n)=w*x1';
    e(n)=x(n+1)-y(n);
    w=w+2*mu*e(n)*x1;
    w1(n-M+1,:)=w(1,:);
end;
J=e.^2;
    %J is the learning curve of the
    %adaptive process;
for n=1:length(x)-5
    Js(n)=(J(n)+J(n+1)+J(n+2))/3;
end;%Js=smoothed out learning curve J;
```

For example, to produce two of the four curves, we used the following Book MATLAB program. To produce the other two curves, you repeat it but with a different step-size factor.

```
>>mu=0.02; x(1)=0; x(2)=0; M=2;
>>for n=1:4000
    x(n+2)=0.6010*x(n+1)-0.7225*x(n)+0.2*randn;
>>end;
>>[w,y,e,J,w1,Js]=lms_one_step_predictor(x,mu,M);
>>plot(w1(:,1),'k');
>>plot(w1(:,2),'k');calabel('n');
```
■

Exercise 11.3.1

Reproduce figures similar to Figure 11.4 with different step-size parameters and variance of the noise.
▲

Example 11.3.2 (Modeling)

Adaptive filtering can also be used to find the coefficients of an unknown filter, as shown in Figure 11.5. The data $x(n)$ were created similar to those in Example 11.3.1. The desired signal is given by $d(n)=x(n)-2x(n-1)+4x(n-2)$. If the output $y(n)$ is approximately equal to $d(n)$, it implies that the coefficients of the LMS filter are approximately equal to those of the unknown system. Figure 11.6 shows the ability of the LMS filter to identify the unknown system. After 1000 iterations, the system is practically identified. For this example, we used $\mu=0.15$ and $M=4$. It is observed that the fourth coefficient is zero. It should be because the system to be identified has only three coefficients and the rest are zero. The Book program to produce Figure 11.6 is given as follows:

BOOK M-FUNCTION FOR SYSTEM IDENTIFICATION: [W,Y,E,J,W1,JS]=URDSP_MODELING_LMS(MU,M)

```
function[w,y,e,J,w1,Js]=urdsp_modeling_lms(mu,M)
x(1)=0; x(2)=0; mu=0.15; M=4;
for n=1:1500
```

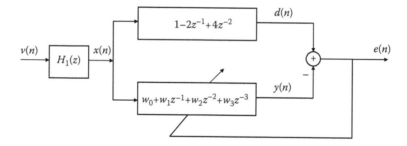

FIGURE 11.5

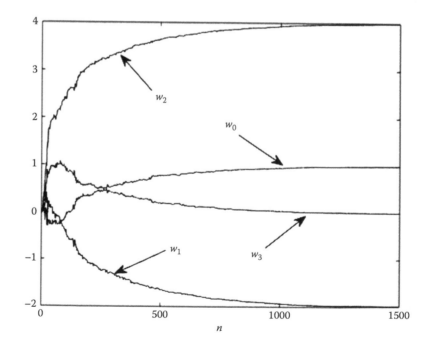

FIGURE 11.6

```
    x(n+2)=0.6010*x(n+1)-0.7225*x(n)+0.2*randn;
end;
for n=1:1500
    d(n+2)=x(n+2)-2*x(n+1)+4*x(n);
end;
[w,y,e,J,w1,Js]=lms1(x,d,mu,M);
plot(w1(:,1),'k'); hold on; plot(w1(:,2),'k');hold on;
plot(w1(:,3),'k'); hold on; plot(w1(:,4),'k'); xlabel('n')
```
■

Example 11.3.3 (Noise Cancelation)

A noise cancellation scheme is shown in Figure 11.7. We introduce in this example the follow-
ing values: $H_1(z)=1$ (or $h(n)=\delta(n)$), $v1(n)=$white noise$=v(n)$, $L=1$(delay), $s(n)=0.99^n \sin(0.2\pi n)$, a
decaying sinusoid. Therefore, the input signal to the filter is $x(n)=s(n-1)+v(n-1)$ and the desired
signal is $d(n)=s(n)+v(n)$. The Book LMS function algorithm **lms1** was used. Figure 11.8 shows the
signal, the signal plus noise, and the outputs of the filter for two different sets of coefficients: $M=4$
and $M=16$. The following noise was used: $v=0.2\text{randn}(1,250)$. The exponent can be written in
MATLAB, for 50 coefficients of a decaying vector, as follows: $0.99.\wedge[1:50]$. ■

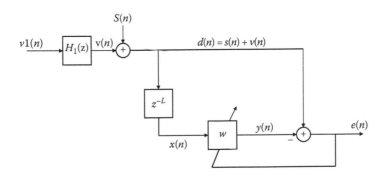

FIGURE 11.7

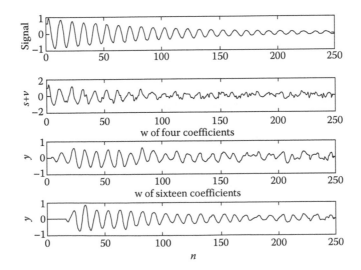

FIGURE 11.8

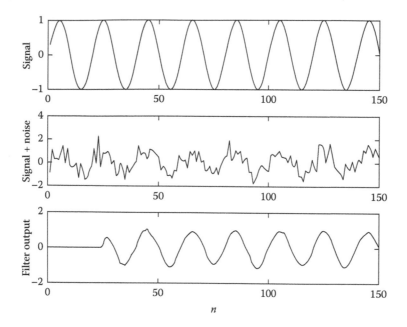

FIGURE 11.9

Example 11.3.4 (Sinusoid Plus Noise)

Assume that the noisy signal is given as $x(n) = a\sin(\omega n) + v(n)$, where $v(n)$ is a white noise sequence. It is desired to get the sinusoid with the smallest amount of noise. Figure 11.9 shows the results. The constants and the signals used were as follows:

```
>>x=sin(0.1*pi*[1:150])+0.5*randn(1,150);
>>dn=sin(0.1*pi*[1:150]);
>>mu=0.02;M=24;
```

To find the results, we used the **lms1** function.

Exercise 11.3.2

Produce and observe similar results using different constants for the step-size constant, noise variance, and number of filter coefficients.   ▲

Example 11.3.5 (Adaptive Channel Estimation)

Figure 11.10 shows a communication noisy channel that is required to be estimated by the adaptive processing technique.

Solution: For this example, we assumed that the channel is represented by a FIR system with four unknown coefficients: $c = [0.6 \quad -0.9 \quad 0.2 \quad -0.1]$. In addition, we used the following values: $M=20$, random channel noise 0.5(rand-0.5), and the signal $s(n)=\sin(0.1\pi n)$. Figure 11.10 presents the noisy channel output signal and the reproduced of the input signal to the channel. It is obvious that we recapture the input to the channel signal. In practice, we can create an adaptive filter by sending a known signal; after the establishment of the proper adaptive filter and assuming that the channel varies slowly, we can use the adoptive filter for a while. Then, periodically, we can find the appropriate adaptive filters.

The Book m-file with which we produced Figures 11.11 and 11.12 is as follows:

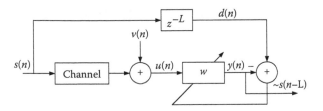

FIGURE 11.10

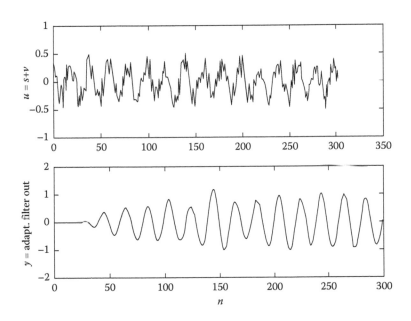

FIGURE 11.11

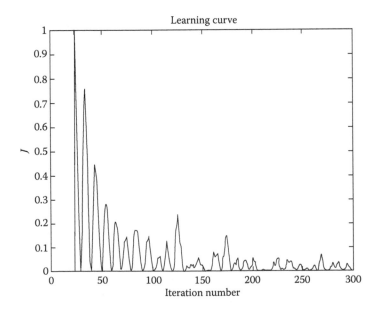

FIGURE 11.12

BOOK M-FUNCTION: [W,E,J,Y,U]=LMS_EX11_3_5(S,MU,C,M)

```
function[w,e,J,y,u]=lms_ex11_3_5(s,mu,c,M)
    %s=signal;mu=step-size coefficient;
    %c=vector with the channel coefficients;
    %M=number of adaptive filter coefficients
w=zeros(1,M);
u=conv(s,c)+0.5*(rand(1,length(conv(s,c)))-0.5);
for n=M:length(u)-length(c)
    u1=u(n:-1:n-M+1);
    y(n)=w*u1';
    e(n)=s(n+1)-y(n);
    w=w+2*mu*e(n)*u1;
end;
J=e.^2;
```

Figure 11.12 shows the learning curve. The reader notes that as the iteration increases, the error tends to decrease and the recaptured signal is less distorted. ∎

Example 11.3.6 (Inverse System Identification)

To find the inverse of an unknown filter, we place the adaptive filter in series with the unknown system as shown in Figure 11.13. The delay is needed so that the system is causal. Figure 11.14 shows the noisy signal, the output from the adaptive filter, and a typical learning curve. In this

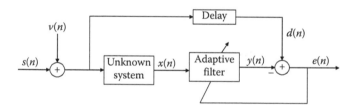

FIGURE 11.13

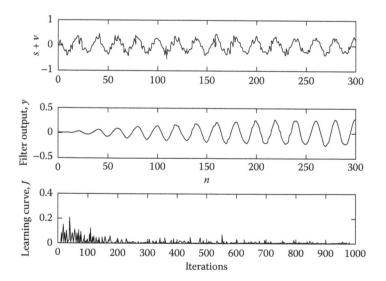

FIGURE 11.14

example, we used four coefficients FIR filter. The input to the unknown system was a sine function with a white Gaussian noise. The following Book script m-file was used:

BOOK SCRIPT M-FILE: EX11_3_6

```
%script file ex11_3_6
xi=sin(0.1*pi*[1:1500])+0.1*randn(1,1500);
x=conv(xi,[0.6 -0.9 0.2 -0.1]);
for n=1:1500
    d(n+1)=x(n);
end;
[w,y,e,J]=lms1(x(1,1:1450),d(1,1:1450),0.02,10);
```

The functions and constants used were as follows: $s(n)=\sin(0.1\pi n)$, $v(n)=0.1$ randn (n), unknown system $c=[0.6 -0.9\ 0.2 -0.1]$, $\mu=0.02$, and $M=10$. ∎

Example 11.3.7 (Adaptive Prediction)

Based on Figure 11.15, which presents an one-tap autoregressive (AR) filter, find the Mean Square Error (MSE) of a single and an ensemble average results.

Solution: The results in Figure 8.16 are found using the Book m-Function that follows.

BOOK M-FUNCTION: [W,Y,E,JAV]=LMS_EX11_3_7(MU,AV,N)

```
function[w,y,e,Jav]=lms_ex11_3_7(mu,av,N)
for m=1:av
    %all quantities are real-valued;
    %x=input data to the filter; d=desired signal;
    %M=order of the filter;
    %mu=step-size factor; x and d must be
    %of the same length;
w(1)=0;x(1)=0;w(2)=0;
for r=1:N+2
    x(r+1)=0.95*x(r)+3*(rand-0.5);
    d(r+1)=x(r+1);
end;

for n=1:N
    y(n+1)=w(n+1)*x(n);
    e(n+1,m)=x(n+1)-y(n);
    w(n+2)=w(n+1)+mu*x(n)*e(n+1,m);
```

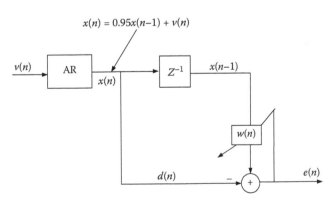

FIGURE 11.15

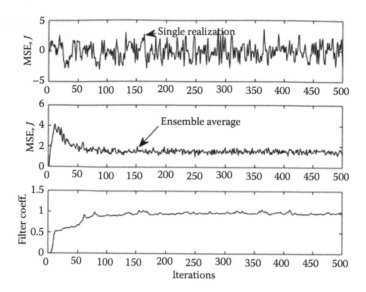

FIGURE 11.16

```
end;
y=zeros(1,N);
end;
Jav=sum((e.^2),2)/av;%Jav is the learning curve;
```

Figure 11.16 was produced using $\mu=0.005$, $N=1000$, and the number ensemble was 100. The third figure shows the creation of the final value of the adaptive filter coefficient versus the number of iterations. ∎

*11.4   PERFORMANCE ANALYSIS OF THE LMS ALGORITHM

By subtracting the Wiener filter w^o (column vector of the optimum adaptive filter) from both sides of Equation 11.3, we obtain the following equation:

$$w(n+1)-w^o = w(n)-w^o +2\mu e(n)x(n) \tag{11.7}$$

The vectors $\xi(n+1)=w(n+1)-w^o$ and $\xi(n)=w(n)-w^o$ are known as the weight errors, which describe on a coordinate system shifted by w^o on the w plane. Therefore, Equation 11.7 becomes

$$\xi(n+1) = \xi(n)+2\mu e(n)x(n) \tag{11.8}$$

$$\begin{aligned}
e(n) &= d(n)-y(n)=d(n)-w^T(n)x(n)=d(n)-x^T(n)w(n)\\
&= d(n)-x^T(n)w(n)-x^T(n)w^o +x^T(n)w^o\\
&= d(n)-x^T(n)w^o -x^T(n)[w(n)-w^o]\\
&= e^o(n)-x^T(n)\xi(n)=e^o(n)-\xi^T(n)x(n)
\end{aligned} \tag{11.9}$$

where

$$e^o(n)=d(n)-x^T(n)w^o \tag{11.10}$$

---

* Means section may be skipped.

is the error when the filter is optimum. Substituting Equation 11.9 in Equation 11.8 and rearranging, we obtain

$$
\begin{aligned}
\xi(n+1) &= \xi(n) + 2\mu[e^o(n) - x^T(n)\xi(n)]x(n) \\
&= \xi(n) + 2\mu x(n)[e^o(n) - x^T(n)\xi(n)] \\
&= [I - 2\mu x(n)x^T(n)]\xi(n) + 2\mu e^o(n)x(n)
\end{aligned}
\tag{11.11}
$$

where

I is the identity matrix (ones in the diagonal and zeros everywhere else)
$e^o(n) - x^T(n)\xi(n)$ is a scalar

Next, we take the expectation of both sides of the preceding equation to obtain:

$$
\begin{aligned}
E\{\xi(n+1)\} &= E\{[I - 2\mu x(n)x^T(n)]\}\xi(n) + 2\mu E\{e^o(n)x(n)\} \\
&= E\{[I - 2\mu x(n)x^T(n)]\}\xi(n)\}
\end{aligned}
\tag{11.12}
$$

Since $e^o(n)$ is orthogonal to all data (see Section 7.7), the last expression is identical to zero. The expression

$$
E\{x(n)x^T(n)\xi(n)\} = E\{x(n)x^T(n)\}E\{\xi(n)\}
\tag{11.13}
$$

is simplified by incorporating the independence assumption, which states that the present observation of the data $\{x(n), d(n)\}$ is independent of the past observations $(x(n-1), d(n-1))$, $(x(n-2), d(n-2))$, …, where

$$
x(n) = [x(n)x(n-1)\cdots x(n-N+1)]
$$

$$
x(n-1) = [x(n-1)x(n-2)\cdots x(n-N)]
$$

Another way to justify the independence assumption is through the following observation: The LMS coefficients $w(n)$ at any given time are affected by the whole past history of the data $\{x(n-1), d(n-1)), (x(n-2), d(n-2)), …\}$. Therefore, for smaller step-size parameter μ, the past N observations of the data have small contribution to $w(n)$, and, thus, we can say that $w(n)$ and $x(n)$ are weakly dependent. This observation clearly suggests the approximation given by substituting Equation 11.13 into Equation 11.12:

$$
\begin{aligned}
E\{\xi(n+1)\} &= [I - 2\mu E\{x(n)x^T(n)\}]E\{\xi(n)\} \\
&= [I - 2\mu R_x]E\{\xi(n)\}
\end{aligned}
\tag{11.14}
$$

The mathematical forms of Equations 11.14 and 10.29 of the steepest-descent method are identical except that the deterministic weight-error vector $\xi(n)$ in Equation 10.29 has been replaced by the average weight-error vector $E\{\xi(n)\}$ of the LMS algorithm. This suggests that, on the average, the present LMS algorithm behaves just like the steepest-descent algorithm. Like the steepest-descent method, the LMS algorithm is controlled by M modes of convergence that are dependent on the eigenvalues of the correlation matrix R_x. In particular, the convergence behavior of the LMS algorithm is directly related to the eigenvalue spread of R_x, and hence to the power spectrum of the input data $x(n)$. The flatter of the power spectrum, the higher speed of convergence of the LMS algorithm is attained.

11.4.1 Learning Curve

In this development, we assume the following: (1) the input signal to LMS filter $x(n)$ is zero-mean stationary process; (2) the desired signal $d(n)$ is zero-mean stationary process; (3) $x(n)$ and $d(n)$ are jointly Gaussian distributed random variables for all n; and (4) at times n, the coefficients $w(n)$ are independent of the input vector $x(n)$ and the desired signal $d(n)$. The validity of (4) is justi-fied for small values of μ (independent assumption). Assumptions (1) and (2) simplify the analysis. Assumption (3) simplifies the final results so that the third-order and higher moments and higher that appear in the derivation are expressed in terms of the second-order moments due to their Gaussian distribution.

If we take the mean square average of the error given by Equation 11.9, we obtain

$$J(n) = E\{e^2(n)\} = E\{[e^o(n) - \boldsymbol{\xi}^T(n)\boldsymbol{x}(n)][e^o(n) - \boldsymbol{x}^T(n)\boldsymbol{\xi}(n)]\}$$
$$= E\{e^{o2}(n)\} + E\{\boldsymbol{\xi}^T(n)\boldsymbol{x}(n)\boldsymbol{x}^T(n)\boldsymbol{\xi}(n)\} - 2E\{e^o(n)\boldsymbol{\xi}^T(n)\boldsymbol{x}(n)\} \tag{11.15}$$

For independent random variables, we have the following equations:

$$E\{xy\} = E\{x\}E\{y\} = E\{xE\{y\}\}$$
$$E\{x^2y^2\} = E\{x^2\}E\{y^2\} = E\{x^2E\{x^2\}\} = E\{xE\{y^2\}x\} \tag{11.16}$$

Based on the preceding two equations, the second term of Equation 11.15 becomes $[\boldsymbol{x}^T(n)\boldsymbol{\xi}(n) = \boldsymbol{\xi}^T(n)\boldsymbol{x}(n) = \text{number}]$:

$$E\{[\boldsymbol{\xi}^T(n)\boldsymbol{x}(n)]^2\} = E\{\boldsymbol{\xi}^T(n)\boldsymbol{x}(n)\boldsymbol{x}^T(n)\boldsymbol{\xi}(n)\} = E\{\boldsymbol{\xi}^T(n)E\{\boldsymbol{x}(n)\boldsymbol{x}^T(n)\}\boldsymbol{\xi}(n)\}$$
$$= E\{\boldsymbol{\xi}^T\boldsymbol{R}_x\boldsymbol{\xi}(n)\} = \text{tr}\{E\{\boldsymbol{\xi}^T(n)\boldsymbol{R}_x\boldsymbol{\xi}(n)\}\}$$
$$= E\{\text{tr}\{\boldsymbol{\xi}^T(n)\boldsymbol{R}_x\boldsymbol{\xi}(n)\}\} = E\{\text{tr}\{\boldsymbol{\xi}(n)\boldsymbol{\xi}^T(n)\boldsymbol{R}_x\}\} \tag{11.17}$$
$$= \text{tr}\{E\{\boldsymbol{\xi}(n)\boldsymbol{\xi}^T(n)\}\boldsymbol{R}_x\} = \text{tr}\{\boldsymbol{K}(n)\boldsymbol{R}_x\}$$

where we used the following properties: (1) the trace of a scalar is the scalar itself; (2) the trace, tr, and the expectation, E, operators are linear and can be exchanged; and (3) the trace of two matrices having $N \times M$ and $M \times N$ dimensions, respectively, is given by

$$\text{tr}\{AB\} = \text{tr}\{BA\} \tag{11.18}$$

The third term in Equation 11.15, due to the independence assumption and due to the fact that $e^o(n)$ is a constant, becomes

$$E\{e^o(n)\boldsymbol{\xi}^T(n)\boldsymbol{x}(n)\} = E\{\boldsymbol{\xi}^T(n)\boldsymbol{x}(n)e^o(n)\} = E\{\boldsymbol{\xi}^T(n)\}E\{\boldsymbol{x}(n)e^o(n)\} = 0 \tag{11.19}$$

The second term is equal to zero due to the orthogonality property (see Section 7.7.1).

Substituting Equations 11.19 and 11.18 in Equation 11.15, we obtain

$$J(n) = E\{e^2(n)\} = J_{\min} + \text{tr}\{\boldsymbol{K}(n)\boldsymbol{R}_x\}, \quad J_{\min} = E\{([e^o(n)]^2)\} \tag{11.20}$$

However, $\boldsymbol{R}_x = \boldsymbol{Q}\boldsymbol{\Lambda}\boldsymbol{Q}^T$, where $\boldsymbol{Q}$ is the eigenvector matrix and $\boldsymbol{\Lambda}$ is the diagonal eigenvalue one. Hence, Equation 11.20 becomes

$$J(n) = J_{min} + \text{tr}\{K(n)Q\Lambda Q^T\} = J_{min} + \text{tr}\{Q^T K(n)Q\Lambda\}$$

$$= J_{min} + \text{tr}\{E\{Q^T \xi(n)\xi^T(n)Q\}\Lambda\} \tag{11.21}$$

$$= J_{min} + \text{tr}\{E\{\xi'(n)\xi'^T(n)\}\Lambda\} = J_{min} + \text{tr}\{K'(n)\Lambda\}$$

where

$$K'(n) = E\{\xi'(n)\xi'^T(n)\} \tag{11.22}$$

Recall that $\xi'(n)$ is the weight-error vector in the coordinate system defined by the basis vectors, which are specified by the eigenvectors of R_x. Since Λ is diagonal, Equation 11.21 becomes

$$J(n) = J_{min} + \sum_{i=0}^{M-1} \lambda_i k'_{ii}(n) = J_{min} + \sum_{i=0}^{M-1} \lambda_i E\{\xi'^2_{ii}(n)\} \tag{11.23}$$

where $k'_{ij}(n)$ is the ijth element of the matrix $K'(n)$.

The learning curve can be obtained by any one of the equations: 11.20, 11.21 and 11.23. It turns out that, on average, the preceding learning curve is similar to the one given by the steepest-descent algorithm.

The general solution of $\xi'(n)$ is given by Equation 10.39, and, hence, Equation 11.23 becomes

$$J(n) = J_{min} + \sum_{i=0}^{M-1} \lambda_i (1 - 2\mu\lambda_i)^{2n} E\{\xi'_{ii}(0)\} \tag{11.24}$$

Example 11.4.1

The filter $H_1(z)$, which produces the desired signal $d(n)=x(n)$, is represented by the difference equation $x(n) + a_1 x(n-1) + a_2 x(n-2) = v(n)$, where a_1 and a_2 are the system coefficients, and $v(n)$ is a zero-mean white noise process of variance σ_v^2. To simulate the system coefficients a_1 and a_2, we use the adaptive predictor (see Figure 11.3a). The LMS algorithm is

$$w(n) = w(n-1) + 2\mu x(n-1)e(n)$$

$$e(n) = d(n) - w^T(n)x(n-1) = x(n) - w^T(n)x(n-1) \tag{11.25}$$

$$J(n) = e^2(n) = \text{learning curve}$$

In this example, we used the following constants: $M=6$, $a_1=-0.96$, $a_2=0.2$, average number=200 and variance $\sigma_x^2 = 0.24$. Figure 11.17 shows the results. The following Book m-Function was used.

BOOK M-FUNCTION: [J,X]=LMS_EX11_4_1(MU,M,AVN)

```
function[J,x]=lms_ex11_4_1(mu,M,avn)
%function[J]=lms_ex8_4_1(mu,M,avn);
%M=number of filter coefficients;
%avn=number of times the MSE (J) to be averaged;

dn(1)=0;dn(2)=0;x(1)=0;x(2)=0;
```

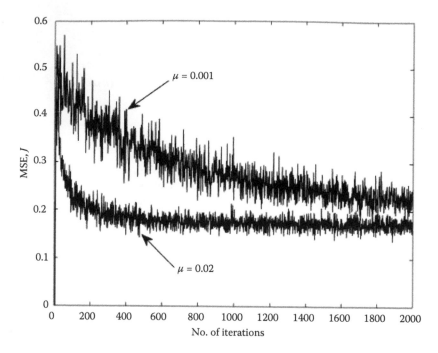

FIGURE 11.17

```
for k=1:avn
    for n=3:2000
        dn(n)=0.96*dn(n-1)-0.2*dn(n-2)+(rand-0.5);
        x(n)=dn(n-1);
    end;
    [w,y,e,J1]=lms1(x,dn,mu,M);
    Jk(k,:)=J1;%this expression creates a
              %matrix avn by 2000;
end;
J=sum(Jk,1)/100;
```

11.4.2   THE COEFFICIENT-ERROR OR WEIGHTED-ERROR CORRELATION MATRIX

Since the MSE is related to the weight-error correlation matrix $K(n)$, the matrix is closely related to the stability of the LMS algorithm. Therefore, $J(n)$ is bounded if the elements of $K(n)$ are also bounded. Since $K(n)$ and $K'(n)$ are related, the stability can be studied by using either one. Multiply both sides of Equation 11.11 by Q^T and use the definitions $\xi'(n) = Q^T \xi(n)$ and $x'(n) = Q^T x(n)$ to obtain

$$\xi'(n+1) = [I - 2\mu x'(n)x'^T(n)]\xi'(n) + 2\mu e^o(n)x'(n) \tag{11.26}$$

Exercise 11.4.2.1

Verify Equation 11.26. ▲

Next, we multiply Equation 11.26 by its transpose and take the ensemble average of both sides to obtain (see Exercise 11.4.2.1):

$$K'(n+1) = K'(n) - 2\mu E\{x'(n)x'^T(n)\xi'(n)\xi'^T(n)\}$$

$$-2\mu E\{\xi'(n)\xi'^T(n)x'(n)x'^T(n)\}$$

$$+4\mu^2 E\{x'(n)\mathrm{x}^T(n)\xi'(n)\xi'^T(n)x'(n)x'^T(n)\}$$

$$+2\mu E\{e^o(n)x'(n)\xi'^T(n)\}$$

$$+2\mu E\{e^o(n)\xi'(n)x'^T(n)\}$$ (11.27)

$$-4\mu^2 E\{e^o(n)x'(n)\xi'^T(n)x'(n)x'^T(n)\}$$

$$-4\mu^2 E\{e^o(n)x'(n)x'^T(n)\xi'(n)x'^T(n)\}$$

$$+4\mu^2 E\{e^{o2}(n)x'(n)x'^T(n)\}$$

Exercise 11.4.2.2

Verify Equation 11.27. ▲

Based on the previous independent assumption, we note the following: (1) $\xi(n)$ is independent of the data $x(n)$ and the desired signal $d(n)$, which is also true for the transformed variables $x'(n)$; (2) $\xi'(n)$ is independent of $x'(n)$ and $d(n)$; (3) $d(n)$ and $x'(n)$ are zero-mean and are jointly Gaussian since $d(n)$ and $x(n)$ have the same properties; (4) applying the orthogonality relationship, we obtain $E\{e^o(n)x'(n)\} = E\{e^o(n)Q^T x(n)\} = Q^T E\{e^o(n)x(n)\} = Q^T \mathbf{0} = \mathbf{0}$; (5) $e^o(n)$ depends only on $d(n)$ and $x(n)$; (6) from (4) $e^o(n)$ and $x'(n)$ are uncorrelated (Gaussian variables are also independent); and (7) $e^o(n)$ has zero mean. With the preceding assumptions in mind, the factors of Equation 11.27 become as follows (see Exercise 11.4.2.3):

$$E\{x'(n)x'^T(n)\xi'(n)\xi'^T(n)\} = E\{x'(n)x'^T(n)\}E\{\xi'(n)\xi'^T(n)\}$$ (11.28)

$$= E\{Q^T x(n)x^T(n)Q\}K'(n) = Q^T R_{xx}QK'(n) = \Lambda K'(n)$$

$$E\{\xi'(n)\xi'^T(n)x'(n)x'^T(n)\} = K'(n)\Lambda$$ (11.29)

$$E\{x'(n)x'^T(n)\xi'(n)\xi'^T(n)x'(n)x'(n)\} = 2\Lambda K'(n)\Lambda + \mathrm{tr}\{\Lambda K'(n)\Lambda\}$$ (11.30)

Exercise 11.4.2.3

Verify Equation 11.30. ▲

Because $e^o(n)$, $x'(n)$, and $\xi'(n)$ are mutually independent, $E\{e^o\} = 0$ and the following are also true:

$$E\{e^o(n)x'(n)\xi'^T(n)\} = E\{e^o(n)\}E\{x'(n)\xi'^T(n)\} = \mathbf{0} \quad (M \times M \text{ matrix})$$ (11.31)

$$E\{e^o(n)\xi'(n)x'^T(n)\} = 0$$ (11.32)

$$E\{e^o(n)x'(n)\xi'^T(n)x'(n)x'^T(n)\} = 0$$ (11.33)

$$E\{e^o(n)x'(n)x'^T(n)\xi'(n)x'^T(n)\} = 0 \tag{11.34}$$

$$E\{e^{o2}(n)x'(n)x'^T(n)\} = E\{e^{o2}(n)\}E\{x'(n)x'^T(n)\} = J_{min}\Lambda \tag{11.35}$$

Substituting Equations 11.28 through 11.35 in Equation 11.27, we obtain

$$K'(n+1) = K'(n) - 2\mu(\Lambda K'(n) + K'(n)\Lambda) + 8\mu^2\Lambda K'(n)\Lambda$$

$$+ 4\mu^2\text{tr}\{\Lambda K'(n)\}\Lambda + 4\mu^2 J_{min}\Lambda \tag{11.36}$$

Concentrating on the *ii*th component of both sides of Equation 11.36, we obtain (see Exercise 11.4.2.4)

$$k'_{ii}(n+1) = (1 - 4\mu\lambda_i + 8\mu^2\lambda_i^2)k'_{ii}(n) + 4\mu^2\lambda_i\sum_{j=0}^{M-1}\lambda_j k'_{jj}(n) + 4\mu^2 J_{min}\lambda_i \tag{11.37}$$

Exercise 11.4.2.4

Verify Equation 11.37.   ▲

Since $K'(n)$ is a correlation matrix, $k'^2_{ij} \le k'_{ii}(n)k'_{jj}(n)$ for all values of i and j, and because the update of $k'_{ii}(n)$ is independent of the off-diagonal elements of $K'(n)$, the convergence of the diagonal elements is sufficient to secure the convergence of all the elements and, thus, guarantees the stability of the LMS algorithm. Therefore, we concentrate on Equation 11.37 with $i=0, 1, 2, \ldots, M-1$.

Equation 11.37 can be written in the following matrix form (see Exercise 11.4.2.4):

$$k'(n+1) = Fk'(n) + 4\mu^2 J_{min}\lambda \tag{11.38}$$

where

$$k'(n) = \begin{bmatrix} k'_{00}(n) & k'_{11} \ldots k'_{M-1,M-1}(n) \end{bmatrix}^T$$

$$\lambda = \begin{bmatrix} \lambda_0 & \lambda_1 \ldots \lambda_{M-1} \end{bmatrix}^T$$

$$F = \text{diag}\begin{bmatrix} f_0 & f_1 \ldots f_{M-1} \end{bmatrix}^T \tag{11.39}$$

$$f_i = 1 - 4\mu\lambda_i + 8\mu^2\lambda_i^2$$

It has been found that if the eigenvalues of F are less than one, the LMS algorithm is stable or equivalently the elements $k'(n)$ remain bounded. An indirect approach to obtain stability is given in Section 11.4.3.

11.4.3 Excess MSE and Misadjustment

We may write the expression in Equation 11.20 as the difference between the MSE and the minimum MSE as follows:

$$J_{ex}(n) = J(n) - J_{min} = \text{tr}[K(n)R_x] \tag{11.40}$$

The steady-state form of Equation 11.40 is

$$J_{ex}(\infty) = J(\infty) - J_{min} = \text{tr}[K(\infty)R_x] \tag{11.41}$$

and it is known as **excess** MSE. Equation 11.23 gives another equivalent form, which is

$$J_{ex}(\infty) = \sum_{i=0}^{M-1} \lambda_i k'_{ii}(\infty) = \xi^T k'(\infty) \tag{11.42}$$

As $n \to \infty$, we set $k'(n+1) = k(n)$, and hence, Equation 11.38 gives another equivalent form, which is

$$k'(\infty) = 4\mu^2 J_{\min} (I-F)^{-1} \lambda \tag{11.43}$$

As a result, Equation 11.42 becomes

$$J_{ex}(\infty) = 4\mu^2 J_{\min} \lambda^T (I-F)^{-1} \lambda \tag{11.44}$$

which indicates that J_{ex} is proportional to $J_{\min}$. The normalized $J_{ex}(\infty)$ is equal to

$$\mathfrak{M} = \frac{J_{ex}(\infty)}{J_{\min}} = 4\mu^2 \lambda^T (I-F)^{-1} \lambda \tag{11.45}$$

which is known is the **misadjustment** factor.

If $A(N \times N)$, $B(M \times M)$, and $C(N \times M)$ are matrices that have an inverse, then

$$(A + CBC^T)^{-1} = A^{-1} - A^{-1}C(B^{-1} + C^T A^{-1} C)^{-1} C^T A^{-1} \tag{11.46}$$

But

$$I - F = \begin{bmatrix} 1-f_0 & 0 & \dots & 0 \\ 0 & 1-f_1 & \dots & 0 \\ \vdots & & & \vdots \\ 0 & 0 & \dots & 1-f_{M-1} \end{bmatrix} = 4\mu^2 \lambda\lambda^T = F_1 + a\lambda\lambda^T \tag{11.47}$$

$$a = -4\mu^2$$

Therefore, Equation 11.45 takes the form (see Exercise 11.4.3.1)

$$\mathfrak{M} = -a\lambda^T (F_1 + a\lambda\lambda^T)^{-1} \lambda = -a\lambda^T \left[F_1^{-1} - \frac{a F_1^{-1} \lambda\lambda^T F_1^{-1}}{1 + a\lambda^T F_1^{-1} \lambda} \right] \lambda$$

$$= a'\lambda^T \left[F_1^{-1} - \frac{a' F_1^{-1} \lambda\lambda^T F_1^{-1}}{1 + a'\lambda^T F_1^{-1} \lambda} \right] \lambda = \frac{\displaystyle\sum_{i=0}^{M-1} \frac{\mu\lambda_i}{1 - 2\mu\lambda_i}}{1 - \displaystyle\sum_{i=0}^{M-1} \frac{\mu\lambda_i}{1 - 2\mu\lambda_i}} \tag{11.48}$$

where in Equation 11.46 we set $C = \lambda$, $B = I$, $a' = -a$, and $A = F_1$. Small $\mathfrak{M}$ implies that the summation on the numerator is small. If, in addition, $2\mu\lambda_i \ll 1$, we obtain the following results:

$$\sum_{i=0}^{M-1} \frac{\mu\lambda_i}{1 - \mu\lambda_i} \cong \mu \sum_{i=0}^{M-1} \mu \, \mathrm{tr}\{R_x\} \tag{11.49}$$

Hence, Equation 11.48 becomes

$$\mathfrak{M} = \frac{\mu\,\mathrm{tr}\{\boldsymbol{R}_x\}}{1 - \mu\,\mathrm{tr}\{\boldsymbol{R}_x\}} \tag{11.50}$$

In addition, for $\mathfrak{M} \ll 0.1$, the quantity $\mu\,\mathrm{tr}|\boldsymbol{R}_x|$ is small and Equation 11.50 becomes $\mu\,\mathrm{tr}\{\boldsymbol{R}_x\}$ Since $r_{xx}(0)$ is the mean square value of the input signal to an M-coefficient filter, we write

$$\mathfrak{M} = \mu M r_{xx}(0) = \mu M E\{x^2(0)\} = \mu M \quad \text{(Power input)} \tag{11.51}$$

The preceding equation indicates that to keep the misadjustment factor small and at a specific desired value as the signal power changes, we must adjust the value of μ.

Example 11.4.3.1

Verify Equation 11.48.   ■

11.4.4 STABILITY

If we set

$$L = \sum_{i=0}^{M-1} \frac{\mu\lambda_i}{1 - \mu\lambda_i} \tag{11.52}$$

then

$$\mathfrak{M} = \frac{L}{1 - L} \tag{11.53}$$

We observe that L and $\mathfrak{M}$ are increasing functions of μ and L, respectively (see Exercise 11.4.4.1). Since L reaches 1 as $\mathfrak{M}$ goes to infinity, this indicates that there is a value $\mu_{\max}$ that μ cannot surpass. To find the upper value of μ, we must concentrate on the expression that can easily be measured in practice. Therefore, from Equation 11.53 we must have

$$\sum_{i=0}^{M-1} \frac{\mu\lambda_i}{1 - \mu\lambda_i} \leq 1 \tag{11.54}$$

Exercise 11.4.4.1

Verify Equations 11.53 and 11.52.   ▲

The preceding equation indicates that the maximum value of μ must make Equation 11.54 an equality. It can be shown that the following inequality holds:

$$\sum_{i=0}^{M-1} \frac{\mu\lambda_i}{1 - \mu\lambda_i} \leq \frac{\mu \sum_{i=0}^{M-1} \lambda_i}{1 - 2\mu \sum_{i=0}^{M-1} \lambda_i} \tag{11.55}$$

Hence, if we set Equation 11.54 as equality and substitute its value to Equation 11.55 as equality, we obtain

$$\mu_{\max} = \frac{1}{3 \sum_{i=0}^{M-1} \lambda_i} = \frac{1}{3\,\mathrm{tr}\{\boldsymbol{R}_x\}} = \frac{1}{3\,\text{(input power)}} \tag{11.56}$$

and hence,

$$0 < \mu < \frac{1}{3\sum\limits_{i=0}^{M-1}\lambda_i} = \frac{1}{3\,\mathrm{tr}\{R_x\}} \tag{11.57}$$

11.4.5 THE LMS AND STEEPEST-DESCENT METHOD

The following similarities and differences exist between the two methods:

1. The steepest-descent method reaches the minimum mean square error J_{min} as $n \to \infty$ and $w(n) \to w^o$.
2. The LMS method produces an error $J(\infty)$ that approaches J_{min} as $n \to \infty$ and remains larger than J_{min}.
3. The LMS method produces a $w(n)$, as the iterations $n \to \infty$, that is close to the optimum w^o.
4. The steepest-descent method has a well-defined learning curve consisting of a sum of decaying exponentials.
5. The LMS learning curve is a sum of noisy decaying exponentials and the noise, in general, decreases the smaller values the step-size parameter μ takes.
6. In the steepest-descent method, the correlation matrix R_x of the data $x(n)$ and the cross-correlation vector $p_{dx}(n)$ are found using ensemble averaging operations from the realizations of the data $x(n)$ and desired signal $d(n)$.
7. In the LMS filter, an ensemble of learning curves is found under identical filter parameters and then averaged point by point.

*11.5 COMPLEX REPRESENTATION OF THE LMS ALGORITHM

In some practical applications, it is mathematically attractable to have complex representation of the underlying signals. For example, baseband signals in quadrature amplitude modulation (QAM) format are written as a summation of two components: real **in-phase** component and imaginary **quadrature** component. Furthermore, signals detected by a set of antennas are also written in their complex form for easy mathematical manipulation. For this reason, we shall present in this section the most rudimentary derivation of the LMS filter assuming that the signals are complex.

In the case where complex-type signals must be processed, we write the output of the adaptive filter in the form:

$$y(n) = \sum_{k=0}^{M-1} w_k^*(n)x(n-k) \tag{11.58}$$

and the error is given by

$$e(n) = d(n) - y(n) \tag{11.59}$$

Therefore, the MSE is

$$J = E\{e(n)e^*(n)\} = E\{|e(n)|^2\} \tag{11.60}$$

Let us define the complex coefficient as follows:

$$w_k = a_k(n) + jb_k(n) \quad k = 0,1,\ldots,M-1 \tag{11.61}$$

---

* Means section may be skipped.

Then the gradient ∇ has the following kth element:

$$\nabla_{w_k} \triangleq \nabla_k = \frac{\partial}{\partial a_k(n)} + j\frac{\partial}{\partial b_k(n)} \quad k = 0,1,\ldots,M-1 \tag{11.62}$$

Which will produce the following kth element of the multielement gradient vector $\nabla_k J$:

$$\nabla_k J = \frac{\partial J}{\partial a_k(n)} + j\frac{\partial J}{\partial b_k(n)} \quad k = 0,1,\ldots,M-1 \tag{11.63}$$

It is noted that the gradient operator is always used to find the minimum points of a function. The preceding equation indicates that a complex constraint must be converted to a pair of real constraints. Hence, we set

$$\frac{\partial J}{\partial a_k(n)} = 0 \quad j\frac{\partial J}{\partial b_k(n)} = 0 \quad k = 0,1,\ldots,M-1 \tag{11.64}$$

The kth element of the gradient vector, using Equation 11.60, is

$$\nabla_k J = E\left\{\frac{\partial e(n)}{\partial a_k(n)}e^*(n) + \frac{\partial e^*(n)}{\partial a_k(n)}e(n) + j\frac{\partial e(n)}{\partial b_k(n)}e^*(n) + j\frac{\partial e^*(n)}{\partial b_k(n)}e(n)\right\} \tag{11.65}$$

Taking into consideration Equations 11.58 and 11.59, we obtain

$$\frac{\partial e(n)}{\partial a_k(n)} = \frac{\partial d(n)}{\partial a_k(n)} - \sum_{k=0}^{M-1}\frac{\partial w_k^*(n)}{\partial a_k(n)}x(n-k)$$

$$= 0 - \sum_{k=0}^{M-1}\frac{\partial[a_k(n) - jb_k(n)]}{\partial a_k(n)}x(n-k) = -x(n-k) \tag{11.66}$$

Similarly, we obtain

$$\frac{\partial e^*(n)}{\partial b_k(n)} = jx(n-k), \frac{\partial e^*(n)}{\partial b_k(n)} = -jx^*(n-k), \frac{\partial e(n)}{\partial b_k(n)} = -jx(n-k) \tag{11.67}$$

Introducing the last three equations into Equation 11.65, we obtain

$$\nabla_k J \equiv \nabla_{w_k} J(w(n)) = -2E\{x(n-k)e^*(n)\} \tag{11.68}$$

and, thus, the gradient vector becomes

$$\nabla_w J(w(n)) = -2E\left\{e^*(n)[x(n)x(n-1)\cdots x(n-M+1)]^T\right\} \tag{11.69}$$

If $w(n)$ is the filter coefficient vector at step n (time), its update value $w(n+1)$ is given by (see Section 10.3)

$$w(n+1) = w(n) + 2\mu e^*(n)x(n) \tag{11.70}$$

Next, we replace the ensemble average in Equation 11.69 by the instantaneous estimate $e^*(n)x(n)$ to obtain

$$w(n+1) = w(n) + 2\mu e^*(n)x(n) \tag{11.71}$$

TABLE 11.5.1

Complex Form of the LMS Algorithm

Parameters	M = number of filter coefficients
	μ = step-size factor
	$x(n) = \left[x(n) \; x(n-1) \cdots x(n-M+1) \right]^T$
Initialization	$w(0) = \mathbf{0}$
Computation	For $n = 0, 1, 2, \ldots$
	1. $y(n) = w^H(n)x(n)$, H = Hermitian = conjugate transpose
	2. $e(n) = d(n) - y(n)$
	3. $w(n+1) = w(n) + 2\mu e^*(n)x(n)$

which is the LMS recursion formula when we are involved with complex-valued processes. The LMS algorithm for complex signals is given in Table 11.5.1.

Example 11.5.1

With the input signal, the desired signal $d(n) = \sin(0.1\pi n)$, $\mu = 0.01$, and number coefficients $M = 16$, we $x(n) = \sin(0.1\pi n) + j1.5(\text{rand} - 0.5)$ obtain the results shown in Figure 11.18. The Book m-Function that produced the results follows.

BOOK M-FUNCTION FOR COMPLEX LMS ALGORITHM

```
function[w,y,e,J,w1]=lms_complex_norm_lms(x,dn,mubar,M,c)
    %function[w,y,e,J,w1]=lms_complex_norm_lms(x,dn,mubar,M,c)
    %x=input data to the filter;dn=desired signal;
    %M=filter order;c=small constant;
    %mubar=step-size equivalent parameter;
    %x and dn must be of the same length;J=learning curve;
N=length(x);
y=zeros(1,N);
```

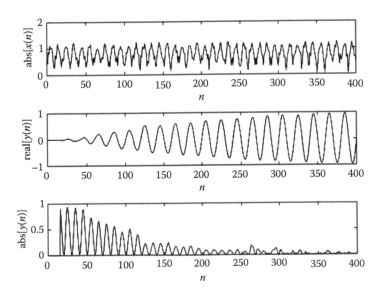

FIGURE 11.18

```
w=zeros(1,M)+j*zeros(1,M);%initialized filter coefficient vector;
for n=M:N
   x1=x(n:-1:n-M+1);%for each n vector x1 is of length
   %M with elements from x in reverse order;
   y(n)=conj(w)*x1';
   e(n)=dn(n)-y(n);
   w=w+(mubar/(c+conj(x1)*x1'))*conj(e(n))*x1;
   w1(n-M+1,:)=w(1,:);
end;
J=e.^2;
   %the columns of the matrix w1 depict the history of the
   %filter coefficients;
```

HINTS, SUGGESTIONS, AND SOLUTIONS OF THE EXERCISES

11.4.2.1

$$Q^T\xi(n+1) = Q^T[\xi(n) - 2\mu x(n)x^T(n)\xi(n)] + 2\mu e^o(n)Q^T x(n) = Q^T\xi(n) - 2\mu Q^T x(n)x^T(n)$$

$$Q^{-T}Q^T\xi(n) + 2\mu e^o(n)x'(n) = \xi'(n) - 2\mu Q^T x(n)\left(Q^T x(n)\right)^T \xi'(n) + 2\mu e^o(n)x'(n) =$$

$$\left[I - 2\mu x'(n)x'^T(n)\right]\xi'(n) + 2\mu e^o(n)x'(n); \quad Q^{-T} = \left(Q^{-1}\right)^T = \left(Q^T\right)^T = Q$$

11.4.2.2

$$\xi'^T(n+1) = \xi'^T(n) - 2\mu\xi'^T(n)x'(n)x'^T(n) + 2\mu e^o(n)x'^T(n)$$

11.4.2.3

Because $x'(n)$ and $\xi'(n)$ are independent $E\{x'(n)x'^T(n)\xi'(n)\xi'^T(n)x'(n)x'^T(n)\}$

$$= E\left\{x(n)x'^T(n)E\left\{\xi'(n)\xi'^T(n)\right\}x'(n)x'^T(n)\right\} = E\left\{x'(n)x'^T(n)K'(n)x'(n)x'^T(n)\right\} \tag{1}$$

$$x'^T(n)K'(n)x'(n) = \sum_{i=0}^{M-1}\sum_{j=0}^{M-1} x_i'(n)x_j'(n)k_{ij}'(n) \tag{2}$$

$$C(n) = M \times M \ matrix = x'(n)x'^T(n)K'(n)x'(n)x'^T(n) \tag{3}$$

$$c_{lm}(n) = x_l'(n)x_m'(m)\sum_{i=0}^{M-1}\sum_{j=0}^{M-1} x_i'(n)x_j'(n)k_{ij}'(n) \tag{4}$$

$$E\{c_{lm}(n)\} = \sum_{i=0}^{M-1}\sum_{j=0}^{M-1} E\{x_l'(n)x_m'(n)x_i'(n)x_j'(n)\}k_{ij}'(n) \tag{5}$$

For Gaussian random variables, we have

$$E\{x_1 x_2 x_3 x_4\} = E\{x_1 x_2\}E\{x_3 x_4\} + E\{x_1 x_3\}E\{x_2 x_4\} + E\{x_1 x_4\}E\{x_2 x_3\} \tag{6}$$

$$E\{x_i'(n)x_j'(n)\} = \lambda_i \delta(i - j) \tag{7}$$

The preceding equation is due to the relation $E\{Q^T x(n)x(n)^T Q\} = Q^T RQ = \Lambda, Rq_i = \lambda_i q_i$

$$E\{x_l'(n)x_m'(n)x_i'(n)x_j'(n)\} = \lambda_l \lambda_i \delta(l-m)\delta(i-j) + \lambda_l \lambda_m \delta(l-i)\delta(m-j) + \lambda_l \lambda_m \delta(l-j)\delta(m-i) \quad (8)$$

Substitute Equation 8 in Equation 5 to find

$$E\{c_{lm}(n)\} = \sum_{i=0}^{M-1}\sum_{j=0}^{M-1} \lambda_l \lambda_i \delta(l-m)\delta(i-j)k_{ij}'(n) + \sum_{i=0}^{M-1}\sum_{j=0}^{M-1} \lambda_l \lambda_m \delta(l-i)\delta(m-j)k_{ij}'(n)$$

$$+ \sum_{i=0}^{M-1}\sum_{j=0}^{M-1} \lambda_l \lambda_m \delta(l-j)\delta(m-i)k_{ij}'(n) = \lambda_l \delta(l-m)\sum_{i=0}^{M-1} \lambda_i k_{ii}'(n) + \lambda_l \lambda_m k_{lm}'(n)$$

$$+\lambda_l \lambda_m k_{ml}'(n) \text{ for } l=0,1,\ldots,M-1 \text{ and } m=0,1,\ldots,M-1. \text{ But } k_{lm}'(n)=k_{ml}'(n) \text{ and } \sum_{i=0}^{M-1}\lambda_i k_{ii}'(n) = tr\{\Lambda K'(n)\}$$

$\lambda_l \lambda_m k_{lm}'(n) + \lambda_l \lambda_m k_{ml}'(n) = 2\lambda_l \lambda_m k_{lm}'(n)$. Based on these results, the desired solution is apparent.

11.4.2.4

$$\begin{bmatrix} k_{00}'(n+1) & k_{01}'(n+1) & \cdots \\ k_{10}'(n+1) & k_{11}'(n+1) & \\ \vdots & & k_{M-1,M-1}'(n+1) \end{bmatrix} = \begin{bmatrix} f_0 + 4\mu^2\lambda_0^2 & 4\mu^2\lambda_1\lambda_0 & \cdots & 4\mu^2\lambda_{M-1}\lambda_0 \\ \lambda_0\lambda_1 & f_1 + 4\mu^2\lambda_1^2 & \cdots & 4\mu^2\lambda_{M-1}\lambda_1 \\ \vdots & \vdots & & \vdots \\ \lambda_0\lambda_{M-1} & \lambda_1\lambda_{M-1} & \cdots & f_{M-1}+4\mu^2\lambda_{M-1}^2 \end{bmatrix}$$

$$\bullet \begin{bmatrix} k_{00}'(n+1) & k_{01}'(n+1) & \cdots \\ k_{10}'(n+1) & k_{11}'(n+1) & \cdots \\ \vdots & & k_{M-1,M-1}'(n+1) \end{bmatrix} + 4\mu^2 J_{min} \begin{bmatrix} \lambda_0 & & & \\ & \lambda_1 & & 0 \\ 0 & & \ddots & \\ & & & \lambda_{M-1} \end{bmatrix}$$

where the iith component of both sides is

$$k_{ii}'(n+1) = f_i k_{ii}'(n) + 4\mu^2\lambda_i[\lambda_0 k_{00}'(n) + 4\mu^2\lambda_1 k_{11}'(n) + \cdots + 4\mu^2\lambda_{M-1}k_{M-1}'(n)] + 4\mu^2 J_{min}\lambda_i$$

which is identical to Equation 11.37.

11.4.3.1

$$a\begin{bmatrix} \lambda_0 & \lambda_1 \end{bmatrix}\begin{bmatrix} \dfrac{1}{1-f_0} & 0 \\ 0 & \dfrac{1}{1-f_1} \end{bmatrix}\begin{bmatrix} \lambda_0 \\ \lambda_1 \end{bmatrix} + \dfrac{a^2\begin{bmatrix} \lambda_0 & \lambda_1 \end{bmatrix}\begin{bmatrix} \dfrac{1}{1-f_0} & 0 \\ 0 & \dfrac{1}{1-f_1} \end{bmatrix}\begin{bmatrix} \lambda_0\lambda_0 & \lambda_0\lambda_1 \\ \lambda_0\lambda_1 & \lambda_1\lambda_1 \end{bmatrix}\begin{bmatrix} \dfrac{1}{1-f_0} & 0 \\ 0 & \dfrac{1}{1-f_1} \end{bmatrix}\begin{bmatrix} \lambda_0 \\ \lambda_1 \end{bmatrix}}{1 - a\begin{bmatrix} \lambda_0 & \lambda_1 \end{bmatrix}\begin{bmatrix} \dfrac{1}{1-f_0} & 0 \\ 0 & \dfrac{1}{1-f_1} \end{bmatrix}\begin{bmatrix} \lambda_0 \\ \lambda_1 \end{bmatrix}}$$

$$
= \frac{a \left[\dfrac{\lambda_0^2}{1-f_0} + \dfrac{\lambda_1^2}{1-f_1} \right] \left[1 - a \left(\dfrac{\lambda_0^2}{1-f_0} + \dfrac{\lambda_1^2}{1-f_1} \right) \right] + 0}{1 - a \left(\dfrac{\lambda_0^2}{1-f_0} + \dfrac{\lambda_1^2}{1-f_1} \right)}
$$

$$
= \left[a \frac{\lambda_0^2}{1-f_0} + \frac{\lambda_1^2}{1-f_1} - a^2 \frac{\lambda_0^4}{(1-f_0)^2} - a^2 \frac{\lambda_1^4}{(1-f_1)^2} - 2a^2 \frac{\lambda_0^2 \lambda_1^2}{(1-f_0)(1-f_1)} - 2a^2 \frac{\lambda_0^2}{(1-f_0)^2} + a^2 \frac{\lambda_0^2 \lambda_1^2}{(1-f_0)(1-f_1)} \right.
$$

$$
\left. + a^2 \frac{\lambda_0^2 \lambda_1^2}{(1-f_0)(1-f_1)} + a^2 \frac{\lambda_1^4}{(1-f_1)^2} \right] \bigg/ \left[1 - a \left(\frac{\lambda_0^2}{1-f_0} + \frac{\lambda_1^2}{1-f_1} \right) \right]
$$

$$
= \frac{a \dfrac{\lambda_0^2}{1-f_0} + \dfrac{\lambda_1^2}{1-f_1}}{1 - 4\mu^2 \dfrac{\lambda_0^2}{1-(1-4\mu\lambda_0 + 8\mu^2 \lambda_0^2)} - 4\mu^2 \dfrac{\lambda_1^2}{1-(1-4\mu\lambda_1 + 8\mu^2 \lambda_1^2)}} = \frac{\displaystyle\sum_{i=0}^{2-1} \frac{\mu\lambda_i}{1-2\mu\lambda_i}}{1 - \displaystyle\sum_{i=0}^{2-1} \frac{\mu\lambda_i}{1-2\mu\lambda_i}}
$$

which confirms Equation 11.48 for $M=2$.

11.4.4.1

Hint: Take derivatives with respect to μ and L, and note the positive values $\partial L / \partial \mu$ and $\partial \mathfrak{M} / \partial L$.

12 Variants of Least Mean Square Algorithm

12.1 THE NORMALIZED LEAST MEAN SQUARE ALGORITHM

Consider the conventional least mean square (LMS) algorithm with the fixed step-size μ replaced with a time-varying $\mu(n)$ as follows (we substitute 2μ with μ for simplicity):

$$w(n+1) = w(n) + \mu(n)e(n)x(n) \tag{12.1}$$

Next, define **a posteriori** error, $e_{ps}(n)$, as

$$e_{ps}(n) = d(n) - w^T(n+1)x(n) \tag{12.2}$$

Substituting Equation 12.1 in Equation 12.2 and taking into consideration the error equation $e(n) = d(n) - w^T(n)x(n)$, we obtain

$$e_{ps}(n) = \left[1 - \mu(n)x^T(n)x(n)\right]e(n), \quad x^T(n)x(n) = \|x(n)\|^2 = \sum_{i=0}^{M-1}|x(n-i)|^2 \tag{12.3}$$

Minimizing $e_{ps}(n)$ with respect to $\mu(n)$ results in (see Exercise 12.1.1)

$$\mu(n) = \frac{1}{\|x(n)\|^2} \tag{12.4}$$

Exercise 12.1.1

Verify Equations 12.3 and 12.4. ▲

Substituting Equation 12.4 in Equation 12.1, we find

$$w(n+1) = w(n) + \frac{1}{\|x(n)\|^2}e(n)x(n) \tag{12.5}$$

However, the most common normalized LMS (NLMS) algorithm is

$$w(n+1) = w(n) + \frac{\mu}{\|x(n)\|^2}e(n)x(n) \tag{12.6}$$

Note that the NLMS is actually a variable step-size algorithm in which this step-size is inversely proportional to the total instantaneous energy of the input signal $x(n)$, estimated over values within the tapped-delay line length.

A modified version of the NLMS algorithm (sometimes called ε-NLMS) is given by

$$
\begin{aligned}
w(n+1) &= w(n) + \frac{\mu}{\varepsilon + \|x(n)\|^2}e(n)x(n) \\
&= w(n) + \frac{\mu}{\varepsilon + \|x(n)\|^2}[d(n) - w^T(n)x(n)]x(n)
\end{aligned}
\tag{12.7}
$$

TABLE 12.1

Some Variants of the LMS Formulas

Algorithm	Recursion
LMS with constant step-size	$w(n+1) = w(n) + \mu\, e(n)x(n);$
	$w(n+1) = w(n) + \mu[d(n) - w^T(n)x(n)]x(n)$
LMS with time-varying step-size	$w(n+1) = w(n) + \mu(n)[d(n) - w^T(n)x(n)]x(n)$
ε-NLMS	$w(n+1) = w(n) + \dfrac{\mu}{\varepsilon + \|x(n)\|^2}[d(n) - w^T(n)x(n)]x(n)$
ε-NLMS with power normalization	$w(n+1) = w(n) + \dfrac{\mu}{\varepsilon + p(n+1)}[d(n) - w^T(n)x(n)]x(n)$
	$p(n+1) = ap(n) + (1-a)|x(n+1)|^2 \quad p(0) = 0$

TABLE 12.2

Normalized Real and Complex LMS Algorithms

Real-Valued Functions		Complex-Valued Functions
Input		
Initialization vector	$\mathbf{w}(n)=\mathbf{0}$	
Input vector	$x(n)$	
Desired output	$d(n)$	
Step-size parameter	μ	
Constant	ε	
Filter length	M	
Output		
Filter output	$y(n)$	
Coefficient vector	$w(n+1)$	
Procedure		
(1) $y(n)=w^T(n)x(n)=w(n)x^T(n)$		(1) $y(n)=w^H(n)x(n)$
(2) $e(n)=d(n)-y(n)$		(2) $e(n)=d(n)-w^H(n)x(n)$
(3) $w(n+1) = w(n) + \dfrac{\mu}{\varepsilon + x^T(n)x(n)}e(n)x(n)$		(3) $w(n+1) = w(n) + \dfrac{\mu}{\varepsilon + x^H(n)x(n)}e*(n)x(n)$
		H=Hermitian, conjugate transpose

In the preceding equation, ε is a small positive number that was introduced to avoid division by zero at a time when the signal becomes zero or extremely small. Tables 12.1 and 12.2 give the LMS and the NLMS formulas.

Example 12.1.1 (Noise Cancellation)

Compare the LMS and NLMS for a line enhancer setup in Figure 12.1.

Solution: The two Book m-functions are given as follows.

BOOK M-FUNCTION FOR UN-NORMALIZED LMS

```
function[w,y,x,J]=lms_ex12_1_1_lms(s,mu,M)
    %s=sinusoidal signal;mu=step-size factor;
    %M=length of adaptive filter;
N=length(s);y=zeros(1,N);w=zeros(1,M);
e=zeros(1,N);
x=[0  s+rand(1,N)-0.5];
    for n=M:N
        x1=x(n:-1:n-M+1);
        y(n)=w*x1';
        d(n+1)=x(n);
        e(n)=d(n+1)-y(n);
        w=w+2*mu*e(n)*x1;
    end;
J=e.^2;
```

BOOK M-FUNCTION FOR NLMS

```
function[w,y,x,J]=lms_ex12_1_1_norm_lms(s,mu,M,c)
    %s=sinusoidal signal;mu=step-size factor;
    %M=length of adaptive filter;c=very small number;
N=length(s);y=zeros(1,N);w=zeros(1,M);
e=zeros(1,N);
d=[0   s+rand(1,N)-0.5];
x=[d(1,2:N)  0];
for n=M:N
    x1=x(n:-1:n-M+1);
    y(n)=w*x1';
    e(n)=d(n)-y(n);
    w=w+((2*mu*e(n)*x1)/(c+x1*x1'));
end;
J=e.^2;
```

Figure 12.2 shows the output of the adaptive filter for the LMS and the NLMS filter. The constants used were $mu=0.01$, $M=16$, $c=0.1$, and $s=\sin(0.1\pi n)$. We observe that the NLMS filter output is closer to the sine signal than the LMS filter. The signal $v(n)$ identifies the white noise. ∎

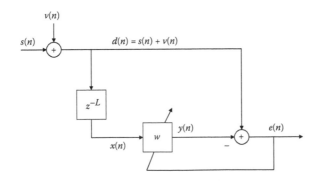

FIGURE 12.1

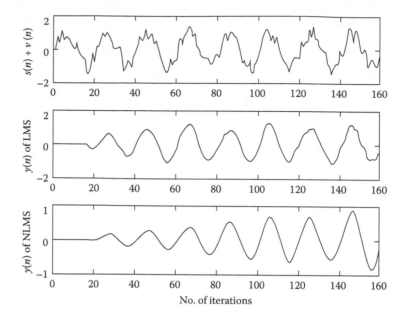

FIGURE 12.2

Example 12.1.2

If we desire to plot the learning function J, we recommend using an average process for a better presentation. The following Book m-Function cancels noise embedded in the signal.

BOOK M-FUNCTION FOR NOISE CANCELATION IN A SINE FUNCTION

```
function[w,jav,js]=lms_denoising_Jav1(N,mu,M,av,a)
es=zeros(1,N);
for m=1:av%av=integer equal to the number of desired
    %averaging;a=multiplier of the random noise,
    %a*randn(1,N);M=number of filter coefficients;
    %N=signal length and noise;mu=step-size factor;
    w=zeros(1,M);
    dn=sin(0.1*pi*[1:N])+a*randn(1,N);
    x=[0 0 dn(1,1:N-2)];%delay 2 units;
    for n=M:N
        x1=x(n:-1:n-M+1);
        y(n)=w*x1';
        e(n)=dn(n)-y(n);
        w=w+mu*e(n)*x1;
    end;
es=es+e;
y=zeros(1,N);
end;
jav=(es/av).^2;
for n=1:N-5
    js(n)=(jav(n)+jav(n+1)+jav(n+2)+jav(n+3))/4;
end;
```

Figure 12.3 was produced using the following constants: $mu=0.01$, $M=16$, $av=1$ and 100, $a=0.5$ and $N=1000$. The signal was a sine function with a white Gaussian additive noise. The learning curves were smoothed for a better understanding of the variations of the two learning curves. The averaging produces a lower mean square error as expected. ∎

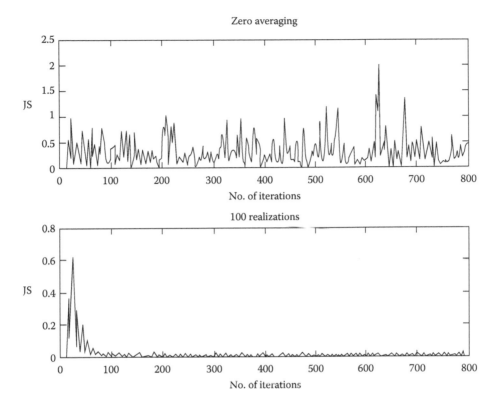

FIGURE 12.3

12.2 POWER NLMS

Observe that the NLMS algorithm can be written in the form (we have substituted $2\mu = \mu$ for convenience)

$$w(n+1) = w(n) + \frac{\mu / K}{\varepsilon / K + \|x(n)\|^2 / K} e(n)x(n) \tag{12.8}$$

The form of the preceding equation suggests a method for approximating the norm square of the data. The method replaces $\|x(n)\|^2 / K$ by a variable $p(i)$, which is updated as follows:

$$p(n+1) = bp(n) + (1-b)|x(n)|^2 \quad p(0) = 0 \tag{12.9}$$

with positive scalar b and values in the range $0 \le b < 1$. Hence, Equation 12.8 takes the form

$$w(n+1) = w(n) + \frac{\mu}{\varepsilon + p(n+1)} e(n)x(n)$$

$$= w(n) + \frac{\mu}{\varepsilon + p(n+1)} [d(n) - w(n)x^T(n)]x(n) \tag{12.10}$$

From the preceding equation, it is understood that the step size of power NLMS (PNLMS) in Equation 12.10 is approximately M times smaller than the step size of the NLMS.

Example 12.2.1 (Channel Equalization)

It is desired to estimate a communication channel, shown in Figure 12.4, so that the noisy output signal from the channel-space system is been de-noised by using a PNLMS.

Solution: At each instant time n, the measured output of the channel, $d(n)$, is compared with the output of the adaptive filter, $y(n)$, and an error signal $e(n)=d(n)-y(n)$, is generated. The error is then used to adjust the filter coefficients according to Equation 12.10. In this case the following constants and functions were used:

$$x(n) = 2\ 0.995^n \sin(0.1\pi n), \mu = 0.004, M = 30,$$

$$b = 0.5, c = 0.0001, \text{Channel} = \text{FIR}\begin{bmatrix} 0.9 & 0.5 & -0.2 \end{bmatrix}$$

The output of the adaptive filter will assume values close to the input to communication channel. Therefore, from an input/output point of view, the adaptive filter behaves as the channel. The top figure of Figure 12.5 shows the output of the channel with additive white noise having zero mean value, $v(n)$=rand−0.5. The middle figure presents the output of the adaptive filter and the third presents the MSE versus iteration. Following is the Book m-function, which was used to create the figure:

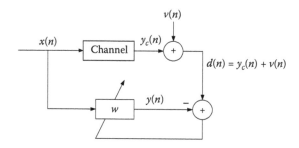

FIGURE 12.4

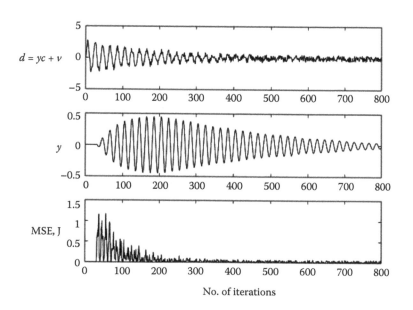

FIGURE 12.5

BOOK M-FUNCTION FOR PNLMS

```
function[w,y,yc,x,J]=lms_power_norm_lms(mu,M,b,c)
    %mu=step-size factor;
    %M=length of adaptive filter;c=very small number;
    %yc=conv(x,[filter coefficients])=channel
    %output;d=yc+v;v=noise;0<b<1;
n=1:1500;p(M)=0;
x=2*0.995.^n.*sin(0.1*pi*n);
N=length(x);y=zeros(1,N);w=zeros(1,M);
yc=conv(x,[0.9 0.5 -0.2]);%conv(x,y)=MATLAB function;
for n=M:N
    x1=x(n:-1:n-M+1);
    p(n+1)=b*p(n)+(1-b)*x1*x1';
    y(n)=w*x1';
    e(n)=(yc(n)+rand-0.5)-y(n);
    w=w+(2*mu*e(n)*x1)/(c+p(n+1));
end;
J=(e/max(e)).^2;
```

We also observe that if *b* in Equation 12.9 is zero, the preceding Book m-function becomes the NLMS m-function.   ■

Exercise 12.2.1

Repeat Example 12.2.1 by introducing different values of the constants and the Finite Impulse Response (FIR) length filter and observe the results.   ▲

Example 12.2.2 (Channel Equalization)

Figure 12.6 shows a baseband data transmission system equipped with an adaptive channel equalizer and a training system. The signal {$s(n)$} transmitted through the communication channel is amplitude or phase modulated pulses. The communication channel distorts the signal, the most important one is the pulse spreading, and results in overlapping of pulses and thus creating the **intersymbol interference** phenomenon. The noise $v(n)$ further deteriorates the fidelity of the signal. It is ideally required that the output of the equalizer is the signal $s(n)$. Therefore, an initialization period is used, during which the transmitter sends a sequence of training symbols that are known to the receiver (**training mode**). This approach is satisfactory if the channel does not change characteristics rapidly in time. However, for slow changes, the output from the channel can be treated as the desired signal for further adaptation of the equalizer so that its variations can be followed (**decision directed mode**).

If the equalization filter is the inverse of the channel filter, $W(z) = 1/H(z)$, the output will be that of the input to the channel assuming, of course, that noise is small. To avoid singularities from

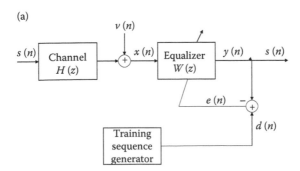

(a)

FIGURE 12.6

the zero of the channel transfer function inside the unit circle, we select an equalizer such that $W(z)H(z) \cong z^{-L}$. This indicates that the output of the filter $W(z)$ is that of the input to the channel shifted by L units of time. Sometimes, more general filters of the form $Y(z)/H(z)$ are used where $Y(z) \neq z^{-L}$. These systems are known as the **partial-response** signaling systems. In these cases, $Y(z)$ is selected such that the amplitude spectra are about equal over the range of frequencies of interest. The result of this choice is that $W(z)$ has a magnitude response of about one, thereby minimizing the noise enhancement.

Figure 12.7 shows a channel equalization problem at training stage. The channel noise $v(n)$ is assumed to be white Gaussian with variance σ_v^2. The equalizer is an M-tap FIR filter and the desired output is assumed to be delayed replica of the signal $s(n)$, $s(n-L)$. The signal $s(n)$ is white, has variance $\sigma_s^2 = 1$, has zero mean value, is made up of random $+1$ and -1, and is uncorrelated with $v(n)$. The channel transfer functions were assumed to be of FIR form. For the present example, we used the following transfer functions:

$$H(z) = H_1(z) = 1$$

$$H(z) = H_2(z) = 0.9 + 0.4z^{-1} - 0.2z^{-2}$$

(12.11)

The following Book m-function produced Figure 12.8.

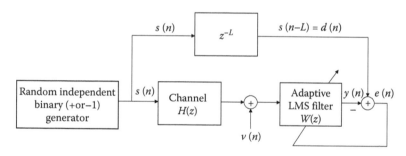

FIGURE 12.7

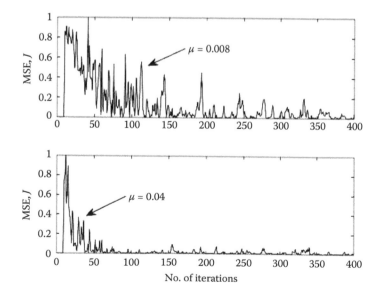

FIGURE 12.8

BOOK M-FUNCTION FOR CHANNEL EQUALIZATION

```
function[Jav,wav,dn,e,x]=lms_equalizer_ex12_2_2(av,M,L,h,N,mu,c)
    %function[Jav,wav,dn,e,x]=
    %lms_equalizer_ex12_2_2(av,M,L,h,N,mu,c)
    %this function solves the example depicted
    %in Fig 12.2.4;av=number of times to average e(error or
    %learning curve); w(filter coefficient);N=length
    %of signal s;L=shift of the signal s to become dn;
    %h=assumed
    %impulse response of the channel system;mu=step-size
    %factor;
    %M=number of adaptive filter coefficients;c=constant
    %multiplier;
wn=zeros(1,M);
J=zeros(1,N);
for i=1:av
    for n=1:N
        v(n)=c*randn;
        s(n)=(rand-0.5);
        if s(n)<=0
            s(n)=-1;
        else
            s(n)=1;
        end;
    end;
        dn=[zeros(1,L)  s(:,1:N-L)];
        ych=filter(h,1,s);
        x=ych(1,1:N)+v;
        [w,y,e,J,w1,Js]=lms1(x,dn,mu,M);
        wn=wn+w;
        J=J+e.^2;
    end;
    Jav=J/av;
    wav=wn/av;
```

The constants used to create Figure 12.8 were as follows: $av=300$, $M=10$, $L=1$, $h=\begin{bmatrix} 0.9 & 0.4 & -0.2 \end{bmatrix}$, $N=2000$, $c=0.05$. ∎

12.3 SELF-CORRECTING LMS FILTER

We can arrange the standard LMS filter in a series form as shown in Figure 12.9. This book proposed configuration LMS filtering permit us to process the signal using filters with fewer coefficients, thus saving in computation. The top of Figure 12.10 shows the desired signal with noise. The middle of Figure 12.10 shows the output of the first stage. The last level of Figure 12.10 shows the output of the third stage of the SCLMS filter, $\{x1(n)\}$. Each stage LMS filter has four coefficients. We also used the following signals and constants: $s(n)=\sin(0.1*[1:1000]*\pi), v(n)=[\text{rand}(1,1000)-0.5], x(n)=dn(n+2)$ (delay by 2 units), $I=3$, $mu=0.005$, $M=4$. Exercise 12.3.1 solves for the self-correcting NLMS filter (SCNLMS).

The output of the ith stage is related to the previous one as follows:

$$x_{i+1}(n) = x_i(n) * w_{i+1}(n)$$

Exercise 12.3.1

Find the output of the first and second stage of an SCNLMS filter that is shown in Figure 12.9. ▲

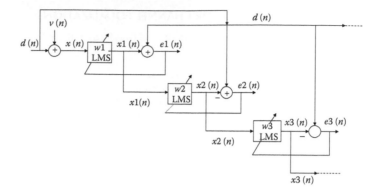

FIGURE 12.9

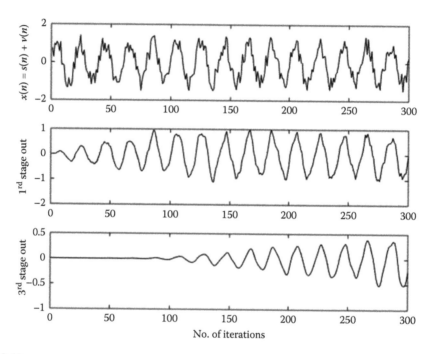

FIGURE 12.10

BOOK M-FUNCTION FOR SCLMS ALGORITHM

```
function[w,y,e,J]=lms_self_correcting_lms(x,dn,mu,M,I)
%function[w,y,e,J]=lms_self_correcting_lms(x,dn,mu,M,I);
[w(1,:),y(1,:),e(1,:)]=lms1(x,dn,mu,M);
for i=2:I%I=number of iterations, I<8-10 is sufficient;
    [w(i,:),y(i,:),e(i,:)]=lms1(y(i-1,:),dn,mu,M);
end;
J=e.^2;
```

12.4 THE SIGN-ERROR LMS ALGORITHM

The sign-error algorithm is defined by

$$w(n+1) = w(n) + \mu \, \text{sign}[e(n)]x(n) \tag{12.12}$$

where

$$\text{sign}(a) = \begin{cases} 1 & a > 0 \\ 0 & a = 0 \\ -1 & a < 0 \end{cases} \qquad (12.13)$$

is the **signum** function. By introducing the signum function and setting μ to a value of power of 2, the hardware implementation is highly simplified (shift add/subtract operation only).

BOOK M-FUNCTION FOR SIGN-ERROR ALGORITHM

```
function[w,y,e,J,w1]=lms_sign_error(x,dn,mu,M)
    %function[w,y,e,J,w1]=lms_sign_error(x,dn,mu,M);
    %all quantities are real-valued;
    %x=input data to the filter;dn=desired signal;
    %M=order of the filter;
    %mu=step size parameter;
    %x and dn must be of the same length;
N=length(x);
y=zeros(1,N);
w=zeros(1,M);%initialized filter coefficient vector;
for n=M:N
    x1=x(n:-1:n-M+1);%for each n the vector x1 is produced
    %of length M with elements from x in reverse order;
    y(n)=w*x1';
    e(n)=dn(n)-y(n);
    w=w+mu*sign(e(n))*x1;
    w1(n-M+1,:)=w(1,:);
end;
J=e.^2;
    %the columns of w1 depict the history of the filter
    %coefficients;
```

Exercise 12.4.1

Compare the results between the two m-functions $[w,y,e,J,w1] = \text{lms1}(x,dn,mu,M)$ and $[w,y,e,J,w1] = \text{lms\_sign\_error}(x,dn,mu,M)$ using identical inputs. Very some constants and observe the differences. ▲

12.5   THE NLMS SIGN-ERROR ALGORITHM

The NLMS sign-error algorithm is

$$w(n+1) = w(n) + \mu \frac{\text{sign}[e(n)]x(n)}{\varepsilon + \|x(n)\|^2} \qquad \|x(n)\|^2 = x(n)^T x(n) \qquad (12.14)$$

The Book m-function for the NLMS sign-error algorithm follows. The m-file `sign()` is a MATLAB function.

BOOK M-FUNCTION FOR NLMS SIGN-ERROR ALGORITHM

```
function[w,y,e,J,w1]=lms_normalized_sign_error(x,dn,mu,M,ep)
    %function[w,y,e,J,w1]=lms_normalized_sign_error(x,dn,mu,M,ep);
    %all quantities are real-valued;
```

```
    %x=input data to the filter;dn=desired signal;
    %M=order of the filter;
    %mu=step size parameter;x and dn must be of the same length;
    %ep=sm
N=length(x);
y=zeros(1,N);
w=zeros(1,M);%initialized filter coefficient vector
for n=M:N
  x1=x(n:-1:n-M+1);%for each n the vector x1 is produced
  %of length M with elements from x in reverse order;
  y(n)=w*x1';
  e(n)=dn(n)-y(n);
  w=w+2*mu*sign(e(n))*x1/(ep+x1*x1');
  w1(n-M+1,:)=w(1,:);
end;
J=e.^2;
    %the columns of w1 depict the history of the filter
    %coefficients;
```

Example 12.5.1

Compare the LMS algorithm and the normalized sign-error algorithms for de-noising a sinusoidal signal as shown in Figure 12.11.

Solution: The results are shown in Figure 12.12 using the following constants and functions: $\mu = 0.01$, $M = 20$, $\varepsilon = 0.0001$; $s = \sin(0.1\pi n)$, $x = \sin(0.1\pi n) + \text{rand}(n) - 0.5$, $x(n) = dn(n+1)$, $n = 1:500$.

We observe that for the same inputs, the NLMS sign-error is superior to the LMS one. You can plot the MSE, J, for both cases and verify your observation with another approach. ∎

Example 12.5.1

Repeat the preceding example and compare the sign-error and the NLMS algorithm. ▲

12.6 THE SIGN-REGRESSOR LMS ALGORITHM

The sign-regressor or data-sign algorithm is given by

$$w(n+1) = w(n) + \mu e(n)\text{sign}[x(n)] \qquad (12.15)$$

where the sign function is applied to $x(n)$ on an element-by-element basis.

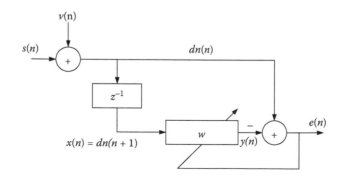

FIGURE 12.11

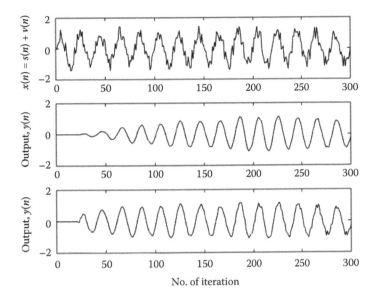

FIGURE 12.12

BOOK M-FUNCTION FOR SIGN-REGRESSOR ALGORITHM

```
function[w,y,e,J,w1]=lms_sign_regressor(x,dn,mu,M)
    %function[w,y,e,J,w1]=lms_sign_error(x,dn,mu,M);
    %all quantities are real-valued;
    %x=input data to the filter;dn=desired signal;
    %M=order of the filter;
    %mu=step size parameter;
    %x and dn must be of the same length;
N=length(x);
y=zeros(1,N);
w=zeros(1,M);%initialized filter coefficient vector;
for n=M:N
    x1=x(n:-1:n-M+1);%for each n the vector x1 is produced
    %of length M with elements from x in reverse order;
    y(n)=w*x1';
    e(n)=dn(n)-y(n);
    w=w+mu*e(n)*sign(x1);
    w1(n-M+1,:)=w(1,:);
end;
J=e.^2;
    %the columns of w1 depict the history of the filter
    %coefficients;
```

12.7   SELF-CORRECTING SIGN-REGRESSOR LMS ALGORITHM

BOOK M-FUNCTION FOR SELF-CORRECTING SIGN-REGRESSOR LMS ALGORITHM

```
function[w,y,e,J]=lms_self_correcting_sign_regressor_lms(x,dn,mu,M,I)
    %function[w,y,e,J]=
    %lms_self_correcting_sign_regressor_lms(x,dn,mu,M,I);
    %x=input data to the filter;dn=desired signal;length(x)=
    %length(dn);
    %y=output of the filter an Ixlength(x) matrix;J=error
```

```
    %function an
    %Ixlength(x) matrix;I=number of stages;
[w(1,:),y(1,:),e(1,:),J(1,:)]=lms_sign_regressor(x,dn,mu,M);
for i=2:I
    [w(i,:),y(i,:),e(i,:),J(i,:)]=lms1(y(i-1,:),dn,mu,M);
end;
J=e.^2;
```

Figure 12.13 was created using the following signals and constants: $s = \sin(0.1\pi n)$, $v = 1.5(\text{rand}(1,1000) - 0.5)$, $x = s + v$, delay 2 units, $mu = 0.005$, $M = 2$, $I = 4$. The reader should observe the power of this Book proposed technique. The adaptive filters have only two coefficients and repeating this process only three times gives excellent results.

12.8   THE NORMALIZED SIGN-REGRESSOR LMS ALGORITHM

The normalized sign-regressor is given by

$$w(n+1) = w(n) + \mu\frac{e(n)\text{sign}(x(n))}{\varepsilon + \|x(n)\|^2}, \quad \|x(n)\|^2 = x(n)^T x(n) \tag{12.16}$$

Book m-Function for Normalized Sign-Regressor LMS Algorithm

```
function[w,y,e,J,w1]=lms_normalized_sign_regressor(x,dn,mu,M)
    %function[w,y,e,J,w1]=lms_sign_error(x,dn,mu,M);
    %all quantities are real-valued;
    %x=input data to the filter;dn=desired signal;
    %M=order of the filter;
    %mu=step size parameter;
    %x and dn must be of the same length;
N=length(x);
y=zeros(1,N);
w=zeros(1,M);%initialized filter coefficient vector;
```

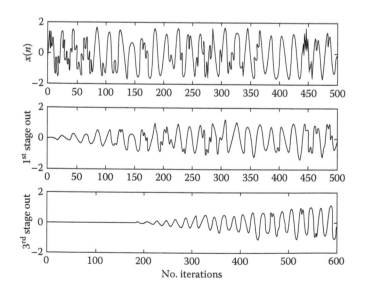

FIGURE 12.13

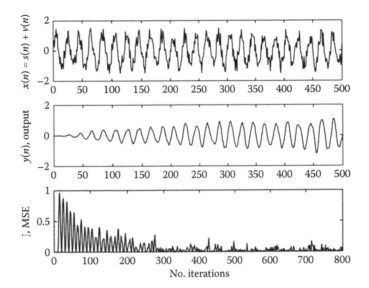

FIGURE 12.14

```
for n=M:N
     x1=x(n:-1:n-M+1);%for each n the vector x1 is produced
     %of length M with elements from x in reverse order;
     y(n)=w*x1';
     e(n)=dn(n)-y(n);
     w=w+mu*e(n)*sign(x1)./(0.0001+x1*x1');
     w1(n-M+1,:)=w(1,:);
end;
J=e.^2;
     %the columns of w1 depict the history of the filter
     %coefficients;
```

Figure 12.14 was created using the following functions and constants: $s(n) = \sin(0.1\pi n)$, $v(n) = 1.2(\text{rand} - 0.5)$, $\mu = 0.005$, $M = 10$.

12.9   THE SIGN–SIGN LMS ALGORITHM

The sign–sign algorithm is defined by

$$w(n+1) = w(n) + \mu \operatorname{sign}[e(n)]\operatorname{sign}[x(n)] \tag{12.17}$$

BOOK M-FUNCTION FOR SIGN–SIGN LMS ALGORITHM

```
function[w,y,e,J,w1]=lms_sign_sign(x,dn,mu,M)
     %function[w,y,e,J,w1]=lms_sign_sign(x,dn,mu,M)
     %all quantities are real-valued;
     %x=input data to the adaptive filter;
     %dn=desired signal;
     %M=order of the filter;
     %mu=step size parameter;x and dn must be of
     %the same length;
N=length(x);
y=zeros(1,N);
w=zeros(1,M);%initialized filter coefficient vector
```

```
for n=M:N
    x1=x(n:-1:n-M+1);%for each n the vector x1 is produced
    %of length M with elements from x in reverse order;
    y(n)=w*x1';
    e(n)=dn(n)-y(n);
    w=w+2*mu*sign(e(n))*sign(x1);
    w1(n-M+1,:)=w(1,:);
end;
J=e.^2;
    %the columns of w1 depict the history of the filter
    %coefficients;
```

Book m-Function for Self-Correcting Sign–Sign LMS Algorithm

```
function[w,y,e,J]=lms_self_correcting_sign_sign_lms(x,dn,mu,M,I)
    %function[w,y,e,J]=lms_self_correcting_sign_sign_lms
    %(x,dn,mu,M,I);
    %x=input data to the filter;y=output
    %data from the filter,
    %y is an Ixlength(x) matrix; J=learning curves,
    %an Ixlength(x)
    %matrix;mu=step-size parameter;M=umber of coefficients;
    %I=number of stages;w=an Ixlength(x) matrix of filter
    %coefficients; dn=desired signal;
[w(1,:),y(1,:),e(1,:),J(1,:)]=lms_sign_sign(x,dn,mu,M);
for i=2:I
    [w(i,:),y(i,:),e(i,:),J(i,:)]=lms1(y(i-1,:),dn,mu,M);
end;
J=e.^2;
```

Figure 12.15 shows the original signal, the output of a sign–sign LMS algorithm, and the bottom picture shows the output of the third stage of a self-correcting LMS algorithm.

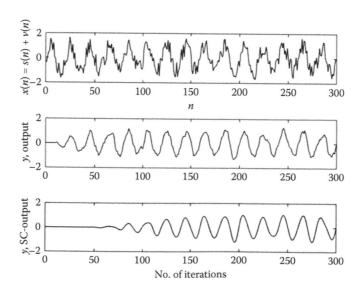

FIGURE 12.15

12.10   THE NORMALIZED SIGN–SIGN LMS ALGORITHM

The normalized sign–sign LMS algorithm is defined by

$$w(n+1) = w(n) + \mu \frac{\text{sign}[e(n)]\text{sign}[x(n)]}{\varepsilon + \|x(n)\|^2} \qquad \|x(n)\|^2 = x'(n)x(n) \qquad (12.18)$$

BOOK M-FUNCTION FOR NORMALIZED SIGN–SIGN LMS ALGORITHM

```
function[w,y,e,J,w1]=lms_normalized_sign_sign(x,dn,mu,M)
    %function[w,y,e,J,w1]=lms_normalized_sign_sign(x,dn,mu,M)
    %all quantities are real-valued;
    %x=input data to the adaptive filter;
    %dn=desired signal;
    %M=order of the filter;
    %mu=step size parameter;x and dn must be of
    %the same length;
N=length(x);
y=zeros(1,N);
w=zeros(1,M);%initialized filter coefficient vector
for n=M:N
    x1=x(n:-1:n-M+1);%for each n the vector x1 is produced
    %of length M with elements from x in reverse order;
    y(n)=w*x1';
    e(n)=dn(n)-y(n);
    w=w+2*mu*sign(e(n))*sign(x1)./(0.0001+x1*x1');
    w1(n-M+1,:)=w(1,:);
end;
J=e.^2;
    %the columns of w1 depict the history of the filter
    %coefficients;
```

To create Figure 12.16 we used the following functions and constants: $s(n) = \sin(0.1\pi n)$, $v(n) = 1.2(\text{rand} - 0.5)$, $x(n) = s(n) + v(n)$; $\mu = 0.005$, $M = 15$.

Note: The book proposed SCLMS filters can be used with any one of the LMS algorithms and their normalized forms.

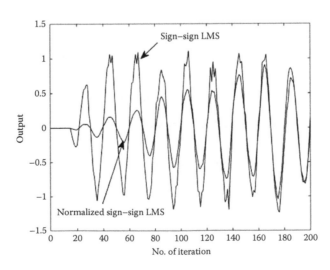

FIGURE 12.16

12.11 VARIABLE STEP-SIZE LMS

The variable step-size LMS (VSLMS) algorithm was introduced to facilitate the conflicting requirements. A large step-size parameter is needed for fast convergence, whereas a small step-size parameter is needed to reduce the miss-adjustment factor. When the adaptation begins and $w(n)$ is far from its optimum value, the step-size parameter should be large in order for the convergence to be rapid. However, as the filter value $w(n)$ approaches the steady-state solution, the step-size parameter should decrease in order to reduce the excess mean-square error (MSE).

To accomplish the variation of the step-size parameter, each filter coefficient is given a separate time-varying step-size parameter such that the LMS recursion algorithm takes the form

$$w_i(n+1) = w_i(n) + \mu_i(n)e(n)x(n-i) \quad i = 0,1,\ldots,M-1 \tag{12.19}$$

where

$w_i(n)$ is the ith coefficient of $w(n)$ at iteration n
$\mu_i(n)$ is its associated step-size

The step sizes are determined in an *ad hoc* manner, based on monitoring sign changes in the instantaneous gradient estimate $e(n)x(n-i)$. Successive changes in the sign of the gradient estimate indicates that the algorithm is close to its optimal solution and, hence, the step-size value must be decreased. The reverse is also true. The decision of decreasing the value of the step-size by some factor c_1 is based on some number m_1 successive changes of $e(n)x(n-i)$. Increasing the step-size parameter by some factor c_2 is based on m_2 successive sign changes. The parameters m_1 and m_2 can be adjusted to optimize performance, as can the factors c_1 and c_2.

The set of update may be written in the matrix form

$$w(n+1) = w(n) + \mu(n)e(n)x(n) \qquad \mu(n) = \begin{bmatrix} \mu_0(n) & & & \mathbf{0} \\ & \mu_1(n) & & \\ & & \vdots & \\ \mathbf{0} & & & \mu_{M-1}(n) \end{bmatrix} \tag{12.20}$$

The VSLMS algorithm is given in Table 12.3.

The following proposed Book m-Function can be used to take care, for example, of a noisy function whose added noise is varying. Figure 12.17 shows the results using the constant value of the step-size value and a varying one. The constants and functions used were

$$N = 1000; n = 1:1000; dn = \sin(0.2\pi n); v(n) = 1.5 \, 0.99^n \, \text{rand}n(1,1000);$$

$mu = 0.04$ for the constant step-size case

BOOK M-FUNCTION FOR VARYING VARIANCE OF THE SIGNAL THAT ADJUSTS APPROXIMATELY THE VALUE OF μ

```
function[w,y,e,J,w1]=lms_varying_mu(x,dn,M,N)
w=zeros(1,M);%N=length of x and dn signals;
for n=M:N
    x1=x(n:-1:n-M+1);
    if var(x1)>=0.6
        mu=0.0005;
```

TABLE 12.3
The VSLMS Algorithm

Input

Initial coefficient vector: $w(0)$

Input data vector: $x(n) = [x(n) \quad x(n-1) \quad \ldots \quad x(n-M+1)]^T$

Gradient term: $g_0(n-1) = e(n-1)x(n-1)$, $g_1(n-1) = e(n-1)x(n-1)$, ... ,

$g_{M-1}(n-1) = e(n-1)x(n-M)$

Step-size parameter: $\mu_0(n-1)$, $\mu_1(n-1)$, ... , $\mu_{M-1}(n-1)$

$a =$ small positive constant

$\mu_{max} =$ positive constant

Outputs

Desired output: $d(n)$

Filter output: $y(n)$

Filter update: $w(n+1)$

Gradient term: $g_0(n)$, $g_1(n)$, ... , $g_{M-1}(n)$

Update step-size parameter: $\mu_0(n)$, $\mu_1(n)$, ... , $\mu_{M-1}(n)$

Execution

(1) $y(n) = w^T(n)x(n)$

(2) $e(n) = d(n) - y(n)$

(3) Weights and step-size parameter adaptation:

For $i = 0,1,2, \ldots ,M-1$

$g_i(n) = e(n)x(n-i)$

$\mu_i(n) = \mu_i(n-1) +$

$\sigma \text{sign}(g_i(n))\text{sign}(g_i(n)) \% \sigma =$ small

positive step-size parameter;

if $\mu_i(n) > \mu_{max}$, $\mu_i(n) = \mu_{max}$

if $\mu_i(n) < \mu_{min}$, $\mu_i = \mu_{min} w_i(n+1) =$

$w_i(n) + \mu_i(n)g_i(n)$

end

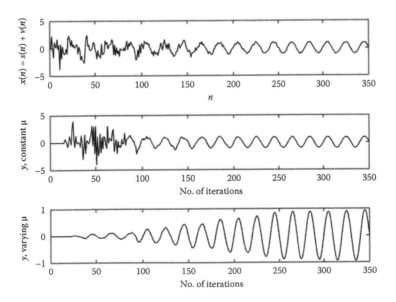

FIGURE 12.17

```
    elseif 0.6<var(x1)<0.2
        mu=0.005;
    elseif var(x1)<=0.2
        mu=0.05;
    end
    y(n)=w*x1';
    e(n)=dn(n)-y(n);
    w=w+mu*e(n)*x1;
    w1(n-M+1,:)=w(1,:);
end;
J=e.^2;
```

12.12 THE LEAKY LMS ALGORITHM

Let us assume that $\{x(n)\}$ and $\{d(n)\}$ are jointly wide-sense stationary processes that determine when the coefficients $w(n)$ converge in the mean to $w^o = R_x^{-1} p_{dx}$. That is,

$$\lim_{n\to\infty} E\{w(n)\} = w^o = R_x^{-1} p_{dx} \tag{12.21}$$

We start by taking the expectation of both sides of the LMS recursion as follows:

$$E\{w(n+1)\} = E\{w(n)\} + \mu E\{d(n)x(n)\} - \mu E\{x(n)x^T(n)w(n)\} \tag{12.22}$$

where

$$y(n) = x^T(n)w(n)$$

Assuming that $\{x(n)\}$ and $\{w(n)\}$ are statistically independent (**independence** assumption), Equation 12.22 becomes

$$E\{w(n+1)\} = E\{w(n)\} + \mu E\{d(n)x(n)\} - \mu E\{x(n)x^T(n)\}E\{w(n)\}$$
$$= (I - \mu R_x)E\{w(n)\} + \mu p_{dx}(n) \tag{12.23}$$

which is similar to steepest-descent method equations, see Section 10.3, where we have set $\mu' = \mu$ for simplicity. The difference in the preceding equation is the presence of the ensemble average symbol. This suggests that the steepest-descent method is applicable to ensemble average $E\{w(n+1)\}$. Re-writing Equation 10.32 in matrix form, we obtain

$$\xi'(n) = (I - \mu\Lambda)^n \xi'(0) \quad k = 0,1,\ldots,M-1 \quad n = 1,2,3,\ldots \tag{12.24}$$

We observe that $w(n) \to w^o$ if $\xi(n) = w(n) - w^o \to 0$ as $n \to \infty$ or when $\xi'(n) = Q^T \xi$ converges to zero. The kth row of Equation 12.24 is

$$\xi_k'(n) = (1 - \mu\lambda_k)^n \xi_k'(0) \tag{12.25}$$

which indicates that $\xi_k'(n) \to 0$ if

$$|1 - \mu\lambda_k| < 1 \text{ or } -1 < 1 - \mu\lambda_k < 1 \text{ or } 0 < \mu < \frac{2}{\lambda_k} \tag{12.26}$$

To have more restrictive condition, we can use the inequality

$$0 < \mu < \frac{1}{\lambda_{\max}} \tag{12.27}$$

If $\lambda_k = 0$, Equation 12.25 indicates that no convergence takes place as n approaches infinity. Since it is possible for these undamped modes to become unstable, it is important for the stabilization of the LMS algorithm to force these modes to zero. One way to avoid this difficulty is to introduce a leakage coefficient γ into the LMS algorithm as follows

$$w(n+1) = (1-\mu\gamma)w(n) + \mu e(n)x(n) \tag{12.28}$$

where

$$0 < \gamma \ll 1$$

The effect of introducing the **leakage coefficient** γ is to force any undamped modes to become zero and to force the filter coefficients to zero if either $e(n)$ or $x(n)$ is zero. (The homogeneous equation $w_i(n+1) = (1-2\mu\gamma)w_i(n)$ has the solution $w_i(n) = A(1-\mu\gamma)^n$, where A is a constant.)

We write Equation 12.28 in the form as follows: $[e(n) = d(n) - x^T(n)w(n)]$.

$$w(n+1) = (1-\mu\gamma)w(n) + \mu[d(n) - x^T(n)w(n)]x(n) = w(n) - \mu\gamma w(n)$$

$$+\mu d(n)x(n) - x^T(n)w(n)]x(n) \tag{12.29}$$

$$= [I - \mu(x(n)x^T(n) + \gamma I)]w(n) + \mu d(n)x(n)$$

We set $x^T(n)w(n)x(n) = x(n)x^T(n)w(n)$ because $x^T(n)w(n)$ is a number. By taking the expected value of both sides of the preceding equation and using the independence assumption, we obtain

$$E\{w(n+1)\} = [I - \mu E\{(x(n)x^T(n)\} + \gamma I)]E\{w(n)\} + \mu E\{d(n)x(n)\}$$
$$= [I - \mu(R_x + \gamma I)]E\{w(n)\} + \mu p_{dx}(n) \tag{12.30}$$

Comparing the preceding equation with Equation 12.23, we observe that the autocorrelation matrix R_x of the LMS algorithm has been replaced with $R_x + \gamma I$. Since the eigenvalues of $R_x + \gamma I$ are $\lambda_k + \gamma$ and since $\lambda_k \geq 0$, all the modes of the leaky LMS algorithm will be decayed to zero. Furthermore, the constraint for the step-size parameter becomes

$$0 < \mu < \frac{1}{\lambda_{max} + \gamma} \tag{12.31}$$

As $n \to \infty, w(n+1) \cong w(n)$ and, hence, Equation 12.30 becomes

$$\lim_{n \to \infty} E\{w(n)\} = [R_x + \gamma I]^{-1} p_{dx} \tag{12.32}$$

which indicates that the leakage coefficient produces a bias into steady-state solution $R_x^{-1} p_{dx}$. For another way to produce the leaky LMS algorithm, see Exercise 12.12.1.

Exercise 12.12.1

Develop the leaky LMS algorithm by minimizing the modified MSE: $J(n) = e(n)^2 + \gamma w^T(n)w(n)$. ▲

BOOK M-FUNCTION FOR LEAKY LMS ALGORITHM

```
function[w,y,e,J,w1]=lms_leaky_lms(x,dn,mu,gama,M)
    %function[w,y,e,J,w1]=lms_leaky_lms(x,dn,mu,gama,M);
    %all signals are real valued;x=input to filter;
    %y=output from the filter;dn=desired signal;
    %mu=step-size factor;gama=gamma factor<<1;
    %M=number of filter coefficients;w1=matrix whose M
    %rows give the history of each filter coefficient;
N=length(x);
```

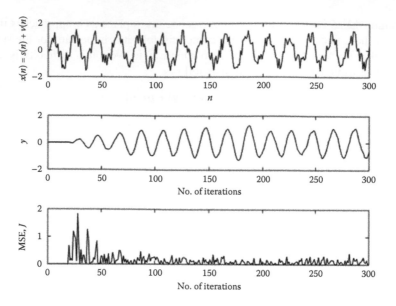

FIGURE 12.18

```
y=zeros(1,N);
w=zeros(1,M);
for n=M:N
    x1=x(n:-1:n-M+1);
    y(n)=w*x1';
    e(n)=dn(n)-y(n);
    w=(1-mu*gama)*w+mu*e(n)*x1;
    w1(n-M+1,:)=w(1,:);
end;
J=e.^2;
```

Figure 12.18 was produced with the following functions and constants: $\sin(0.1\pi n)$, $v(n) = 1.2[\text{rand}(1,1000) - 0.5]$, $dn(n) = s(n) + v(n)$, delay of dn was 2, $dn(n+2) = x(n)$, $mu = 0.005$, gama $= 0.01$ and $M = 20$.

Exercise 12.12.2

Compare the leaky LMS and the normalized leaky LMS.   ▲

12.13   THE LINEARLY CONSTRAINED LMS ALGORITHM

In all previous analysis of the Wiener filtering problem, the steepest-descent method, Newton's method, and the LMS algorithm, no constrain was imposed on the solution of minimizing the mean-square error. However, in some applications, there might be some mandatory constraints that must be taken into consideration in solving optimization problems. For example, the problem of minimizing the average output power of a filter while the frequency response must remain constant at specific frequencies. In this section, we discuss the filtering problem of minimizing the mean-square error subject to a general constraint.

The error between the desired signal and the output of the filter is

$$e(n) = d(n) - w^T(n)x(n) \tag{12.33}$$

We then minimize this error in the mean-square sense subject to the constant

$$c^T w = a \tag{12.34}$$

where

    a is a constant
    c is a constant vector

Using the Lagrange multiplier method (see Appendix 4), we write

$$J_c = E\left\{e^2(n)\right\} + \lambda\left(c^T w - a\right)$$
$$\lambda = \text{Lagrange multiplier} \tag{12.35}$$

Therefore, the following relations must be satisfied simultaneously: in

$$\nabla_w J_c = 0 \qquad \frac{\partial J_c}{\partial \lambda} = 0 \tag{12.36}$$

the second term produces the constraint (Equation 12.34). Next, we substitute the error $e(n)$ in Equation 12.35 to obtain

$$J_c = J_{\min} + \xi^T R_x \xi + \lambda\left(c^T \xi - a'\right) \tag{12.37}$$

where

$$\xi(n) = w(n) - w^o, \quad w^o = R_x^{-1} p_{dx}, \quad R_x = E\{x(n)x^T(n)\} \tag{12.38}$$

and

$$p_{dx} = E\{d(n)x^T(n)\} \quad a' = a - c^T w^o \tag{12.39}$$

Exercise 12.13.1

Verify Equation 12.37.   ▲

Exercise 12.13.2

We can derive the NLMS subject to a constraint: minimize $\min_w \|w(n+1) - w(n)\|^2 = \min_w \left\{ \left[\|w(n+1) - w(n)\|^T \|w(n+1) - w(n)\| \right] \right\}$ subject to the constraint $d(n) = w^T(n+1)x(n)$.   ▲

Exercise 12.13.3

Study the application of the LMF algorithm by making changes to different constants and input signals with varying noise strength.   ▲

The solution has now changed to $\nabla_\xi J_c = 0$ and $\partial J_c / \partial \lambda = 0$. Hence, from Equation 12.37, we obtain (see also Appendix 2)

$$\nabla_\xi J_c = \begin{bmatrix} \dfrac{\partial J_c}{\partial \xi_1} \\ \vdots \\ \dfrac{\partial J_c}{\partial \xi_M} \end{bmatrix} = \begin{bmatrix} 2\xi_1 r_1 + 2\xi_2 r_2 + & \cdots & +2\xi_M r_M \\ \vdots & & \vdots \\ 2\xi_1 r_M + 2\xi_2 r_{M-1} + & \cdots & +2\xi_M r_1 \end{bmatrix} + \lambda \begin{bmatrix} c_1 \\ \vdots \\ c_M \end{bmatrix} = \mathbf{0} \qquad (12.40)$$

or in matrix form

$$2 R_x \xi_c^o + \lambda c = \mathbf{0}, \; \xi_c^o \text{ is the constraint optimum of the vector } \xi \qquad (12.41)$$

In addition, the constraint gives the relation

$$\frac{\partial J_c}{\partial \lambda} = c^T \xi_c^o - a' = 0 \qquad (12.42)$$

Solving the of the last two equations for λ and ξ_c^o, we obtain

$$\lambda = -\frac{2a'}{c^T R_x^{-1} c} \quad \xi_c^o = \frac{a' R_x^{-1} c}{c^T R_x^{-1} c} \qquad (12.43)$$

Substituting the value of λ in Equation 12.37, we obtain the minimum value of J_c as

$$J_c = J_{\min} + \frac{a'^2}{c^T R_x^{-1} c} \qquad (12.44)$$

But $w(n) = \xi(n) + w^o$ and, hence, using Equation 12.43, we obtain the equation

$$w_c^o = w^o + \frac{a' R_x^{-1} c}{c^T R_x^{-1} c} \qquad (12.45)$$

Note: *The second term of Equation 12.44 is the excess MSE produced by the constraint.*

To obtain the recursion relation subject to constraint $c^T w = a$, we must proceed in two steps as follows:

$$\text{Step 1} : w'(n) = w(n) + \mu e(n) x(n) \qquad (12.46)$$

$$\text{Step 2} : w(n+1) = w'(n) + \eta(n) \qquad (12.47)$$

where $\eta(n)$ is chosen so that $c^T w(n+1) = a$ while $\eta^T(n)\eta(n)$ is minimized. In other words, we choose the vector $\eta(n)$ so that the constraint $c^T w = a$ after Step 2, while the perturbation introduced by $\eta(n)$ is minimized. The problem can be solved using the Lagrange multiplier method, which gives

$$\eta(n) = \frac{a - c^T w'(n)}{c^T c} c \qquad (12.48)$$

Thus we obtain the final form of Equation 12.47 as

$$w(n+1) = w'(n) + \frac{a - c^T w'(n)}{c^T c} c \qquad (12.49)$$

The constraint algorithm is given in Table 12.4. Figure 12.19 shows the results of a constrained LMS filter with the following data: $s(n) = \sin(0.1\pi n)$; $v(n) = \text{noise} = 1.2(\text{rand} - 0.5)$; $dn(n) = s(n) + v(n)$; x delayed by 1 from dn; $c = 0.6*\text{ones}(1,35)$; $a = 0.8$; $\mu = 0.002$; $M = 35$. For c, you can also use the vector $c = 0.98.^{\wedge}[1:32]$ with the same results.

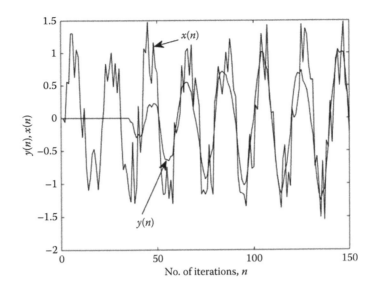

$y(n), x(n)$

No. of iterations, n

FIGURE 12.19

TABLE 12.4

Linearly Constrained LMS Algorithm

Input	Initial coefficient vector, $w(0)=\mathbf{0}$
	Input data, $x(n)$
	Desired output, $d(n)$
	Constant vector, c
	Constraint constant, a
Output	Filter output, $y(n)$
Procedure	$y(n)=w^{T}(n)x(n)$
	$e(n)=d(n)-y(n)$
	$w'(n)=w(n)+\mu e(n)x(n)$
	$w(n+1) = w'(n) + \dfrac{a - c^{T} w'(n)}{c^{T} c} c$

Note: *Solving the constrained optimization problem, using the Lagrange multiplier method, the NLMS can be obtained (see Exercise 12.13.2).*

BOOK CONSTRAINT M-FUNCTION: $[W,E,Y,J,W2]$ = CONSTRAINED_LMS(X,DN,C,A,MU,M)

```
function[w,e,y,J,w2]=constrained_lms(x,dn,c,a,mu,M)
    %x=data vector; dn=desired vector of equal length
    %with x;c=constant row vector of length M;a=constant
    %e.g. a=0.8;mu=step-size parameter;M=number of filter
    %coefficients;w2=matrix whose columns give the history
    %of each coefficient;
    w=zeros(1,M);
    N=length(x);
    for n=M:N
        y(n)=w*x(n:-1:n-M+1)';
        e(n)=dn(n)-y(n);
        w1=w+2*mu*e(n)*x(n:-1:n-M+1);
```

```
        w=w1+((a-c*w1')*c/(c*c'));
        w2(n-M+1,:)=w(1,:);
end;
J=e.^2;
```

12.14   THE LEAST MEAN FOURTH ALGORITHM

The optimal weight vector w^o that solves $\min_{w} E\left|d(n) - x^T(n)w(n)\right|^4$ (see Sayed 2008) can be approximated iteratively by the recursion

$$w(n+1) = w(n) + \mu x(n)e(n)e(n)^2$$

$$w(n+1) = w(n) + \mu x*(n)e(n)\left|e(n)\right|^2 \text{ (complex case)}$$

(12.50)

For the application of the least mean forth (LMF) algorithm, see Exercise 12.13.1 (Figure 12.20).

12.15   THE LEAST MEAN MIXED NORMAL (LMMN) LMS ALGORITHM

The optimal weight vector w^o that solves $\min_{w} E\left[\delta|e|^2 + \frac{1}{2}(1-\delta)|e|^4\right], e = d - x^T w$ for some constant $0 \leq \delta \leq 1$ can be approximated iteratively using the recursion (see Sayed 2008)

$$w(n+1) = w(n) + \mu x(n)e(n)(\delta + (1-\delta)|e(n)|^2)$$

(12.51)

For the application of the present algorithm and its comparison with the normalized one, see Exercise 12.15.1.

Exercise 12.15.1

Study the application of the LMMN LMS algorithm by varying constants and input signals and with varying noise strengths. Compare the results between the standard and the normalized algorithm.

▲

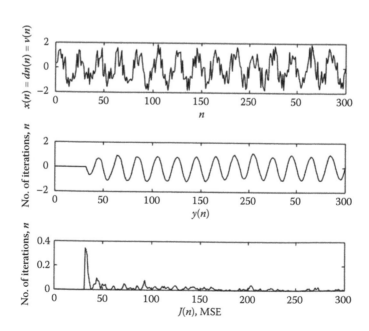

FIGURE 12.20

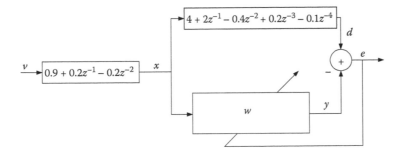

FIGURE 12.21

12.16 SHORT-LENGTH SIGNAL OF THE LMS ALGORITHM

Below is proposed a Book program that alleviates the problem of using a short-length signal present. To test the proposed approach, we proceed to identify an unknown system as shown in Figure 12.21.

BOOK M-FILE FOR SHORT-TIME SIGNALS

```
%short_length_signal_lms
M=8;mu=0.08;
v=0.5*[rand(1,200)-0.5];
x=conv(v,[0.9 0.2 -0.2]);
d=conv(x,[4 2 -0.4 0.2 -0.1]);
x=x(1,1:200);
d=d(1,1:200);
ve=[v v v v v v v v v v v v];%extended signal;
xe=conv(ve,[0.9 0.2 -0.2]);
dn=conv(xe,[4 2 -0.4 0.2 -0.1]);
xe=xe(1,1:2400);
de=dn(1,1:2400);
[w,y,e,j,w1]=lms_short_signal(x,d,mu,M);
[we,ye,ee,je,w1e]=lms_short_signal(xe,de,mu,M);
```

BOOK M-FUNCTION FOR SHORT-TIME SIGNALS

```
function[w,y,e,J,w1]=lms_short_signal(x,dn,mu,M)
N=length(x);w=0.02*(rand(1,M)-0.5);
for n=M:N
    x1=x(n:-1:n-M+1);
    y(n)=w*x1';
    e(n)=dn(n)-y(n);
    w=w+mu*e(n)*x1;
    w1(n-M+1,:)=w(1,:);
end;
J=e.^2;
```

The results of the short signal with 200 time units are as follows:

| 1.0289 | 0.5579 | −0.2093 | −0.0763 | −0.0882 | −0.130 | −0.0238 | −0.1382 |

The results of the extended signal with 2400 time units are as follows:

| 3.8568 | 1.9526 | −0.4496 | 0.1432 | −0.1393 | −0.0355 | −0.0116 | −0.0828 |

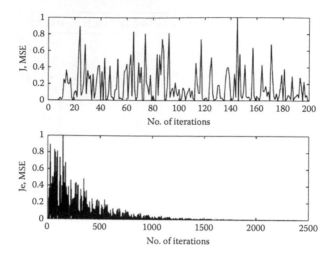

FIGURE 12.22

Figure 12.22 shows the learning curve for the two cases. It is interesting to note that this approach is successful for stationary random input signals.

12.17   THE TRANSFORM DOMAIN LMS ALGORITHM

The implementation of the LMS filter in the frequency domain can be accomplished simply by taking the Discrete Fourier Transform (DFT) of both the input data, $\{x(n)\}$, and the desired signal, $\{d(n)\}$. The advantage of doing this is due to the fast processing of the signal using the Fast Fourier Transform (FFT) algorithm. However, this procedure requires a block-processing strategy, which results in storing a number of incoming data in buffers, and thus some delay is unavoidable.

The block-diagram approach for the transform domain LMS (TDLMS) algorithm is shown in Figure 12.23. The signals are processed by block-by-block format; that is, $\{x(n)\}$ and $\{d(n)\}$ are sequenced into blocks of length M so that

$$x_i(n) = x(iM+n), d_i(n) = d(iM+n) \quad n = 0,1,\ldots,M-1, \quad i = 0,1,2,\ldots \tag{12.52}$$

The values of the ith block of the signals $\{x_i(n)\}$ and $\{d_i(n)\}$ are Fourier transformed using the DFT to find $X_i(k)$ and $D_i(k)$, respectively. Due to DFT properties, the sequences $X_i(k)$ and $D_i(k)$ have M complex elements corresponding to frequency indices ('bins') $k=0, 1, 2,\ldots, M-1$. Hence,

$$X_i(k) = \text{DFT}\{x_i(n)\} = \sum_{n=0}^{M-1} x_i(n) e^{-j\frac{2\pi nk}{M}} \qquad k = 0,1,\ldots,M-1 \tag{12.53}$$

$$D_i(k) = \text{DFT}\{d_i(n)\} = \sum_{n=0}^{M-1} d_i(n) e^{-j\frac{2\pi nk}{M}} \qquad k = 0,1,\ldots,M-1 \tag{12.54}$$

During the ith block processing, the output of each frequency bin of the adaptive filter is computed as follows:

$$Y_i(k) = W_{i,k} X_i(k) \qquad k = 0,1,2,\ldots,M-1 \tag{12.55}$$

where
$W_{i,k}$ is the kth frequency bin corresponding to the ith update (corresponding to the ith block data).

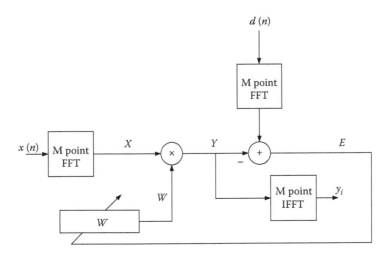

FIGURE 12.23

The error in the frequency domain is

$$E_i(k) = D_i - Y_i(k) \quad k = 0,1,2,\dots,M-1 \tag{12.56}$$

The system output is given by

$$y_i(n) = y(iM+n) = \text{IDFT}\{Y_i(k)\} = \frac{1}{M}\sum_{k=0}^{M-1} Y_i(k)e^{j\frac{2\pi nk}{M}} \qquad n = 0,1,\dots,M-1 \tag{12.57}$$

To update the filter coefficients we use, by analogy of the LMS recursion, the following recursion is used:

$$\boldsymbol{W}_{i+1} = \boldsymbol{W}_i + \mu \boldsymbol{E}_i \bullet \boldsymbol{X}_i^*$$

$$\boldsymbol{W}_{i+1} = \begin{bmatrix} W_{i+1,0} & W_{i+1,1} & \cdots & W_{i+1,M-1} \end{bmatrix}^T$$

$$\boldsymbol{W}_i = \begin{bmatrix} W_{i,0} & W_{i,1} & \cdots & W_{i,M-1} \end{bmatrix}^T \tag{12.58}$$

$$\boldsymbol{E}_i = \begin{bmatrix} E_i(0) & E_i(1) & \cdots & E_i(M-1) \end{bmatrix}^T$$

$$\boldsymbol{X}_i^* = \begin{bmatrix} X_i^*(0) & X_i^*(1) & \cdots & X_i^*(M-1) \end{bmatrix}^T$$

The dot (•) in Equation 12.58 implies element-by-element multiplication and the asterisk (*) stands for complex conjugate. If we set $\boldsymbol{X}_i^*$ in the form

$$\boldsymbol{X}_i = \text{diag}\{X_i(0) \quad X_i(1) \quad \cdots \quad X_i(M-1)\}$$

$$= \begin{bmatrix} X_i(0) & 0 & \cdots & 0 \\ 0 & X_i(1) & \cdots & 0 \\ \vdots & \vdots & \cdots & 0 \\ 0 & 0 & \cdots & X_i(M-1) \end{bmatrix} \tag{12.59}$$

Then Equation 12.58 becomes

$$W_{i+1} = W_i + \mu X_i^* E_i \tag{12.60}$$

Therefore, Equations 12.52 through 12.58 constitute the frequency domain of the LMS algorithm. The Book m-function gives the adaptive filter coefficient after I blocks (or iteration) is given as follows.

Book m-Function for the Frequency Domain of the LMS

```
function[A]=lms_FT_lms(x,d,M,I,mu)
%function[A]=lms_FT_lms(x,d,M,I,mu);
wk=zeros(1,M);
for i=0:I          %I=number of iterations (or blocks);
   if I*M>length(x)-1
      ('error:I*M<length(x)-1')
   end;
                   %M=number of filter coefficients;
      x1=x(M*(i+1):-1:i*M+1);
      d1=d(M*(i+1):-1:i*M+1);
   xk=fft(x1);
   dk=fft(d1);
   yk=wk.*xk;
   ek=dk-yk;
   wk=wk+mu*ek.*conj(xk);
   A(i+1,:)=wk;
end;
    %all the rows of A are the wk's at an increase order
    %of iterations(blocks);
    %to filter the data, wk must be inverted in the time
    %domain, convolve with the data x and then plot the
    %real part of the output y, e.g. wn4=the forth iteration
    %=ifft(A(4,:)),yn4=filter(wn4/4,1,x) for even M;
```

Example 12.17.1

The following Book m-file (or script) produces the output of the Figure 12.24. To produce the figure in the command window, we just write ex12_17_1.

BOOK M-FILE (SCRIPT FILE): EX12_17_1

```
%ex12_17_1
M=40;I=12;mu=0.01;
n=0:999;
d=sin(0.1*pi*n);v=1.5*rand(1,1000);
x=d+v;
[A]=lms_FT_lms(x,d,M,I,mu);
wn10=ifft(A(10,:))/M;%inverse FT of row 10 of
    %matrix A;
yn10=filter(wn10/I,1,x)/M;
subplot(2,1,1);plot(x(1,1:200),'k');
xlabel('n');ylabel('x(n)');
subplot(2,1,2);plot(real(yn10(1,1:200)),'k');
xlabel('n');ylabel('y(n)');
```

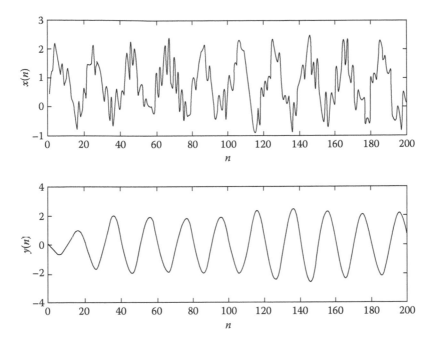

FIGURE 12.24

*12.17.1 CONVERGENCE

Let the signals $\{x(n)\}$ and $\{y(n)\}$ by jointly WSS processes and the initial filter coefficient be zero, $W_o = 0$. Based on Equations 12.60 and 12.56 (see Exercise 12.17.1.1), we obtain

$$W_{i+1,k} = \left(1 - \mu|X_i|^2\right)W_{i,k} + \mu D_i(k)X_i^*(k) \tag{12.61}$$

Exercise 12.17.1.1

Verify Equation 12.61. ▲

The expected value of Equation 12.61, assuming W_{ik} and $X_i(k)$ are statistically independent, is given by

$$E\{W_{i+1,k}\} = \left(1 - \mu E\left\{|X_i(k)|^2\right\}\right)E\{W_{i,k}\} + \mu E\{D_i(k)X_i^*(k)\} \tag{12.62}$$

Because $x(n)$ and $x(n)$ are stationary, their statistical characteristics do not change from block to block and, therefore, the ensembles $E\{|X_i(k)|^2\}$ and $E\{D_i(k)X_i^*(k)\}$ are independent of i but depend on k. Taking the z-transform of Equation 12.62 with respect to i of the dependent variable $W_{i,k}$, we find the relation (see the z-transform table in Chapter 3 and Exercise 12.17.1.2)

$$W_k(z) = -\mu E\left\{|X_i(k)|^2\right\}\frac{W_k(z)}{z-1} + \frac{\mu E\{D_i(k)X_i^*(k)\}}{z-1} \tag{12.63}$$

Exercise 12.17.1.2

Verify Equation 12.62. ▲

---

* Means section may be skipped.

Applying the final value theorem (see Table 3.1 and Exercise 12.17.1.2), we obtain the steady-state value for the filter coefficients:

$$E\{W_k^\infty\} = \frac{E\{D_i(k)X_i^*\}}{E\{|X_i(k)|^2\}} \tag{12.64}$$

Let the mean filter coefficient error $E_i(k)$ be defined by

$$E_i(k) = E\{W_{i,k}\} - E\{W_k^\infty\} \tag{12.65}$$

Then, using Equations 12.64 and 12.62, we find (see Exercise 12.17.1.3)

$$E_{i+1} = \left(1 - \mu E\{|X_i|^2\}\right)E_i(k) \quad k = 0,2,3,\ldots,M-1 \tag{12.66}$$

Exercise 12.17.1.3

Verify Equation 12.66.   ▲

12.18   THE ERROR NORMALIZED STEP-SIZE LMS ALGORITHM

The variable error normalized step size (ENSS) depends on the optimization problem as follows:
Define a **posteriori** error $e_p(n)$ as

$$e_p(n) = d(n) - \mathbf{w}^T(n+1)\mathbf{x}(n) \tag{12.67}$$

Then, we can define the following constraint optimization problem (Sayed 2008):

$$\min_{\mathbf{w}} \|\mathbf{w}(n) - \mathbf{w}(n-1)\|^2 \text{ subject } e_p(n) = \left(1 - \mu \frac{\|\mathbf{x}(n)\|^2}{1 + \mu\|e_L(n)\|^2}\right)e(n) \tag{12.68}$$

where

$$\|\mathbf{e}(n)\|^2 = \sum_{i=0}^{L-1}|e(n-i)|^2 \tag{12.69}$$

By defining the difference as

$$\Delta\mathbf{w}(n) = \mathbf{w}(n+1) - \mathbf{w}(n) \tag{12.70}$$

and multiplying both sides of Equation 12.70 by $\mathbf{x}^T(n)$ from left, we obtain

$$\mathbf{x}^T(n)\Delta\mathbf{w}(n) = \mathbf{x}^T(n)\mathbf{w}(n+1) - \mathbf{x}^T(n)\mathbf{w}(n) \tag{12.71}$$

By adding and subtracting $d(n)$ from the right side of Equation 12.71, we have

$$\mathbf{x}^T(n)\Delta\mathbf{w}(n) = e(n) - e_p(n) \tag{12.72}$$

Substituting Equation 12.68 in 12.72, we obtain

$$\mathbf{x}^T(n)\Delta\mathbf{w}(n) = \mu e(n)\frac{\|\mathbf{x}(n)\|^2}{1 + \mu\|e_L(n)\|^2}\frac{dy}{dx} \tag{12.73}$$

There are infinite solutions to Equation 12.73. The solution that results in a minimum value of $\Delta w(n)$ in the Euclidian second norm sense is that one taken directly from Equation 12.73, (Sayed, 2008); that is

$$\Delta w(n) = \mu e(n) \frac{x(n)}{1 + \mu \|e_L(n)\|^2}$$

$$\left(x^T(n)x(n) = \|x(n)\|^T \right)$$

(12.74)

Equation 12.74 can be proved by assuming $(\Delta w(n) + z)$ is any other solution, where z is a column vector. The Euclidian vector norm of this solution can be written as

$$\|\Delta w(n) + z\|^2 = (\Delta w(n) + z)^T (\Delta w(n) + z)$$

$$= \|\Delta w(n)\|^2 + \|z\|^2 + \Delta w^T(n)z + z^T \Delta w(n)$$

(12.75)

Substituting Equation 12.74 in 12.75 yields

$$\|\Delta w(n) + z\|^2 = \|\Delta w(n)\|^2 + \|z\|^2 + \mu e(n) \frac{x^T(n)z}{1 + \mu \|e_L(n)\|^2} + \mu e(n) \frac{z^T x(n)}{1 + \mu \|e_L(n)\|^2}$$

(12.76)

Since $\Delta w(n)$ is a solution, we conclude from Equation 12.74 that $x^T(n)z = 0 = z^T x(n)$ and hence Equation 12.76 becomes

$$\|\Delta w(n) + z\|^2 = \|\Delta w(n)\|^2 + \|z\|^2$$

(12.77)

The right side of Equation 12.77 is larger than $\|\Delta w(n)\|^2$ for any nontrivial vector z. Thus the minimum value of $\|\Delta w(n) + z\|^2$ is $\|\Delta w(n)\|^2$ (that is, when $z = 0$), which proves Equation 12.74. Now, substituting Equation 12.74 in 12.70, we obtain the error normalized step-size (ENSS) LMS algorithm:

$$w(n+1) = w(n) + \frac{\mu x(n)}{1 + \mu \|e_L(n)\|^2} x(n)e(n)$$

(12.78)

The fraction quantity in Equation 12.78 represents the variable step size of the algorithm; that is

$$\mu(n) = \frac{\mu}{1 + \mu \|e_L(n)\|^2} \quad \text{or} \quad \mu(n) = \frac{1}{1 + \frac{1}{\mu} \|e_L(n)\|^2}$$

(12.79)

The preceding equation shows that $\mu(n)$ is an increasing function of the step-size μ. In a stationary environment, the best choice of L is $L = n$, which in turn makes $\mu(n)$ a monotonic decreasing function of μ. Clearly, large values of μ should be used in this case to increase $\mu(n)$ and thus obtaining fast rate of convergence mainly at the early stages of adaptation, where the error value is large. In stationary environments, the length of the error vector should be constant (L = constant value) such that the algorithm will adapt to statistical input data. In this case, as n increases, $\|e_L(n)\|^2$ decreases and $\mu(n)$ increases to a maximum value μ, which is the step-size of the LMS algorithm.

The proposed Book algorithm performs better than many other algorithms that have been examined in stationary environments for both small and large filter length.

Example 12.18.1

Compare the error normalized step-size LMS algorithm with the LMS and NLMS algorithms. To check the algorithms, we used the identification of an unknown system as shown in Figure 12.25. The results are shown in Figure 12.26.

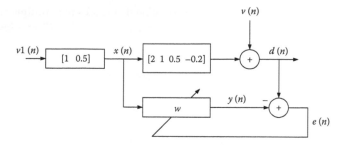

FIGURE 12.25

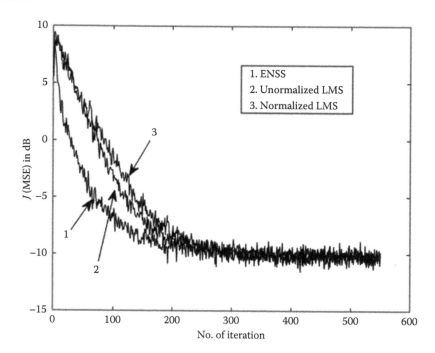

FIGURE 12.26

Solution: The following Book m-functions and Book program were used.

BOOK M-FUNCTIONS FOR ENSS APPLICATION

```
function[J,w]=lms_error_normalized_SS(h,avn,N,mu1,M)
J=zeros(1,N);
for i=1:avn
    x=filter([1 0.5],1,randn(1,N));y=zeros(1,N);
    w=zeros(1,M);e=zeros(1,N);X=zeros(1,M);
    d=filter(h,1,x);v=0.3*randn(1,N);
    D=0;
        %x=input to the adaptive filter and known filter h;
        %y=output from the adaptive filter;
        %avn=number of averaging;d=output from
        %the unknown filter plus system noise;
        %M=number of adaptive filter coefficient;
    for k=1:N
        X=[x(k) X(1:M-1)];
        den=X*X'+0.0001;
```

```
        y=w*X';
        e(k)=d(k)+v(k)-y;
        D=D+e(k)^2;
        den1=D*mul+1;
        w=w+(mul/den1)*e(k)*X;
        J(k)=J(k)+abs(e(k))^2;
    end
end
J=J/avn;


function[J,w]=lms_average_normalized_lms(h,avn,N,mul,M)
J=zeros(1,N);
for i=1:avn
    x=filter([1 0.5],1,randn(1,N));y=zeros(1,N);
    w=zeros(1,M);e=zeros(1,N);X=zeros(1,M);
    d=filter(h,1,x);v=0.3*randn(1,N);
        %J=learning curve (MSE);h=known filter;
        %w=adaptive filter coefficients;v=internal
        %system noise;avn=number of times been
        %averaged;M=number of adaptive filter
        %coefficients;N=number of data;
    for k=1:N
        X=[x(k) X(1:M-1)];
        den=X*X'+0.0001;
        y=w*X';
        e(k)=d(k)+v(k)-y;
        w=w+(mul/den)*e(k)*X;
        J(k)=J(k)+abs(e(k))^2;
    end
end
J=J/avn;


function[J,w]=lms_average_unnormalized_lms(h,avn,N,mul,M)
J=zeros(1,N);
    %for explanation see the above two functions;
for i=1:avn
    x=filter([1 0.5],1,randn(1,N));y=zeros(1,N);
    w=zeros(1,M);e=zeros(1,N);X=zeros(1,M);
    d=filter(h,1,x);v=0.3*randn(1,N);
    for k=1:N
        X=[x(k) X(1:M-1)];
        y=w*X';
        e(k)=d(k)+v(k)-y;
        w=w+mul*e(k)*X;
        J(k)=J(k)+abs(e(k))^2;
    end
end
J=J/avn;
```

BOOK MATLAB PROGRAM

```
>>[J,w]=lms_error_normalized_SS([2  1  0.5  -0.2], 200, 1000, …
0.8, 6);
>>[J1,w1]=lms_average_normalized_lms([2  1  0.5  -0.2], 200, …
1000, 0.06, 6);
>>[J2,w2]=lms_average_unnormalized_lms([2  1  0.5  -0.2], 200, …
1000, 0.01,6);
>>plot(10*log10(J(1,1:550)),'k'); hold on; >>plot(10*log10(J1(1,1:550)),'k');
>>hold on; plot(10*log10(J2(1,1:550)),'k'); xlabel('No. of >>iterations');
>>ylabel('J (MSE) in dB')
```

12.19 THE ROBUST VARIABLE STEP-SIZE LMS ALGORITHM

Based on regularization Newton's recursion (Sayed 2008), we write

$$w(n+1) = w(n) + \mu(n)[\varepsilon(n)I + R_x]^{-1}[p - R_x w(n)] \tag{12.80}$$

where n is the iteration number, w is an $M \times 1$ vector of adaptive filter weights, $\varepsilon(n)$ is an iteration-dependent regularization parameter, $\mu(n)$ is an iteration-dependent step-size, and I is an $M \times M$ identity matrix, $p(n) = E\{d(n)x(n)\}$ is the cross-correlation vector between the desired signal $d(n)$ and the input signal $\{x(n)\}$, and $R_x(n)=E\{x(n)x^T(n)\}$ is the autocorrelation matrix of $x(n)$.

Writing Equation 12.80 in the LMS form by replacing p and R_x by their instantaneous approximation $[d(n)x(n)]$ and $[x(n)x^T(n)]$, respectively, with the appropriate proposed weights, we obtain

$$w(n+1) = w(n) + \mu\|e_L(n)\|^2 \times \left[\alpha\|e_L(n)\|^2 I + \gamma x(n)x^T(n)\right]^{-1} x(n)e(n) \tag{12.81}$$

where μ is a positive constant step-size, α and γ are positive constant, $e(n)$ as the system output error defined by

$$e(n) = d(n) - x^T(n)w(n) \tag{12.82}$$

$$\|e(n)\|^2 = \sum_{i=0}^{n-1}|e(n-i)|^2 \tag{12.83}$$

$$\|e_L(n)\|^2 = \sum_{i=0}^{L-1}|e(n-i)|^2 \tag{12.84}$$

Equation 12.83 is the square norm of the error vector, $e(n)$, estimated over its entire updated length n, and Equation 12.84 is the squared norm of the error vector, $e(n)$, estimated over the last L values.

Expanding Equation 12.81 and applying the matrix inversion formula (see Appendix 2)

$$[A + BCD]^{-1} = A^{-1} - A^{-1}B\left[C^{-1} + DA^{-1}B\right]^{-1}DA^{-1} \tag{12.85}$$

with $A = \alpha\|e(n)\|^2 I, B = x(n), C = \gamma$ and $D = x^T(n)$, we obtain

$$\left[\alpha\|e(n)\|^2 I + \gamma x(n)x^T(n)\right]^{-1}$$
$$= \alpha^{-1}\|e(n)\|^{-2} I - \alpha^{-1}\|e(n)\|^{-2} Ix(n)\frac{x^T(n)\alpha^{-1}\|e(n)\|^{-2}}{\gamma^{-1} + x^T(n)\alpha^{-1}\|e(n)\|^{-2} x(n)} \tag{12.86}$$

Multiplying both sides of Equation 12.86 by $x(n)$ from the right and rearranging the equation, we have (see Exercise 12.19.1)

$$\left[\alpha\|e(n)\|^2 I + \gamma x(n)x^T(n)\right]^{-1} x(n) = \frac{x(n)}{\alpha\|e(n)\|^2 + \gamma\|x(n)\|^2} \tag{12.87}$$

Excise 12.19.1

Verify Equation 12.87.

▲

Substituting Equation 12.87 in 12.81, we obtain a new proposed robust variable size step size (RVSS):

$$w(n+1) = w(n) + \frac{\mu\|e_L(n)\|^2}{\alpha\|e(n)\|^2 + (1-\alpha)\|x(n)\|^2} x(n)e(n) \qquad (12.88)$$

where we replaced γ by $(1-\alpha) \geq 0$ in Equation 12.88 without loss of generality. The fractional quantity in Equation 12.88 may be viewed as a time-varying step-size, $\mu(n)$, of the proposed RVSS algorithm. Clearly, $\mu(n)$ is controlled by normalized of both error and input data.

The parameters α, L, and μ are appropriately chosen to achieve the best trade-off between rate of convergence and low final mean-squared error. A small constant ε could be added to the denominator of Equation 12.88 to prevent instability of the algorithm if the denominator becomes too small. The quantity $\|e_L(n)\|^2$ is large at the beginning of adaptation and it decreases as n increases, while $\|x(n)\|^2$ fluctuates depending on the recent values of the input signal. On the other hand, $\|e(n)\|^2$ is an increasing function of n since $e(n)$ is a vector of increasing length. To compute Equation 12.83 with minimum computational complexity, the error value produced in the first iteration is squared and stored. The error value in the second iteration is squared and added to the previous stored value. Then, the result is stored in order to be used in the next iteration, and so on.

Example 12.19.1

Use the unknown system shown in Figure 12.25, study the RVSS algorithm with respect to normalized and ENSS algorithm. Figure 12.27 shows the desired results.

Solution: The following Book m-function and program were used.

BOOK M-FUNCTION FOR RVSS ALGORITHM

```
function[J,w]=lms_robust_variable_SS(avn,N,mul,h,M)
J=zeros(1,N);
    %M>length(h);J=MSE-learning curve;avn=number of averaging
    %the learning curve; N=number of inputs;M=number
    %of adaptive filter coefficients; h=unknown system;
    %mul=constant step-size;v=internal system noise;
```

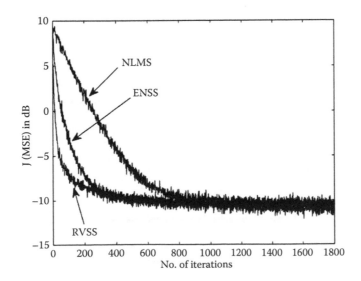

FIGURE 12.27

```
for i=1:avn
    y=zeros(1,N);w=zeros(1,M);e=zeros(1,N);
    X=zeros(1,M);D=zeros(1,M);
    x=filter([1 0.5],1,0.3*randn(1,N));v=0.1*randn(1,N);
    for k=1:N
        d=filter(h,1,x)+v;
        X=[x(k) X(1,1:M-1)];
        den=X*X'+0.0001;
        y=w*X';
        e(k)=d(k)-y;
        D=[e(k) D(1,1:M-1)];
        denx=D*D';
        a=0.5;
        mu=(mu1*denx)/((a*den+(1-a)*e*e'));
        w=w+mu*e(k)*X;
        J(k)=J(k)+abs(e(k))^2;
    end;
end;
J=J/avn;
```

BOOK MATLAB PROGRAM

```
>>[J,w]=lms_robust_variable_SS(300,2000,0.07,[2 1 0.5 -0.2],6);
>>[J1,w1]=lms_error_normalized_SS([2 1 0.5 - 0.2],300,2000,0.15,6);
>>[J2,w2]=lms_average_normalized_lms([2 1 0.5 - 0.2],300,2000,0.02,6);
>>plot(10*log10(J(1,1:1800)),'k');hold on;plot(10*log10(J1(1,1:1800)),'k');
>>hold on;plot(10*log10(J2(1,1:1800)),'k');xlabel('No. of iterations');
>>ylabel(J, MSE in dB');
```

The reader will observe that RVSS adjust faster than the other algorithm.   ∎

Example 12.19.2

Use NLMS to study the results for an abrupt change in the unknown system, from *h* becomes *-h*, which must be identified. The unknown system is the one shown in Figure 12.25.

Solution: The following Book m-functions and Book program was used to produce Figures 12.28 and 12.29. The value of **w** found with the constants used was
1.9988   0.9973   0.5573   −0.2052   0.0055   0.0519.

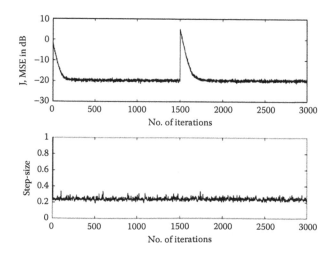

FIGURE 12.28

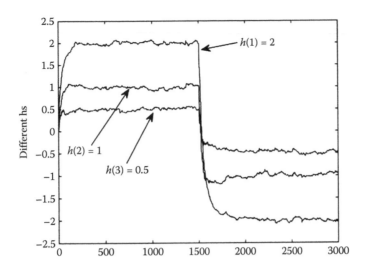

FIGURE 12.29

```
function[J,mukav]=lms_abrupt_change_normalized_lms(avn,N,M,mu1)
    %N must be even;muk=gives the variation of step-size;
    %N=number of data;M>=length(h);avn=number of averaging;
    %v=internal system noise;mu1=step-size constant;J=MSE;
J=zeros(1,N);muk=zeros(1,N);
for i=1:avn
    h=[2 1 0.5 -0.2];
    y=zeros(1,N);w=zeros(1,M);e=zeros(1,N);
    X=zeros(1,M);D=zeros(1,M);
    x=filter([1 0.5],1,0.3*randn(1,N));
    v=0.1*randn(1,N);
    for k=1:N
        d=filter(h,1,x)+v;
        X=[x(k) X(1:M-1)];
        den=X*X'+0.0001;
        y=w*X';
        e(k)=d(k)-y;
        if k==N/2;%N must be even;
            h=-h;
        end;
        w=w+(mu1/den)*e(k)*X;
        muk(k)=muk(k)+(mu1/den);
        J(k)=J(k)+abs(e(k))^2;
    end;
end;
mukav=muk/avn;
J=J/avn;

function[w1,w]=lms_abrupt_change_normalized_lms_w(N,M,mu1)
    %N must be even;muk=gives the variation of step-size;
    %N=number of data;M>=length(h);avn=number of averaging;
    %v=internal system noise;mu1=step-size constant;J=MSE;
    h=[2 1 0.5 -0.2];
    y=zeros(1,N);w=zeros(1,M);e=zeros(1,N);
    X=zeros(1,M);D=zeros(1,M);
    x=filter([1 0.5],1,0.3*randn(1,N));
    v=0.1*randn(1,N);
    for k=1:N
        d=filter(h,1,x)+v;
        X=[x(k) X(1:M-1)];
        den=X*X'+0.0001;
```

```
        y=w*X';
        e(k)=d(k)-y;
        if k==N/2;%N must be even;
            h=-h;
        end;
        w=w+(mul/den)*e(k)*X;
        w1(k,:)=w(1,:);
    end;
```

BOOK MATLAB PROGRAM

```
>>[J,mukav]=lms_abrupt_change_normalized_lms(200,3000,6,0.1);
>>[w1,w]=lms_abrupt_change_normalized_lms_w(3000,6,0.1);
```

For Figure 12.28, we used the following commands:

```
>>subplot(2,1,1);plot(10*log10(J),'k');
>>xlabel('No. of iterations');
>>ylabel('J, MSE in dB');subplot(2,1,2);plot(mukav,'k');
>>axis([0  3000 0 1]);
>>ylabel('Step-size');xlabel('No. of iterations');
```

For Figure 12.29, we used the following commands:

```
>>plot(w1(:,1),'k');hold on;plot(w1(:,2),'k');
>>hold on;plot(w1(:,5),'k');
>>xlabel('No. of iteration');ylabel('Different hs');                    ∎
```

12.20 THE MODIFIED LMS ALGORITHM

From Figure 12.30, the desired signal is the output of the linear regression model system and it is given by (Haykin 2001)

$$d(n) = w_{un}^T x(n) + v(n) \tag{12.89}$$

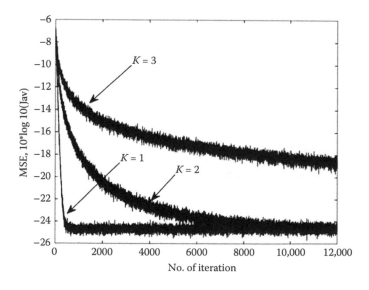

FIGURE 12.30

where w_{un} is the coefficients vector of the unknown system, $x(n)$ is the input vector (regressor) and $v(n)$ is a nonmeasurable white noise with variance σ_v^2. The LMS FIR filter is chosen to minimize the following index of performance:

$$J(w,K) = E\left\{e^{2K}(n)\right\} \quad K = 1,2,3,\ldots \tag{12.90}$$

From the preceding equation and using the instantaneous gradient vector, we find the **modified** LMS filter (Haykin 2001)

$$w(n+1) = w(n) + \mu K x(n) e^{2K-1}(n) \tag{12.91}$$

The following Book m-function simulates the preceding equation.

BOOK M-FUNCTION FOR MODIFIED LMS FILTER

```
function[w,e,Jav]=lms_modified_lms(mu,av,a,b,N,M,K)
for m=1:av%av=integer equal to the number of desired
    %averaging;a=multiplier of the random noise,
    %a*randn(1,N);
    w=zeros(1,M);
    x=a*(rand(1,N)-0.5);
    d=filter([0.9 0.4  -0.2],1,x)+b*(rand(1,N)-0.5);
            %multiplier of multi-linear regressor model;
    for n=M:N
        x1=x(n:-1:n-M+1);
        y(n)=w*x1';
        e(n,m)=d(n)-y(n);%e(n,m) becomes a Nxav
                        %matrix;
        w=w+mu*K*e(n,m)^(2*K-1)*x1;
    end
y=zeros(1,N);
end
Jav=sum((e.^2),2)/av;%e.^2 squares each matrix element;
```

Figure 12.30 shows the results with the following inputs: $mu=0.04$, $av=300$ (ensemble number), $a=1.5$, $b=0.2$, $N=15{,}000$, $M=6$. It is obvious that the modified LMS filter is not as efficient as the standard LMS filter. ∎

12.21 MOMENTUM LMS ALGORITHM

Another LMS type filter has been proposed, known a **momentum** *LMS*, which is given by

$$w(n+1) = w(n) + (1-g)[w(n) - w(n-1)] + ag\mu e(n)x(n) \tag{12.92}$$

Where $0 < g < 1$ and $a > 1$.

BOOK M-FUNCTION FOR MOMENTUM LMS

```
function[w,y,e,J,w1]=lms_momentum_lms(x,d,mu,g,a,M)
    %0<g<1,
    %function[w,y,e,J,w1]=lms_momentum_lms(x,dn,mu,M);
    %all quantities are real-valued;
    %x=input data to the filter; d=desired signal;
    %M=order of the filter;
    %mu=step-size factor; x and d must be
```

```
%of the same length, in the program they are
%reversed;
%w1=is a matrix of dimensions: length(x)xM,
%each column represents the variation of
%each filter coefficient;
N=length(x);w(2,1:M)=zeros(1,M);w1=zeros(1,M);
w(1,1:M)=zeros(1,M);
xr=fliplr(x);dr=fliplr(d);
for n=1:N-M
    x1=xr(n:1:M+n-1); %for each n the vector x1 is
            %of length M;
    y(n)=w(n+1,1:M)*x1';
    e(n)=dr(n)-y(n);
    w(n+2,:)=w(n+1,:)+(1-g)*(w(n+1,:)-w(n,:))+g*a*mu*e(n)*x1;
    w1(n,:)=w(1,:);
end;
J=e.^2;%J is the learning curve of the adaptation;
```

If we set mu=0.06 and $M=6$ for a LMS filter and compare the MSE with the results of a momentum LMS filter with the constants mu=0.06, $M=6$, $g=0.4$ and $a=4$, we observe that the momentum one is superior.

12.22   THE BLOCK LMS ALGORITHM

We can subdivide a set of data L sections and process the data sequentially. The following Book m-function process the data in a block-by-block format.

BOOK M-FUNCTION FOR BLOCK LMS PROCESSING

```
function[w,y,e,J,w1]=lms_block_lms(x,d,mu,M,L)
    %all quantities are real-valued;
    %x=input data to the filter; d=desired signal;
    %M=order of the filter;
    %mu=step-size factor; x and d must be
    %of the same length, in the program they are
    %reversed;N/L must be an integer;
    %w1=is a matrix of dimensions: length(x)x(M+1),
    %each column represents the variation of
    %each filter coefficient;
N=length(x);w(2,1:M+1)=zeros(1,M+1);w1=zeros(1,M+1);
w(1,1:M+1)=zeros(1,M+1);
xr=fliplr(x);dr=fliplr(d);
for m=1:L-1
    if m*(N/L)+1>N
    end
for n=((m-1)*(N/L)+1):1:m*(N/L)
    x1=xr(n:1:M+n); %for each n the vector x1 is
            %of length M+1;
    y(n)=w(n+1,1:M+1)*x1';
    e(n)=dr(n)-y(n);
    w(n+2,:)=w(n+1,:)+(mu/L)*e(n)*x1;
    w1(n,:)=w(1,:);
end;
end;
J=e.^2;%J is the learning curve of the adaptation;
```

Using the identification of a FIR system with impulse response $h = \begin{bmatrix} 0.9 & 0.4 & -0.1 \end{bmatrix}$ and with a zero mean white noise as input, we obtain the learning curve shown in Figure 12.31. The constants used were $N = 12,000$, $L=4$, $M=6$, $mu=0.08$, $d = \text{filter}([\ 0.9 \quad 0.4 \quad -0.1\ ],1,x)$. The values of the adaptive filter were at the 3,000 iteration $w(100,:) = 0.7402 \quad 0.3215 \quad -0.0938 \quad 0.0320 \quad 0.0141 \quad -0.0016$; at the 300 iteration $w(300,:) = 0.8963 \quad 0.3971 \quad -0.1020 \quad -0.0019 \quad -0.0011 \quad -0.0011$; at the 500 iteration $w(500,:) = 0.9000 \quad 0.4000 \quad -0.1000 \quad -0.0000 \quad -0.0000 \quad 0.0000$.

12.23   THE COMPLEX LMS ALGORITHM

The complex LMS algorithm is given in the Table 12.5.

Figure 12.32 shows the two learning curves, one for un-normalized and the other for normalized complex LMS adaptive filter. It shows that, after the first about 100 iterations, the normalized filter performs better. The input for both the unknown system and the adaptive filter was

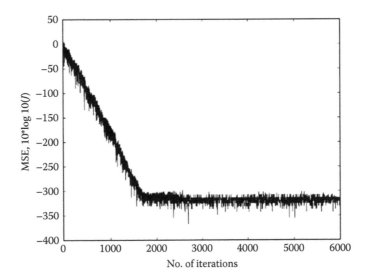

FIGURE 12.31

TABLE 12.5

Complex LMS Algorithm

Un-normalized

$x(0) = w(0) = [\ 0 \quad 0 \quad 0 \quad \cdots \quad 0\ ]^T$

For $n \geq 0$

$e(n) = d(k) - w^H(n)x(n)$

$w(n+1) = w(n) + \mu e^*(n)x(n)$

Normalized

For $n \geq 0$

$e(n) = d(k) - w^H(n)x(n)$

$w(n+1) = w(n) + \mu e^*(n)x(n) / (0.0001 + x(n)x^H(n))$

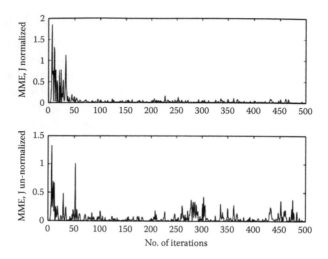

FIGURE 12.32

$x = \text{rand}n(1,12000) + j0.2\,\text{filter}([\quad 0.9 \quad 0.2 \quad],1,\text{rand}(1,12000)-0.5)$; The output from the unknown filter was $d = \text{filter}([\quad 0.9 \quad 0.4 \quad -0.1 \quad],1,x)$ and $M=5$. To obtain the unknown filter, we write $\text{real}(w) = \quad 0.9081 \quad 0.4043 \quad -0.1040 \quad 0.002 \quad 0.0040$.

Book m-Function for Normalized Complex LMS Filter

```
function[w,y,e,J,w1,Js]=lms_complex_normalized_lms(x,dn,mu,M)
    %x=input data to the filter; dn=desired signal;
    %M=order of the filter;
    %mu=step-size factor; x and dn must be
    %of the same length;
    %Js=smooths the learning curve;
    %w1=is a matrix of dimensions: length(x)xM,
    %each column represents the variation of
    %each filter coefficient;
N=length(x);w=zeros(1,M);w1=zeros(1,M);
for n=M:N
    x1=x(n:-1:n-M+1); %for each n the vector x1 is
                %of length M with elements from x in
                %reverse order;
    y(n)=conj(w)*x1';
    e(n)=dn(n)-y(n);
    w=w+2*mu*conj(e(n))*x1/(0.00001+x1*x1');
    w1(n-M+1,:)=w(1,:);
end;
J=conj(e).*e;%J is the learning curve of the adaptation;
for n=1:length(x)-5
    Js(n)=(J(n)+J(n+1)+J(n+2))/3;
end
```

Book m-Function for Unnormalized Complex LMS Filter

```
function[w,y,e,J,w1,Js]=lms_complex_lms(x,dn,mu,M)
    %x=input data to the filter; dn=desired signal;
    %M=order of the filter;
```

```
%mu=step-size factor; x and dn must be
%of the same length;
%Js=smooths the learning curve;
%w1=is a matrix of dimensions: length(x)xM,
%each column represents the variation of
%each filter coefficient;
N=length(x);w=zeros(1,M);w1=zeros(1,M);
for n=M:N
    x1=x(n:-1:n-M+1); %for each n the vector x1 is
                %of length M with elements from x in
                %reverse order;
    y(n)=conj(w)*x1';
    e(n)=dn(n)-y(n);
    w=w+2*mu*conj(e(n))*x1;
    w1(n-M+1,:)=w(1,:);
end;
J=conj(e).*e;%J is the learning curve of the adaptation;
for n=1:length(x)-5
    Js(n)=(J(n)+J(n+1)+J(n+2))/3;
end
```

12.24 THE AFFINE LMS ALGORITHM

We first create the signal matrix, which is of the form

$$
X(n) = \begin{bmatrix}
x(n) & x(n-1) & \cdots & x(n-N+1) & x(n-N) \\
x(n-1) & x(n-2) & \cdots & x(n-N) & x(n-N-1) \\
\vdots & \vdots & & \vdots & \vdots \\
x(n-M) & x(n-M-1) & \cdots & x(n-N-M+1) & x(n-N-M)
\end{bmatrix} \quad (12.93)
$$

The output vector from the adaptive filter is given by

$$
y(n) = X^T(n)w(n) = \begin{bmatrix} y_0(n) \\ y_1(n) \\ \vdots \\ y_N(n) \end{bmatrix} \quad (12.94)
$$

The desired and error signals are given by

$$
d(n) = \begin{bmatrix} d(n) \\ d(n-1) \\ \vdots \\ d(n-N) \end{bmatrix}, \quad
e(n) = \begin{bmatrix} e_0(n) \\ e_1(n) \\ \vdots \\ e_N(n) \end{bmatrix} = \begin{bmatrix} d(n)-y_0(n) \\ d(n-1)-y_1(n) \\ \vdots \\ d(n-N)-y_N(n) \end{bmatrix} = d(n) - y(n) \quad (12.95)
$$

The affine algorithm is given in Table 12.6.

BOOK M-FUNCTION FOR SYSTEM IDENTIFICATION

```
function[w,y,e,J]=lms_affine_lms(x,d,mu,M,g)
    %all quantities are real-valued;0<g<1
    %x=input data to the filter; dn=desired signal;
    %M=order of the filter;
```

TABLE 12.6
The Affine Projection Algorithm

Initialization

$$x(0) = w(0) = [\ 0 \quad 0 \quad \cdots \quad 0\ ]^T$$

γ = small constant

For $n \geq 0$

$$e(n) = d(n) - X^T(n)w(n)$$

$$w(n+1) = w(n) + \mu X(n)\left(X^T(n)X(n) + \gamma I\right)^{-1} e(n)$$

```
    %mu=step-size factor; x and dn must be
    %of the same length;
    %Js=smooths the learning curve;
    %w1=is a matrix of dimensions: length(x)xM,
    %each column represents the variation of
    %each filter coefficient;
N=length(x);w=zeros(1,M)';
xr=fliplr(x)';dr=fliplr(d)';
for k=1:M
    X(k,:)=xr(k:1:N-M+k);%Mx(N-M+1)
end
for n=1:N-M+1

    y=X'*w(:,n);%((N-M+1)xM)x(Mx1)=(N-M+1)x1
    e=dr(1:(N-M+1))-y;
    w(:,n+1)=w(:,n)+mu*X*inv(X'*X+g*eye(N-M+1))*e;

end;
J=e.^2;%J is the learning curve of the adaptation;
```

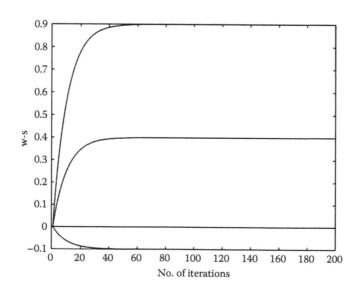

FIGURE 12.33

The evolving of the first four coefficients versus iteration numbers is shown in Figure 12.33. The fifth coefficient is also zero for all the iteration numbers.

12.25   THE COMPLEX AFFINE LMS ALGORITHM

The complex affine LMS algorithm is presented in Table 12.7.

We have tested the preceding algorithm by using the following constants and vectors to identify an unknown system with $h = [\ \ 0.9 \quad 0.4 \quad -0.1\ \ ]$. The input vectors and constants were

$$x = \text{filter}([\ \ 0.9 \quad 0.2\ \ ], 1, 0.5\,\text{rand}\,n(1,200) + j(\text{rand}(1,200) - 0.5))$$

$$d = \text{filter}([\ \ 0.9 \quad 0.4 \quad -0.1\ \ ] 1, x)$$

$$\mu = 0.05$$

$$\gamma = 0.2$$

$$M = 5$$

Based on the preceding, we obtained the following approximations:

$$\text{real}\big(w(:,20)\big) = 20\text{th iteration} = [\ \ 0.5590 \quad 0.2487 \quad -0.0621 \quad 0.0000 \quad -0.0000\ \ ]^T$$

$$\text{real}\big(w(:,100)\big) = 100\text{th iteration} = [\ \ 0.8943 \quad 0.3975 \quad -0.0994 \quad 0.0000 \quad -0.0000\ \ ]^T$$

$$\text{real}\big(w(:,150)\big) = 150\text{th iteration} = [\ \ 0.8996 \quad 0.3998 \quad -0.1000 \quad 0.0000 \quad -0.0000\ \ ]^T$$

To obtain the preceding results, we used the following Book m-function.

BOOK M-FUNCTION FOR COMPLEX AFFINE ALGORITHM

```
function[w,y,e,J]=lms_complex_affine_lms(x,d,mu,M,g)
    %all quantities are real-valued;0<g<1
    %x=input data to the filter; d=desired signal;
    %M=order of the filter;
    %mu=step-size factor; x and d must be
    %of the same length;
N=length(x);w=zeros(1,M)';
xr=fliplr(x)';dr=conj(fliplr(d)');
for k=1:M
    X(k,:)=xr(k:1:N-M+k);%Mx(N-M+1)
end
for n=1:N-M+1
```

TABLE 12.7
Complex Affine Algorithm

$$x(0) = w(0) = [\ \ 0 \quad 0 \quad \cdots \quad 0\ \ ]^T$$

$\gamma = $ small constant

For $n \geq 0$

$$e*(n) = d*(n) - X^H(n)w(n)$$

$$w(n+1) = w(n) + \mu X(n)\big(X^H(n)X(n) + \gamma I\big)^{-1} e*(n)$$

```
y=X'*w(:,n);%((N-M+1)xM)x(Mx1)=(N-M+1)x1
e=dr(1:(N-M+1))-y;
w(:,n+1)=w(:,n)+mu*X*inv(X'*X+g*eye(N-M+1))*e;

end;
J=e.*conj(e);%J is the learning curve of the adaptation;
```

HINTS, SOLUTIONS, AND SUGGESTIONS OF THE EXERCISES

12.1.1

$$w(n+1) = w(n) + \mu(n)e(n)x(n) \tag{1}$$

$$e_{ps}(n) = d(n) - w^T(n+1)x(n), \tag{2}$$

$$(1) \text{ in } (2) \Rightarrow$$

$$e_{ps}(n) = d(n) - [w^T(n) + \mu(n)e(n)x^T(n)]x(n) = d(n) - w^T(n)x(n) - \mu(n)e(n)x^T(n)x(n)$$

$$= e(n) - \mu(n)e(n)\|x(n)\|^2, \frac{\partial e_{ps}^2(n)}{\partial \mu(n)} = 2\Big(e(n) - \mu(n)e(n)\|x(n)\|^2\Big) = 0 \Rightarrow \mu(n) = \frac{1}{\|x(n)\|^2}$$

12.3.1

Use the following constants and signals: $\mu=0.01$, $M=6$, $dn(n)=0.995^n \cos(0.1\pi n)$, $x(n)=0.995^n \cos(0.1\pi n)+0.2\text{rand}n(1,1500)$, $n=1:1500$.

12.12.1

Taking the gradient of $J(n)$, we obtain

$$\nabla_w J(n) = -e(n)x(n) + \gamma w(n) \Rightarrow w(n+1) = w(n) - \mu\nabla_w J(n)\{\text{steepest-descent algorithm}\}$$

$$= w(n) - \mu[-e(n)x(n) + \gamma w(n)] = (1-\mu\gamma)w(n) + \mu e(n)x(n)$$

12.12.2

The normalized leaky LMS algorithm is

```
function[w,y,e,J,w1]=lms_normalized_leaky_lms(x,dn,mu,gama,M)
    %function[w,y,e,J,w1]=lms_normalized_leaky_lms(x,dn,mu,gama,M);
    %all signals are real valued;x=input to filter;
    %y=output from the filter;dn=desired signal;
    %mu=step-size factor;gama=gamma factor<<1;
    %M=number of filter coefficients;w1=matrix whose M
    %rows give the history of each filter coefficient;
N=length(x);
y=zeros(1,N);
w=zeros(1,M);
for n=M:N
    x1=x(n:-1:n-M+1);
    y(n)=w*x1';
    e(n)=dn(n)-y(n);
    w=(1-mu*gama)*w+mu*e(n)*x1/(0.0001+x1*x1');
    w1(n-M+1,:)=w(1,:);
end;
J=e.^2;
```

The reader will find that the normalized perform better.

12.13.1

$$J = E\left\{\left[d(n) - w^T(n)x(n)\right]\left[d(n) - x^T(n)w(n)\right] + \lambda\left(c^T w(n) - a\right)\right\} = E\left\{d^2(n) - d(n)w^T(n)x(n)\right.$$

$$-d(n)x^T(n)w(n) + w^T(n)x(n)x^T(n)w(n) + \lambda c^T w(n) - \lambda a\right\} = \sigma_d^2 - w^T(n)E\{d(n)x(n)\}$$

$$-wE\{d(n)x(n)\} + w^T(n)E\{x(n)x^o(n)\}w(n) + \lambda c^T w(n) - \lambda a = \sigma_d^2 - 2w^T(n)p_{dx} + w^T(n)R_x w(n)$$

$$+\lambda c^T w(n) - \lambda a = \sigma_d^2 - 2w^T(n)p_{dx} + (w(n) - w^o)^T R_x(w(n) - w^o) + w^{oT}R_x w(n) + w^T(n)R_x w^o$$

$$+w^{oT}R_x w^o + \lambda\left[c^T\left(w(n) - w^o\right) - \left(a - c^T w^o\right)\right] = \sigma_d^2 - w^{oT}p_{dx} + \xi^T R_x \xi + \lambda\left(c^T \xi - a'\right)$$

where $R_x w^o = p_{dx}$, $(w^{oT}R_x w(n))^T = w^T(n)R_x w^o$ (R_x = symmetric)

12.3.2

We write the cost function

$$J(n) = [w(n+1) - w(n)]^T[w(n+1) - w(n)] + \lambda[d(n) - w(n+1)x(n)] \tag{1}$$

Differentiate (1) with respect to $w(n+1)$ (see also Appendix 2) to obtain

$$\frac{\partial J(n)}{\partial w(n+1)} = 2[w(n+1) - w(n)] - \lambda x(n) = 0 \Rightarrow w(n+1) = w(n) + \frac{1}{2}\lambda x(n) \tag{2}$$

Substitute (2) in the constraint $d(n) = w^T(n+1)x(n)$ to obtain

$$d(n) = \left(w(n) + \frac{1}{2}\lambda x(n)\right)^T x(n) = w^T(n)x(n) + \frac{1}{2}\lambda\|x(n)\|^2 \tag{3}$$

But $e(n) = d(n) - w^T(n)x(n)$ and (3) becomes

$$\lambda = \frac{2e(n)}{\|x(n)\|^2} \tag{4}$$

Substitute (4) in (2), we obtain

$$w(n+1) = w(n) + \frac{1}{\|x(n)\|^2}e(n)x(n) \tag{5}$$

From (5), we obtain the final form by introducing the step-size factor μ to control the change in weight vector.

12.13.3

The Book m-Function for the LMF Algorithm

```
function[w,y,e,J,w1]=lms_least_mean_fourth(x,dn,mu,M)
N=length(x);w=zeros(1,M);
```

```
for n=M:N
    x1=x(n:-1:n-M+1);
    y(n)=w*x1';
    e(n)=dn(n)-y(n);
    w=w+mu*x1*e(n)^3;
    w1(n-M+1,:)=w(1,:);
end;
J=e.^2;
```

12.15.1

Book m-function for the least-mean-mixed-norm LMS algorithm:

```
function[w,y,e,J,w1]=lms_least_mean_mixed_norm(x,dn,mu,M,delta)
N=length(x); w=zeros(1,M);%0<delta<1;
for n=M:N
    x1=x(n:-1:n-M+1);
    y(n)=w*x1';
    e(n)=dn(n)-y(n);
    w=w+mu*x1*e(n)*(delta+(1-delta)*e(n)^2);
    w1(n-M+1,:)=w(1,:);
end
J=e.^2;
```

To obtain the normalized form of the preceding algorithm, divide the last term of line 7 with the expression $(\varepsilon + x1 * x1')$.

12.17.1.1

The kth value is

$$W_{i+1,k} = W_{i,k} + \mu X_i^* E_i = W_{i,k} + \mu X_i^*[D_i(k) - Y_i(k)] = W_{i,k} + \mu X_i^*[D_i(k) - W_{i,k} X_i(k)]$$

$$= W_{i,k} + \mu X_i^* D_i(k) - \mu W_{i,k}|X_i|^2 = \left(1 - \mu|X_i|^2\right)W_{i,k} + \mu X_i^* D_i(k)$$

12.17.1.2

The z-transform and the ensemble operator are linear operations and can be interchanged. Therefore, the z-transform is

$$zW_k(z) - zW_{0,k} = (1 - \mu E\{|X_i(k)|^2\})W_k(z) + \left[\mu E\{D_i(k)X_i^*(k)\} / (1 - z^{-1})\right] \tag{1}$$

where $W_{0,k} = 0$ since it was assumed that the initial conditions have zero values and $W_k(z) = \sum_{i=0}^{\infty} E\{W_{i,k}\}z^{-i}$. Multiplying (1) by z^{-1} and $(z-1)$ and applying the final value theorem (see z-transform properties in Chapter 3), we obtain.

$$E\{W_k^{\infty}\} = \lim_{z \to 1}\{z-1)W_k(z)\} = \lim_{z \to 1} \frac{\mu E\{D_i(k)X_i^*(k)\}}{1 - \left(1 - \mu E\{|X_i(k)|^2\}\right)z^{-1}} = \frac{E\{D_i(k)X_i^*(k)\}}{E\{|X_i(k)|^2\}} \tag{2}$$

12.17.1.3

$$E_{i+1}(k) = E\{W_{i+1,k}\} - E\left\{W_k^\infty\right\} = \left(1 - 2\mu E\left\{\left|X_i(k)\right|^2\right\}\right) E\{W_{i,k}\} + 2\mu E\{D_i(k)X_i^*(k)\}$$

$$-\frac{E\{D_i(k)X_i^*(k)\}}{E\left\{\left|X_i(k)\right|^2\right\}} = \left(1 - 2\mu E\left\{\left|X_i(k)\right|^2\right\}\right) E\{W_{i,k}\} - \left(1 - 2\mu E\left\{\left|X_i(k)\right|^2\right\}\right)\frac{E\{D_i(k)X_i^*(k)\}}{E\{\left|X_i(k)\right|^2\}}$$

$$= \left(1 - 2\mu E\left\{\left|X_i(k)\right|^2\right\}\right) E_i(k)$$

13 Nonlinear Filtering

13.1 INTRODUCTION

Nonlinear filtering techniques remove unknown interference to one-dimensional signals $f(n)$ or to two-dimensional signals (images) $f(m,n)$, where the integers m and n have, in practice, finite images: $0 \leq m \leq M-1, 1 \leq n \leq N-1$. The basic problem is to use the received (or detected) signal $x = [x_0 \quad x_1 \quad \cdots \quad x_{N-1}]^N$, which is corrupted by noise, and try to extract the original signal $s = [s_0 \quad s_1 \quad \cdots \quad s_{N-1}]$, which is another random vector. Although it is rather difficult to find a useful solution, by making simplifications, we may be able to solve the problem with the available methods at hand. Complexity of the reduced solution depends on the model of the underlying signal s, the nature of the additive noise (corruption) to the signal, and the accuracy of the solution with respect to the preceding assumptions.

When the noise is Gaussian process and linearly added to the signal, the theory of the linear filtering gives optimum solutions. In this case, the mean square error is used as the criterion of accuracy.

In some cases, more pronounced in images, the intensity of a pixel is due to the signal (desired image) plus scattering light from different parts of the environment during its formation. In such a situation, the white noise additive model seldom holds. The intensity of each pixel of an image created by an acquisition system is usually a **multiplicative** process with respect to background illumination. This is equivalent to say that the ith pixel intensity is equal $s_i v_i$, where s_i is the signal intensity and v_i is the noise intensity. Another important noise is the **impulsive noise**, which is recognized if the pixel values do not change at all or change slightly and some pixel values change "enormously;" that is, their change is highly visible. Because this type of noise is hard to handle with linear filters, the use of **nonlinear filters** is more appropriate. In this chapter, we only deal with order-based filters, which have been employed successfully to pass the desired signal (image) structures while suppressing noise.

13.2 STATISTICAL PRELIMINARIES

13.2.1 SIGNAL AND NOISE MODEL-ROBUSTNESS

The simplest and most important case of signal and noise is the additive white noise model where the signal s and noise v are assumed independent. The white indicates that v_i's are independent or at most uncorrelated. For this model, we write

$$x = s + v \tag{13.1}$$

As we mentioned in the previous section, the multiplicative noise model is expressed as

$$x = sv \tag{13.2}$$

or, in the expanded form,

$$[x_0 \quad x_1 \quad \cdots \quad x_{N-1}] = [s_0 v_0 \quad s_1 v_1 \quad \cdots \quad s_{N-1} v_{N-1}] \tag{13.3}$$

where s_i's are independent of v_i's.

Another type of noise is the **impulsive noise**, also known as **outliers**. An outlier can be defined as an observation that appears to be inconsistent with the remaining data.

13.2.2 POINT ESTIMATION

In a typical case, we are faced with the problem of extracting **parameter** values from a discrete-time waveform or a data set. For example, let us have an N-point data set $\{x(0) \quad x(1) \quad \cdots \quad x(N-1)\}$ which depends on an unknown parameter θ. Our aim is to determine θ based on the data or define an **estimator**

$$\hat{\theta} = g(x(0), x(1), \ldots, x(N-1)) \tag{13.4}$$

where g is some function (statistic) of the data, and its numerical value is called an estimate of θ.

To be able to determine a good estimator, we must model mathematically the data. Because the data are inherently random, we describe them by the probability density function (PDF) or $p(x(0),x(1),\ldots,x(N-1);\theta)$, which is **parametrized** by the unknown parameter θ. This type of dependence is denoted by a semicolon.

As an example, let $x(0)$ be a random variable from a Gaussian population and with mean value $\theta = \mu$. Hence, the PDF is

$$p(x(0);\theta) = \frac{1}{\sqrt{2\pi\sigma^2}} e^{-\frac{1}{2\sigma^2}(x(0)-\theta)^2} \tag{13.5}$$

The plots of $p(x(0);\theta)$ for different values of θ are shown in Figure 13.1. From the figure, we can **infer** that, if the value of $x(0)$ is positive, it is doubtful that $\theta = \theta_2$ or θ_3 and, hence, the value $\theta = \theta_1$ is more probable. In the area of point estimation, the specification of the appropriate PDF is critical in determining a good estimator. In the case when the PDF is not given, which is found more often than not, we must try to choose one that is consistent with the constraints of any prior knowledge of the problem, and furthermore that is mathematically tractable.

Once the PDF has been selected, we want to determine an optimal estimator (a function of the data). The estimator may be thought of as a rule that assigns a value to θ for each realization of the sequence $\{x\}_N$. The estimate of θ is the **value** of θ obtained at a particular realization $\{x\}_N$.

13.2.3 ESTIMATOR PERFORMANCE

Let the set $\{x(0),x(1),\ldots,x(N-1)\}$ of random data be the sum of a constant c and a zero-mean white noise $v(n)$:

$$x(n) = c + v(n) \tag{13.6}$$

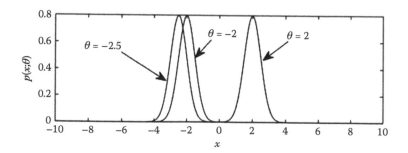

FIGURE 13.1

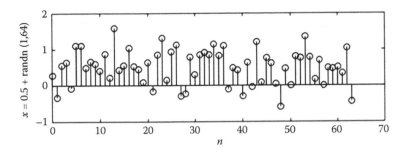

FIGURE 13.2

Intuitively, we may set as an estimate of c the sample mean of the data

$$\hat{c} = \frac{1}{N} \sum_{n=0}^{N-1} x(n) \tag{13.7}$$

From Figure 13.2, we find that $x(0) = 0.2837$ and we may accept the random variable $x(0)$ as another estimate of the mean:

$$\tilde{c} = x(0) \tag{13.8}$$

The basic question is as follows: which of these two estimators will produce the more accurate mean value? Instead of repeating the experiment a large number of times, we proceed to prove that sample mean is a better estimator than $x(0)$. To do this, we first look at their mean value (expectation):

$$E\{\hat{c}\} = E\{\frac{1}{N} \sum_{n=0}^{N-1} x(n)\} = \frac{1}{N} \sum_{n=0}^{N-1} E\{x(n)\}$$

$$= \frac{1}{N} E\{x(0) + x(1) + \cdots + x(N-1)\} = \frac{N\mu}{N} = \mu \tag{13.9}$$

$$E\{\tilde{c}\} = E\{x(0)\} = \mu \tag{13.10}$$

which indicates that on average, both estimators produce the true mean value of the population. Next, we investigate their variances, as shown in Exercise 13.2.3.1.

Exercise 13.2.3.1

Verify Equation 13.11. ▲

$$\text{var}\{\tilde{c}\} = \text{var}\left\{\frac{1}{N} \sum_{n=0}^{N-1} x(n)\right\} = \frac{\sigma^2}{N} \tag{13.11}$$

$$\text{var}\{\tilde{c}\} = \text{var}\{x(0)\} = \sigma^2 \tag{13.12}$$

The preceding results show that $\text{var}\{\tilde{c}\} > \text{var}\{\hat{c}\}$, and the $\text{var}\{\hat{c}\}$ approaches zero as $N \to \infty$. To prove Equation 13.11, we assume that $v(n)$'s are independent identical distributed (IID) and have the same variance σ^2.

13.2.4   BIASED AND UNBIASED ESTIMATOR

An estimator that **on average** yields the true value of the unknown parameter is known as unbiased one and, mathematically, is given by

$$E\{\hat{\theta}\} = \theta \qquad a < \theta < b \tag{13.13}$$

where (a, b) denotes the range of possible values of θ (see Exercise 13.2.4.1).

Exercise 13.2.4.1

Let $v(n)$ be a white Gaussian noise (WGN) with zero mean and equal variance. If the observation is $x(n)=c+v(n)$ for $n=0, 1, \ldots, N-1$, find the estimate of the parameter c.   ▲

An unbiased estimator is given mathematically by the relation (see Exercise 13.2.4.2):

$$E\{\hat{\theta}\} = \theta + b(\theta) \tag{13.14}$$

where the biased of the estimator is given by

$$b(\theta) = E\{\hat{\theta}\} - \theta \tag{13.15}$$

Exercise 13.2.4.2

Show that $\hat{c} = (1/4N)\sum_{n=0}^{N-1} x(n)$, with $x(n)=c+v(n)$ (see also Exercise 13.2.4.1) is a biased estimator.   ▲

13.2.5   CRAMER–RAO LOWER BOUND

It is helpful to be able to place a lower bound on the variance of any unbiased estimator. The Cramer–Rao lower bound (CRLB) is the appropriate measure. It assures us if the estimator is the minimum variance unbiased estimator (MVUE) or provides us with a benchmark to compare the performance of the estimator.

Theorem 13.1: Cramer–Rao Lower Bound

It is assumed that the pdf $p(\mathbf{x};\theta)$ satisfies the regular condition

$$E\left\{\frac{\partial \ln p(\mathbf{x};\theta)}{\partial \theta}\right\} = 0 \quad \text{for all } \theta \tag{13.16}$$

where the expectation is taken with respect to $p(\mathbf{x};\theta)$. Then, the variance of any unbiased estimator $\hat{\theta}$ must satisfy

$$\text{var}\{\hat{\theta}\} \geq \frac{1}{-E\left\{\dfrac{\partial^2 \ln p(\mathbf{x};\theta)}{\partial \theta^2}\right\}} \tag{13.17}$$

$$E\left\{\frac{\partial^2 \ln p(\mathbf{x};\theta)}{\partial \theta^2}\right\} = \int \frac{\partial^2 \ln p(\mathbf{x};\theta)}{\partial \theta^2} p(\mathbf{x};\theta)\, d\mathbf{x}$$

where the derivative is evaluated at the true value of θ and the expectation is taken with respect to $p(x;\theta)$. Furthermore, an unbiased estimator may be found that contains the bound for all θ if and only if

$$\frac{\partial \ln p(x;\theta)}{\partial \theta} = I(\theta)(g(x) - \theta) \tag{13.18}$$

for some function of g and I. The estimator, which is the MVUE, is $\theta = g(x)$, and the minimum variance (MV) is $1/I(\theta)$. ∎

Let us consider a number of observations with

$$x(n) = c + v(n) \quad n = 0,1,\ldots,N-1 \tag{13.19}$$

where $v(n)$ is a WGN with variance σ^2. To determine the CRLB for c (v's are IID), we find the PDF

$$p(x;c) = \prod_{n=0}^{N-1} \frac{1}{(2\pi\sigma^2)^{1/2}} \exp\left(-\frac{1}{2\sigma^2}(x(n)-c)^2\right)$$

$$= \frac{1}{(2\pi\sigma^2)^{N/2}} \exp\left(-\frac{1}{2\sigma^2} \sum_{n=0}^{N-1}(x(n)-c)^2\right) \tag{13.20}$$

Taking the first derivative, we obtain

$$\frac{\partial \ln p(x;\theta)}{\partial c} = \frac{\partial}{\partial c}\left(-\ln[(2\pi\sigma^2)^{N/2}] - \frac{1}{2\sigma^2} \sum_{n=0}^{N-1}(x(n)-c)^2\right)$$

$$= \frac{1}{\sigma^2} \sum_{n=0}^{N-1}(x(n)-c) = \frac{N}{\sigma^2}(\bar{x}-c); \quad \bar{x} = \text{sample mean} \tag{13.21}$$

Differentiating once again, we obtain

$$\frac{\partial^2 \ln p(x;\theta)}{\partial c^2} = -\frac{N}{\sigma^2} \tag{13.22}$$

Since the second derivative is constant, from Equation 13.17, we obtain the CRLB

$$\text{var}(\hat{c}) \geq \frac{1}{N/\sigma^2} = \frac{\sigma^2}{N} \tag{13.23}$$

Comparing Equation 13.21 with Equation 13.18, we find the following corresponding relations:

$$I = \frac{N}{\sigma^2} \quad \text{and} \quad (g(x)-\theta) = (\bar{x}-c) \tag{13.24}$$

The relations give us the MV and the MVUE.

13.2.6 Mean Square Error Criterion

When we seek the minimum mean square estimator, we can use a simplified form of the estimator where the statistic is a linear combination of the random data set $\{x(0),x(1),\ldots,x(N-1)\}$. Hence, we need to determine $a_0,a_1,\ldots,a_{N-1}$ such that

$$\text{mse}(\hat{\theta}) = E\{(\hat{\theta}-\theta)^2\} = E\{[a_0x(0)+a_1x(1)+\cdots+a_{N-1}x(N-1)]^2\} \tag{13.25}$$

is minimized. Since expectation is a linear operation, the preceding equation becomes

$$\text{mse}(\overline{\theta}) = E\{\theta^2\} - 2\sum_{n=0}^{N-1} a_i E\{\theta x(n)\} + \sum_{m=0}^{N-1}\sum_{n=0}^{N-1} a_m a_n E\{x(m)x(n)\} \tag{13.26}$$

The above equation is quadratic in a_n's and setting the derivatives equal to zero, we obtain the following set of equations:

$$r_{00}a_0 + r_{01}a_1 + \cdots + r_{0N-1}a_{N-1} = r(\theta,x(0)) \triangleq r_{\theta x}(0)$$

$$r_{10}a_0 + r_{11}a_1 + \cdots + r_{1N-1}a_{N-1} = r(\theta,x(1)) \triangleq r_{\theta x}(1)$$

$$\vdots$$

$$r_{N-1,0}a_0 + r_{N-1,1}a_1 + \cdots + r_{N-1,N-1}a_{N-1} = r(\theta,x(N-1)) \triangleq r_{\theta x}(N-1)$$

or

$$\boldsymbol{Ra} = \boldsymbol{r_{\theta x}} \tag{13.27}$$

where

$$r_{mn} = E\{x(m)x(n)\}$$

$$r_{\theta x} = E\{\theta x(m)\}$$

Exercise 13.2.6.1

Verify Equation 13.27. ▲

From Equation 13.25, we can also proceed as follows:

$$\text{mse}(\hat{\theta}) = E\{[(\hat{\theta}-E\{\hat{\theta}\})+(E\{\hat{\theta}\}-\theta)]^2\}$$

$$= E\{[(\hat{\theta}-E\{\hat{\theta}\}]^2\} + E\{[E\{\hat{\theta}\}-\theta]^2\} + 2E\{[\hat{\theta}-E\{\hat{\theta}\}][E\{\hat{\theta}\}-\theta]\} \tag{13.28}$$

$$= \text{var}(\hat{\theta}) + [E\{\hat{\theta}\}-\theta]^2 = \text{var}(\hat{\theta}) + b^2(\theta)$$

which shows that the mean square error is the sum of the error due to the variance of the estimator as well as its bias. Constraining the bias to zero, we can find the estimator, which minimizes its variance. Such an estimator is known as the **minimum variance unbiased estimator** (MVUE).

13.2.7 Maximum Likelihood Estimator

Very often, MVUEs may be difficult or impossible to determine. For this reason, in practice, many estimators are found using the **maximum likelihood function** (MLE) principle. Besides being easy to implement, its performance is optimal for large numbers of data. The basic idea is to find a statistic

$$\hat{\theta} = g(x(0), x(1), \ldots, x(N-1)) \tag{13.29}$$

so that if the random variables $x(m)$'s take the observed experimental value $x(m)$'s, and then the number $\hat{\theta} = g(x(0), x(1), \ldots, x(N-1))$ will be a good estimate of θ.

Definition 8.1: Likelihood Function

Let $x = \begin{bmatrix} x(0) & x(1) & \cdots & x(N-1) \end{bmatrix}^T$ be a random vector with density function:

$$p(x(0), x(1), \ldots, x(N-1); \theta), \theta \in \Theta$$

The function

$$l(\theta; x(0), x(1), \ldots, x(N-1)) = p(x(0), x(1), \ldots, x(N-1); \theta) \tag{13.30}$$

is considered as a function of the parameter $\theta = \begin{bmatrix} \theta_0 & \theta_1 & \cdots & \theta_{N-1} \end{bmatrix}^T$, which is called the **likelihood function** (*l* identifies the likelihood function with one parameter (scalar)). ▲

The random variables $x(0)$, $x(1)$,..., $x(N-1)$ are IID with a density function $p(x; \theta)$; the likelihood function is

$$L(\theta; x(0), x(1), \ldots, x(N-1)) = \prod_{n=0}^{N-1} p(x(n); \theta) \tag{13.31}$$

Example 13.2.7.1

Let $x(0)$, $x(1)$,..., $x(N-1)$ be the random sample from the normal distribution $N(\theta, 1), -\infty < \theta < \infty$ ($\theta \triangleq \mu =$ mean of the population). Using Equation 13.31, we write

$$L(\theta; x(0), x(1), \ldots, x(N-1)) = \left(\frac{1}{\sqrt{2\pi}} \right)^N \exp\left(-\sum_{n=0}^{N-1} (x(n) - \theta)^2 / 2 \right) \tag{13.32}$$

Since the likelihood function $L(\theta)$ and its logarithm $\ln\{L(\theta)\}$ are maximized for the same value of the parameter θ, we can use either $L(\theta)$ or $\ln\{L(\theta)\}$. Therefore,

$$\frac{\partial \ln\{L(\theta; x(0), x(1), \ldots, x(N-1))\}}{\partial \theta} = \frac{\partial}{\partial \theta} \left(N \ln\{1/\sqrt{2\pi}\} - \sum_{n=0}^{N-1} (x(n) - \theta)^2 / 2 \right)$$

$$= -\sum_{n=0}^{N-1} (x(n) - \theta) = 0 \tag{13.33}$$

and the solution of Equation 13.33 for θ is

$$\hat{\theta} = g(x(0), x(1), \ldots, x(N-1)) = \frac{1}{N} \sum_{n=0}^{N-1} x(n) \tag{13.34}$$

The preceding equation shows that the estimator maximizes $L(\theta)$. Therefore, the preceding statistic $g(.)$ (the sample mean value of the data) is the maximum likelihood estimator of the mean, $\theta \triangleq \mu$. Since $E\{\hat{\theta}\} = \frac{1}{N} \sum_{n=0}^{N-1} E\{x(n)\} = \frac{N\theta}{N} = \theta$, the estimator is unbiased one. ∎

Definition 8.2

If we choose a function $g(x) = g(x(0), x(1), \ldots, x(N-1))$ such that θ is replaced by $g(x)$, the likelihood function L is maximum. That is, $L(g(x); x(0), \ldots, x(N-1))$ is at least as great as $L(\theta; x(0), x(1), \ldots, x(N-1))$ for all $\theta \in \Theta$, or in mathematical form:

$$L(\hat{\theta}; x(0), x(1), \ldots, x(N-1)) = \sup_{\theta \in \Theta} \; L(\theta; x(0), x(1), \ldots, x(N-1)) \tag{13.35}$$

∎

Definition 8.3

Any statistic whose expectation is equal to a parameter θ is called an unbiased estimator of the parameter θ. Otherwise, the statistic is said to be biased.

$$E\{\hat{\theta}\} = \theta \quad \text{unbiased estimator} \tag{13.36}$$

∎

Definition 8.4

Any statistic that converges stochastically to a parameter θ is called a consistent estimator of that parameter. Mathematically, we write $\lim_{N \to \infty} \Pr\{|\hat{\theta} - \theta| > \varepsilon\} = 0$. ∎

If as $N \to \infty$ the relation

$$\hat{\theta} \to \theta$$

holds, the estimator $\hat{\theta}$ is said to be a **consistent estimator**. If, in addition, as $N \to \infty$, the relation

$$E\{\hat{\theta}\} \to \theta$$

holds, the estimator $\hat{\theta}$ is said to be **asymptotically unbiased**. Furthermore, if, as $N \to \infty$, the relation

$$\mathrm{var}\{\hat{\theta}\} \to \text{lowest value for all } \theta$$

holds, it is said that $\hat{\theta}$ is **asymptotically efficient**.

In Example 13.2.7.1,

Let the observed data $x(n)$ be the set

$$x(n) = c + v(n) \quad n = 0, 1, \ldots, N-1 \tag{13.37}$$

where c is an unknown constant greater than 0 and $v(n)$ is a WGN with zero mean and with unknown variance c. The PDF is

$$p(x; c) = \frac{1}{(2\pi c)^{N/2}} \exp\left[-\frac{1}{2c} \sum_{n=0}^{N-1} (x(n) - c)^2 \right] \tag{13.38}$$

Considering the preceding equation as a function of c, it becomes a likelihood function $L(c; x)$. Differentiating its natural logarithm, we obtain

$$\frac{\partial \ln[(2\pi)^{N/2} p(\mathbf{x};c)]}{\partial c} = \frac{\partial}{\partial c}\left[-\frac{N}{2}\ln c - \frac{1}{2c}\sum_{n=0}^{N-1}(x(n)-c)^2 \right]$$

$$= -\frac{N}{2}\frac{1}{c} + \frac{1}{c}\sum_{n=0}^{N-1}(x(n)-c) + \frac{1}{2c^2}\sum_{n=0}^{N-1}(x(n)-c)^2 = 0 \qquad (13.39)$$

where a multiplication of the PDF by a constant does not change the maximum point. From Equation 13.39, we obtain

$$\hat{c}^2 + \hat{c} - \frac{1}{N}\sum_{n=0}^{N=1}x^2(n) = 0$$

Solving for $\hat{c}$ and keeping the positive sign of the quadratic root, we find

$$\hat{c} = -\frac{1}{2} + \sqrt{\frac{1}{N}\sum_{n=0}^{N-1}x^2(n) + \frac{1}{4}} \qquad (13.40)$$

Note $\hat{c} > 0$ for all values of the summation under the square root. Since

$$E\{\hat{c}\} = E\left\{-\frac{1}{2} + \sqrt{\frac{1}{N}\sum_{n=0}^{N-1}x^2(n) + \frac{1}{4}}\right\} \neq -\frac{1}{2} + \sqrt{E\left\{\frac{1}{N}\sum_{n=0}^{N-1}x^2(n) + \frac{1}{4}\right\}} \qquad (13.41)$$

it implies that the estimator is biased. From the law of the large numbers as $N \to \infty$

$$\frac{1}{N}\sum_{n=0}^{N-1}x^2(n) \to E\{x^2(n)\} = \mathrm{var}(x) + E\{x^2\} = c + c^2 \qquad (13.42)$$

$$(E\{(x-\bar{x})^2\} \triangleq \mathrm{var}(x) = E\{x^2\} - 2E\{x\}\bar{x} + \bar{x}^2 = E\{x^2\} - \bar{x}^2)$$

and, therefore, Equation 13.40 gives

$$\left(\hat{c} + \frac{1}{2}\right)^2 = \left(c + c^2 + \frac{1}{4}\right) \text{ or } \bar{c}^2 + \frac{1}{4} + \bar{c} = c + c^2 + \frac{1}{4}$$

which indicates that

$$\bar{c} \to c \qquad (13.43)$$

Hence, the estimator is a consistent estimator. ∎

Example 13.2.7.2

It is required to find the maximum likelihood estimator for the mean m and variance σ^2 of a set of data $\{x(n)\}$ provided by a Gaussian random generator.

The Gaussian PDF $p(x; m, \sigma^2)$ for one random variable (RV) is

$$p(x; m, \sigma^2) = (2\pi\sigma^2)^{-1/2} \exp\left[-\frac{1}{2}\left(\frac{x-m}{\sigma}\right)^2\right] \tag{13.44}$$

Its natural logarithm is

$$\ln p(x; m, \sigma^2) = -\frac{1}{2}\ln(2\pi\sigma^2) - \frac{1}{2}\left(\frac{x-m}{\sigma}\right)^2 \tag{13.45}$$

The likelihood function for the data (IID) is given by

$$L(m, \sigma^2) = p(x(0); m, \sigma^2) p(x(1); m, \sigma^2) \cdots p(x(N-1); m, \sigma^2) \tag{13.46}$$

and its logarithm is

$$\ln L(m, \sigma^2) = \sum_{n=0}^{N-1} \ln p(x(n); m, \sigma^2) \tag{13.47}$$

Substituting Equation 13.45 in Equation 13.47, we obtain

$$\ln L(m, \sigma^2) = \sum_{n=0}^{N-1}\left[-\frac{1}{2}\ln(2\pi\sigma^2) - \frac{1}{2}\left(\frac{x(n)-m}{\sigma}\right)^2\right]$$

$$= -\frac{N}{2}\ln(2\pi\sigma^2) - \frac{1}{2\sigma^2}\sum_{n=0}^{N-1}(x(n)-m)^2 \tag{13.48}$$

There are two unknown in the logs of the likelihood function. Differentiating Equation 13.48 with respect to the mean and variance, we obtain

$$\frac{\partial \ln L}{\partial m} = \frac{1}{\sigma^2}\sum_{n=0}^{N-1}(x(n)-m); \quad \frac{\partial \ln L}{\partial(\sigma^2)} = -\frac{N}{2}\frac{1}{\sigma^2} + \frac{1}{2\sigma^4}\sum_{n=0}^{N-1}(x(n)-m)^2 \tag{13.49}$$

Equating the partial derivatives to 0, we obtain

$$\frac{1}{\hat{\sigma}^2}\sum_{n=0}^{N-1}(x(n)-\hat{m}) = 0 \tag{13.50}$$

$$-\frac{N}{2}\frac{1}{\hat{\sigma}^2} + \frac{1}{2\hat{\sigma}^4}\sum_{n=0}^{N-1}(x(n)-\hat{m})^2 = 0 \tag{13.51}$$

For the estimate variance to be different than 0, Equation 13.50 reduces to

$$\sum_{n=0}^{N-1}(x(n)-\hat{m}) = 0 \quad \text{or} \quad \hat{m} = \frac{1}{N}\sum_{n=0}^{N-1}x(n) = \text{sample mean} \tag{13.52}$$

Therefore, the MLE of the mean of a Gaussian population is equal to the sample mean, which indicates that the sample mean is an optimal estimator.

Multiplying Equation 13.51 by $2\hat{\sigma}^4$ leads to

$$-N\hat{\sigma}^2 + \sum_{n=0}^{N-1}(x(n)-\hat{m})^2 = 0 \text{ or } \hat{\sigma}^2 = \frac{1}{N}\sum_{n=0}^{N-1}(x(n)-\hat{m})^2 \tag{13.53}$$

Therefore, the MLE of the variance of a Gaussian population is equal to the sample variance. ∎

Let the PDF of the population be the Laplacian (a is a positive constant):

$$p\left(x;\theta\right) = \frac{a}{2}\exp(-a|x-\theta|) \tag{13.54}$$

Then, the likelihood function corresponding to the preceding PDF for a set of data $\{x(n)\}$ (IID) is

$$L(\theta) = \ln\left\{\left(\frac{a}{2}\right)^N \prod_{n=0}^{N-1}\exp(-a|x(n)-\theta|)\right\}$$

$$= -a\sum_{n=0}^{N-1}|x(n)-\theta| + N\ln\{a\} - N\ln\{2\} \tag{13.55}$$

Before we proceed further, we must define the term **median** for a set of random variables. Median is the value of an RV of the set when half of the RVs of the set have values less than the median and half have higher values (odd number of terms). Hence, the med{1, 4, 2, 3, 5, 9, 8, 7, 11}=med{1, 2, 3, 4, 5, 7, 8, 9, 11}=5.

Definition 8.5

Take an RV whose CDF (distribution) is F_x. The point x_{med} is the median of x if

$$F_x(x_{\mathrm{med}}) = \frac{1}{2} \tag{13.56}$$

∎

From Equation 13.55, we observe that the derivative of $-\sum_{n=0}^{N-1}|x(n)-\theta|$ with respect to θ is negative if θ is larger than the sample median, and positive if it is less than the sample median (remember that $|x(n)-\theta| = [(x(n)-\theta)(x(n)-\theta)]^{1/2}$ for real values). Therefore, the estimator

$$\hat{\theta} = \mathrm{med}\left\{\begin{array}{cccc} x(0) & x(1) & \cdots & x(N-1)\end{array}\right\} \tag{13.57}$$

which maximizes $L(\theta)$ for the Laplacian PDF likelihood function.

Note: (1) From the previous discussion, we have observed that the minimization of $L(\theta)$ created from a Gaussian distribution is equivalent in minimizing $\sum_{n=0}^{N-1}(x(n)-\theta)^2$. The minimization result in finding the estimator $\hat{\theta} = \frac{1}{N}\sum_{n=0}^{N-1}x(n)$, which is the sample mean of the data $x(n)$. (2) Similarly, the minimization of the likelihood function $L(\theta)$ created from a Laplacian distribution is equivalent in minimizing $\sum_{n=0}^{N-1} |x(n)-\theta|$. This minimization results in finding the estimator

$$\hat{\theta} = \text{med}\left\{ \begin{array}{cccc} x(0) & x(1) & \cdots & x(N-1) \end{array} \right\}$$

Exercise 13.2.7.1

Let $x(n)=s(n)+v(n)$ for $n=0$, 1, 2,..., $N-1$, where $v(n)$ are IID with zero mean and variance σ_n^2 for $n=0$, 1, 2,..., $N-1$. Find the as in Equation 13.27.   ▲

Exercise 13.2.7.2

Let the RVs $x(0)$, $x(1)$,..., $N(n-1)$ denoting a random sample, been IID and having PDF

$$p(x(m)) = \left\{ \begin{array}{ll} \theta^{x(m)}(1-\theta)^{1-x(m)} & x(m)=0,1,0 \le \theta \le 1 \\ 0 & \text{elsewhere} \end{array} \right.$$

Find the maximum likelihood function θ.   ▲

Exercise 13.2.7.3

A set of data is given by $x(n)=c+v(n)$ for $n=0$, 1,..., $N-1$, where $v(n)$ is a WGN with zero mean and known variance σ^2. Show that the estimator $\hat{c}$ is an efficient estimator.   ▲

13.3 MEAN FILTER

Let us consider the following signal:

$$x(n) = s(n)+v(n) \quad n=0,1,...,N-1 \tag{13.58}$$

where $v(n)$ is a WGN with zero mean and variance about one. For the deterministic signal, we have selected a constant, a ramp, and a step function. The reader can easily plot them using MATLAB.

For simplicity in our study, we shall use filters with odd numbers of elements and, hence, $K=2k+1$. Under these circumstances, the output of a **mean** filter at time n is given by

$$y(n) = \frac{1}{2k+1}\sum_{j=-k}^{k} x(n+j) \tag{13.59}$$

Because in the preceding equation for any n, we average only $2k+1$ values of the signal sample from $-k$ to $+k$ with the middle point n, the process is equivalent in multiplying the original signal with a window called the **filter window**. Because this window is moving with each new n, it is also known as the **moving filter window**. The signal samples that contribute to the output at time n are $x(n-k),x(n-k+1),...,x(n),...,x(n+k-1),x(n+k)$. For example, with filter length 5 and signal length 50, the output signal identically distributed (ID) is given by

$$y(n) = \frac{1}{5}\sum_{j=-2}^{2} x(n+j) \quad n=3,4,...,48 \tag{13.60}$$

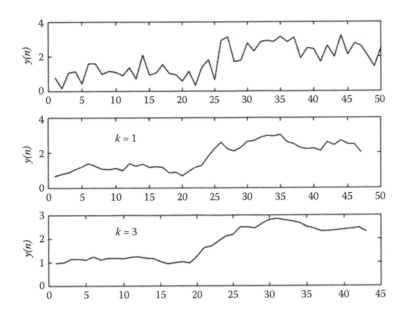

FIGURE 13.3

Figure 13.3 shows the noisy signal $x(n)$ and its filtered version for the mean filter of $k=1$ and $k=3$. The signal $s(n)$ is a step function with values 1 and 2.5. The random noise $v(n)$ is a WGN with mean zero. From the figure, we note that the output becomes **smoother** as the filter window increases, and the step change of the signal becomes a **ramp**. This indicates that the edges in images (abrupt change from white to black, for example) will be smoothed out, like looking at the image through a lens that is slightly out of focus.

BOOK MATLAB FUNCTION FOR ONE-DIMENSIONAL MEAN FILTER

```
function[yo1]=urdsp_one_dimens_mean_filter(x,k)
%filter length 2k+1;x=s+v=noisy signal
for n=1:length(x)-(2*k+1)
    for j=0:2*k
        y(j+1)=x(n+j);
    end;
    yo1(n)=sum(y)/(2*k+1);
end;
```

Example 13.3.1

The three-dimensional signal used in this example is the sum of a two-dimensional step function plus white noise as shown in the upper part of Figure 13.4. The lower part of Figure 13.4 shows the output of a two-dimensional mean filter with sliding window dimensions 7×7. The Book MATLAB functions and programs that produce Figure 13.4 follow.

BOOK MATLAB FUNCTION PRODUCING THE NOISY STEP SIGNAL

```
function[z]=urdsp_2d_step_signal(N,M,i)
    %N=points in y direction;M=points in x direction;
    %i=controlling intensity of the noise;
    %z=matrix presenting the step function and noise;
xy=1+i*rand(N,M);
xys=2*ones(N,M)+i*rand(N,M);
z=[xy xys];
```

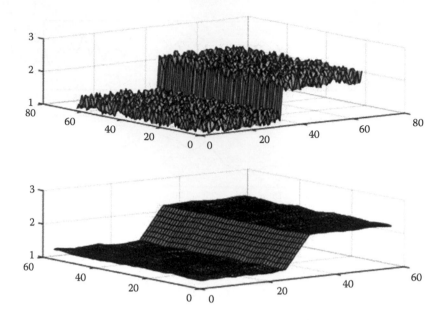

FIGURE 13.4

BOOK MATLAB FUNCTION THAT PRODUCES OUTPUT FROM A MEAN FILTER

```
function[yo2]=urdsp_2d_mean_filter(z,k)
    %z=2-dimensional signal with noise, matrix;
    %2k+1=width and length of sliding window;
for i=1:length(z(1,:))-(2*k+1)
    for j=1:length(z(:,1))-(2*k+1)
        zw=z(i:i+2*k,j:j+2*k);
        yo2(i,j)=sum(sum(zw))/((2*k+1)^2);
    end;
end;
```

TO PRODUCE FIGURE 13.4, WE USED THE FOLLOWING MATLAB PROGRAM

```
>>N=64;M=32;i=0.5;
>>z=ssp_2d_step_signal(N,M,i);
>>k=3;
>>yo=ssp_2d_mean_filter(z,k);%yo=57x57 matrix;
>>[X1,Y1]=meshgrid(1:64);[X,Y]=meshgrid(1:57);
>>subplot(2,1,1);
>>surfl(X1,Y1,z);
>>subplot(2,1,2);
>>surfl(X,Y,yo);
```
■

13.4 MEDIAN FILTER

When a single impulse exists in the signal, the mean filter spreads the impulse and reduces the amplitude. On the other hand, the **median** filter eliminates totally the impulse, provided that the pulse width is **less** than $k+1$ of a window length $N=2k+1$. Figure 13.5 shows the effect of mean and median filter on a one-dimensional signal. Note the different effect on the input these two figures have.

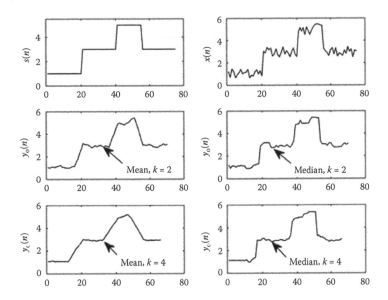

FIGURE 13.5

Book MATLAB One-Dimensional Median Filter Algorithm

```
function[yo2]=urdsp_2d_median_filter(x,k)
    %x=matrix (image);(2k+1)(2k+1)=number of
    %matrix elements of sliding window;
for i=1:length(x(1,:))-(2*k+1)
    for j=1:length(x(:,1))-(2*k+1)
        xw=x(i:i+2*k,j:j+2*k);
        cvxw=xw(:);
        yo2(i,j)=median(cvxw);
    end;
end;
```

Book MATLAB Two-Dimensional Median Filter Algorithm

```
function[yo2]=urdsp_2d_median_filter(x,k)
    %x=matrix (image);(2k+1)(2k+1)=number of
    %matrix elements of sliding window;
for i=1:length(x(1,:))-(2*k+1)
    for j=1:length(x(:,1))-(2*k+1)
        xw=x(i:i+2*k,j:j+2*k);
        cvxw=xw(:);
        yo2(i,j)=median(cvxw);
    end;
end;
```

Figure 13.6 shows in the upper part the two-dimensional step signal with noise and a pulse. The middle figure shows the output of a median filter with $k=2$. The bottom figure is the output of the median filter with $k=4$. Note the complete disappearance of the pulse in the noisy signal. To produce the figure, we used the following Book MATLAB program. First, we had to produce a step function with a pulse present. This is accomplished with the following Book MATLAB function with a pulse 6×6 and $N=64$, $M=32$, and $i=0.5$.

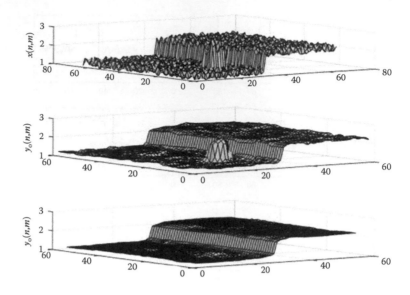

FIGURE 13.6

BOOK M-FUNCTION WITH PULSE PRESENT

```
function[z]=urdsp_2d_step_pluspulse_signal(N,M,i)
    %N=points in y direction;M=points in x direction;
    %i=controlling intensity of the noise;
    %z=matrix presenting the step function and noise;
    %assume M=N=60;
xy=i*rand(N,M);
xys=2.5*ones(N,M)+i*rand(N,M);
z=[xy xys];
z(20:25,20:25)=2.5+(rand(6,6)-0.5);
```

THE FOLLOWING BOOK PROGRAM WAS USED TO PRODUCE FIGURE 13.6

```
>>N=64;M=32;i=0.5;
>>z=urdsp_2d_step_pluspulse_signal(N,M,i);
>>subplot(3,1,1);
>>surfl(z);zlabel('z(n,m)');
>>subplot(3,1,2);
>>yo2k2=urdsp_2d_median_filter(z,2);
>>surfl(yo2k2);zlabel('y_0(n,m)');
>>subplot(3,1,3);
>>yo2k4=urdsp_2d_median_filter(z,4);
>>surfl(yo2k4);zlabel('y_0(n,m)');
```

13.5 TRIMMED-TYPE MEAN FILTER

13.5.1 $(r-s)$-FOLD TRIMMED MEAN FILTERS

It can be easily shown by simulation that the mean filter is more efficient in deleting Gaussian noise than the median filter. However, the mean filter is less efficient in removing impulse noise. As long as the length of the impulse is less than $k+1$, where $2k+1$ is the width of the sliding window, the median filter completely eliminates the impulse noise. When both Gaussian and impulse noise are present, the **trimmed mean filter** becomes a compromise between mean and median filter.

One of the forms of the trimmed mean filter is the **(r−s)-fold trimmed mean filter**, which is obtained by sorting the samples, omitting a total $r+s$ samples: $x_{(1)}, x_{(2)}, \ldots, x_{(r)}$ and $x_{(N-s+1)}, x_{(N-s+2)}, \ldots, x_{(N)}$ then average the remaining ones. The subscript with parentheses indicates and ascending random values, for example, $x_{(1)} \leq x_{(2)} \leq, \cdots, \leq x_{(r)}$. Hence, we write

$$\text{trim-mean } \{x(1), x(2), \ldots, x(N); r, s\} = \frac{1}{N-r-s} \sum_{i=r+1}^{N-s} x_{(i)} \qquad (13.61)$$

where $N = 2k+1$ is the width of the sliding window.

The following Book m-function executes the operation of one-dimensional trimmed mean filter.

BOOK M-FUNCTION FOR ONE-DIMENSIONAL TRIMMED MEAN FILTER

```
function[yo1]=urdsp_1d_trimmed_mean_filter(x,r,s,k)
    %x=input noisy signal;r=integer;s=integer;
    %r+s<N=2k+1=width of sliding window;
for n=1:length(x)-(2*k+1)
    for j=0:2*k
        y(j+1)=x(n+j);
    end;
    ys=sort(y);%sort(.)=MATLAB function;
    ystr=ys(r+1:2*k+1);
    yo1(n)=sum(ystr)/(2*k+1-r-s);
end;
```

For Figure 13.7, we built the signal x as follows:

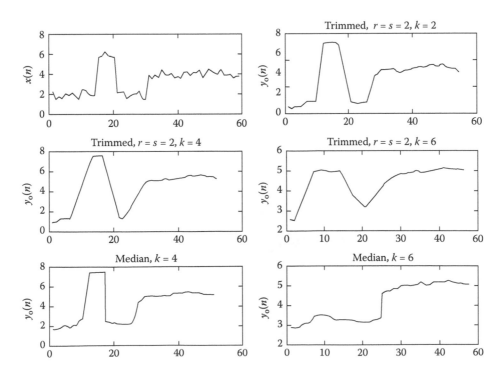

FIGURE 13.7

```
>>x1=2*ones(1,30)+(rand(1,30)-0.5);
>>x=[x1 4*ones(1,30)+(rand(1,30)-0.5)];
>>x(15:20)=6+(rand(1,6)-0.5);
>>              %substitutes the points 15 to 20 of x
>>              %with values of magnitude 6 plus noise;
```

The following book m-function executes the two-dimensional trimmed mean filter.

BOOK M-FUNCTION FOR THE TWO-DIMENSIONAL MEAN TRIMMED FILTER

```
function[yo2]=urdsp_2d_trimmed_mean_filter(x,r,s,k)
    %x=input 2d signal (matrix);r=integer;s=integer;
    %(2k+1)x(2k+1)=the number of elements inside the
    %sliding window;r+s<2k+1;yo2=output matrix;
for i=1:length(x(1,:))-(2*k+1)
    for j=1:length(x(:,1))-(2*k+1)
        w=x(i:i+2*k,j:j+2*k);
        ws=sort(w(:));%sort(.) is a MATLAB function;
        wstr=ws(r+1:2*k+1-s);
        yo2(i,j)=sum(wstr)/(2*k+1-r-s);
    end;
end;
```

Figure 13.8 shows at the top the input signal to the filter. The middle Figure 13.8 shows the output of a mean filter with $k=4$ and the bottom Figure 13.8 shows the output of a trimmed-mean filter with $r=2$, $s=2$ and $k=4$.

To plot, for example, the two-dimensional signal x (a 64×64 matrix) with dark color (black), we write the program: >> [X,Y]=meshgrid(1:64); surfl(X,Y,x); colormap(gray);%x is a two-dimensional matrix;.

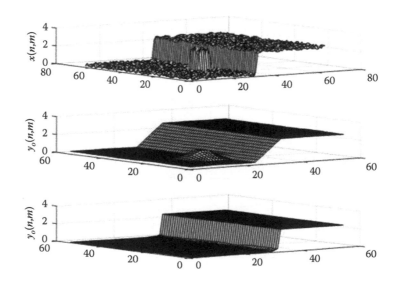

FIGURE 13.8

13.5.2   (*r,s*)-Fold Winsorized Mean Filter

A modification of the (*r,s*)-fold trimmed mean filter is accomplished by substituting the values of the *r* smallest samples with the value $x_{(r+1)}$, and the values of the *s* largest samples are replaced by $x_{(N-s)}$, yielding

$$\text{Winsorized mean filter } \{x(1),\ldots,x(N); r,s\} = \frac{1}{N}\left(rx_{(r+1)} + \sum_{i=r+1}^{N-s} x_{(i)} + sx_{(N-s)}\right) \qquad (13.62)$$

where $N = 2*k+1$.

Book m-Function for One-Dimensional (r,s)-Fold Winsorized Mean File

```
function[yo1]=urdsp_1d_winsorized_mean_filter(x,r,s,k)
    %x=input signal;r=integer;s=integer;r+s<N=2k+1=
    %width of sliding window;
for n=1:length(x)-(2*k+1)
    for j=0:2*k
        y(j+1)=x(n+j);
    end;
    ys=sort(y);%sort(.) is a MATLAB function;
    ystr=ys(r+1:2*k+1-s);
    yo1(n)=(r*ys(r+1)+sum(ystr)+s*ys(2*k+1-s))/(2*k+1);
end;
```

Book m-Function for Two-Dimensional (r,s)-Fold Winsorized Mean Filter

```
function[yo2]=urdsp_2d_winsorized_mean_filter(x,r,s,k)
    %x=input 2d signal (image)=matrix;r=integer;s=integer;
    %r+s<(2k+1)(2k+1)=number of elements inside the
    %2d window;yo2=output matrix;
for i=1:length(x(1,:))-(2*k+1)
    for j=1:length(x(:,1))-(2*k+1)
        w=x(i:i+2*k,j:j+2*k);
        ws=sort(w(:));
        wstr=ws(r+1:2*k+1-s);
        yo2(i,j)=sum(wstr)/(2*k+1-s-r);
    end;
end;
```

13.5.3   Alpha-Trimmed Mean Filter and Alpha-Winsorized Mean Filter

In both cases, the trimmed elements are assumed to be equal number, $r = s$. The trimmed number, often specified by a proportion denoted by $\alpha = j/N$, $0 \le j \le N/2$, is an integer. Hence, the number of samples trimmed at each side are αN. Because we have accepted N to be an odd number, we can select alpha to be even percentage such that αN is an even integer. The α-**trimmed mean filter** is given by

$$\text{Alpha-trimmed mean filter } \{x(1),\ldots,x(N); \alpha\} = \sum_{i=\alpha N+1}^{N-\alpha N} x_{(i)} \frac{1}{N-2\alpha N} \qquad (13.63)$$

BOOK M-FUNCTION FOR ONE-DIMENSIONAL ALPHA-TRIMMED MEAN FILTER

```
function[yo1]=urds_1d_alpha_trimmed_mean_filter(x,a,k)
    %x=data;a=proportion of trimmed elements;N=2k+1=
    %number of elements of the sliding window;
N=2*k+1;
for n=1:length(x)-N
    for j=0:2*k
        y(j+1)=x(n+j);
    end;
    ys=sort(y);
    ystr=ys(a*N+1:N-a*N);
    yo1(n)=sum(ystr)/(N-2*a*N);
end;
```

BOOK M-FUNCTION FOR TWO-DIMENSIONAL ALPHA-TRIMMED MEAN FILTER

```
function[yo2]=urdsp_2d_alpha_trimmed_mean_filter(x,a,k)
    %x=data;a=portion of trimmed elements;NxN=(2k+1)(2k+1)=
    %number of elements inside the sliding window;yo2=
    %filter output (matrix);a*N*N=must be an integer;
N=2*k+1;
for i=1:length(x(1,:))-N
    for j=1:length(x(:,1))-N
        w=x(i:i+2*k,j:j+2*k);
        ws=sort(w(:));
        wstr=ws(a*N*N+1:N*N-a*N*N);
        yo2(i,j)=sum(wstr)/(N*N-2*a*N*N);
    end;
end;
```

13.5.4 ALPHA-TRIMMED WINSORIZED MEAN FILTER

The alpha-trimmed Winsorized mean filter is given by the relationship:

$$\{x(1),x(2),\ldots,x(N); \alpha\} = \frac{1}{N}(\alpha N x_{(\alpha N+1)}$$

$$+ \sum_{i=\alpha N+1}^{N-aN} x_{(i)} + \alpha N x_{(N-\alpha N)})$$

(13.64)

BOOK M-FUNCTION FOR ONE-DIMENSIONAL ALPHA-TRIMMED WINSORIZED MEAN FILTER

```
function[yo1]=urdsp_1d_alpha_tr_wins_mean_filter(x,a,k)
    %x=data;a=proportion of trimmed elements;N=k+1=number
    %of elements inside the sliding window;a*N=must
    %be an integer;
N=2*k+1;
for n=1:length(x)-N
    for j=0:2*k
        y(j+1)=x(n+j);
    end;
    ys=sort(y);
    ystr=ys(a*N*N+1:N-a*N);
    yo1(n)=(sum(ystr)+a*N*ys(a*N+1)+a*N*ys(N-a*N))/N;
end;
```

BOOK M-FUNCTION FOR TWO-DIMENSIONAL ALPHA-TRIMMED WINSORIZED MEAN FILTER

```
function[yo2]=urdsp_2d_alpha_tr_wins_mean_filter(x,a,k)
    %x=data (matrix);a=portion of trimmed elements;
    %NxN=(2k+1)(2k+1)=number of elements inside the
    %sliding window;a*N*N=must be integer;
    %yo2=output (matrix);
N=2*k+1;
for i=1:length(x(1,:))-N
    for j=1:length(x(:,1))-N
        w=x(i:i+2*k,j:j+2*k);
        ws=sort(w);
        wstr=ws(a*N*N+1:N*N-a*N*N);
        yo2(i,j)=sum(wstr)/(N*N-2*a*N*N);
    end;
end;
```

13.6   L-FILTERS

L-estimators are useful and widely used because, by varying the associated constants, we obtain many useful estimators.

Definition 8.6

The L-estimators are of the form:

$$\text{L-estimator} \quad \hat{\theta} = \sum_{i=1}^{N} a_i x_{(i)} \quad a_i\text{'s} = \text{constants} \tag{13.65}$$

▲

The L-filters are also known as **order statistic filters**. They have been used by the statisticians for a long time since they have robust and often optimal properties for estimating population parameters of IID random variables. Furthermore, these filters are a compromise between nonlinear and linear operation since they include ordering and weighting. The L-filter is given by

$$\text{L-filter} \quad \{x(1), x(2), \ldots, x(N); \boldsymbol{a}\} = \sum_{i=1}^{N} a_i x_{(i)} \quad \boldsymbol{a} = \begin{bmatrix} a_1 & a_2 & \cdots & a_N \end{bmatrix}^T \tag{13.66}$$

If, in addition,

$$\sum_{i=1}^{N} a_i = 1 \tag{13.67}$$

the L-filter is known as the smooth L-filter. The great advantage of these filters is the ability to choose appropriate weighting factor to optimize the filtering in the mean square sense.

BOOK M-FUNCTION FOR ONE-DIMENSIONAL L-FILTER

```
function[yo1]=urdsp_1d_L_filter(x,a,k)
    %x=input vector data;a=input weighting vector,
    %length(a)=N;N=2k+1=length of sliding window;
N=2*k+1;
for n=1:length(x)-N
    for j=0:2*k
        y(j+1)=x(j+n);
    end;
    ys=sort(y);
    yo1(n)=sum(ys.*a);
end;
```

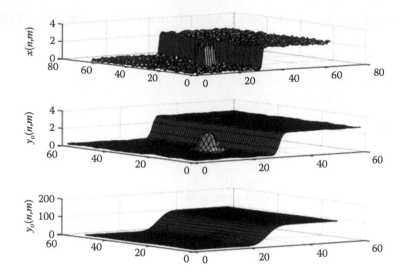

FIGURE 13.9

BOOK M-FUNCTION FOR TWO-DIMENSIONAL L-FILTER

```
function[yo2]=urdsp_2d_L_filter(x,a,k)
    %x=input signal, matrix;a=input weighting matrix (NxN);
    %NxN=(2k+1)(2k+1)=number of elements of the two-
    %dimensional window;
N=2*k+1;
for i=1:length(x(1,:))-N
    for j=1:length(x(:,1))-N
        w=x(i:i+2*k,j:j+2*k);
        ws=sort(w(:));
        yo2(i,j)=sum(a(:)'.*ws');
    end;
end;
```

The top of Figure 13.9 shows the step function with noise and a noisy pulse. The middle Figure 13.9 was produced with 2D L-filter and with $k=2$, which means that a=hamming(5)*hamming(5)′. The bottom Figure 13.9 was produced using the 2D L-filter and with $k=6$, which means that a=hamming(13)*hamming(13)′.

13.7 RANK-ORDER STATISTIC FILTER

The rank-order filters have been used by statisticians for a long time and are simple modifications of the median filter. Mathematically, we write

$$\text{Rank-order filter } \{x(1),x(2),\ldots,x(N);r\}= x_{(r)} \tag{13.68}$$

which is the rth order statistic of the sample $\{x(1),x(2),\ldots,x(N)\}$. If $r=k+1$, we obtain $x_{(k+1)}$, which acts as a median filter. If $r<k+1$, a bias towards lower values is accomplished. If $r>k+1$, a bias toward higher values is accomplished.

BOOK M-FUNCTION FOR ONE-DIMENSIONAL RANK-ORDER FILTER

```
function[yo1]=urdsp_1d_ranked_order_filter(x,r,k)
    %x=input data;r=integer<=N=2k+1;N=length
    %of sliding window;
N=2*k+1;
for n=1:length(x)-N
    for j=0:2*k
        y(j+1)=x(n+j);
    end;
    yst=sort(y);
    yo1(n)=yst(r);
end;
```

BOOK M-FUNCTION FOR TWO-DIMENSIONAL RANK-ORDER FILTER

```
function[yo2]=urdsp_2d_ranked_order_filter(x,r,k)
    %x=input data (matrix);r=integer<NxN;
    %NxN=number of elements inside the sliding
    %two-dimensional window; N=2k+1;
N=2*k+1;
for i=1:length(x(1,:))-N
    for j=1:length(x(:,1))-N
        w=x(i:i+2*k,j:j+2*k);
        ws=sort(w(:));
        yo2(i,j)=ws(r);
    end;
end;
```

The top part of Figure 13.10 is the step function with a pulse and random noise. The middle Figure 13.10 was created with $r=k=2$ and the bottom one was created with $r=2$ and $k=6$.

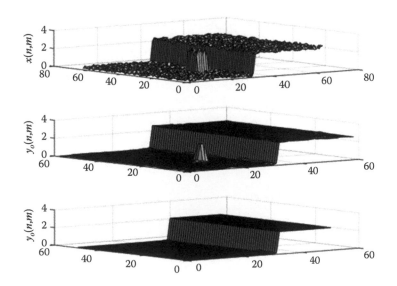

FIGURE 13.10

13.8 EDGE-ENHANCEMENT FILTERS

One type of edge-enhancement filters is the **comparison and selection** filter. Because the output of the median filter is smaller than the value of the output of the mean filter at the beginning of the signal edge, values of some smaller sample than the median is retained. Similarly, since the output of the median filter is larger than the output of the mean filter at the end of the signal edge, values of some larger sample than the median are retained. Formally, for the **comparison selection** filter, we write

$$\{x(1), x(2), \ldots, x(N); j\} = \{x_{(k+1-j)}, \text{mean}\{x(1), \ldots, x(N)\} \ge med\{x(1), \ldots, x(N)\} \quad (13.69)$$

Book m-Function for One-Dimensional Comparison Selection Filter

```
function[yo1]=urdsp_1d_compar_selection_filt(x,j,k)
    %x=data;N=2k+1=length of sliding window;
    %j=integer; 1<=j<k
N=2*k+1;
for n=1:length(x)-N
    for m=0:2*k
        y(m+1)=x(n+m);
    end;
    ys=sort(y);
    ym=mean(y);%mean()=MATLAB function;
    ymed=median(y);%median()=MATLAB function;
    if ym>=ymed
        yo1(n)=ys(k+1-j);
    else
        yo1(n)=ys(k+1+j);
    end;
end;
```

Book m-Function Two-Dimensional Comparison Selection Filter

```
function[yo2]=urdsp_2d_comparison_select_filt(x,m,k)
    %x=data (matrix); NxN=(2k+1)(2k+n1)=dimensions
    %of two-dimensional sliding window;m=integer,
    %1<=m<k;
N=2*k+1;
for i=1:length(x(1,:))-N
    for j=1:length(x(:,1))-N
        w=x(i:i+2*k,j:j+2*k);
        ws=sort(w(:));
        wm=mean(ws);
        wmed=median(ws);
        if wm>=wmed
            yo2(i,j)=ws(k+1-m);
        else
            yo2(i,j)=ws(k+1+m);
        end;
    end;
end;
```

Figure 13.11 shows the effect of increasing sharpness of the edge of a one-dimensional filter with the increase of the parameter *j*.

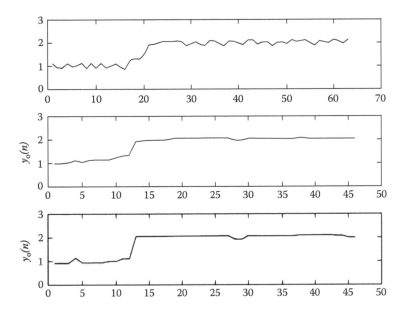

FIGURE 13.11

13.9 R-FILTERS

R-filters are the result of R-estimators that are **robust** and are based on rank **test**. The rank of an observation x_i is denoted by $R(x_i)$ and is given by

$$x_i = x_{(R(x_i))} \quad i = 1, 2, \ldots, N \tag{13.70}$$

which means that the rank of x_i is in the ordered sequence. Next, we assign the weights

$$w_{jk} = \frac{d_{N-k+1}}{\displaystyle\sum_{i=1}^{N} id_i} \tag{13.71}$$

to each of the $n(n+1)/2$ averages $(x_{(j)} + x_{(k)})/2$ for $j \le k$. Then, the R-estimator is the median of the discrete distribution that assigns the probability w_{jk} to each average $(x_{(j)} + x_{(k)})/2$ for $j \le k$. For example, if $d_1 = d_2 = \cdots = d_N = 1$, the distribution assigns the weights $2/(N(N+1))$ to each of the averages $(x_{(j)} + x_{(k)})/2$ for $j \le k$ and, thus,

$$\text{Hodges–Lehmann } \{x_1, \ldots, x_N\} \triangleq \hat{\theta} = \text{med}\{(x_{(j)} + x_{(k)})/2 : 1 \le j \le k \le N\} \tag{13.72}$$

known as the Hodges–Lehmann estimator.

Example 13.9.1

Let the input data be the vector $x = [4 \quad 5 \quad 16 \quad 9 \quad 6]$. Then we sort the vector to obtain $\{x\} = [4 \quad 5 \quad 6 \quad 9 \quad 16]$. The output of the Hodges-Lehmann estimator is $\text{med}\{(4+16)/2, (5+9)/2, (6+6)/2\} = \text{med}\{10, 7, 6\} = 7$. ∎

Book m-Function One-Dimensional Hodges–Lehmann Filter

```
function[yo1]=urdsp_1d_hodges_lehmann_filter(x,k)
    %x=data;N=2k+1=length of sliding window;
N=2*k+1;
for n=1:length(x)-N
    for m=0:2*k
        y(m+1)=x(n+m);
    end;
    ys=sort(y);
    for i=1:k+1
        aws(i)=(ys(i)+ys(N-i+1))/2;
    end;
    yo1(n)=median(aws);
end;
```

Book m=Function Two-Dimensional Hodges–Lehmann Filter

```
function[yo2]=urdsp_2d_hodges_lehmann_filter(x,k)
    %x=data matrix;NxN=(2k+1)(2k+1)=dimensions of
    %two dimensions sliding window;
N=2*k+1;
for n=1:length(x(1,:))-N
    for m=1:length(x(:,1))-N
        for m=1:length(x(:,1))-N
            w=x(n:n+2*k,m:m+2*k);
            ws=sort(w(:));
            for i=1:k+1
                aws(i)=(ws(i)+ws(N-i+1))/2;
            end;
            yo2(n,m)=median(aws);
        end;
    end;
end;
```

Figure 13.12 shows the result of the Hodges–Lehmann on a noisy two-dimensional signal with $k=2$ and $k=4$.

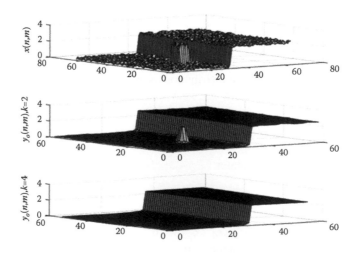

FIGURE 13.12

ADDITIONAL EXERCISES

1. Show that the mean filter is linear.
2. Find the output of a median filter if the input sequence is
   $x = \{ 1 \quad 9 \quad 4 \quad 3 \quad 8 \quad 7 \quad 7 \}$ and $k=1$.
3. Find the output of a trimmed mean filter if the input sequence is:
   $x = \{ 2 \quad 6 \quad 1 \quad 5 \quad 7 \quad 9 \quad 8 \}$ with $r=1$, $s=1$, and $k=2$.
4. Find the output of a (r,s)-fold Winsorized mean filter if the input is
   $x = \{ 2 \quad 6 \quad 1 \quad 5 \quad 7 \quad 9 \quad 8 \}$ and $r=1$, $s=1$, and $k=2$.
5. Find the output of an alpha-trimmed mean filter if the input is
   $x = \{ 2 \quad 6 \quad 1 \quad 5 \quad 7 \quad 9 \quad 8 \}$ with $\alpha=0.2$ and $k=2$.
6. Find the output of an L-filter if
   $x = \{ 2 \quad 1 \quad 6 \quad 4 \quad 5 \quad 9 \quad 8 \quad 7 \quad 3 \}$, $a = [0.4 \quad 0.2 \quad 0.4]$, and $k=1$.
7. Find the output of a rank-order filter with $r=2$ and $k=2$. The input data are
   $x = \{ 5 \quad 1 \quad 3 \quad 9 \quad 8 \quad 7 \quad 6 \quad 2 \quad 4 \}$.

PROBLEMS, SOLUTIONS, SUGGESTIONS, AND HINTS

13.2.3.1

$$\text{var}\{\hat{c}\} = \text{var}\left\{\frac{1}{N}\sum_{n=0}^{N-1}x(n)\right\} = E\{[\hat{c}-E\{\hat{c}\}]^2\} = E\left\{\left[\frac{1}{N}\sum_{n=0}^{N-1}x(n)-c\right]^2\right\}$$

$$= E\left\{\left(\frac{1}{N}\sum_{n=0}^{N-1}x(n)\right)^2 + c^2 - 2c\frac{1}{N}\sum_{n=0}^{N-1}x(n)\right\} = E\left\{\left(\frac{1}{N}\sum_{n=0}^{N-1}x(n)\right)^2\right\} + c^2 - 2cE\left\{\frac{1}{N}\sum_{n=0}^{N-1}x(n)\right\}$$

$$= E\left\{\left(\frac{1}{N}\sum_{n=0}^{N-1}x(n)\right)^2\right\} - c^2 \text{ (since } E\left\{\frac{1}{N}\sum_{n=0}^{N-1}x(n)\right\} = \frac{1}{N}\sum_{n=0}^{N-1}E\{x(n)\} = \frac{Nc}{N} = c)$$

$$= E\left\{\frac{1}{N^2}\sum_{n=0}^{N-1}x^2(n) + \frac{1}{N}\sum_{n\neq m}x(n)x(m)\right\} - c^2 = \frac{1}{N^2}\sum_{n=0}^{N-1}E\{x^2(n)\} + \frac{1}{N}\sum_{n=0}^{N-1}c^2 - c^2$$

$$= \frac{1}{N^2}\sum_{n=0}^{N-1}\text{var}\{x(n)\} + \frac{Nc}{N} - c^2 = \frac{N\sigma^2}{N^2} = \frac{\sigma^2}{N}; \ x(n),x(m) \text{ are independent]}$$

13.2.4.1

If we set $\hat{c} = \frac{1}{N}\sum_{n=0}^{N-1}x(n) = g(x(0),x(1),\ldots,x(N-1))$ (1), then

$$E\{\hat{c}\} = \frac{1}{N}\sum_{n=0}^{N-1}E\{x(n)\} = \frac{1}{N}\sum_{n=0}^{N-1}E\{c+v(n)\} = \frac{1}{N}\sum_{n=0}^{N-1}(c+E\{v(n)\}) = \frac{Nc}{N} = c$$

for all c. This implies that the estimator (1) is unbiased.

13.2.4.2

$$E\{\hat{c}\} = \frac{1}{4N} \sum E\{x(n)\} = \frac{1}{4}c$$

which indicates that if $c=0$, then $E\{\hat{c}\}=c$, and if $c \neq 0$, then $E\{\hat{c}\} \neq c$, and, thus, $\hat{c}$ is an unbiased estimator.

13.2.6.1

Taking the partial derivative of Equation 13.26 with respect to a_m, we find

$$\partial \text{mse}(\hat{\theta})/\partial a_m = -2E\{\theta x(m)\} + \frac{\partial}{\partial a_m} \left[\sum_{m=0}^{N-1} \sum_{n=0}^{N-1} a_m a_n E\{x(m)x(n)\} \right] = -2r_{\theta x}(m) + 2\sum_{n=0}^{N-1} a_n r_{mn} = 0 \quad (1) \quad \text{since}$$

$a_m a_n$ appears twice in the double summation expansion. Therefore, (1) is the mth equation of Equation 13.27.

13.2.7.1

$$r_{mn} = E\{x(m)x(n)\} = E\{(s(m)+v(m))(s(n)+v(n))\}:$$
$$= E\{(s(m)s(n)\} + E\{v(m)s(n)\} + E\{s(m)v(n)\} + E\{v(m)v(n)\} \quad (1)$$

If

$$m \neq n \ r_{mn} = \sigma_s^2 \quad (2)$$

and if

$$m = n \ r_{mm} = \sigma_s^2 + \sigma_v^2 \quad (3)$$

Also

$$r_{sx}(m) = E\{s(m)s(m) + s(m)v(m) = \sigma_s^2 + E\{s(m)v(m)\} = \sigma_s^2 \quad (4)$$

Hence $Ra = r\theta_x$, where

$$\mathbf{R} = \begin{bmatrix} \sigma_s^2 & \sigma_s^2 & \cdots & \sigma_s^2 \\ \sigma_s^2 & \sigma_s^2 & \cdots & \sigma_s^2 \\ & & \vdots & \\ \sigma_s^2 & \sigma_s^2 & \cdots & \sigma_s^2 \end{bmatrix} + \begin{bmatrix} \sigma_0^2 & 0 & \cdots & 0 \\ 0 & \sigma_1^2 & \cdots & 0 \\ & & \vdots & \\ 0 & 0 & \cdots & \sigma_{N-1}^2 \end{bmatrix}; \mathbf{a} \triangleq \begin{bmatrix} a_0 \\ \vdots \\ a_{N-1} \end{bmatrix} = \mathbf{R}^{-1} \begin{bmatrix} \sigma_s^2 \\ \vdots \\ \sigma_s^2 \end{bmatrix}$$

13.2.7.2

The likelihood function is

$$L(\theta; x(0), \ldots, x(N-1)) = \theta^{\sum x(m)} (1-\theta)^{N-\sum x(m)} \quad (1)$$

where $x(m)$ are 0 or $1 \leq$ for $m=0, 1, \ldots, N-1$.

The natural logarithm of (1) is $\ln L(\theta) = \sum_{m=0}^{N-1} x(m) \ln \theta + N - \sum_{m=0}^{N-1} x(m) + \ln(1-\theta)$. Hence,

$$\partial \ln L(\theta) / \partial \theta = \sum_{m}^{N-1} x(m)(1/\theta) - (N - \sum_{m}^{N-1} x(m)) / (1-\theta) = 0 \tag{2}$$

provided that θ is not zero or one. Solving (2) for θ, we obtain the maximum likelihood estimator $\hat{\theta} = \sum_{m=0}^{N-1} x(m) / N$, if we set $N=3$, $x(0)=0$, $x(1)=1$, $x(2)=1$ then $L(\theta) = \theta^2(1-\theta)^{3-2} = \theta^2(1-\theta)$ and $\hat{\theta} = 2/3$.

13.2.7.3

The PDF is

$$p(\mathbf{x};c) = p(x(0);c)p(x(1);c)\cdots p(x(N-1);c) = [1/(2\pi\sigma^2)^{N/2}]\exp\left[-\frac{1}{2\sigma^2}\sum_{n=0}^{N-1}(x(n)-c)^2\right] \tag{1}$$

The derivative of the logarithm of the likelihood function is

$$\partial \ln p(\mathbf{x};c) / \partial c = \frac{1}{\sigma^2}\sum_{n=0}^{N-1}(x(n)-c) = 0 \tag{2}$$

which yields the estimator

$$\frac{1}{\sigma^2}\sum_{n=0}^{N-1}x(n) = \frac{1}{\sigma^2}N\hat{c} \text{ or } \hat{c} = \frac{1}{N}\sum_{n=0}^{N-1}x(n) \tag{3}$$

From (2), we also obtain $\dfrac{\partial^2 \ln p(\mathbf{x};c)}{\partial c^2} = -\dfrac{N}{\sigma^2}$, which shows that the second derivative is a constant. From (2), we also obtain the mean value to be $E\{\hat{c}\} = \dfrac{1}{N}\sum_{n=0}^{N-1}E\{x(n)\} = (1/N)Nc = c$, which shows that the estimator is **unbiased**. Since the Cramer–Rao formula gives the relation $1/E\{-\partial^2 \ln p(x;c)/\partial c^2\} = \sigma^2/N$ indicates that the variance of the estimator is given in the formulas: $\mathrm{var}(\hat{c}) = \sigma^2/N$, which attains the lowest value, known as the *Cramer–Rao lower bound*, as $N \to \infty$. Hence the estimator is **asymptotically efficient**.

ADDITIONAL EXERCISES

1.
   The linearity is defined by $f(ax+by) = af(x)+bf(y)$, where a and b are real or complex constants. From the definition, we obtain

$$y(n) = \frac{1}{2k+1}\sum_{j=-k}^{k}x(n+j) = \frac{1}{2k+1}\sum_{j=-k}^{k}[s(n+j)+v(n+j)]$$

$$= \frac{1}{2k+1}\sum_{j=-k}^{k}s(n+j) + \frac{1}{2k+1}\sum_{j=-k}^{k}v(n+j) = (\text{filtered signal}) + (\text{filtered noise})$$

2. $N = 2*1+1 = 3$, $\mathrm{yo1} = \Big\{ \mathrm{med}\big\{ 1 \quad 9 \quad 4 \big\} \, \mathrm{med}\big\{ 9 \quad 4 \quad 3 \big\} \, \mathrm{med}\big\{ 4 \quad 3 \quad 8 \big\}$
   $\mathrm{med}\big\{ 3 \quad 8 \quad 7 \big\} \Big\} = \big\{ 4 \quad 4 \quad 4 \quad 7 \big\}$

3. $xs = \{1\ 2\ 5\ 6\ 8\ 9\}$, $\mathrm{yo1} = \left\{ \dfrac{2+5+6}{5-1-1}, \dfrac{5+6+7}{3}, \dfrac{6+7+8}{3} \right\} = \{4.3\ 6\ 7\}$

4. $xs = \big\{ 1 \quad 2 \quad 5 \quad 6 \quad 7 \quad 8 \quad 9 \big\}$, $\mathrm{yo1}(1) = (1\times 2 + (2+5+6) + 1\times 6)/5 = 4.2$, $\mathrm{yo1}(2) = 6$

5. $\mathrm{yo1}(1) = 1\times 0.4 + 2\times 0.2 + 6\times 0.4 = 3.2$, $\mathrm{yo1}(2) = 1\times 0.4 + 4\times 0.2 + 6\times 0.4 = 3.6$, $\mathrm{yo1}(3) = 5.0$,
   $\mathrm{yo1}(4) = 6.2$, $\mathrm{yo1}(5) = 7.2$, $\mathrm{yo1}(6) = 8.0$

6. $\mathrm{yo1}(1) = 1\times 0.4 + 2\times 0.2 + 6\times 0.4 = 3.2$, $\mathrm{yo1}(2) = 1\times 0.4 + 4\times 0.2 + 6\times 0.4 = 3.6$, $\mathrm{yo1}(3) = 5.0$,
   $\mathrm{yo1}(4) = 6.2$, $\mathrm{yo1}(5) = 7.2$, $\mathrm{yo1}(6) = 8.0$

7. $N = 5$, $\mathrm{yo1s}(1) = \big\{ 1 \quad 3 \quad 5 \quad 8 \quad 9 \big\}$, $\mathrm{yo1}(1) = x_{(2)} = 3$, $\mathrm{yo1}(2) = \big\{ 1 \quad 3 \quad 6 \quad 7 \quad 8 \quad 9 \big\}$,
   $\mathrm{yo1}(2) = x_{(2)} = 3$, $\mathrm{yo1s}(3) = \big\{ 2 \quad 6 \quad 7 \quad 8 \quad 9 \big\}$, $\mathrm{yo1}(3) = x_{(2)} = 6$, etc.

Appendix 1: Suggestions and Explanations for MATLAB Use

The reader who does not have a lot of experience with MATLAB should review this appendix and try to execute the presented material in MATLAB before reading the text.

CREATING A DIRECTORY

We suggest that, for a particular project, you create your own directory where your own developed MATLAB m-files are stored. However, whenever you need any of these files, you must include the directory in the MATLAB path. Assume that you have the following directory path: c:\ap\sig-syt\ ssmatlab. You can use the following two approaches:

```
>>cd 'c:\ap\sig-syst\ssmatlab'
```

or

```
>>path(path,'c:\ap\sig-syst\ssmatlab')%remember to introduce the path any
% time you start new MATLAB operations; the symbol % is necessary
% for the MATLAB to ignore the explanations;
```

The MATLAB files are included in "ssmatlab" directory.

HELP

If you know the name of a MATLAB function and you would like to know how to use it, write the following command in the command window:

```
>>help sin
```

or

```
>>help exp
```

and so on.
  To look for a keyword, write

```
>>look for filter
```

SAVE AND LOAD

When you are in the command window and you have created many variables and, for example, you would like to save two of them in a particular directory and in a particular file, proceed as follows:

```
>>cd 'c:\ap\matlabdata'
>>save data1 x dt %it saves in the matlabdata directory the
                  %file data1 having the two variables x and dt;
```

Assume now that you want to bring these two variables in the working space to use them. You first change directory, as you did previously, and then you write in the command window:

```
>>load data1
```

Then, the two variables will appear in the working space ready to be used.

MATLAB AS CALCULATOR

```
>>pi^pi-10;
>>cos(pi/4);
>>ans*ans %the result will be (√2 / 2) × (√2 / 2) = 1 / 2 because the
          %first output is eliminated, only the last output is
          %kept in the form of ans;
```

VARIABLE NAMES

```
>>x=[1  2  3  4  5];
>>dt=0.1;
>>cos(pi*dt);%since no assignment takes place there is no
             %variable;
```

COMPLEX NUMBERS

```
>>z=3+j*4;%note the multiplication sign;
>>zs=z*z;%or z^2 will give you the same results;
>>rz=real(z);iz=imag(z):%will give rz=3, and iz=4;
>>az=angle(z); abz=abs(z);%will give az=0.9273 rad, and abz=5;
>>x=exp(-z)+4;%x=3.9675+j0.0377;
```

ARRAY INDEXING

```
>>x=2:1:6;%x is an array of the numbers {2, 3, 4, 5, 6};
>>y=2:-1:-2:%y is an array of the numbers {2, 1, 0, -1, -2};
>>z=[1  3  y];%z is an array of the numbers {1, 3, 2, 1, 0, -1, -2};
             %note the required space between array numbers;
>>xt2=2*x;%xt2 is an array of numbers of x each one multiplied by 2;
>>xty=x.*y;%xty is an array of numbers which are the result of
           %multiplication of corresponding elements, that is
           %{4, 3, 0, -5, -12};
```

EXTRACTING AND INSERTING NUMBERS IN ARRAYS

```
>>x=2:1:6;
>>y=[x zeros(1,3)];%y is an array of the numbers {2, 3, 4, 5, 6,
                   %0, 0, 0};
>>z=y(1,3:7);%1 stands for row 1 which y is and 3:7 instructs to
             %keep columns
             % 3 through 7 the result is the array {4, 5, 6, 0, 0};
lx=length(x);%lx is the number equal to the number of columns of
             %the row vector x, that is lx=5;
x(1,2:4)=4.5*(1:3);%this assignment substitutes the elements of x
                   %at column
                   %positions 2,3 and 4 with the numbers 4.5*[1 2 3]=4.5, 9,
                   % and 13.5, note the columns of 2:4 and 1:3 are the same;
```

```
x(1,2:2:length(x))=pi;% substitutes the columns 2 and 4 of x with
        %the value of pi, hence the array is {2, 3.1416, 4,3.1416 6}
```

VECTORIZATION

```
>>n=0:0.2:1;
>>s=sin(0.2*pi*n);% the result of these two commands gives the
        %signal s (sine function) at times (values of n) 0, 0.2,
        %0.4, 0.6, 0.4, 1;
```

This approach is preferable since MATLAB executes faster the vectorization approach rather than the loop approach, which is

```
>>s=[];% initializes all values of vector s to zero;
>>for n=0:5% note that the index must be integer;
>>s(n+1)=sin(0.2*pi*n*0.2);% since we want values of s every 0.2
            %seconds we must multiply n by 0.2; note also that
            %for n=0 the variable becomes s(1) and this
            %because the array in MATLAB always starts
            %counting columns from 1;
>>end;
```

The results are identical to the previous one.

WINDOWING

The following windows are used for correcting, to some extent, the effects of truncating a vector.

@bartlett	-	Bartlett window
@barthannwin	-	Modified Bartlett–Hanning window
@blackman	-	Blackman window
@blackmanharris	-	Minimum four-term Blackman–Harris window
@bohmanwin	-	Bohman window
@chebwin	-	Chebyshev window
@flattopwin	-	Flat Top window
@gausswin	-	Gaussian window
@hamming	-	Hamming window
@hann	-	Hann window
@kaiser	-	Kaiser window
@nuttallwin	-	Nuttall defined minimum four-term Blackman–Harris window
@parzenwin	-	Parzen (de la Valle–Poussin) window
@rectwin	-	Rectangular window
@tukeywin	-	Tukey window
@triang	-	Triangular window

Example:
```
>>N = 65;
>>w = window(@blackmanharris,N);
>>w1= window(@hamming,N);
>>w2= window(@gausswin,N,2.5);
>>plot(1:N, [w,w1,w2]); axis([1N 0 1]);
>>legend('Blackman-Harris','Hamming','Gaussian');
```

MATRICES

If a and b are matrices such that a is a 2x3 and b is 3x3, c=a*b is a 2×3 matrix.

```
>>a=[1 2; 4 6]; %a is a 2x2 matrix
```
$\begin{bmatrix} 1 & 2 \\ 4 & 6 \end{bmatrix}$;

```
>>b=a';%b is a transposed 2x2 matrix of a and is
```
$\begin{bmatrix} 1 & 4 \\ 2 & 6 \end{bmatrix}$;

```
>>da=det(a);%da is a number equal to the determinant of a, da=-2;
>>c=a(:);%c is a vector which is made up of the columns of a,
        %c=[1  4  2  6];
>>ia=inv(a); %ia is a matrix which is the inverse of a;
>>sa1=sum(a,1);%sa1 is a row vector made up of the sum of the
              %rows,sa1=[5  8];
>>sa2=sum(a,2);%sa2 is a column vector made up by the sum of the
              %columns,sa2=[3  10]';
```

PRODUCE A PERIODIC FUNCTION

```
>>x=[1  2  3  4];
>>xm=x'*ones(1,5);%xm is 4x5 matrix and each of its column is x';
>>xp=xm(:)';% xp is a row vector, xp=[x  x  x  x  x];
```

SCRIPT FILES

Script files are m-files that, when you introduce their names in the command window, you receive the results. You must, however, have the directory that includes the file in the MATLAB search directories. You can modify the file any desired way and get new results. Suppose that any time you ask for the file pexp.m, the magnitude and angle of the exponential function $e^{j\omega}$ are plotted. To accomplice this, you first go to the command window and open a new m-file. At the window, you type the file as follows. As soon as you finished typing, you click *Save As* and save the file in, say, c:\ap\ssmatlab. If you want to see the results, at the command window, write pexp and hit the Enter key.

SCRIPT FILE PEXP.M

```
>>w=0:pi/500:pi-pi/500;%they are 500 at pi/500 apart;
>>x=exp(j*w);ax=abs(x);anx=angle(x);
>>subplot(2,1,1);plot(w,ax,'k')%'k' means plot line in black;
>>xlabel('\omega rad/s');ylabel('Magnitude');
>>subplot(2,1,2);plot(w,anx,'k');
>>xlabel('\omega rad/s');ylabel('Angle');
```

If you have the function $\dfrac{2e^{j\omega}}{e^{j\omega} - 0.5}$ and want to plot the results as previously, substitute in the script file the function x with the function:

```
x=2*exp(j*w)./(exp(j*w)-0.5);
```

In the preceding MATLAB expression, note the dot before the slash. This instructs MATLAB to operate at each value of w separate and, thus, give results at each frequency point.

FUNCTIONS

Here is an example of how to write functions. The reader should also study the functions that are presented throughout the book. In Fourier series, for example, you have to plot functions of the form

$$s(t) = \sum_{n=0}^{N} A_n \cos n\omega_0 t$$

and you want to plot this sum of cosines, each one having different amplitude and frequency. Let $A = \begin{bmatrix} 1 & 0.6 & 0.4 & 1 \end{bmatrix}$, $\omega_0 = 2$ and $0 \le t \le 4$. You approach this solution by vectorizing the summation. The MATLAB function is of the form

```
function[s]=sumofcos(A,N,w0,rangeoft)
n=0:N-1;
s=A*cos(w0*n'*rangeoft)
```

```
%when we want to use this function at the command window to find s we
%write for example:
```

```
>>A=[1   0.6   0.4   0.1];N=4;w0=2;rangeoft=0:0.05:6;
```

```
>> [s]=sumofcos(A,N,w0,rangeoft);
```

At the enter key click the vector s is one of the variables in the command window and it can be plotted at the wishes of the reader; you must secure that the directory in which sumofcos function exists is in the MATLAB %path; after you type the function in the editing window you save as.. in the directory, for example, c:\ap\ssmatlab and filename: sumofcos.m

It is recommended that the reader set small numbers for N ($N=4$) and range of t (0:0.2:1) and produce first the matrix `cos(w0*n'*t)` and, then, see the result `A*cos(w0*n'*t)`.

COMPLEX EXPRESSIONS

You can produce results by writing, for example,

```
>>x=[1 3 1 5 3 4 5  8];
>>plot(abs(fft(x,256)),'r');%will plot in red color the spectrum of the
                        %vector x of 256 points;
```

AXES

```
>>axis([xmin xmax ymin ymax]);%sets the max and min values of the axes;
>>grid on;%turns on grid lines in the graph;
```

2D GRAPHICS

To plot a sine and a cosine signal

```
>>x=linspace(0,2*pi,40);%produces 40 equal spaced points between 0 and
                        %2π;
>>y=sin(x);plot(x,y,'r');%will plot the sine signal with color red;
>>y1=cos(x);plot(x,y1,'g');%will plot the cosine signal with color
                        %green;
```

For other color lines: 'y'=yellow,'c'=cyan,'b'=blue,'w'=white,'k'=black

Type of lines: 'g:'=green dotted line,'r--'=red dashed line,'k--x'=black dashed line with x's,'k-.'=black dash-dot line,'+'=plus sign,'ko'=black circles

Add Greek Letters: \omega=will produce Greek lower case omega,\Omega= will produce capital case Greek omega. The same is true for the rest of the Greek letters. For example if we want to write the frequency in a figure under the x-axis, in the command window we write:>>xlabel('\omega rad/s');. For an omega with a subscript 01 we write: >>xlabel('\omega_{01} rad/s');

Add grid lines: >>grid on;%this is done after the command plot;

Adjusting axes: >>axis square;%sets the current plot to be square rather %than the default rectangle;

>>axis off;%turn off all axis labeling, grid, and tick %marks;leave the title and any labels placed by the 'text' %and 'gtext' commands;

>>axis on;%turn on axis labeling, tick marks and grid;

>>axis([xmin xmax ymin ymax]);%set the maximum and %minimum values of the axes using values given in the row %vector;

Subplots (Example): >>n=0:100;x=sin(n*pi*n);y=cos(n*pi*n);z=x.*y;w=x+y; %subplot(2,2,1);plot(n,x);subplot(2,2,2);plot(n,y); %subplot(2,2,3);plot(n,z);subplot(2,2,4);plot(n,w);

Log plotting: >>semilogx(x);%will plot the vector x in log scale in x-axis %and linear scale in y-axis;

>>semilogy(x);%will plot the vector x in log scale in y-%direction and linear scale in the x-axis;

>>loglog(x);%will plot the vector x in log scale both axes;

Histogram: >>x=randn(1,1000);hist(x,40);colormap([0 0 0]);%will plot a Gaussian histogram of 40 bars white;if instead we entered the vector [1 1 1] the bars would be black;the vector [1 0 0] will give red and the vector [0.5 0.5 0.5] will produce gray;

>>x=-3.0:0.05:3;y=exp(-x.*x);bar(x,y);colormap([.5 .5 .5]); %will produce bar-figure of the bell curve with gray color; >>**stairs**(x,y,'k');%will produce a stair-like black curve;

Add words: >>gtext('the word');

After the return, the figure and a crosshair appear. Move the center to the point in the figure where the word must start and click.

Add legend: >>plot(x1,y1,'+',x2,y2,'*');%there will be two curves in the %graph; >>legend('Function 1','Function 2');

The following rectangle will appear in the figure:

```
+ Function 1
* Function 2
```

3D PLOTS

MESH-TYPE FIGURES

If, for example, you desire to plot the function $f(x) = e^{-(x^2+y^2)}$ in the ranges $-2 \leq x \leq 2$, $-2 \leq y \leq 2$, you proceed as follows:

```
>>x=-2:0.1:2;y=-2:0.1:2;[X,Y]=meshgrid(x,y);
>>f=exp(-(X.*X+Y.*Y));mesh(X,Y,f);colomap([0 0 0]);
```

The preceding commands will produce a mesh-type 3D figure with black lines.

GENERAL-PURPOSE COMMANDS

MANAGING COMMANDS AND FUNCTION

help   Online help for MATLAB functions and m-files, for example, >>help plot

path   Shows the path to MATLAB directories that are available at the command window

MANAGING VARIABLES AND THE WORKPLACE

clear   Removes all the variables and items in the memory. Let us assume that the memory contains the variables x,y,z, then ≫clear x z; only y will remain in the memory

length   A number that gives the length of a vector. ≫x=[1 3 2 5]; then ≫length(x); will give the number 4. If you write ≫y=length(x); the variable y is equal to 4.

size   Array dimensions. ≫x=[1 3 2 5]; then size(x) will give the numbers 1 4 which means one row and four columns. Let us write ≫x=[1 2; 3 5; 6 4]; then size(x) will give the numbers 3 2, which means that x is a matrix of 3×2 dimensions

who   Produces a list of the variables in the memory

format   This command is used as follows for display: ≫format short,pi; will produce the number 1.1416, ≫format long,pi; will produce the number 3.14159265358979, and ≫format long,single(pi); will produce the number 3.1415927

OPERATORS AND SPECIAL CHARACTERS

OPERATORS AND SPECIAL CHARACTERS

+   Plus

-         Minus

*   Number and matrix multiplications

.*   Array multiplication. >>x=[1 2 3];y=[2 3 4]; z=x.*y; hence, z=[2 6 12]

.^   Array power. >>x=[2 3 4]; y=x.^3; hence y=[8 27 64]. >>x=[2 4;1 5]; y=x.^2; hence y=[4 16;1 25]

/   Right division

./   Array division. >>x=[2 4 6]; >>y=[4 4 12]; z=x./y; hence z=[0.5 1 0.5]

:   Colon. >>x=[1 3 6 2 7 8]; y=x(1,3:6); hence y=[6 2 7 8]

.   Decimal point

...   Continuation. >>x=[1 4 6 7 8 9 ... >> 2 5 8 1]; The vector x is interpreted by MATLAB
           as a row vector having 10 elements

%   Comments. >>x=[1 4 2 6];%this is a vector. MATLAB ignores "this is a vector."

           Transpose of a matrix or vector. >>x=[2 6 3]; y=x'; will have $y = \begin{bmatrix} 2 \\ 6 \\ 3 \end{bmatrix}$

&   Logical AND

| Logical OR

~   Logical NOT

xor   Logical exclusive (XOR)

CONTROL FLOW

for   Repeat statements a specific number of times.
           >>for n=0:3;
           >> x(n+1)=sin(n*pi*0.1);%observe the n+1 , if the +1 was not there x(0)
           >>end;   %was not defined by MATLAB

           Then x=[0 0.3090 0.5878 0.8090]

           >>for n=0:2
           >>   for m=0:1
           >>   x(n+1,m+1)=n^2+m^2;
           >>   end;
           >>end;

           Then $x = \begin{bmatrix} 0 & 1 \\ 1 & 2 \\ 4 & 5 \end{bmatrix}$

while   Repeat statements an indefinite times of times.

           >>a=1;num=0;
           >>while (1+a)<=2 & (1+a)>=1.0001
           >>a=s/2;
           >>num=num+1;
           >>end;
           We obtain a=0.0001, and num=14

if   Conditionally execute statements.

           if expression
                   commands evaluated if true
           else
                   commands evaluated if false
           end

       If there is more than one alternative, the if-else-end statement takes the form

elseif
           if expression1
                   commands evaluated if expression1 is true

>   *elseif expression2*
>   *commands evaluated if expression2 is true*
>   *elseif ....*
>   .
>   .
>   .
>   *else*
>   *commands evaluated if no other expression is true*
>   *end*

ELEMENTARY MATRICES AND MATRIX MANIPULATION

ELEMENTARY MATRICES AND ARRAYS

eye(n,n)	Identity matrix (its diagonal elements are ones and all the others are zeros)
linspace	linspace(x1,x2) generates 100 equally spaced points between x1 and x2. Linspace(x1,x2,N) generates N equally spaced points between x1 and x2
ones	ones(1,5) generates a row vector with its elements only ones ones(2,4) generates a 2×4 matrix with all its elements ones
rand	Uniformly distributed random numbers. >>x=rand(1,5); x is a row vector of 5 elements of random numbers. >>x=rand(2,3); x is a 2×3 matrix whose elements are random numbers
randn	Normally distributed random numbers. Applications are similar to rand
zeros	Creates arrays and matrices of all zeros. >>x=zeros(1,4); x is a row vector of four elements all with zero value. >>x=zeros(3,4); x is a 3×4 matrix with all of its elements zero
: (colon)	Regularly spaced vector. >>x=[1 4 2 5 8 3]; y=x(1,3:6); hence, y=[2 5 8 3];
eps	Floating-point relative accuracy. To avoid NA response in case there is a zero over zero expression at a point, as in the sinc function, you, for example, write >> n=-4:4;x=sin(n*pi*.1)./((n*pi+eps);
i,j	Imaginary unit
pi	Ratio of a circle's circumference to its diameter

MATRIX MANIPULATION

diag	Diagonal matrices and diagonals of a matrix. >>x=[1 3 5; 2 6 9; 4 7 0]; y=diag(x); will give a column vector y=[1 6 0]^T. >>y1=diag(x,1); will give a column vector y1=[3 9]^T which is the diagonal above the main diagonal. >>y2=diag(x,-1); will give the column vector y2=[2 7] which is the diagonal just below the main diagonal y3=diag(diag(x)); will give a 3×3 matrix with the diagonal 1,6,0 and the rest of the elements zero
fliplr	Flips vectors and matrices left and right
flipud	Flip matrices and vectors up and down
tril	Lower triangular part of a matrix including the main diagonal and the rest are zero. If x=[1 3 5; 2 6 9; 4 7 0] then y=tril(x) is the matrix [1 0 0; 3 6 0; 4 7 0]
triu	Upper triangular part of a matrix
toeplitz	Produces a Toeplitz matrix given a vector. >>x=[1 5 2];y=toeplitz(x) produces the matrix y=[1 5 2; 5 1 5; 2 5 1]

ELEMENTARY MATHEMATICS FUNCTION

ELEMENTARY FUNCTIONS

abs	Absolute value of a number and the magnitude of a complex number
acos, acosh	Inverse cosine and inverse hyperbolic cosine
acot, acoth	Inverse cotangent and inverse hyperbolic cotangent
acsc, acsch	Inverse cosecant and inverse hyperbolic cosecant
angle	Phase angle of a complex number. angle(1+j)=0.7854
asec, asech	Inverse secant and inverse hyperbolic secant
asin, asinh	Inverse sine and inverse hyperbolic sine
atan, atanh	Inverse tangent and inverse hyperbolic tangent
ceil	Round toward infinity, for example, ceil(4.22)=5
conj	Complex conjugate, for example, conj(2+j*3)=2-j*3
cos, cosh	Cosine and hyperbolic cosine
cot, coth	Cotangent and hyperbolic cotangent
csc, csch	Cosecant and hyperbolic cosecant
exp	Exponential, for example, exp(-1)=1/e=0.3679
fix	Rounds towards zero, for example, fix(-3.22)=-3
floor	Round towards minus infinity, for example, floor(-3.34)=-4 and floor(3.65)=3
imag	Imaginary part of a complex number, for example, imag(2+j*5)=5
log	Natural logarithm, for example, log(10)=2.3026
log2	Based 2 logarithm, for example, log2(10)=3.3219
log10	Common (base 10) logarithm, for example, log10(10)=1
mod	Modulus (signed remainder after division), for example, mod(10,3)=1 and mod(10,4)=2. In general, mod(x,y)=x-n*y
real	Real part of complex number
rem	Remainder after division, for example, rem(10,3)=1, rem(10,5)=0, and rem(10,4)=2
round	Round to the nearest integer, for example, round(3.22)=3 and round(3.66)=4
sec, sech	Secant and hyperbolic secant
sign	Signum function. sign(x)=0 for x=0, sign(x)=1 for x>0 and sign(x)=-1 for x<1
sin, sinh	Sine and hyperbolic sine
sqrt	Square root, for example, sqrt(4)=2
tan, tanh	Tangent and hyperbolic tangent
erf, erfc	Error and co-error function
gamma	Gamma function. for example, gamma(6)=120 or 1*2*3*4*(6-1)=120

NUMERICAL LINEAR ALGEBRA

MATRIX ANALYSIS

det	Matrix determinant >>a=[1 2; 3 4]; det(a)=1×4-2×3=-2
norm	The norm of a vector, for example, norm(v)=sum(abs(v).^2)^(1/2)
rank	Rank of a matrix. rank(A) provides the number of independent columns or rows of matrix A
trace	Sum of the diagonal elements, for example, trace([1 3; 4 12])=13

eig   Eigenvalues and eigenvectors. >>[v,d]=eig([1 3; 5 8]); Therefore

$$v = \begin{bmatrix} -0.8675 & -0.3253 \\ 0.4974 & -0.9456 \end{bmatrix}, \quad d = \begin{bmatrix} -0.7202 & 0 \\ 0 & 9.7202 \end{bmatrix}$$

inv   Matrix inversion, for example, >>A=[1 3; 5 8]; B=inv(A); Therefore,

$$B = \begin{bmatrix} -1.1429 & 0.4286 \\ 0.7143 & -0.1429 \end{bmatrix}, \quad A*B = \begin{bmatrix} 1.0000 & 0 \\ 0 & 1.0000 \end{bmatrix}$$

DTA ANALYSIS

BASIC OPERATIONS

max   Maximum element of an array. >>v=[1 3 5 2 1 7];x=max(v); Therefore, x=7

mean   Average or mean value of an array, for example, mean([1 3 5 2 8])=19/5=3.8

median   Median value of an array, for example, median([1 3 5 2 8])=3

min   Minimum element of an array

sort   Sorts elements in ascending order, for example, sort([1 3 5 2 8])=[1 2 3 5 8]

std   Standard deviation

sum   Sum of an array elements, for example, sum([1 3 5 2 8])=19

FILTERING–CONVOLUTION

conv   Convolution and polynomial multiplication, for example,conv([1 1 1])=[1 2 3 2 1], if you have to multiply these two polynomials (x2+2x+1)*(x+2), you convolve their coefficients conv([1 2 1],[1 2])=[1 4 5 2]; therefore, you write the polynomial x^3+4x^2+5x+2

conv2   Two-dimensional convolution

filter   Filter data with infinite impulse response or finite impulse response (FIR) filter. Let the FIR filter be given by y(n)=0.8x(n)+0.2x(n-1)-0.05x(n-2). Let the input data be x=[0.5 -0.2 0.6 0.1]. Hence, a=[1], b=[0.5 0.2 -0.05] and the output is given by y=filter(a,b,x). The result is y=[0.6250 -0.4063 0.8906 -0.1230].

FOURIER TRANSFORMS

abs   Absolute value and complex magnitude, for example, abs(4+j*3)=5, abs([-0.2 3.2])=[0.2 3.2]

angle   Phase angle, for example, angle(4+j*3)=0.6435 in radians

fft   One-dimensional fast Fourier transform. >>x=[1 1 1 0]; y=fft(x); Hence, y=[3 0-1.0000i 1.0000 0+1.0000i]. If you had written z=fft(x,8), you would have obtained z=[3 1.7071-1.7071i 0-1.0000i 0.2929+0.2929i 1 0.2929-0.2929i 0+1.0000i 1.7071+1.7071i]

fft2   Two-dimensional fast Fourier transform

fftshift   Shift DC component of fast Fourier transform to the center of spectrum. For example, you write in the command window: >>x=[1 1 1 1 1 ⋯ 0]; y=fft(x,256); Then the command plot(abs(fftshift(y))) will center the spectrum. You can also write plot(abs(fftshift(fft(x,256))))

ifft   Inverse one-dimensional fast Fourier transform

ifft2   Inverse two-dimensional fast Fourier transform

TWO- AND THREE-DIMENSIONAL PLOTTING

TWO-DIMENSIONAL PLOTS

plot	Linear plot. If you have three vectors of equal length such as x with numbers of equal distance, y and z, you can create the following simple plots: plot(y) will plot the values of y at numbers 1, 2, ... in the x-direction; plot(x,y) will plot the y values versus the equal-distance values of the vector x in the x direction; and plot(x,y,x,z) will plot both vectors y and z on the same graph. You can plot the two vectors by writing >>plot(x,y); hold on;plot(x,z); If you would like the second graph to have different color, you write plot(x,z,'g') for green color
loglog	Log-log scale plot. For example, loglog(y) will produce the plot
semilogx	Semilog scale plot. The log scale will be on the x-axis and the linear scale on the y-axis. The plot is accomplished by writing semilogx(y)
semilogy	Semilog scale plot. The log scale will be on the y-axis and the linear scale on the x-axis. The plot is accomplished by writing semilogy(y)
axis	Controls axis scaling. For example, if you want the axes to have specific ranges, you write after you created a plot using the MATLAB default axis([minx maxx miny max])
grid	Grid lines. After you created the plot, you write: grid on
subplot	Create axes in tiled positions. For example, when you write subplot(3,1,1), you expect 3×1 plots in one page starting plot one. Next you write subplot(3,1,2) and then proceed to plot the second plot, and so on. If you write subplot(3,2,1), you expect 3×2=6 plots on the page. After you write subplot(3,2,1), you proceed to plot the first of the 3×2 matrix format plots. For example, if you write subplot(3,2,2) and proceed to plot the figure, you create a plot at line two and the second plot.
legend	Graph legend. For example, if you have two lines on the plot, one red and one green, and write legend('one','two'), a rectangle frame will appear on the graph with a red line and the letters one and under a green line with the letters two.
title	Graph title. For example, if you write title('This is a graph'), the script in parentheses will appear on the top of the graph.
xlabel	X-axis label. For example, if you write xlabel('n time'), the n time will appear under the x-axis.
gtext	Place text with mouse. After you have created a plot, if you write in the command window gtext('this is the 1st graph') at the return, a crosshair will appear on the graph. At the click, the phrase in parentheses will appear on the graph.

Appendix 2: Matrix Analysis*

A2.1 DEFINITIONS

Let A be an $m \times n$ matrix with elements a_{ij}, $i=1,2, \ldots, m$; $j=1,2, \ldots, n$. A shorthand description of A is

$$[A]_{ij} = a_{ij} \tag{A2.1}$$

The **transpose** of A, denoted by A^T, is defined as the $n \times m$ matrix with elements a_{ji} or

$$\left[A^T\right]_{ij} = a_{ji} \tag{A2.2}$$

Example A2.1.1

$$A = \begin{bmatrix} 1 & 2 \\ 4 & 9 \\ 3 & 1 \end{bmatrix}; \quad A^T = \begin{bmatrix} 1 & 4 & 3 \\ 2 & 9 & 1 \end{bmatrix}$$

A **square** matrix is one for which $m=n$. A square matrix is **symmetric** if $A^T=A$.

The **rank** of a matrix is the number of linearly independent rows or columns, whichever is less. The **inverse** of a square $n \times n$ matrix A^{-1} for which

$$A^{-1}A = AA^{-1} = I \tag{A2.3}$$

where I is the identity matrix

$$I = \begin{bmatrix} 1 & 0 & \cdots & 0 \\ 0 & 1 & \cdots & 0 \\ \vdots & \vdots & & \vdots \\ 0 & 0 & \cdots & 1 \end{bmatrix} \tag{A2.4}$$

A **matrix** A is singular if its inverse does not exist.

The determinant of a square $n \times n$ matrix is denoted by $\det\{A\}$ and it is computed as

$$\det\{A\} = \sum_{j=1}^{n} a_{ij}C_{ij} \tag{A2.5}$$

where

---

* In this appendix, uppercase letters represent matrices and lowercase letters without subscripts, excluding identifiers, indicate vectors. If lowercase letters have subscripts and indicate vectors, they will be written in a bold-faced format.

$$C_{ij} = (-1)^{i+j} M_{ij} \tag{A2.6}$$

and M_{ij} is the determinant of the submatrix A obtained by deleting the ith row and jth column and is called the **minor** of a_{ij}. C_{ij} is the **cofactor** of a_{ij}.

Example A2.1.2

$$A = \begin{bmatrix} 1 & 2 & 4 \\ 3 & -3 & 9 \\ -1 & -1 & 6 \end{bmatrix}, \det\{A\} = (-1)^{1+1} \begin{vmatrix} -3 & 9 \\ -1 & 6 \end{vmatrix} + (-1)^{1+2} 2 \begin{vmatrix} -4 & 9 \\ -1 & 6 \end{vmatrix}$$

$$+ (-1)^{1+3} 4 \begin{vmatrix} 4 & -3 \\ -1 & -1 \end{vmatrix}$$

$$= C_{11} + C_{12} + C_{13}$$

$$= (18 + 9) + [-2(24 + 9)] + [4(-4 - 3)]$$

Any choice of i will yield the same value for the $\det\{A\}$.
A **quadratic form** Q is defined as

$$Q = \sum_{n=1}^{n} \sum_{j=1}^{n} a_{ij} x_i x_j \tag{A2.7}$$

In defining the quadratic form, we assume that $a_{ji} = a_{ij}$. This entails no loss in generality because any quadratic functions may be expressed in this manner. Q may also be expressed as

$$Q = x^T A x \tag{A2.8}$$

where $x = \begin{bmatrix} x_1 & x_2 & x_n \end{bmatrix}^T$ and A is a square $n \times n$ matrix with $a_{ji} = a_{ij}$ (symmetric matrix).

Example A2.1.3

$$Q = \begin{bmatrix} x_1 & x_2 \end{bmatrix} \begin{bmatrix} a_{11} & a_{12} \\ a_{21} & a_{22} \end{bmatrix} \begin{bmatrix} x_1 \\ x_2 \end{bmatrix} = \begin{bmatrix} a_{11}x_1 + a_{21}x_2 & a_{12}x_1 + a_{22}x_2 \end{bmatrix} \begin{bmatrix} x_1 \\ x_2 \end{bmatrix}$$

$$= a_{11}x_1^2 + a_{21}x_1x_2 + a_{12}x_1x_2 + a_{22}x_2^2$$

A square $n \times n$ matrix A is **positive semidefinite** if A is symmetric and

$$x^T A x \geq 0 \tag{A2.9}$$

for all $x \neq 0$. If the quadratic form is strictly positive, matrix A is called **positive definite**. If a matrix is positive definite or positive semidefinite, it is automatically assumed that the matrix is symmetric.

The **trace** of a **square** matrix is the sum of the diagonal elements or

$$tr\{A\} = \sum_{i=1}^{n} a_{ii} \tag{A2.10}$$

A partitioned $m \times n$ matrix A is one that is expressed in terms of its submatrices. An example is the 2×2 partitioned

$$A = \begin{bmatrix} A_{11} & A_{12} \\ A_{21} & A_{22} \end{bmatrix}, \quad \begin{bmatrix} k \times l & k \times (n-l) \\ (m-k) \times l & (m-k) \times (n-l) \end{bmatrix} \tag{A2.11}$$

MATLAB Functions

```
B=A';% B is the transpose of A
B=inv(A);%B is the inverse of A
a=det(A);% a is the determinant of A
I=eye(n);%I is an nxn identity matrix
a=trace(A);%a is the trace of A
```

A2.2   SPECIAL MATRICES

A **diagonal** matrix is a square $n \times n$ matrix with $a_{ij} = 0$ for $i \neq j$. A diagonal matrix has all the elements off the principal diagonal equal to zero. Hence

$$A = \begin{bmatrix} a_{11} & 0 & \cdots & 0 \\ 0 & a_{22} & \cdots & 0 \\ \vdots & \vdots & & \vdots \\ 0 & 0 & \cdots & a_{nm} \end{bmatrix} \tag{A2.12}$$

$$A^{-1} = \begin{bmatrix} a_{11}^{-1} & 0 & \cdots & 0 \\ 0 & a_{22}^{-1} & \cdots & 0 \\ \vdots & \vdots & & \vdots \\ 0 & 0 & \cdots & a_{nn}^{-1} \end{bmatrix} \tag{A2.13}$$

A generalization of the diagonal matrix is the square $n \times n$ **block diagonal** matrix

$$A = \begin{bmatrix} A_{11} & 0 & \cdots & 0 \\ 0 & A_{22} & \cdots & 0 \\ \vdots & \vdots & & \vdots \\ 0 & 0 & \cdots & A_{kk} \end{bmatrix} \tag{A2.14}$$

where all A_{ii} matrices are square and the submatrices are identically zero. The submatrices may not have the same dimensions. For example, if $k=2$, A_{11} may be a 2×2 matrix and A_{22} might be a scalar. If all A_{ii} are nonsingular, then

$$A^{-1} = \begin{bmatrix} A_{11}^{-1} & 0 & \cdots & 0 \\ 0 & A_{22}^{-1} & \cdots & 0 \\ \vdots & \vdots & & \vdots \\ 0 & 0 & \cdots & A_{kk}^{-1} \end{bmatrix} \qquad (A2.15)$$

and

$$\det\{A\} = \prod_{i=1}^{n} \det\{A_{ii}\} \qquad (A2.16)$$

A square $n \times n$ matrix is orthogonal if

$$A^{-1} = A^T \qquad (A2.17)$$

Example A2.2.1

$$A = \begin{bmatrix} \dfrac{2}{\sqrt{5}} & \dfrac{1}{\sqrt{5}} \\ -\dfrac{1}{\sqrt{5}} & \dfrac{2}{\sqrt{5}} \end{bmatrix}, \quad A^{-1} = \dfrac{1}{\det\{A\}} \begin{bmatrix} \dfrac{2}{\sqrt{5}} & \dfrac{1}{\sqrt{5}} \\ -\dfrac{1}{\sqrt{5}} & \dfrac{2}{\sqrt{5}} \end{bmatrix}^T = \begin{bmatrix} \dfrac{2}{\sqrt{5}} & -\dfrac{1}{\sqrt{5}} \\ \dfrac{1}{\sqrt{5}} & \dfrac{2}{\sqrt{5}} \end{bmatrix} = A^T$$

A matrix is **orthogonal** if its columns (and rows) are orthonormal. Therefore, we must have

$$A = \begin{bmatrix} \mathbf{a_1} & \mathbf{a_2} & \cdots & \mathbf{a_n} \end{bmatrix}$$

$$\mathbf{a_i^T a_j} = \begin{cases} 0 & \text{for } i \neq j \\ 1 & \text{for } i = j \end{cases} \qquad (A2.18)$$

An idempotent matrix is a square $n \times n$ matrix, which satisfies the relations

$$A^2 = A$$
$$A^m = A \qquad (A2.19)$$

Example A2.2.2

The **projection** matrix $A = H(H^TH)^{-1}H^T$ becomes $A^2 = H(H^TH)^{-1}H^TH$
$(H^TH)^{-1}H^T = H(H^{-1}H^{-T}H^TH(H^TH)^{-1})H^T = H(H^{-1}IH(H^TH)^{-1})H^T = H(H^TH)^{-1}H^T$
Hence, it is an idempotent matrix.

A **Toeplitz** square matrix is defined as

$$[A]_{ij} = a_{i-j} \tag{A2.20}$$

$$A = \begin{bmatrix} a_0 & a_{-1} & a_{-2} & \cdots & a_{-(n-1)} \\ a_1 & a_0 & a_{-1} & \cdots & a_{-(n-2)} \\ \vdots & \vdots & \vdots & & \vdots \\ a_{n-1} & a_{n-2} & a_{n-3} & \cdots & a_0 \end{bmatrix} \tag{A2.21}$$

Each element along the northwest-to-southeast diagonals is the same. If in addition $a_{-k}=a_k$, then A is **symmetric Toeplitz**.

MATLAB Functions

```
A=diag(x);%creates a diagonal matrix A with its diagonal the vector x;
A=toeplitz(x);%A is a symmetric Toeplitz matrix;
A=toeplitz(x,y)%x, and y must be of the same length, the main diagonal
        % will be the first element of x, the first element of y is not used;
```

A2.3 MATRIX OPERATION AND FORMULAS

ADDITION AND SUBTRACTION

$$A + B = \begin{bmatrix} a_{11} & a_{12} & \cdots & a_{1n} \\ a_{21} & a_{22} & \cdots & a_{2n} \\ \vdots & & & \\ a_{m1} & a_{m2} & \cdots & a_{mn} \end{bmatrix} + \begin{bmatrix} b_{11} & b_{12} & \cdots & b_{1n} \\ b_{21} & b_{22} & \cdots & b_{2n} \\ \vdots & & & \\ b_{m1} & b_{m2} & \cdots & b_{mn} \end{bmatrix}$$

$$= \begin{bmatrix} a_{11}+b_{11} & a_{12}+b_{12} & \cdots & a_{1n}+b_{1n} \\ a_{21}+b_{21} & a_{22}+b_{22} & \cdots & a_{2n}+b_{2n} \\ \vdots & & & \\ a_{m1}+b_{m1} & a_{m2}+b_{m2} & \cdots & a_{mn}+b_{mn} \end{bmatrix} \tag{A2.22}$$

Both matrices must have the same dimension.

MULTIPLICATION

$$AB(m \times n \times n \times k) = C(m \times k)$$

$$c_{ij} = \sum_{j=1}^{n} a_{ij} b_{ji} \tag{A2.23}$$

Example A2.3.1

$$AB = \begin{bmatrix} a_{11} & a_{12} \\ a_{21} & a_{22} \\ a_{31} & a_{32} \end{bmatrix} \begin{bmatrix} b_{11} & b_{12} \\ b_{21} & b_{22} \end{bmatrix}$$

$$= \begin{bmatrix} a_{11}b_{11} + a_{12}b_{21} & a_{11}b_{12} + a_{12}b_{22} \\ a_{21}b_{11} + a_{22}b_{21} & a_{21}b_{12} + a_{22}b_{22} \\ a_{31}b_{11} + a_{32}b_{21} & a_{31}b_{12} + a_{32}b_{22} \end{bmatrix}; \quad 3 \times 2 \times 2 \times 2 = 3 \times 2 \tag{A2.24}$$

TRANSPOSITION

$$(AB)^T = B^T A^T \tag{A2.25}$$

INVERSION

$$\left(A^T\right)^{-1} = \left(A^{-1}\right)^T \tag{A2.26}$$

$$(AB)^{-1} = B^{-1} A^{-1} \tag{A2.27}$$

$$A^{-1} = \frac{C^T}{\det\{A\}} \quad (A \equiv n \times n \text{ matrix}) \tag{A2.28}$$

$$c_{ij} = (-1)^{i+j} M_{ij} \tag{A2.29}$$

$M_{ij} \equiv$ minor of a_{ij} obtained by deleting the ith

row and jth column of A

Example A2.3.2

$$A^{-1} = \begin{bmatrix} 2 & 4 \\ -1 & 5 \end{bmatrix}^{-1} = \frac{1}{10+4} \begin{bmatrix} 5 & 1 \\ -4 & 2 \end{bmatrix}^T = \frac{1}{14} \begin{bmatrix} 5 & -4 \\ 1 & 2 \end{bmatrix};$$

$$AA^{-1} = \begin{bmatrix} 2 & 4 \\ -1 & 5 \end{bmatrix} \frac{1}{14} \begin{bmatrix} 5 & -4 \\ 2 & 2 \end{bmatrix} = \frac{1}{14} \begin{bmatrix} 14 & -8+8 \\ -5+5 & 4+10 \end{bmatrix} = \begin{bmatrix} 1 & 0 \\ 0 & 1 \end{bmatrix} = I$$

DETERMINANT (SEE EQUATION A2.5)

$A = n \times n$ matrix; $B = n \times n$ matrix

$$\det\{A^T\} = \det\{A\} \tag{A2.30}$$

$$\det\{cA\} = c^n \det\{A\} \tag{A2.31}$$

$$\det\{AB\} = \det\{A\}\det\{B\} \tag{A2.32}$$

$$\det\{A^{-1}\} = \frac{1}{\det\{A\}} \tag{A2.33}$$

TRACE (SEE EQUATION A2.10)

$A = n \times n$ matrix; $B = n \times n$ matrix

$$\text{tr}\{AB\} = \text{tr}\{BA\} \tag{A2.34}$$

$$\text{tr}\{A^T B\} = \sum_{i=1}^{n}\sum_{j=1}^{n} a_{ij}b_{ij} \tag{A2.35}$$

$$\text{tr}\{xy^T\} = y^T x; \quad x, y = \text{vectors} \tag{A2.36}$$

MATRIX INVERSION FORMULA

$A = n \times n$; $B = n \times m$; $C = m \times m$; $D = m \times n$

$$(A + BCD)^{-1} = A^{-1} - A^{-1}B(DA^{-1}B + C^{-1})^{-1}DA^{-1} \tag{A2.37}$$

$$(A + xx^T)^{-1} = A^{-1} - \frac{A^{-1}xx^T A^{-1}}{1 + x^T A^{-1}x}, \quad x = n \times 1 \text{ vector} \tag{A2.38}$$

PARTITION MATRICES

Examples of 2×2 partition matrices are as follows:

$$AB = \begin{bmatrix} A_{11} & A_{12} \\ A_{21} & A_{22} \end{bmatrix}\begin{bmatrix} B_{11} & B_{12} \\ B_{21} & B_{22} \end{bmatrix} = \begin{bmatrix} A_{11}B_{11} + A_{12}B_{21} & A_{11}B_2 + A_{12}B_{22} \\ A_{21}B_{11} + A_{22}B_{21} & A_{21}B_{12} + A_{22}B_{22} \end{bmatrix} \tag{A2.39}$$

$$\begin{bmatrix} A_{11} & A_{12} \\ A_{21} & A_{22} \end{bmatrix}^T = \begin{bmatrix} A_{11}^T & A_{12}^T \\ A_{21}^T & A_{22}^T \end{bmatrix} \tag{A2.40}$$

$$A = \begin{bmatrix} A_{11} & A_{12} \\ A_{21} & A_{22} \end{bmatrix} = \begin{bmatrix} k \times k & k \times (n-k) \\ (n-k) \times k & (n-k) \times (n-k) \end{bmatrix}$$

$$A^{-1} = \begin{bmatrix} (A_{11} - A_{12}A_{22}^{-1}A_{21})^{-1} & -(A_{11} - A_{12}A_{22}^{-1}A_{21})^{-1}A_{12}A_{22}^{-1} \\ -(A_{22} - A_{21}A_{11}^{-1}A_{12})^{-1}A_{21}A_{11}^{-1} & (A_{22} - A_{21}A_{11}^{-1}A_{12})^{-1} \end{bmatrix} \tag{A2.41}$$

$$\det\{A\} = \det\{A_{12}\}\det\{A_{11} - A_{12}A_{22}^{-1}A_{21}\} = \det\{A_{11}\}\det\{A_{22} - A_{21}A_{11}^{-1}A_{12}\} \qquad (A2.42)$$

IMPORTANT THEOREMS

1. A square matrix A is singular (invertible) if and only if its columns (or rows) are linearly independent or, equivalently, if its $\det\{A\} \neq 0$. If this is true, A is of **full rank**. Otherwise, it is singular.
2. A square matrix A is positive definite if and only if
   (a)

$$A = CC^T \qquad (A2.43)$$

   where C is a square matrix of the same dimension as A and it is of full rank (invertible), or
   (b) The principal minors are all positive. (The ith principal minor is the determinant of the submatrix formed by deleting all rows and columns with an index greater than i.) If A can be written as in Equation A2.43, but C is not full rank or the principal minors are only nonnegative, then A is positive definite.
3. If A is positive definite, then

$$A^{-1} = (C^{-1})^T C^{-1} \qquad (A2.44)$$

4. If A is positive definite and B $(m \times n)$ is of full rank $(m \leq n)$, then BAB^T is positive definite.
5. If A is positive definite (or positive semi-definite), then the diagonal elements are positive (nonnegative).

A2.4   EIGENDECOMPOSITION OF MATRICES

Let λ denotes an **eigenvalue** of the matrix A $(n \times n)$, then

$$Av = \lambda v \qquad (A2.45)$$

where v is the **eigenvector** corresponding to the eigenvalue λ. If A is symmetric, then

$$Av_i = \lambda_i v_i , Av_j = \lambda_j v_j \quad (\lambda_i \neq \lambda_j)$$

and

$$v_j^T A v_i = \lambda_i v_j^T v_i \qquad (a)$$

$$v_i^T A v_j = \lambda_j v_i^T v_j \quad \text{or} \quad v_j^T A v_i = \lambda_j v_j^T v_i \qquad (b)$$

Subtracting (a) from (b), we obtain $(\lambda_i - \lambda_j)v_j^T v_i = 0$. But $\lambda_i \neq \lambda_j$ and hence $v_j^T v_i = 0$, which implies that the eigenvectors of a symmetric matrix are orthogonal. We can proceed and normalize them, producing orthonormal eigenvectors.

From Equation A2.45, we write

$$A[\; v_1 \quad v_2 \quad \cdots \quad v_n \;] = [\; \lambda_1 v_1 \quad \lambda_2 v_2 \quad \cdots \quad \lambda_n v_n \;]$$

or

$$\qquad (A2.46)$$

$$AV = V\Lambda$$

where

$$\Lambda = \begin{bmatrix} \lambda_1 & 0 & 0 & \cdots & 0 \\ 0 & \lambda_2 & 0 & \cdots & 0 \\ & & \vdots & & \\ 0 & 0 & 0 & \cdots & \lambda_n \end{bmatrix} \tag{A2.47}$$

Because v_i are mutually orthogonal, $v_i^T v_j = \delta_{ij}$ makes V a **unitary** matrix, $V^T V = I = V V^T$. Postmultiply Equation A2.46 by V^T, we obtain

$$A = V \Lambda V^T \sum_{i=1}^{n} \lambda_i v_i v_i^T \tag{A2.48}$$

which is known as **unitary decomposition** of A. We also say that A is **unitary similar** to the diagonal Λ because a unitary matrix V takes A to diagonal form: $V^T A V = \Lambda$.

If $\Lambda = I$, from Equation A2.48, $A = V V^T = I$ and, hence

$$I = V V^T = \sum_{i=1}^{n} v_i v_i^T \tag{A2.49}$$

Each of the terms in the summation is of rank 1 projection matrix:

$$P_i^2 = v_i v_i^T v_i v_i^T = v_i v_i^T = P_i \quad \left(v_i^T v_i = 1 \right) \tag{A2.50}$$

$$P_i^T = v_i v_i^T = P_i \tag{A2.51}$$

Hence, we write (see Equations A2.48 and A2.49)

$$A = \sum_{i=1}^{n} \lambda_i P_i \tag{A2.52}$$

$$I = \sum_{i=1}^{n} P_i \tag{A2.53}$$

INVERSE

Because V is unitary matrix, $V V^T = I$ or $V^T = V^{-1}$ or $V = (V^T)^{-1}$ and, therefore,

$$A^{-1} = (V^T)^{-1} \Lambda^{-1} V^{-1} = V \Lambda^{-1} V^T = \sum_{i=1}^{n} \frac{1}{\lambda_i} v_i v_i^T \tag{A2.54}$$

DETERMINANT

$$\det\{A\} = \det\{V\} \det\{\Lambda\} \det\{V^{-1}\} = \det\{\Lambda\} = \prod_{i=1}^{n} \lambda_i \tag{A2.55}$$

A2.5   MATRIX EXPECTATIONS

$$E\{x\} = m_x \tag{A2.56}$$

$$E\{(x - m_x)(x - m_x)^T\} = R_{xx} \tag{A2.57}$$

$$E\{tr\{A\}\} = tr\{E\{A\}\} \tag{A2.58}$$

$$E\{Ax + b\} = Am_x + b \tag{A2.59}$$

$$E\{xx^T\} = R_{xx} + m_x m_x^T \tag{A2.60}$$

$$E\{xa^T x\} = (R_{xx} + m_x m_x^T)a \tag{A2.61}$$

$$E\{(x + a)(x + a)^T\} = R_{xx} + (m_x + a)(m_x + a)^T \tag{A2.62}$$

$$E\{x^T x\} = tr\{R_{xx}\} + m_x^T m_x = tr\{R_{xx} + m_x m_x^T\} \tag{A2.63}$$

$$E\{x^T a x^T\} = a^T \{R_{xx} + m_x m_x^T\} \tag{A2.64}$$

$$E\{x^T A x\} = tr\{AR_{xx}\} + m_x^T A m_x = tr\{A(R_{xx} + m_x m_x^T)\} \tag{A2.65}$$

A2.6   DIFFERENTIATION OF A SCALAR FUNCTION WITH RESPECT TO A VECTOR

$$x = [\ x_1 \quad x_2 \quad \cdots \quad x_n\ ]^T; \quad \frac{\partial}{\partial x} = \left[\ \frac{\partial}{\partial x_1} \quad \frac{\partial}{\partial x_2} \quad \cdots \quad \frac{\partial}{\partial x_n}\ \right]^T$$

$$\frac{\partial}{\partial x}(y^T x) = \frac{\partial}{\partial x}(x^T y) = y \tag{A2.66}$$

$$\frac{\partial}{\partial x}(x^T A) = A \tag{A2.67}$$

$$\frac{\partial}{\partial x}(x^T) = I \tag{A2.68}$$

$$\frac{\partial}{\partial x}(x^T x) = 2x \tag{A2.69}$$

$$\frac{\partial}{\partial x}(x^T A y) = Ay \tag{A2.70}$$

$$\frac{\partial}{\partial x}(y^T A x) = A^T y \tag{A2.71}$$

$$\frac{\partial}{\partial x}(x^T A x) = (A + A^T)x \tag{A2.72}$$

$$\frac{\partial}{\partial x}(x^T A x) = 2Ax \quad \text{if } A \text{ is symmetric} \tag{A2.73}$$

$$\frac{\partial}{\partial x}(a^T A x x^T) = (A + A^T)xx^T + x^T A x I \tag{A2.74}$$

Appendix 3: Mathematical Formulas

A3.1 TRIGONOMETRIC IDENTITIES

$$\cos(-a) = \cos a$$

$$\sin(-a) = -\sin a$$

$$\cos\left(a \pm \frac{\pi}{2}\right) = \mp \sin a$$

$$\sin\left(a \pm \frac{\pi}{2}\right) = \pm \cos a$$

$$\cos(a \pm \pi) = -\cos a$$

$$\sin(a \pm \pi) = -\sin a$$

$$\cos^2 a + \sin^2 a = 1$$

$$\cos^2 a - \sin^2 a = \cos 2a$$

$$\cos(a \pm b) = \cos a \cos b \mp \sin a \sin b$$

$$\sin(a \pm b) = \sin a \cos b \pm \cos a \sin b$$

$$\cos a \cos b = \frac{1}{2}[\cos(a-b) + \cos(a+b)]$$

$$\sin a \sin b = \frac{1}{2}[\cos(a-b) - \cos(a+b)]$$

$$\sin a \cos b = \frac{1}{2}[\sin(a-b) + \sin(a+b)]$$

$$c \cos a + d \sin a = \sqrt{a^2 - b^2} \, \cos[a - \tan^{-1}(d/c)]$$

$$\cos^2 a = \frac{1}{2}(1 + \cos 2a)$$

$$\cos^3 a = \frac{1}{4}(3\cos a + \cos 3a)$$

$$\cos^4 a = \frac{1}{8}(3 + 4\cos 2a + \cos 4a)$$

$$\sin^2 a = \frac{1}{2}(1 - \cos 2a)$$

$$\sin^3 a = \frac{1}{4}(3\sin a - \sin 3a)$$

$$\sin^4 a = \frac{1}{8}(3 - 4\cos 2a + \cos 4a)$$

$$e^{\pm ja} = \cos a \pm j\sin a$$

$$\cosh a = \cos ja$$

$$\cos a = \frac{1}{2}\left(e^{ja} + e^{-ja}\right)$$

$$\sin a = \frac{j}{2}\left(e^{-ja} - e^{ja}\right)$$

$$\sinh a = -j\sin ja$$

$$\tanh a = -j\tan ja$$

A3.2   ORTHOGONALITY

$$\sum_{n=0}^{N-1}\cos\frac{2\pi k}{N}n\cos\frac{2\pi l}{N}n = 0 \qquad 1 \le k,l \le N-1, \quad k \ne l$$

$$\sum_{n=0}^{N-1}\sin\frac{2\pi k}{N}n\sin\frac{2\pi l}{N}n = 0 \qquad 1 \le k,l \le N-1, \quad k \ne l$$

$$\sum_{n=0}^{N-1}\sin\frac{2\pi k}{N}n\cos\frac{2\pi l}{N}n = 0 \qquad 1 \le k,l \le N-1, \quad k \ne l$$

$$\sum_{n=0}^{N-1}\cos^2\frac{2\pi k}{N}n = \begin{cases} N/2 & 1 \le k,l \le N-1, k \ne l \\ N & k = 0, N/2 \end{cases}$$

$$\sum_{n=0}^{N-1}\sin^2\frac{2\pi k}{N}n = \begin{cases} N/2 & 1 \le k,l \le N-1, \quad k \ne l \\ 0 & k = 0, N/2 \end{cases}$$

The preceding formulas are correct if all k, l, and n are replaced by $k \bmod N$ and $l \bmod N$.

A3.3   SUMMATION OF TRIGONOMETRIC FORMS

$$\sum_{n-0}^{N-1}\cos\frac{2\pi k}{N}n = \begin{cases} 0 & 1\le k\le N-1 \\ N & k=0,N \end{cases}$$

$$\sum_{n-0}^{N-1}\sin\frac{2\pi k}{N}n = \begin{cases} 0 & 1\le k\le N-1 \\ N & k=0,N \end{cases}$$

(k, l, and n are integers)

A3.4   SUMMATION FORMULAS

FINITE SUMMATION FORMULAS

$$\sum_{k=0}^{n}a^k = \frac{1-a^{n+1}}{1-a}, \quad a\ne 1$$

$$\sum_{k=1}^{n}ka^k = \frac{a\left(1-(n+1)a^n+na^{n+1}\right)}{(1-a)^2}, \quad a\ne 1$$

$$\sum_{k=1}^{n}k^2a^k = \frac{a[(1+a)-(n+1)^2 a^n+(2n^2+2n-1)a^{n+1}-n^2 a^{n+2}]}{(1-a)^3}, \quad a\ne 1$$

$$\sum_{k=1}^{n}k = \frac{n(n+1)}{2}$$

$$\sum_{k=1}^{n}k^2 = \frac{n(n+1)(2n+1)}{6}$$

$$\sum_{k=1}^{n}k^3 = \frac{n^2(n+1)^2}{4}$$

$$\sum_{k=0}^{2n-1}(2k+1) = n^2$$

INFINITE SUMMATION FORMULAS

$$\sum_{k=0}^{\infty}a^k = \frac{1}{1-a} \quad |a|<1$$

$$\sum_{k=0}^{\infty} ka^k = \frac{a}{(1-a)^2} \quad |a| < 1$$

$$\sum_{k=0}^{\infty} k^2 a^k = \frac{a^2 + a}{(1-a)^3} \quad |a| < 1$$

A3.5 SERIES EXPANSIONS

$$e^a = 1 + a + \frac{a^2}{2!} + \frac{a^3}{3!} + \cdots$$

$$\ln(1+a) = a - \frac{a^2}{2} + \frac{a^3}{3} - \frac{a^4}{4} + \cdots \quad |a| < 1$$

$$\sin a = a - \frac{a^3}{3!} + \frac{a^5}{5!} - \frac{a^7}{7!} + \cdots$$

$$\cos a = 1 - \frac{a^2}{2!} + \frac{a^4}{4!} - \frac{a^6}{6!} + \cdots$$

$$\tan a = a + \frac{a^3}{3} + \frac{2a^5}{15} + \frac{17a^7}{315} + \cdots \quad |a| < \frac{\pi}{2}$$

$$\sinh a = a + \frac{a^3}{3!} + \frac{a^5}{5!} + \frac{a^7}{7!} + \cdots$$

$$\cosh a = 1 + \frac{a^2}{2!} + \frac{a^4}{4!} + \frac{a^6}{6!} + \cdots$$

$$\tanh a = a - \frac{a^3}{3} + \frac{2a^5}{15} - \frac{17a^7}{315} + \cdots \quad |a| < \frac{\pi}{2}$$

$$(1+a)^n = 1 + na + \frac{n(n-1)}{2!}a^2 + \frac{n(n-1)(n-2)}{3!}a^3 + \cdots \quad |a| < 1$$

A3.6 LOGARITHMS

$$\log_b N = \log_a N \log_b a = \frac{\log_a N}{\log_a b}$$

A3.7 SOME DEFINITE INTEGRALS

$$\int_0^{\infty} x^2 e^{-ax} \, dx = \frac{2}{a^3}$$

$$\int_0^\infty x^n e^{-ax}\, dx = \frac{n!}{a^{n+1}} \quad a > 0$$

$$\int_0^\infty e^{-a^2 x^2}\, dx = \frac{\sqrt{\pi}}{2a} \quad a > 0$$

$$\int_0^\infty x e^{-a^2 x^2}\, dx = \frac{1}{2a^2} \quad a > 0$$

$$\int_0^\infty \frac{e^{-ax}}{x} \sin mx\, dx = \tan^{-1} \frac{m}{a} \quad a > 0$$

$$\int_0^\infty \frac{\sin mx}{x}\, dx = \frac{\pi}{2}$$

Appendix 4: MATLAB Functions

[m,z] = hist(x,a)	m = counts in each bin; z = coordinates of the a bins;
	w = max(x)/length(z); bp = m/(w*a) = probability per bin;
[r,p,k] = residue(nu,de)	example: $F(z) = z^2/(z^2-0.04)$, nu = [1 0 0], de = [1 0 −0.04],
[r] = xcorr(x,y,'biased')	produces a biased cross-correlation; if x = y r = auto-correlation; the r is a 2N−1 symmetric function with respect to zero
[r] = xcorr(x,y,'unbiased')	produces a an unbiased cross-correlation; if x = y r = auto-correlation
\omega	will produce the lower case omega, the same for the rest of the Greek letters
\Omega	will produce the upper case omega, the same for the rest Greek letters
	'w' = white, 'k' = black
A(:,1)	will give $[1\ 4]^T$ for matrix A = [1 2;4 3], all rows of the first column
A(2,:)	will give [2 4] for the matrix A = [1 2;4 3], all columns for the second row
a'*a	$a^T \cdot b^T$ is a row vector $c = (a_1b_1,\ a_2b_2,\ldots,a_nb_n)$
abs()	gives the absolute value of the complex expression in the parenthesis
acos()	will produce the inverse of the cosine
angle()	gives the angle of a complex number or vector
asin()	will produce the inverse of the sine
atan()	will produce the inverse of the tangent
axis([0 2 −2 5])	will create axes with length 0 to 2 in x-axis and −2 to 5 in the y-axis
bar(z,p)	plots bars at the x-coordinates of z and heights p
	bar(z,pb) = plots the probability distribution with z bars
color designation	'y' = yellow, 'm' = magenta, 'c' = cyan, 'r' = red, 'g' = green, 'b' = blue,
colormap(gray)	produces black figure; see also meshgrid()
conj()	gives the complex conjugate of a complex number or expression
conj(z)	gives the conjugate value of the complex number z, conj(3−j*5) = 3.0000 + 5.0000i
	contains the rows 4–6 and columns 8–15
contour(X,Y,Z,15)	will produce 15 contours (see also meshgrid() below);
contour(X,Y,Z,V,'g')	V = [a b c] is a vector with values a, b and c which can be any
contour(Z,[1.6 3.1 5])	produces contours of the performance surface at heights 1.6, 3.1 and 5
	contour(Z,[2.5 3.1 5]) will produce 3 contours at the heights 2.5, 3.1 and 5
contour(Z,30)	create 30 contours of the performance surface
conv(**x,y**)	convolution between two vectors **x** and **y**; the output vector signal
det(A)	gives the determinant of A; we can also write det([1 2;4 3]) for 2x2 matrix
dimpulse(nu,de,5)	inverts the z-transform; example: $z^2/(z^3 + 0.1)$, nu = [0 1 0 0],de = [1 0 0 0.1], gives 5 terms of the time function
dimpulse(num,den,a)	gives a numbers of the time function $f(n)$, num = vector with the
dot	$a. * b$ is a column vector with elements $c = (a_1b_1,\ a_2b_2,\ldots,a_nb_n)'$
eig(R)	[Q,D] = eig(R), R = correlation matrix, Q = each one of its columns equal to an eigenvector, D = is a diagonal matrix with the eigenvalues of the correlation matrix
	empty circle on top at points $\{x\}$
eye(4)	crates a 4x4 identity matrix, its main diagonal have unit value
	$f(n) = 0.1(0.2)^n - 0.1(-0.2)^n + \delta(n)$; r = residues, p = poles
	$F(z) = 0.1(z/(z-0.2)) - 0.1(z/(z + 0.2)) + 1$, the inverse is
fft(x,N)	takes the DFT of the sequence (vector) x having N elements
figure(2)	creates a second figure and the next plot will introduce insert the plot at this new figure

filter(b,a,x)	b = vector of the input coefficients; a = vector of the output coefficients; x = vector of the input to the system; y(n) = output;
fliplr(x)	flips a vector from left to right, x = vector
floor(a)	gives the number without the decimal, floor(5.68) = 5, flor(−2.6) = −3
freqz(); [H,W] = freqz(A,B,N)	B = vector of the b's coefficients of the numerator of the transfer function H(e^{jw}); A = vector of the a's coefficients of the denominator of H(e^{jw}); N = number of bins in the frequency domain; W = frequency vector in rad/sample of the filter H(e^{jw}) = transfer function
grid on	after plot() we right *grid on* and a grid will appear with horizontally and vertically lines
hist(*a*,α)	creates a histogram with α number of bins from the elements of the vector *a*
hold on	we have made a plot first; then we write *hold on* and then we proceed for the second plot on the same figure
ifft(X,N)	takes the inverse DFT (IDFT) of the frequency sequence (vector) {*X*} with *N* elements
	inv([1 2; 4 3]) to invert a 2x2 matrix
inv(**A**)	gives the inverse of matrix **A, A**$^{-1}$
legend('FT')	will produce a legend identifying with the type of plotting, here FT
length(*a*)	gives the number of the elements of the vector *a*
log(a$_1$)	gives the natural logarithm of the number a$_1$
log10(a$_1$)	gives the logarithm of the number a$_1$ with base 10
matrix: A(4:6,8:15)	produces a 3x8 matrix from the elements of A that
matrix: A.^n	every element is raised in the power of n
mean(*a*)	gives the mean value of the elements of the vector *a*
meshgrid(x,y)	example: x = 0:0.1:4;y = 0:0.1:4;[X,Y] = meshgrid(x,y);
norm(x)	gives the norm of x, the square root of the sum of the square of the vector x coefficients
	number; the contours will have color green
plot(x,y)	plots a blue color graph x versus y
plot(x,y,'k')	black color graph, 'g' = green, 'r' = red
	r = [0.1000 −0.1000], p = [0.2000 −0.2000], k = 1, hence
rand(1,n)	produces a vector of n elements which are white noise
randn(1,n)	produces a vector of n elements which are normally distributed
rank(A)	gives the rank of the matrix A, the number of independent columns of the matrix
	roots(a) will give the roots of the above polynomial
roots(a)	a is a vector of the coefficients of the polynomial e.g.
sign()	examples: sign(0.2) = 1, sign(−2) = −1, sign(0) = 0,
	sign([−2 0.1 0 −0.3 −5]) = [−1 1 0 −1 −1]
sqrt()	takes the square root of a number
std(*x*)	standard deviation = square root of the variance, *x* = data vector
stem(x,y)	produces a set of lines of magnitude of the set {*y*} with
stem(x,y,'k','filled')	black color graph with the top circles filled with black color
subplot(4,1,3)	produces a subplot on the third row (3) of a figure that is made of four (4) rows and one column(1);
subplot(5,2,3)	a figure of 5 rows, 2 columns and plots the 3rd subplot
sum(*a*)	sums the elements of a vector *a*
sum(X,1)	sums all rows of the matrix X
sum(X,2)	sums all the columns of the matrix X
surf(X,Y,Z)	produces three dimensional figure; see also meshgrid()
	surf(X,Y,Z); colormap(gray);%black figure; see contour above
	the number 5 asks to produce 5 values of the time function
title('Figure')	will produce the title, Figure, on the top of the figure

trace(A)	sums the main diagonal of the matrix A
transpose()	transposes a matrix
	values of the numerator coefficients of the z-function in descending order; den = same as for num but for the denominator
var(a)	gives the variance of the elements of vector a
var(x)	variance of the data vector x
	will have length: length(x) + length(y)−1
	with 0.5 mean value; rand(1,n)−0.5 is a vector of n elements which are uniformly distributed and with zero mean
xcorr(x,y)	gives two-sided correlation of the two vectors; if x = y the one-sided correlation starts from length(x) to length(x) + length(y)−1; if x<y MATLAB adds zeros to the smaller, the one-sided starts from length(x) and ends at 2length(y)−1
xlabel('x_{cD}')	on the x-axis will appear x_{cD}
xlabel('x-axis')	on the x-axis will appear x-axis
ylabel('y-label')	on the y-axis will appear y-axis
	$Z = 4*X.^2 + Y.^3$; %(the function we want to produce);

Bibliography

Astola, J. and P. Kuosmanen, *Fundamentals of Nonlinear Digital Filtering*, CRC Press, Boca Raton, FL, 1997.

Farhang-Boroujeny, B., *Adaptive Filters: Theory and Applications*, John Wiley & Sons, New York, 1999.

Haykin, S., *Adaptive Filter Theory*, Prentice Hall, Upper Saddle River, NJ, 2001.

Hays, M. H., *Statistical Digital Signal Processing and Modeling*, John Wiley & Sons, New York, 1996.

Hogg, R. V. and A. T. Craig, *Introduction to Mathematical Statistics*, 4th Edition, Macmillan Publishing Co., New York, 1978.

Kay, S. M., *Modern Spectrum Estimation: Theory and Applications*, Prentice Hall, Englewood Cliffs, NJ, 1988.

Kay, S. M., *Statistical Signal Processing*, Prentice Hall, Upper Saddle River, NJ, 1993.

Manolakis, D. G., V. K. Ingle and S. T. Kogon, *Statistical and Adaptive Signal Processing*, McGraw-Hill, New York, 2000.

Marple, S. L. Jr., *Digital Spectral Analysis with applications*, Prentice Hall, Englewood Cliffs, NJ, 1987.

Pitas, I. and A. N. Venetsanopoulos, *Nonlinear Digital Filters: Principles and Applications*, Kluwer Academic Publishers, Boston, MA, 1990.

Poularikas, A. D. and Z. M. Ramadan, *Adaptive Filtering Primer with MATLAB*, CRC Taylor & Francis, Boca Raton, FL, 2006.

Poularikas, A. D., *Signals and Systems Primer with MATLAB*, CRC Taylor & Francis, Boca Raton, FL, 2007.

Sayed, A. H., *Adaptive Filters*, Wiley Interscience, Hoboken, NJ, 2008.

Shiavi, R., *Introduction to Applied Statistical Signal Analysis*, IRWIN, Boston, MA, 1991.

Stoica, P. and R. Moses, *Introduction to Spectral Analysis*, Prentice Hall, Upper Saddle River, NJ, 1997.

Index

Milton Keynes UK
Ingram Content Group UK Ltd.
UKHW050456071024
449327UK00015B/403